KU-207-138

Frontiers in Geographical Teaching

The Madingley Lectures for 1963

Frontiers in Geographical Teaching

THE MADINGLEY LECTURES FOR 1963

EDITED BY

Richard J. Chorley and Peter Haggett

COURSE SECRETARY

J. M. Y. Andrew

METHUEN & CO LTD

LONDON

First published 1965
by Methuen & Co Ltd
11 New Fetter Lane, London EC4

Reprinted with corrections 1967
Printed in Great Britain by
The Camelot Press Ltd
London and Southampton

Contents

PART II TECHNIQUES

Foreword

In 1951 Madingley Hall, which had recently been acquired by the University, first became available to the Board of Extra-mural Studies in vacations for the arrangement of residential courses. In the intervening years there have been many and various courses attended by a wide range of adult students. At an early stage it became evident that there was a need for intensive and highly specialized courses designed largely to bring teachers and like persons into the University, there to encounter and discuss recent developments and advances in their subjects. In these enterprises the Board have received unstinting help from many teaching officers of the University.

Although geography was not amongst the first subjects to be dealt with in this way, it was apparent that there was a strong latent demand, so that the course of study on which this volume is based was held in Madingley Hall in the summer of 1963. The Board were fortunate in that Mr R. J. Chorley and Mr P. Haggett, University Lecturers in Geography, were willing to direct the course, which was attended mainly by specialist teachers of the subject in Training Colleges, Grammar Schools, and similar institutions. There was no doubt that this opportunity to keep abreast of a subject which, like others, is changing and growing was highly appreciated. It is hoped that this volume, the first to be published arising specifically out of a course in Madingley Hall, will be equally welcomed.

The thanks of the Board are due to Mr Chorley and Mr Haggett not only for having directed this successful course but also for editing this book. I would like to add an expression of my own gratitude to them and also to my colleagues, Mr J. M. Y. Andrew and Dr R. E. Pahl, for their vital part in the initiation and daily conduct of this course.

G. F. HICKSON,
Secretary of the Board of Extra-mural Studies
University of Cambridge.

Acknowledgements

The authors gratefully acknowledge the permissions by the following individuals and organizations to reproduce figures: The Warden, Madingley Hall, Cambridge, for the print used for the dust-jacket; The Director, U.S. Geological Survey, for figures 2.1, 2.2, and 8.2; G. B. Masefield, Esq., M.A., Editor, *Tropical Agriculture*, for figures 3.3, 3.5, 3.6, 3.7, and 3.10; Professor A. N. Strahler, Columbia University, New York, for figure 8.1; The Editor, Royal Geographical Society, for figures 9.7, 13.2, 13.4, 13.6, 13.7, 13.10, and 17.4; Her Majesty's Stationery Office (Crown Copyright Reserved), for figures 13.1, 13.5, 13.8, 13.9, 13.12, 13.13, and 13.14; Longmans, Green & Co. Ltd., for figures 17.1, 17.3, 17.5, and 17.6; The Petroleum Information Bureau, for figure 17.2; The Editor, Institute of British Geographers, for figure 13.3; The Editor, *Town Planning Review*, for figure 13.11.

PART ONE

CONCEPTS

CHAPTER ONE

Changes in the Philosophy of Geography

E. A. WRIGLEY

Lecturer in Geography, University of Cambridge

It is perhaps a platitude that a perennial problem of geography, true of many subjects but felt very acutely in geography, has been to find a satisfactory way of organizing the welter of observational material with which geographers commonly deal. It may prove helpful to look back to some solutions to this problem offered in the past as a background to contemporary developments, and to pay especial attention to the writings of Vidal de la Blache, since both his successes and his final failure are very instructive.

'CLASSICAL' GEOGRAPHY

Modern geography is often said to begin with two important early nineteenth-century German scholars, von Humboldt and Ritter.[1] These two men saw eye to eye in many things and agreed in pouring scorn on their predecessors because they dealt with geographical information in such a haphazard and unsystematic fashion. They gave clear expression to a view of geographical methodology which remained dominant for most of the nineteenth century. It was still the guiding principle of Friedrich Ratzel, at least in his earlier writings in the 1880's, and was shared by many who were not geographers. It is indeed a pointless exercise at this comparatively early date to distinguish between those who were geographers and those who were not. Humboldt would not fit any label conveniently. Buckle in the early chapters of his *History of Civilisation in England* (1857–61) gives a succinct account of the new attitude.

This conception of the organization of geography, which may

[1] This claim is conventional and perhaps convenient, but it is quite possible to argue the case for a later date; for example for the work of Vidal de la Blache.

conveniently be labelled the 'classical'[1] view since it held sway during the formative period of modern geography in the nineteenth century, was straightforward and simple. Men like Humboldt and Ritter considered the writings of earlier geographers to be defective because they were largely descriptive and were in their view very ill-organized.[2] They considered the scientific organization of knowledge to be a two-stage affair: a first stage which consisted of the careful assembly of detailed and accurate factual material; and a second in which the material was given coherence and made intelligible by being subsumed under a number of laws which should express the relationships of cause and effect to be found in the phenomena as simply and concisely as possible.[3] The vital feature of any science was this second stage. Without it any branch of learning was simply pigeon-holing and antiquarianism. This was a first main characteristic of 'classical' geography. Status in the world of learning depended on the successful formulation of laws enabling the material to be organized and made intelligible. If geography were to be worthy of ranking with the sciences it must succeed in establishing such laws. It must go on, as Ratzel put it, airing his Latin tags, *rerum cognoscere causas*,[4] to know the causes of things.

A second chief prop of the 'classical' view of geography was the

[1] The term 'classical' is used here in a different sense from that employed by Hartshorne for whom the deaths of Humboldt and Ritter in 1859 mark the end of the 'classical' period.

[2] E.g. 'A systematic organization of material is seldom to be found in them (the older type of geographies). . . . They contain at bottom only an arbitrary, unorganized and unsystematic compilation of all sorts of note-worthy phenomena, which in the different parts of the globe appear to be especially striking. . . . The description of Europe is begun with either Portugal or Spain because Strabo began his narration in this order. The facts are arranged like the pieces of a patchwork quilt, now one way, now another, as if each disconnected piece could stand by itself' (Ritter, 1862, pp. 21–2).

[3] E.g. 'In proportion as laws admit of more general application, and as sciences mutually enrich each other, and by their extension become connected together in more numerous and more intimate relations, the developments of general truths may be given with conciseness devoid of superficiality. On being first examined, all phenomena appear to be isolated, and it is only by the result of a multiplicity of observations, combined with reason, that we are able to trace the mutual relations existing between them' (Humboldt, 1849, Vol. 1, p. 29).

[4] 'But this deeper conception (of geography) cannot possibly remain content with description but must, following the irresistible example of all natural sciences, within whose sphere it developed in the most intimate relationship, go on from description to the higher task of "Rerum cognoscere causas"' (Ratzel, 1882, Vol. 1, p. 5).

conviction[1] that in the final analysis there was no difference methodologically between what would now be called the social and the physical sciences. In both cases the ultimate aim was the formulation of laws expressing the universal operation of cause and effect.[2] It was widely agreed that the subject matter was vastly more diverse and complicated in the study of societies than, say, in physics and that it might be much longer before satisfactory formulations of laws could be made, but the Newtonian model was assumed, though often implicitly rather than explicitly, to be appropriate. The denial of the methodological unity of all knowledge, and particularly the assertion of a special position for the study of social change and functioning in the writings of men like Dilthey and Max Weber, still lay in the future. This is important because it largely obviated a difficulty which has been much more keenly felt by geographers in the last two generations, namely the problem of running in harness, as it were, physical geography and social geography. Since the methodology of the social sciences is now very commonly held to be different from that of the physical sciences (for example, the possibility of formulating universally valid laws descriptive of social functioning is often denied, and Weberian 'ideal' types and middle-order generalizations advocated in their stead), there are clearly problems in asserting the unity of geographical knowledge and particularly of geographical methodology. During the period of 'classical' geography no difficulties could arise on this score because the general methodology of all branches of the subject could be held to be the same.

A third general point on which the writings of the 'classical' geographers show agreement in the main is that a prime object of geographical study is to investigate the ways in which the physical environment affects the functioning and development of societies.[3]

[1] Shared of course by many contemporaries, for example A. Comte and J. S. Mill.

[2] 'In regard to nature, events apparently the most irregular and capricious have been explained, and have been shown to be in accordance with certain fixed and immutable laws. This has been done because men of ability and, above all, men of patient, untiring thought have studied natural events with the view of discovering their regularity: and if human events were subjected to a similar treatment, we have every right to expect similar results' (Buckle, 1857, Vol. 1, p. 6).

[3] 'Only having a firm methodological principle can protect it (geography) from going astray: the clear commitment to the central theme of the relationship between the forms of terrestrial phenomena and mankind' (Ritter, 1862, p. 28).

The *Erdkunde* gained Humboldt's warm approbation because of its success

This is not to suggest that this was the main theme of all their works. There were many in which it appeared rarely; some of a specialist nature in which it did not appear at all; and in some works of geographical methodology it was firmly rejected. But, from a general view, this appears to be the third necessary support of the 'classical' attitude to geography.[1] It gave point to all the subsidiary lines of investigation and it made it perfectly clear why geography must have both a physical and social side. Some knowledge of both is obviously vital to any attempt to understand this matter. Given the first two main characteristics of 'classical' geography, the belief in the necessity of formulating laws expressing relationships of cause and effect and the conviction that the basic methodology of both social and physical sciences is the same, it is very reasonable to feel that one of the most promising lines of investigation to pursue is the explanation of social change and function by reference to features of the physical environment. In the early decades after the publication of the *Origin of Species* this type of work, while it might be modified, was also encouraged and strengthened by a flow of new concepts, as may be seen in much of Ratzel's large output.

The story of the decline and fall of the 'classical' conception of geography is a most interesting chapter in intellectual history. By the end of the century it was widely attacked for its rigidity, because it was wedded to what came to be called geographical determinism (although this was not an accurate criticism to level at the major figures of the group), and because the new ideas about methodology in the social sciences made its whole approach to the understanding of social action and social change seem unrewarding.[2] 'Classical' geography has left a considerable legacy. It can be seen, for example, in the continuing arguments about what are usually termed 'possibilism' and 'probabilism', which are rooted in the 'classical' attitude

in showing the influence of the environment 'on the migrations, laws, and manners, of nations, and on all the principal events enacted upon the face of the earth' (Humboldt, 1849, Vol, 1, p. 28).

Ratzel subtitled his great work *Anthropo-geographie: oder Grundzüge der Anwendung der Erdkunde auf die Geschichte.*

[1] It is interesting that Hettner once wrote to Joseph Partsch that what first attracted him to geography was the idea of the dependence of man on nature and described his surprise at discovering how little this entered into his teaching when he went to study under Kirchoff at his first university, Halle (*Heidelberger Geographische Arbeiten*, 1960, p. 77).

[2] It is interesting to note that Hettner knew Max Weber in the years when both men were at Heidelberg.

to the subject, and are open to much the same objections as the 'classical' system. Or again, it can be seen in the layout of many textbooks which begin with such things as solid geology and climate and progress through vegetation and soils to settlement, agriculture, industry and transport—a perfectly logical sequence of exposition in 'classical' terms,[1] but less so if the 'classical' view is abandoned.

'REGIONAL' GEOGRAPHY

The reaction against 'classical' geography took many forms: some essentially a development from it, like the writings of Hettner; some in opposition to it, like Brunhes' ideas;[2] some developed along new lines without close reference to it. The work of Vidal de la Blache falls best perhaps in the last category.[3] It is of the greatest importance and may serve as an introduction to the 'post-classical' world, and to what may be termed the 'regional' view of the nature of geography.

Vidal saw that it made little sense to set the physical and social environments of man over against one another, as it were, and examine the way the former influenced the latter, still less to do this in a systematic fashion in the hope that general laws describing the relationships could be discovered. Instead he propounded a different idea. Whereas in the 'classical' view the study of the physical environment and the study of society were riveted to each other because a

[1] One great strength and beauty of the 'classical' system was that the sequence was at one and the same time a coherent method of conveying information and of explaining it stage by stage. Description and explanation were so neatly bound up together that the second could hardly be distinguished from the first.

[2] Brunhes admired Ratzel's work very greatly, considering that he had conceived a fresh and fruitful approach to human geography, but his views on method stand in strong contrast with the 'classical' school. 'Between the facts of physical nature there are sometimes causal relations; between those of human geography there are really only relations of connexion. To force, as it were, the bond which connects phenomena to each other is to produce a work of false science; and a critical attitude is very necessary to allow one to specify with good judgement those complex cases where interconnexion (*connexité*) does not in the least imply causation' (Brunhes, 1925, Vol. 2, p. 877).

[3] This is not, of course, to say that Vidal de la Blache was not acquainted with the works of his predecessors and contemporaries. On the contrary, few men have drawn with such facility from the whole corpus of geographical literature from Greek times onwards.

main purpose of geographical study was to investigate the conditioning of society by environment, in Vidal's scheme they were linked because they were inseparable. Any physical environment in which a society settles is greatly affected by the presence of man, the more so if the society has an advanced material culture. The plant and animal life of France, to take an obvious example, was vastly different in the nineteenth century from what it would have been if there had been no settlement there by man. Equally, the adjustment of each society to the peculiarities of the local physical environment, taking place over many centuries, produces local characteristics in that society which are not to be found elsewhere. Man and nature become moulded to one another over the years rather like a snail and its shell. Yet the connexion is more intimate even than that, so that it is not possible to disentangle influences in one direction, of man on nature, from those in the other, of nature on man. The two form a complicated amalgam. Vidal often reiterated that he was not studying a people but a landscape, yet he chose to do this in a way which came close to denying the distinction between the two. The area within which an intimate connexion between man and land had grown up in this way over the centuries formed a unit, a region, which was a proper object for geographical study. Since each region was so much a product of local circumstance, both social and physical, what was significant in one area might prove to be irrelevant in another. His conception of the subject weighted it strongly in favour of the regional and against the systematic treatment of material.

It was essential to the best flowering of Vidal's method that the society living in an area should be 'local' and that it should be basically rural. It must be local in the sense that the bulk of the materials used as food, for building, for fuel, in the manufacture of tools and machines and so on, should be of local origin. Each small region might conduct a trade with other areas in special commodities but the basic stuff of life and work was to be local. This naturally gave rise to typical regional foods and dishes, styles of domestic and farm architecture, clothes, and so on. Equally, it must be rural, rooted in the land, with the bulk of the population either working on the land or servicing those who did. Even the bourgeoisie and the local landed gentry might find it difficult to break free from local, rural patterns of life. The peasant was deeply imbedded in them (see e.g. Vidal de la Blache, 1911, pp. 384–5). Hence Vidal's interest in the minutiae of the material culture of each small area. He dwelt at some length on the importance to geographers of ethnological museums in this connexion. Within them might be preserved the whole range of tools with which a society went about its daily business, not merely the instruments used for grinding corn, of working wood, or ploughing the soil, but also the houses, the types of clothing, the methods of heating and lighting which were or had been in use (Vidal de la Blache, 1922, pp. 119–21). These would all be made from local materials and designed to overcome local difficulties or take advantage of local opportunities. With such a display before him a geographer should be able to read off many both of the main and the more detailed characteristics of the environment in which they were developed.

To emphasize the continuity of traits in the life of societies of this sort Vidal made use of an arresting image. He reminded his readers that the surface of a pond may be ruffled by a passing breeze so that the eye of the watcher is unable any longer to see the bottom of the pond through the clear water, but that as soon as the wind dies down and the waters are again calm the bottom is once more visible and all the old contours may again be seen undisturbed by the movements of the surface water (Vidal de la Blache, 1911, p. 386). In just the same way the advent of war, pestilence, famine, rebellion, may appear to disrupt the life of a region and throw its steady routines of action into chaos, but once the crisis is over the same long-established pattern of life, of working and holding the land, of building, of clothing, of feeding, even of trade, will reappear. Change may come in such communities. Vidal showed much change in Alsace and Lorraine in *La France de l'Est* in the centuries before the French Revolution, but it was change within a continuing dialectic of man and land, the elaboration of a pattern set by the exigencies of life and soil.

All this was rooted in a very apt appreciation of a truth of great importance. The communities of Europe throughout the medieval and early modern period were rural, were local, and were the result of a long interplay between man and land. There were important local variations in material culture which gave to each region characteristic styles of domestic architecture, clothing and food. In some cases and for some classes in the community there were national rather than local characteristics, but Vidal's was a powerful and legitimate vision of the functioning of societies during most of European history. It was, however, ironically, a vision of things past or about to pass, not a vision of things present or to come. The method developed by Vidal de la Blache is admirably suited to the historical geography of Europe before the Industrial Revolution, or indeed to the limited and shrinking areas of the world today whose economies are still based on peasant agriculture and local self-sufficiency in most of the material things of life, but it is not applicable to a country which has undergone industrial revolution.

What gives to the work of Vidal de la Blache its special interest is not just the fact that he conceived a new attitude to the organization of geographical material and founded a very influential 'school', but also that the rigour of his argument led him to recognize that his method could not cope with the aftermath of the Industrial Revolution. This can be seen clearly in the *France de l'Est* (1917), perhaps his most original and important work. The *France de l'Est* is devoted to the study of the formation of the landscapes and rural societies of Alsace and Lorraine. It covers a period of two millenia during which the landscapes and societies emerged. As a result a large fraction of the book is taken up with the period before 1789. The method of treatment is chronological in the main and there comes a point in the later stages of the work where the finely developed dialectic between man and land[1] which he had been at such pains to pursue over many centuries suddenly begins to fail to comprehend and make sense of the course of events. In the middle of the nineteenth century – Vidal actually found a date, 1846, for it (1917, p. 126) – the waters of the pond were ruffled by something much more disturbing than the storms of the past, something which did not leave the contours on the bottom of the pond as they had always been but caused them to adopt quite a new configuration. The advent of the steam-engine, the railways, the coal-carrying canals and of the new Alsatian cotton industry did not mean the superimposition of a few new strands upon an old-established pattern, but was the first stage in the dissolution of the traditional, rural, local, regional pattern of life. Vidal was too shrewd and conscientious not to recognize this. Whereas the industries of the past were easily assimilated into the model he proposed, the new industries represented a new type of society. The new society was able to produce industrial goods on a vastly bigger scale, and was formed round cheap and speedy communications. Food, clothing, building materials, tools, all soon ceased to be locally made and different in one region from its neighbour. Instead, from the

[1] 'A people, great or small, possesses a personality, whose appearance, like anything else, must submit to the erosion of time, but yet it keeps through the ages the fundamental traits which it acquired as it developed in the region where it settled. . . . It is in conjunction with time, and face to face with the soil, that their traits became fixed once and for all, and thus a personality became established which cannot but be noticed and deserves respect' (Vidal de la Blache, 1917, p. 43).

mid-nineteenth century onwards in Alsace and Lorraine, and from a much earlier date in England, the leading characteristics of the traditional society and economy crumbled slowly away. As a result geographical methodology has been obliged to abandon the 'regional' concept of the subject, just as half a century earlier it had to abandon the 'classical' scheme. The changes in society and economy which produced this situation did not of course come overnight, nor did they affect all areas equally or equally quickly, but the Industrial Revolution began a process which today leaves only one man in twenty on the land in this country[1] and a small and falling fraction on the land in all other materially advanced countries; which has left almost no typical local foodstuffs, clothes or house types; which leaves us all dependent upon a network of communications covering the whole globe. The basic stuff out of which Vidal fashioned his analyses is no longer to be found in Europe and North America, though it is still to be seen in parts of Africa, Asia and South America. Advanced communities are no longer local, no longer fundamentally rural, no longer characterized in their material culture by a host of features which are not to be seen elsewhere.

Vidal de la Blache regretted what he could not help but observe. He considered that much that was best in the life of France arose out of the range and balance of regional communities to be found there. He considered, like many of his French contemporaries, that the moral qualities of rural life were important to the nation and feared their decay. It was only natural also that he should regret the onset of a train of events which was in time to make nonsense of his life's work, but it is a measure of his stature as a scholar that he not only saw that the change was coming, but also suggested how sense would come to be made of the new order of things. He noted in the *France de l'Est*, for example, that the organizing principle of economic life in the future would be the relationship of an area to the metropolitan centre to which it was subservient, that is a relationship born of ease of access to an urban centre, rather than a rural relationship between man and land such as had been so long the case.[2]

The work of Vidal de la Blache was, of course, only one of a range

[1] In December 1963 there were 862,000 persons employed in agriculture, forestry and fishing out of a national total of 24,234,000, or about 3·6 per cent.

[2] 'The idea of the region in its modern form is a conception to do with industry; it is associated with that of the industrial metropolis' (Vidal de la Blache, 1917, p. 163).

of variants within the general 'regional' view of geography. This view gained widest currency between the wars under the umbrella title of landscape or *Landschaft* geography.[1] At the end of the inter-war period Dickinson (1939), argued that landscape geography was the main recent development in the subject and that Britain was out of step in being much less fully committed than the continent to this view of the subject whose acceptance he strongly urged. Landscape geography developed many forms, and it is necessarily hazardous to generalize, but in outline the starkest form of landscape geography consisted in the examination of all that was visible on the surface of the earth (Brunhes' use of the idea of surveying the surface of the earth from a balloon illustrates both what was intended and the vintage of the concept), and the investigation of the characteristic associations of phenomena which existed there. A part of that which is visible consists of the works of man, part may be chiefly natural, a great deal is, as it were, an admixture of the two – fields, crops, animals, afforested areas, etc. It was frequently remarked that landscape geography provided the discipline with its own peculiar subject matter. This 'classical' geography had not possessed. 'Moreover, a science cannot be defined on the basis of particular causal relationships; it must have a definite body of material for investigation. The recognition of this fact, and the direction of research to the study of landscape and society on the lines here presented, is the most significant and widespread trend of post-war geography in other countries' (Dickinson, 1939, p. 8). Landscape geography developed a scholasticism of its own so that there were arguments about whether a house as a more permanent and fixed feature of the landscape was more 'geographical' than a man; and in a broader context there was much argument about how far it was proper to take into account things which were not visible. Landscape geographers of a purist inclination either excluded economic geography altogether or wished to see it relegated to the status of an associated subject rather than an integral part of the discipline, because it was by nature systematic and involved the consideration of abstract and general principles rather than the concrete reality of the landscape.[2] Landscape

[1] The terms are used here loosely and in a very general sense. Hartshorne devotes a chapter to the problems of the correct use of the terms; see Hartshorne, *The Nature of Geography* (1939), pp. 149–74.

[2] See, e.g. M. A. Lefèvre, *Principes et problèmes de géographie humaine* (1945), esp. pp. 29–30 and 195–6. The book in general is an interesting and typical example of landscape geography of a rather strict type.

geography had its strengths – the inter-war generation of continental scholars produced much of interest and value in conformity with its precepts – but it suffered from many weaknesses. Above all it suffered from the same fundamental shortcoming which afflicted Vidal's work. All variants of the 'regional' view of geography are at their best when dealing with areas of rural, local economies. All are ill at ease when dealing with areas thoroughly caught up in the Industrial Revolution. Yet the 'regional' period of geographical methodology, like the 'classical', has left many traces, some of which will perhaps prove permanent, on the methods used in organizing and presenting geographical material. Any discipline is both the product and the victim of its own past successes and these were two of the most important successes thrown up by geographical scholarship.[1]

WHAT REPLACES 'REGIONAL' GEOGRAPHY?

The view that the study of the region and regional life was the peculiar crown and peak of geographical work, that which held the subject together, that which solved most of the methodological difficulties which had become apparent in 'classical' geography by the turn of the present century, is no longer tenable. There has been, however, little, if any, retreat from regional geography, if by that one means the study of things in association in area, which still affords endless opportunities for *ad hoc* studies. The regional method thus remains the means for much geographical work but is no longer its end. One may say that much geography is still regional, but no longer that geography is about the region. What we have seen is a concept overtaken by the course of historical change. 'Regional' geography in the great mould has been as much a victim of the Industrial Revolution as the peasant, landed society, the horse and the village community, and for the same reason.

Granted that the 'regional' concept of geographical methodology has lost its general appeal because the western world has changed so rapidly, one might argue that it is possible to adopt two attitudes to

[1] These are not, of course, the *only* two general views of the subject which have gained wide currency. There is, for example, the inversion of the 'classical' system, the view that geography is the study of the effect of man on land, of the effect of the material culture of societies on local ecological, hydrological and other systems. This has supported and still supports a very interesting literature.

the question of the place of regional analysis in geographical work when dealing with communities which have been drawn into the Industrial Revolution. One may hold either that with the final disappearance of the old local, rural, largely self-sufficient way of life the centrality of regional work to geography has been permanently affected – that in one respect we are back with the 'classical' early nineteenth-century position where regional study was important but on the whole less important than systematic study. Or, secondly, one may argue that what departed with an older type of economic life was only one type of regional study and that the general significance of the region to geographical work remains unchanged. On this view all that has happened is that there are now different fish to catch in the sea, fish which escape the older sort of net which Vidal made so well, but which are worth catching and can be had with different types of net or methods of fishing.[1]

On the whole the first of the two alternatives seems the truer. The line of intellectual descent which began with von Thünen and J. G. Kohl and leads down through Alfred Weber, Christaller, Lösch and Isard has perhaps supplied the most fruitful of the ideas which have enabled geographers to tackle the question of the regional ordering and functioning of economy and society in post-industrial communities, and their thinking is, of course, systematic in nature, though very flexible for use in special studies. It is true that in a sense the Industrial Revolution has made possible a degree of regional differentiation of economic activity which was not possible earlier and in this way brought out regional distinctiveness with a sensitivity not previously seen. In the days of substantial local self-sufficiency, for example, corn was grown for local consumption in many parts of England where today the land is largely down to grass. Each agricultural area of Britain can realize advantages of site and soil which remained potential only as long as transport was expensive and uncertain and markets little developed. Or again, to contrast the agricultural economy of California, with its specialization on a range of crops peculiarly suited to its climate and irrigation possibilities, with the agricultural economy of an area physically similar, such as central Chile, where a wider range of basic foodstuffs is grown, is to see the implication of modern high-speed, cheap communications and easy access to great consuming centres. But these changes and the

[1] Either view is consistent with the assertion that regional and systematic studies are both necessary. This was recognized in the main by both 'classical' and 'regional' geographers. The difference is a matter of emphasis.

contemporary pattern of regional specialization are only intelligible in terms not of one region but of a whole congerie of interlocked economies. Furthermore, the great bulk of employment in modern industrial countries is to be found in secondary and tertiary occupations, not on the land, and in these days when in almost all industries the most important locating factor is accessibility to the major markets, this means that a systematic treatment alone holds out hope of understanding.

It is notoriously difficult to be clear about the trend of contemporary events, but the question which naturally suggests itself at this point in the argument is, of course, whether, granted that the two great earlier conceptions of the subject are inadequate, some new over-arching conception of the nature and methodology of the subject is emerging. Since the situation is alive and changing, any observation about this is likely to contain an element of advocacy as well as analysis. This section of the essay will have served its purpose, therefore, if it helps to stimulate further thought.

One may begin by remarking that geography and geographers do not live in intellectual isolation. Both 'classical' and 'regional' methodologies of geography were closely related to the general intellectual history of their day. The 'classical' view of geography was a natural avenue to explore given the contemporary assumptions about the nature of scientific knowledge and the way in which it should be organized. Equally the 'regional' idea in the hands of a man like Brunhes has many affinities with 'functional' social anthropology[1] and with the whole argument at the turn of the last century about the methodology appropriate to the social as opposed to the physical sciences. In the same way many contemporary developments in geographical technique and in ideas about geographical methodology are linked to thought in the social and physical sciences in a wider context.

One aspect of this which appears to be of singular importance is the application of statistical concepts and devices to many new areas of study. This is true of social sciences like economics where in the last generation econometrics had greatly flourished; of the biological sciences; and of more utilitarian branches of study like operational

[1] The meaning which Brunhes attached to the word *connexité* and its importance in his organization of material make this clear. There are many points of resemblance between what he was attempting to do for geography and what men like Malinowski or Raymond Firth wished to do for social anthropology.

research where the use of statistical techniques and computers has made possible rational planning of such things as the holding of stocks or the optimal use of machinery.

The use of statistical techniques has spread rapidly in geography in the last generation. Statistical methods are now commonly used in dealing with questions like the testing of regional boundaries; the spacing, size and areas of influence of settlements; locational theory; migratory movements; characteristic crop combinations and plant associations; and a host of geomorphological and hydrological questions. Sometimes such studies make use of available statistical techniques; sometimes modifications are used designed to help especially with the measurement of association in area.[1] Although geographers have been rather slow to make use of the opportunities offered, statistical techniques are peculiarly well suited to geographical problems for two reasons. In the first place some statistical techniques are capable of bringing into a meaningful relationship to one another a large number of variables which may be only rather weakly correlated with one another, or which may be significantly related only when combined and considered in groups. This characteristic is a godsend to geography because geographers have often wished to hold within the focus of their attention a large number of rather disparate factors, all of which are thought to be of some importance, but they have commonly been unable to establish the nature or degree of intensity of relationship between the many elements in the situation. Attempts to overcome the difficulty by intuitive assessments have often been strikingly unsuccessful, leaving the impression that more had been bitten off than could conveniently be chewed. If the welter of possible interconnections can be examined statistically and an accurate measurement of correlation made a much firmer foundation for analysis is available. It needs to be said repeatedly that statistics is an aid to good judgement rather than a substitute for it, but it is a very powerful aid and enables many problems to be examined again or for the first time which could not be tackled previously, or tackled only perfunctorily without statistical aid.

Secondly, statistical techniques are likely to be attractive to geography because they help with one of the subject's most intractable methodological problems of the last three-quarters of a century since the decay of the 'classical' school, the recurring worry about the best way to accommodate in a single discipline a physical and a social

[1] See Chapter 9.

side.[1] This issue was always at its most sensitive when the question of the influence of the physical environment on social change and functioning was raised, but arose in other connexions also. Experience made men wary of treating these questions in an '*A* caused *B*' way yet geographers continued to be interested in questions which demanded a knowledge both of society and environment. It was in this connexion that Brunhes advocated the idea of *connexité*, interconnection, functional relationship. The use of statistical techniques permits greater precision in these matters and, if properly used, helps in avoiding some pitfalls frequently visited by the unwary in the past. In short both the practical and methodological problems of geography are such that the use of statistical techniques is likely to exercise a strong attraction.

It may be remarked that it is one thing to say that statistical techniques are very useful and another to say that this is a development equivalent to the rise of the 'classical' or 'regional' conceptions of the subject; that statistical techniques are only a tool; that they cannot be a methodology in themselves; that they cannot supply a general vision of the subject or give it the sort of unity that the older conceptions, whatever their defects, provided. This is true. Geographical writing and research work has in recent years lacked any generally accepted, overall view of the subject even though techniques have proliferated. This, where it has been recognized, has been widely regarded as a bad thing. A unifying vision is a very comforting thing, but one may perhaps question whether it is as vital a thing as is sometimes supposed. Without it there is always a danger of a slow drifting apart of the congerie of interests which together make up a subject. With it, on the other hand, there is also danger from rigidity and from the creation of an orthodoxy. At all events it is arguable that the best sign of health is the production of good research work rather than the manufacture of general methodologies, though perhaps the two together are to be preferred to either singly if they form a fruitful harmony, as in the work of Vidal de la Blache.

Perhaps the most sensible attitude now as at other times to adopt towards the question of method in geography is to be eclectic – to use whichever method of analysis, Blachian or systematic, landscape or Löschian, appears to offer the best hope of dealing with the problem

[1] This is one of the most fundamental questions in the recent history of geographical methodology and deserves study on a much larger scale. It is perhaps the best point of departure for a history of geographical writing in the last hundred years since it is central to so much else.

in hand. There is no reason, for example, why a study similar to that of the *France de l'Est* or others done by the followers of Vidal should not be written for many parts of the underdeveloped world today where life is still essentially local and rural. However, for reasons already touched upon, geographers tend to find it hard to leave it at that. Both the older 'classical' school, because of the question to which they addressed themselves, and the followers of Vidal de la Blache, because of the way in which they conceived of the region, held the physical and social halves of geography very firmly together, indeed in the second case they they were held to form a seamless robe which could not be perceived except as a unity. If both the view of geography as the investigation of the effects of the physical environment upon social functioning and development, and the view of it as the examination of the region in the manner of Vidal de la Blache are rejected there is an evident danger that the links between the two halves of the subject will be weakened. This makes little or no difference to individual pieces of research. There are now and have always been since the days of Humboldt and Ritter, and indeed much earlier, particular pieces of work in which a knowledge of both social and physical geography have been essential, and others which were purely physical or social. But in the general sense, viewed overall, the connexion is weakened.

It would be mistaken to suppose that this is a new situation, that we are moving away from a period when there were no uncertainties, seeing the breaking up of something which in times past was a firm whole. Even at the beginning of modern geography the same problem was present. When Ritter wrote the introduction to the *Erdkunde* he intended to survey each of the earth's continents in turn. In the event he was able to deal only with Africa and parts of Asia. This is a pity because he also said that he considered Europe to be different from other continents (following Hegel) because hers was not a static but a developing civilization, and he suggested that whereas the study of the local physical environment might provide many clues to the functioning of the static civilizations of other continents it would be of much less utility in Europe.[1] In other words he foresaw that when dealing with Europe there would be difficulty in holding the two sides of the subject in the same close conjunction that was possible elsewhere. From Ritter's day to the present this and similar questions have

[1] See e.g. Ritter, *Allgemeine Erdkunde* (1862, p. 229). Buckle has much of interest to say on this theme: see especially *Civilization in England*, Vol. 1, Chap. 2.

given rise to discussion, at some times desultory, at others urgent. The degree of overlap between the two halves of the subject has always varied from topic to topic. This made the existence of an overall vision the more important if a clear connexion between the two were judged of supreme importance. In general the more backward in material culture, and the more rural in nature, the closer the evident connexion: while the connexion is least obvious in the industrial and urban countries of today. An eclectic attitude towards geographical method in work on the 'advanced' countries today will tend to underline the comparatively slight degree of overlap which exists.[1]

Handsome is as handsome does. The final test of the value of any intellectual labour is its ability to help men to understand questions in which they are interested. Men such as Ritter and Vidal de la Blache in their day succeeded notably in this. Questions of method in geography in the future as in the past will be decided by the quality of the work produced by men of different methodological persuasions. Progress lies in rejecting conceptions which are no longer fruitful in favour of those which can help the understanding. All such schema are provisional: in time they will be replaced by others which meet contemporary needs better. Intellectual development is a continuing process of modification, rejection, addition and replacement of conceptual tools. The more fully past experience in geography is digested the more likely it is that contemporary discussions will be productive. *Reculer pour mieux sauter* is good advice here as in other connexions. Only when the merits of long-standing methods are seen in their original setting can their present utility be adequately judged. The most complete prisoners of the past are those who are unconscious of it.

[1] Forde contributed an interesting article to the little burst of methodological writing which occurred in the *Scottish Geographical Magazine* in 1939, and in dealing with a similar issue came to the conclusion that 'Actually the human geographer stands in need of a knowledge of physical environment to precisely the same degree as do the archaeologist, the ethnographer, and the economic historian' (Forde, 1939, pp. 229–30). And the published work of many human geographers bears him out in his contention.

References

BRUNHES, J., 1925, *La Géographie humaine*; 2 vols, 3rd ed. (Paris).

BUCKLE, H. T., 1857–61, *History of Civilization in England*; 2 vols (London).

DICKINSON, R. E., 1939, 'Landscape and Society', *Scot. Geog. Mag.*, **55**, 1–15.

FORDE, C. D., 1939, 'Human Geography, History and Sociology', *Scot. Geog. Mag.*, **55**, 217–35.

HARTSHORNE, R., 1939, *The Nature of Geography* (Lancaster, Pa.).

HEIDELBERGER GEOGRAPHISCHE ARBEITEN, 1960, *Alfred Hettner: Gedenkschrift zum 100 Geburtstag* (Heidelberg).

HUMBOLDT, A. VON, 1849, *Cosmos*; Transl. by E. C. Otté (London).

LEFEVRE, M. A., 1945, *Principes et problèmes de géographie humaine* (Brussels).

RATZEL, F., 1882–91, *Anthropogeographie*; 2 vols (Stuttgart).

RITTER, K., 1862, *Allgemeine Erdkunde* (Berlin).

VIDAL DE LA BLACHE, P., 1911, *Tableau de la géographie de la France* (Paris).

— 1917, *La France de l'Est* (Paris).

— 1922, *Principes de la géographie humaine* (Paris).

CHAPTER TWO

A Re-evaluation of the Geomorphic System of W. M. Davis

R. J. CHORLEY

Lecturer in Geography, University of Cambridge

As the work of William Morris Davis recedes further into the past and becomes more and more identified with the basic structure of what might be termed 'classical geomorphology' its outlines blur and the impression which it commonly produces is one of intangible strength. His voluminous, repetitive, but often subtly-modulated writings appear to us as through a haze of secondary interpretation producing unreal optical effects, not the least striking of which is that the cycle of erosion concept seems large enough to embrace the whole of geomorphological reality. An added distortion results from Davis' avowedly prime position as a teacher, for the achievement of a teacher must be judged largely by the effects which his teachings produce. Thus, an evaluation of the cycle of erosion theory requires that one distinguishes between the stated intention of Davis as conveyed by his writings, the implicit and often unstated assumptions underlying his work, the sometimes distorted interpretations of Davis' work stemming from his students, and, finally, the effect of his teaching on succeeding generations of geomorphologists.

Increasingly, however, since the death of Davis in 1934, criticisms of the cyclic approach to landform study have been intensifying, but the vague and confused form taken by these criticisms and, in particular, the form in which they have reached teachers has not permitted the true character of these objections to appear clearly. It is true to recognize also that the obvious teaching qualities possessed by the cycle concept have hardened the resistance of teachers to these criticisms, and it must be stressed at the outset that modern objections to the Davisian approach have not developed because the cycle has been found to be a totally inappropriate vehicle for geomorphic thought or teaching, but because its restrictive and highly-specialized, built-in characteristics have been high-lighted by recent investigations.

The cycle of erosion is thus now being recognized as merely one framework within which geomorphology may be viewed, wherein those aspects of landforms which are susceptible to progressive, sequential and irreversible change through time are especially stressed (just as the system of Euclid is now considered as merely one of many 'geometries'). The cycle is no more a complete and exclusive definition of geomorphic reality than the pronouncement by the proverbial Indian blind man on feeling an elephant's leg that the animal is like a tree. What has happened in the last thirty years or so is that, to continue the metaphor, other blind geomorphologists have been feeling the geomorphological elephant's trunk and sides and are variously describing it as being like a snake or a wall. It is understandable that the equally-blind 'onlookers' should have become confused and vaguely resentful, particularly because they have been brought up to believe that trees are much more rational, beautiful and believable things than either snakes or walls!

THE MODEL

The strongest and most compelling feature of the cycle of erosion concept is that it presents many of the features of a theoretical model (Chorley, 1964). As distinct from classification, which merely involves the dissection and categorizing of information in some convenient manner, model-building requires the identification and association of some supposedly significant aspects of reality into a working system which seems to possess some special properties of intellectual stimulation. This stimulative quality, often resulting from the special juxtaposition of information which is the very foundation of the structure (one is almost tempted to say the 'artistic form') of the model, finds expression in an enlargement of what is thought of as 'reality' (i.e. involving the kind of scientific prediction which resulted from the construction of Newton's model). It was thus very characteristic of Davis' intellectual achievement that after he developed his cyclic model he was able to say that he could think of many more landforms than he could find examples of in the field! A moment's reflection on the magnitude of possible combinations of structure, process and stage in landforms shows exactly what he meant. It is therefore in the bringing together of certain aspects of the 'web of reality', stripped of other considerations, into a clear-cut theoretical model that much of the intellectual attractiveness and teaching strength of the cycle lies. Davis (1909, pp. 253–4) knit certain aspects of landforms together into a meaningful association both in space within a given landscape and in time throughout an assumed evolutionary history. In order to understand many of the special characteristics of the cyclic theory, and in particular to recognize both its strengths and limitations, it is profitable to consider it in the light of three of the properties common to all such models – their essentially theoretical character, the inherent need for the discarding or 'pruning' of much information, and the fact that no part of reality can be uniquely and completely built into any one model. Davis (1909, p. 281) himself recognized and defended the theoretical nature of his model, writing '. . . the scheme of the cycle is not meant to include any actual examples at all, because it is by intention a scheme of the imagination and not a matter for observation; yet it should be accompanied, tested, and corrected by a collection of actual examples that match just as many of its elements as possible'. Rather than being surprised by such a statement, one should recognize this as a very characteristic state of mind for the model builder in a natural science, where the subject matter of even a small part of reality has usually to be accepted in uncontrollable mutual associations. The result is that there is an attendently large 'elbow room' within which the researcher may select, organize and interpret his material, introducing a subjective bias into all work – good or bad. This brings one to the second model property, that much possible information relating to even a small part of reality has to be rejected in order that the rest (i.e. that information and those relationships which appear especially significant or interesting) may be presented in sharp outline. All models caricature reality by this pruning, but the most successful ones (e.g. that of Newton) are those wherein that which remains still retains some observable or testable significance as far as the 'real world' is concerned. Davis' cyclic model is heavily pruned, such that changes in the geometry of erosional landforms through time emerge as the central theme. When he excluded the possible effects of climatic change or of progressive movements of base level from his scheme, Davis was not (as he asserted, e.g. 1909, p. 283) doing so to facilitate and simplify his *explanation* but to make the cyclic scheme *possible at all.* One can only imagine what would have remained of the cycle if Davis had permitted the possible effects of continuous movements of base level to have been superimposed upon those associated with the progressive subaerial degradation of the landmass. This is, in fact, just what Walther Penck attempted to do, and is the reason why his model is much more confused and unsatisfactory than the cycle. The third property of models which I think is appropriate here follows directly from the second. If discrimination and selection operate in terms of the building of information into a model structure then no single model can form a universally appropriate approximation to a segment of reality. It is interesting that it was one of Davis' most faithful supporters, Nevin Fenneman (1936), who, almost inadvertently one feels, stated this most cogently: '. . . the cycle itself is not a physical process but a philosophical conception. It contemplates erosion in one of its aspects, that of changing form. But erosion does not always and everywhere present this aspect. . . .' 'Cycles have parts and the parts make wholes, and the wholes may be counted like apples. Non-cyclic erosion can only be measured like cider. There is neither part nor whole, only much or little.' It is around the essentially non-cyclic model of Grove Karl Gilbert (Gilbert, 1877, Chapter 5; Chorley, 1962) that much modern thought is centring.

THE DOGMA OF PROGRESSIVE, IRREVERSIBLE AND SEQUENTIAL CHANGE

Another feature of the cycle of erosion concept, one which is basic to the whole reasoning underlying it, is the tacit assumption that the amount of energy available for the transformation of landforms is a simple and direct function of relief or of angle of slope. This unformulated, but nevertheless real, assumption on the part of Davis (and one which seems so logical in the abstract as to be unquestionably accepted as an axiom) is that rates of mass transfer by all agencies are greater on steeper slopes than on less steep ones. From this assumption many others are deduced – some apparently true and others, often not so apparently, untrue. The ideas, for example, that steep slopes are eroded faster than less steep ones and that stream velocity is solely dependent on bed slope, derive from this axiom, and lead inevitably to the conclusion (Davis, 1909, pp. 255–6) that rates of change of landforms, as well as their geometrical magnitude, are direct functions of local relief. It follows, therefore, that the progressive changes of relief during the consumption of a landmass by erosion are held to be universally associated with a progressive landscape evolution wherein the geometry of individual landforms and the rates of their erosional change are both subject to sequential transformations through time. Considering individual valley-side slope elements and stream reaches, for example, this reasoning leads to the assumption that they are progressively transformed into lower and lower energy (i.e. gradient) forms as the general relief is reduced following late youth (Davis, 1909, pp. 268–9). The study of change is therefore the guiding purpose of the 'geographical cycle', wherein a sequence of sketchily-treated changes leading to ill-defined conditions of 'grade' (Chorley, 1962) in stream channels and slopes, is followed by a progressive, sequential and irreversible transformation of virtually all aspects of landforms as the potential energy (i.e. relief) of the system is dissipated. Although it is obvious that the general reduction of a landmass by erosion must be associated with broad changes involving in the long run the replacement of steeper slopes by less steep ones, modern research is indicating that the relationship between gradient and rate of mass transfer is more partial and complex than Davis assumed (this word is deliberately employed in that Davis never attempted to test this axiom), such that some aspects of landscape geometry may be relatively unchanging throughout large segments of 'cyclic time', whereas the detailed pattern of change of others may be neither progressive nor continual. Recent works relating gradient and process by Leopold (1953) and Young (1960) have respectively demonstrated, for example, that both mean velocity and bed velocity of rivers *increase* downstream (the increasing depth of flow more than offsetting the decrease of bed slope, as embodied in the eighteenth-century Chézy flow formula), and that the rate of soil creep on some slopes seems to be more strongly controlled by the frequency and amount of moisture changes than by the slope angle.

In order to illustrate both how the built-in sequential assumptions of Davis inevitably lead to highly stylized and restricted concepts of change of form through time and how modern research is questioning these concepts, we can profitably examine three aspects of landscape geometry – drainage density, erosional slopes and river meanders.

A cyclic interpretation of the development of drainage density (the total length of stream channels in a given area divided by the area) was given by Glock (1931; see also Wooldridge and Morgan, 1959, p. 173) wherein drastic changes through time were inferred, but Melton (1957) and Strahler (1958) have shown that the factors which seem to exercise the most important control over drainage density are those of rainfall characteristics, infiltration, surface resistance to erosion and runoff intensity, rather than those (e.g. relief) which might dictate significant or progressive changes of drainage density through time. It seems that, virtually independent of relief or 'stage' in any cyclic scheme, drainage density (which is probably the most important single parameter of landscape geometry) is most characteristic of the rainfall/infiltration characteristics of a region and thus may be relatively unconnected with stage throughout long periods of erosional history.

Valley-side slope (which combines with drainage density, relief and upper slope curvature to virtually define the geometry of erosional landscapes) was treated by Davis (1909, pp. 266–9) as beginning steep and irregularly covered with coarse debris in youth and as getting progressively less steep with age as it becomes composed of a thickening mantle of finer and finer debris. As has been mentioned above, there can be no dispute as to the decrease of *average* erosional slope angle as relief is lowered, but modern work is shedding some interesting light on the detailed pattern of the development of slope elements within the broad framework of surface degradation (which is, after all, a concept which predates Davis by several hundreds of years). It is now patently apparent that it is completely unrealistic to hold rigid views as to any unique pattern of slope development, even within a given region of uniform structure, lithology or climate. The geometry of valley-side slopes is controlled by a number of interlocked variables which may operate in very different combinations and magnitudes. Thus, within a single climatic environment some slopes may recline whereas others retreat parallel (Schumm, 1956 B); in the same lithological environment and 'stage' of dissection adjacent recline and parallel retreat can be deduced (Strahler, 1950 A, p. 804); and in some limiting cases a steepening of slopes can be deduced through part of their history (Carter and Chorley, 1961).

Davis (1909, p. 265) associated the development of meanders with the practical cessation of downcutting at 'grade' when the continuance of outward cutting changes the curves of youth into systematic meanders of radius proportional to the river's discharge, which increase in size progressively as the gradient of the flood plain lowers during the subsequent progress of the cycle. Thus the existence of meanders, together with their magnitude, is viewed as having some time-significance in terms of 'stage' within the cycle. The vexed question of grade has been touched on elsewhere (Chorley, 1962), but it is profitable here to examine briefly how modern research bears on Davis' interpretation of meanders. At the outset it is important to recognize two elements of the problem which are interconnected but quite distinct – the question of the initiation of meanders and that of the relationship between meander form and fluvial processes once the meanders have developed. Obviously the first question is the more difficult one, but it is now reasonably certain that the regularly spaced pools and shallows (riffles) of meandering rivers are comparable with similar features in straight streams (Leopold and Wolman, 1957; Leopold, Wolman and Miller, 1964), that the deposition of these riffles probably occurs during the time of falling discharge (when the decreasing thread of water becomes more and more deflected), and that meandering can develop in streams which are in the broadest sense, aggrading, degrading, or 'poised' (Matthes, 1941).

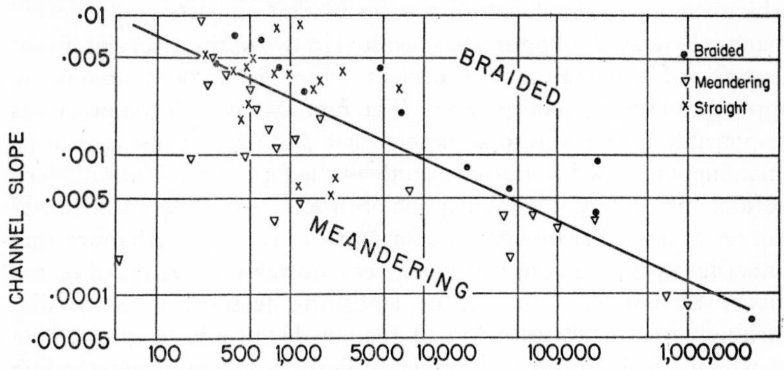

FIG. 2.1. *The control over the meandering or braided condition of rivers exercised by bankfull discharge and channel slope (after Leopold and Wolman, 1957).*

The relationships between meander geometry and discharge have been exhaustively examined in actual rivers, irrigation canals and in model flumes, but some recent work by Leopold and Wolman (1957), involving observations directed towards the second of the questions identified above has had an important bearing on the first question. Leopold and Wolman found that whether streams meander or braid seems to depend on a fairly simple multiple relationship between bankfull discharge and channel slope (the latter being partly a function of the calibre of the bed material) (Figure 2.1). It is thus apparent that whether a stream develops a sinuous meandering course on unconsolidated flood plain material is really determined by *chance* relationships between discharge and slope (calibre) and does not *per se* have any time or 'stage' significance. Having said this, it is only fair to add that with the passage of time the decrease of bed calibre and channel slope makes the *chance* of obtaining a meandering condition (as distinct from braiding or straightness) *more likely*, but this is a rather different statement from that of Davis. In short, many geomorphic form changes are now being viewed from a more *ad hoc* standpoint than they were by Davis.

When one compares the assumptions of Davis with the findings of more recent workers regarding the three examples given above, an outstanding feature of Davis' reasoning becomes abundantly clear. This is that given aspects of landscape result simply from a small number of given causes – often from one single cause – which operate through time in a progressive and sequential manner. This reasoning, which was often employed in a more blatant and less sophisticated manner by Davis' followers, has resulted in two of the most significant features of the reasoning processes which have been commonly applied to cyclic geomorphology. The first is that both the processes responsible for a given topographic form and its past history can be unambiguously deduced from a study of the form itself. Thus modern notions that different combinations of processes or different histories can result in similar topographic forms are at variance with much that underlies cyclic thinking. In its extreme forms this disregard of the highly ambiguous character of landscape features leads to disturbingly uncritical treatments of geomorphic problems, an example of which will be given in Chapter 8. Perhaps the most outstanding instance of this is the naïve view that river terraces and nickpoints are almost invariably associated with movements of base level, and one has only to turn to the work of Lewis (1944) and Yatsu (1955) – the former showing that terraces can be produced by the varying of stream load/discharge relationships, and the latter that breaks of stream slope can be associated with changes in calibre of the bed load – to see the limitations of this assumption. The second result of the lack of a multivariate view of reality nourished by the cyclic approach is that geomorphic processes tend often to be viewed in an over-simplified manner. Everyone is familiar with such arguments as 'more rain means more erosion' (i.e. desert wadis must have been excavated during more pluvial past conditions) or that 'the intense rainstorms of desert areas are associated with high rates of erosion'. Recent work by Langbein and Schumm (1958) has indicated that rates of sediment yield and erosion tend to be at a maximum in climates having a rainfall of about 12 inches per year (i.e. where rainfall is high enough to cause substantial erosion and vegetation not dense enough to prevent it) and possibly only rising to another maximum at rainfalls exceeding 50 or 60 inches where the impeding effect of vegetation cannot be increased by further increase of precipitation (Figure 2.2).

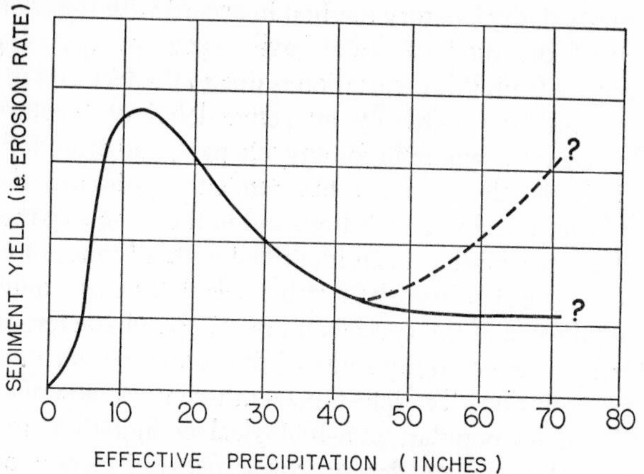

FIG. 2.2. *A schematic suggestion of variations in sediment yield (i.e. erosion rate) associated with variations in precipitation (after Langbein and Schumm, 1958, and Fournier).*

THE INTELLECTUAL SETTING OF THE 'CYCLE'

A proper understanding of the specialized character of the cycle of erosion concept can be achieved only by considering it in the context of nineteenth-century thought in the natural sciences, for, despite the deceptively youthful appearance which its teaching facility has maintained, the cyclic notion was first mentioned by Davis as long ago as 1884 and first stated in fairly complete terms only five years later (Davis, 1889). At this time the writings of Herbert Spencer and others were extending the concept of evolution from the biological into the physical, social and mental spheres such that it seemed to form a basic organizational framework for the whole world of experience. Although prominent Harvard philosophers of the time were resisting this extension (Leighly, 1955, p. 312), there is no doubt that the idea of organic evolution was one of the most important mainsprings of the cycle of erosion theory. In his first statement of the cycle notion Davis (1884) termed it a *cycle of life* in which, as he later wrote, 'land forms, like organic forms, shall be studied in view of their evolution', (Davis, 1909, p. 279), such that the cyclic concept has the 'capacity to set forth the reasonableness of land forms and to replace the arbitrary, empirical methods of description formerly in universal use, by a rational, explanatory method in accord with the evolutionary philosophy of the modern era' (Davis, 1922, p. 594). The main problem arising from this association is due to the fact that in the later nineteenth century the highly attractive label of 'evolution' had practically become a synonym for any 'change', and often for 'history' in general. This identification has tended to obscure the *special character* of the concept of evolution, and in the same way the concept of the cycle of erosion has been identified with all types of change in landforms and with landform history in general. Thus it is only possible to understand some of the special and restrictive characteristics of the cyclic idea by understanding some of the contemporary implications of the term 'evolution'. The late-nineteenth-century view of evolution, particularly in its popular, non-biological connotation, implied an inevitable, continuous and irreversible process of change producing an orderly sequence of transformations, wherein earlier forms could be considered as stages in a sequence leading to later forms. 'Time' thus became, at least for many of those concerned with adapting the evolutionary notion to wider fields, almost synonymous with 'development' and 'change', such that it was viewed not merely as a temporal framework within which events occur but *a process itself*. It was in this sense that Davis employed the concept of evolution as a basis for the cycle of erosion, and it is easy to see why what Fenneman termed 'non-cyclic erosion' seems just as inconsequential to the cyclic concept as the lack of sequential development of certain biological organisms through long time periods seemed to the theory of evolution.

In other ways, too, Davis' synthesis was typical of nineteenth-century scholarship in general and of geographical scholarship in particular. The emphasis upon historical sequences rather than functional associations, the reconnaissance and artistic basis of his field work, and the stress laid upon causal description are features of Davis' work which make it appear most antique to the modern student of landforms. Davis followed Ritter in his concept of the scope and nature of geography, wherein human activities were subordinated to, and largely based upon, the main features of the physical environment. There are many overtones of the 'landschaft' concept of geography implicit in Davis' reasoning, in that it is assumed that the landscape features contain within themselves the unambiguous evidence of their origin. While stressing the Victorian character of much of Davis' work it is only fair to note that he departed from the characteristic standards of much nineteenth-century work in the natural sciences in three important particulars; his lack of detailed field observations, his unconcern with details of processes prompting change, and the entirely qualitative nature of his methods.

This last characteristic leads us to a further feature of the geographical cycle which militates against its popularity with modern workers – its highly dialectical and semantic quality. Anyone at all familiar with the voluminous writings of Davis cannot but be struck with the essentially verbal logic which he employed, characterized by his obsessive concern over terminology. Much of this resulted, of course, from the theoretical basis of his work, but in its extreme form this preoccupation resulted in 'research by debate' (e.g. Symposium, 1940). Speaking of the reaction of the Davisian geomorphologist confronted by a radically different approach to the subject, Bryan (1940, p. 254) wrote . . . 'Slightly bemused by long, though mild intoxication on the limpid prose of Davis' remarkable essays, he wakes with a gasp to realize that in considering the important question of slope he has always substituted words for knowledge, phrases for critical observation.' Again, to be fair to Davis, it must be recognized that he never intended his cyclic theory to be 'scientific' – at least in the sense that the term is currently employed, but there can be no doubt that the long-term effect of his work was to take a whole branch of natural science often intimately concerned with mass, force, resistance, rates of change, and many of the other basic parameters of physics and to effectively divorce it from the main stream of scientific thought. Until the Second World War geomorphology developed very much like a private game, played by comparatively few initiates most of whom were unable or unwilling to draw upon the general body of scientific experience. Thus, although the avowed aim of the cyclic approach to landforms was to provide a general view of the degradational succession of erosional forms, its effect (particularly when yoked to the concept of denudation chronology) was to throw the emphasis upon historical studies of special regions. This idiographic attitude has always found favour with geographers since the collapse of nineteenth-century determinism, but its application in geomorphology successfully isolated the subject from every other science except a small segment of historical geology. Largely deprived of the stimulus of cross-fertilization, geomorphology in the half century after 1890 developed by in-breeding into a highly-stylized discipline wherein the keen edge of research was blunted, and lacking the active professional echelon concerned with practical problems which during the same period, for example, transformed the sister science of meteorology.

THE DEVELOPMENT OF THE CYCLE

In the academic life cycle of William Morris Davis old age was succeeded by rejuvenation and, even as atrophy was setting in with regard to the subject in general, he was during the period 1920–34 revising many of his earlier views. The irony is that what is most easily available to students today as the 'essential' teaching of Davis are certain of his essays written prior to 1909 and the writings of his most influential students. Pre-eminent in both these respects was the editor of *Geographical Essays*, Douglas Wilson Johnson, the Professor of Geomorphology at Columbia University. In his teaching Johnson, who until his death in 1944 held a foremost – almost dictatorial – position in American geomorphology, followed in detail the approach to the subject adopted by Davis in his middle years (Strahler, 1950 B), differing from him only on the spelling of 'peneplain' (peneplane). Davis, however, showed remarkable versatility after the age of seventy, modifying his views on peneplanation and the youthful stage (1922), recognizing the lack of real differences between many humid and arid landforms (1930 A), and acknowledging after setting up permanent residence in California in 1928 the difficulty of applying simple cyclical notions to an area of active orogeny.

However, the scheme of the cycle represented such a compelling geomorphic framework that in a few decades it was extended into all branches of geomorphology, usually much less satisfactorily than had been its application to 'normal' fluvial features. Davis was variously responsible for these extensions: the arid cycle was wholly worked out by Davis (1905, 1909, pp. 296–322), as was the cycle of upland glacial erosion (Davis, 1900; 1909, pp. 658–66); and the suggestion by Davis (1896; 1909, p. 709) regarding the stages of shoreline development was inflated by Johnson (1919) into cycles of submergence and emergence. In contrast, however, he viewed karstic features as developing only as an early mature stage in the normal cycle (1930 B) although a whole cycle of karst development had been deduced in the meantime (Cvijić, 1918, following Sawički and others). The reasons for the lack of current popularity for these cyclic extensions are precisely those which account for the decline of the 'normal' cycle, although in the former instances they are usually more obvious than in the latter. The most important reason has been the specialized and restrictive nature of the initial assumptions, for example, that the arid cycle is referred to block-faulted basin and range structures and that the destruction of glacial mountains is almost entirely attributed to cirque-cutting. The other problem is the one to which I have already referred with regard to the normal cycle, that of the misinterpretations regarding the successive development of topographic forms attendant upon a very imperfect knowledge of process. Thus Shepard (1960) has shown, for example, that barrier beaches (offshore bars), far from being criteria for coastal emergence, are now being observed to develop in association with stationary and rising sea-levels.

DENUDATION CHRONOLOGY AND THE CYCLE

Of all the apparent developments of the cyclic approach to the study of landforms none has been more important than that relating to 'denudation chronology'. During the first half of the twentieth century studies in the sequential development of erosional forms referred to changes in baselevel formed the mainstream of geomorphological work in Britain, the United States and France, under the influence of S. W. Wooldridge, D. W. Johnson and H. Baulig, although there was a different emphasis on either side of the Atlantic regarding the eustatic or diastrophic nature of the baselevel changes (Chorley, 1963). The relationship of the concept of the cyclic development of landforms to denudation chronology is at once complex and ambiguous, such that the two are commonly confused. It is important to realize that studies of denudation chronology, using the term in precisely the same way as it is currently employed, preceded or accompanied the formulation of the cyclical approach by Davis. In the 1860's and 70's Jukes and Ramsay proposed sequences of landscape development involving changes of baselevel; in the 1880's in the United States Joseph Le Conte interpreted breaks in stream profiles as indicative of the discontinuous lowering of baselevel; and one year before the first really important statement of the cycle McGee (1888) developed an erosional chronology for that part of the

Appalachians later to be made classic by Davis (1889), Johnson (1931) and many others. The notion of cyclic change fitted so well into the interpretation of landforms directed pre-eminently towards an evaluation of baselevel changes (which is the real aim of students of denudation chronology) that the two approaches merged, and it is with surprise that we now realize that, in terms of research (as distinct from teaching) all that remains of the cyclic concept can be measured largely in terms of its reinforcement of denudation chronology. The closeness of this association tells us a great deal both about the character of denudation chronology and the cycle. The former, relying often upon highly ambiguous evidence, assumed like the cycle the character of a highly stylized game indulged in by a freemasonry who after commiting themselves to certain basic initial steps of faith (e.g. topographic flat means stillstand; higher is older and lower is younger; uplift is generally discontinuous, etc.) reached conclusions which seem often to be more a product of the means of analysis rather than a physical reality. To adapt an expression of Sauer's (1925, p. 52), many studies of denudation chronology look like the products of men set out to 'bag their own decoys'. However, the best and most convincing studies of this type (e.g. Wooldridge and Linton, 1955) rely for their conviction and satisfaction upon evidence provided by deposits of known origin and date. This permits the true character of denudation chronology to emerge, and it becomes patently apparent that it represents, as McGee stated quite clearly, a branch of *historical geology* in which the central theme is the interpretation of past forms rather than the full understanding of the present landscape. Of course, these two aims may occasionally coincide, particularly when landforms are changing at a slow rate, but an historical pre-occupation for its own sake cannot provide a universal basis for the discipline of geomorphology. Despite Davis' protestations as to the essentially geographical nature of his cyclic approach, the same preoccupation with the deduction of past forms is apparent, and the expulsion of geomorphology from American geography some forty years ago was largely due to this feature of Davis' emphasis. It is significant that, according to Johnson (1929, p. 209), in his last years Davis came to the view that most of his writings were not strictly geographical in character.

THE 'GEOGRAPHICAL CYCLE' AND GEOGRAPHY

This brings us to the last, and undoubtedly the most presently appropriate question which must be asked: What is the relevance of the cyclic basis of geomorphology to current geographical teaching? It has, I think, become apparent that the supposed geographical significance of the cycle assumed by Davis was based upon his essentially antique view of the nature of the subject. In terms of the modern, man-oriented geographical synthesis the explanatory description of landforms as a function of deduced origins has never been satisfactorily assimilated, despite the efforts of S. W. Wooldridge. Even the most sophisticated treatments of geographical 'regions' or 'landscapes' have not satisfactorily circumvented this difficulty, while the standard geographical text commonly presents the reader with a ritualistic introduction of dead, undigested and largely irrelevant physical information. Even when this is presented within a cyclic framework the relevance of this material does not increase (usually the reverse!), and this is largely the result of the difference in the time scales which are involved. Commonly the 'yesterday' of historical geography is still the 'today' of geomorphology. Russell (1949) put some of the difficulties of integrating a past-oriented geomorphology into the body of geography when he wrote: 'Geographers ordinarily find difficulty in discovering useful information in the conclusions of the pure morphologist. That a particular river is a consequent stream with an obsequent extension, or that some part of a river is superimposed rather than antecedent, or that a windgap suggests a cause of stream piracy, really means little to the person working on the problems of some specific cultural landscape. . . . The geomorphologist may concern himself deeply with questions of structures, process, and time, but the geographer wants specific information along the lines of what, where, and how much' (1949, pp. 3–4). Neither can it be contended, however, that the modern 'quantitative' approach to geomorphology which I shall treat in a later chapter is any more 'geographical' in character than its predecessor but, almost paradoxically, these studies which have been attacked as 'removing geomorphology from geography' are in the process of providing *as a by-product* just that basically relevant geographical material called for by Russell. A geographer does not want to know that a stream is 'mature' but what discharges have been recorded for it, not that a river terrace may possibly represent an interglacial event but information regarding its dimensions and composition, not that a slope form may indicate the poly-cyclic origin of the valley but details of its geometry and soil characteristics. This is not to degrade either geography or geomorphology, for both are quite distinct disciplines and are each proceeding to higher and different syntheses. It is no more the exclusive aim of the geomorphologist to provide physical data for the geographer, than it is for the geographer to content himself with a mass of physical information unintegrated into his human theme.

References

BRYAN, K., 1940, 'The Retreat of Slopes', *Ann. Assn. Amer. Geog.*, **30**, 254–68.

— 1941, 'Physiography 1888–1938', *Geol. Soc. Amer.*, *50th Anniversary Vol.*, 1–15.

CARTER, C. S. and CHORLEY, R. J., 1961, 'Early Slope Development in an Expanding Stream System', *Geol. Mag.*, **98**, 117–30.

CHORLEY, R. J., 1962, 'Geomorphology and General Systems Theory', *U.S. Geol. Survey, Prof. Paper*, *500–B*, 10 pp.

— 1963, 'Diastrophic Background to Twentieth-century Geomorphological Thought', *Bull. Geol. Soc. Amer.*, **74**, 953–70.

— 1964, 'Geography and Analogue Theory', *Ann. Assn. Amer. Geog.*, **51**, 127–37.

CVIJIĆ, J., 1918, 'Hydrographie souterraine et évolution morphologique du karst', *Rec. des Trav. de l'Inst. de Géog. alpine* (Grenoble), 6(4), 56 pp.

DAVIS, W. M., 1884, 'Geographic Classification, Illustrated by a Study of Plains, Plateaus and their Derivatives', *Proc. Amer. Assn. Adv. Sci.*, **33**, 428–32.

— 1889, 'The Rivers and Valleys of Pennsylvania', *Nat. Geog. Mag.*, **1**, 183–253 (also *Geographical Essays*).

— 1896, 'The Outline of Cape Cod', *Proc. Amer. Acad. Arts and Sciences*, **31**, 303–32 (also *Geographical Essays*).

— 1899, 'The Geographical Cycle', *Geog. Jour.*, **14**, 481–504 (also *Geographical Essays*).

— 1900, 'Glacial Erosion in France, Switzerland and Norway', *Proc. Boston Soc. Nat. Hist.*, **29**, 273–322 (also *Geographical Essays*).

— 1904, 'Complications of the Geographical Cycle', *Proc. Eighth Int. Geog. Congr.* (Washington), 150–63 (also *Geographical Essays*).

— 1905, 'The Geographical Cycle in an Arid Climate', *Jour. Geol.*, **13**, 381–407 (also *Geographical Essays*).

— 1909, *Geographical Essays* (Boston), 777 pp.

DAVIS, W. M., 1922, 'Peneplains and the Geographical Cycle', *Bull. Geol. Soc. Amer.*, **33**, 587–98.

— 1930 A, 'Rock Floors in Arid and Humid Climates', *Jour. Geol.*, **38**, 1–27 and 136–58.

— 1930 B, 'Origin of Limestone Caverns', *Bull. Geol. Soc. Amer.*, **41**, 475–628.

FENNEMAN, N. M., 1936, 'Cyclic and Non-Cyclic Aspects of Erosion', *Science*, **83**, 87–94.

GILBERT, G. K., 1877, *The Geology of the Henry Mountains*, U.S. Dept. of the Interior (Washington) (Chapter 5, Land Sculpture).

GLOCK, W. S., 1931, 'The Development of Drainage Systems', *Geog. Rev.*, **21**, 475–82.

JOHNSON, D. W., 1919, *Shore Processes and Shoreline Development* (New York), 584 pp.

— 1929, 'The Geographic Prospect', *Ann. Assn. Amer. Geog.*, **19**, 167–231.

— 1931, *Stream Sculpture on the Atlantic Slope* (New York), 142 pp.

LANGBEIN, W. B. and SCHUMM, S. A., 1958, 'Yield of Sediment in Relation to Mean Annual Precipitation', *Trans. Amer. Geophys. Union*, **39**, 1076–84.

LEIGHLY, J., 1955, 'What has happened to Physical Geography?', *Ann. Assn. Amer. Geog.*, **45**, 309–18.

LEOPOLD, L. B., 1953, 'Downstream Change of Velocity in Rivers', *Amer. Jour. Sci.*, **251**, 606–24.

LEOPOLD, L. B. and WOLMAN, M. G., 1957, 'River Channel Patterns: Braided, Meandering and Straight', *U.S. Geol. Survey, Prof. Paper, 282–B*, 39–85.

LEOPOLD, L. B., WOLMAN, M. G. and MILLER, J. P., 1964, *Fluvial Processes in Geomorphology* (San Francisco), 522 pp.

LEWIS, W. V., 1944, 'Stream Trough Experiments and Terrace Formation', *Geol. Mag.*, **81**, 241–53.

MCGEE, W. J., 1888, 'Three Formations on the Middle Atlantic Slope', *Amer. Jour. Sci.*, 3rd Ser., **35**, 120–43, 328–30, 367–88 and 448–66.

MATTHES, G. H., 1941, 'Basic Aspects of Stream Meanders', *Trans. Amer. Geophys. Union*, Pt. 3, 632–6.

MELTON, M. A., 1957, 'An Analysis of the Relations among Elements of Climate, Surface Properties, and Geomorphology', *Office of Naval Research Project NR 389–042*, Tech. Rept. 11, Dept. of Geol., Columbia Univ., New York, 102 pp.

RUSSELL, R. J., 1949, 'Geographical Geomorphology', *Ann. Assn. Amer. Geog.*, **39**, 1–11.

SAUER, C. O., 1925, 'The Morphology of Landscape', *Univ. of Calif. Pubs. in Geog.*, **2**, 19–53.

SCHUMM, S. A., 1956 A, 'Evolution of Drainage Systems and Slopes in Badlands at Perth Amboy, New Jersey', *Bull. Geol. Soc. Amer.*, **67,** 597–646.

— 1956 B, 'The Role of Creep and Rainwash on the Retreat of Badland Slopes', *Amer. Jour. Sci.*, **254,** 693–706.

SHEPARD, F. P., 1960, 'Gulf Coast Barriers', in 'Recent Sediments, North-west Gulf of Mexico', ed. by F. P. Shepard, F. B. Phleger and T. H. Van Andel, *Amer. Assn. Petroleum Geologists*, Tulsa, 197–220.

STRAHLER, A. N., 1950 A, 'Equilibrium Theory of Erosional Slopes, approached by Frequency Distribution Analysis', *Amer. Jour. Sci.*, **248,** 673–96 and 800–14.

— 1950 B, 'Davis' Concepts of Slope Development viewed in the Light of Recent Quantitative Investigations', *Ann. Assn. Amer. Geog.*, **40,** 209–13.

— 1952, 'Dynamic Basis of Geomorphology', *Bull. Geol. Soc. Amer.*, **63,** 923–38.

— 1958, 'Dimensional Analysis applied to Fluvially Eroded Landforms', *Bull Geol. Soc. Amer.*, **69,** 279–300.

SYMPOSIUM, 1940, 'Walther Penck's Contribution to Geomorphology', *Ann. Assn. Amer. Geog.*, **30,** 219–80.

WOOLDRIDGE, S. W., 1958, 'The Trend of Geomorphology', *Trans. Inst. Brit. Geog.*, No. 25, 29–35.

WOOLDRIDGE, S. W. and LINTON, D. L., 1955, *Structure, Surface and Drainage in South-east England*, 2nd Edn. (London), 176 pp.

WOOLDRIDGE, S. W. and MORGAN, R. S., 1959, *An Outline of Geomorphology*, 2nd. Edn. (London), 409 pp.

YATSU, E., 1955, 'On the Longitudinal Profile of a Graded River', *Trans. Amer. Geophys. Union*, **36,** 655–63.

YOUNG, A., 1960, 'Soil Movement by Denudational Processes on Slopes', *Nature*, **188,** 120–2.

CHAPTER THREE

Some Recent Trends in Climatology

R. P. BECKINSALE

Senior Lecturer in Geography, University of Oxford

Since 1945 all branches of meteorology have made great progress and have stimulated corresponding advances in climatology, although allowance must be made for the usual long educational time-lag. As a rule meteorological progress has been associated with advances in physics, chemistry, observational techniques and in the use of mathematical models, all of which tend to move meteorology outside the sphere of simple climatology or at least to remove many meteorological findings, unless grossly over-simplified, beyond the scope of non-scientific studies. Fortunately, since meteorology is based mainly on scientific principles, many of the narrower aspects of climatology can also be based securely on a simple knowledge of physics and chemistry. Unfortunately, much of traditional climatology demands geographical correlations and global or regional generalizations which at best are unscientific and at worst are little short of incorrect. Thus generalizations which are intended to form a fundamental prop for non-scientific students may become anathema to students with more than a nodding acquaintance with simple atmospheric physics and aerodynamics. Whereas the meteorologist writes for specialists in a specialized way, the climatologist often has the unenviable task of making meteorological data intelligible and useful to non-specialists. The climatologist today has to balance intelligibility and accuracy, and not infrequently it seems better to be intelligible and semi-accurate than unintelligible and accurate. The cardinal rule is that elementary principles of physics and chemistry should not be flouted.

The climatologist usually has the advantage of acute geographical knowledge. He can often come down to earth, though there seems no need always to parachute when a gentler descent from the attractive refuge above the bright blue sky would often lead to a happier landing. Modern meteorology has opened the way to greater climatic reality: climate has never been so real, nor so complicated. An attempt to demonstrate this will be made from recent advances in five

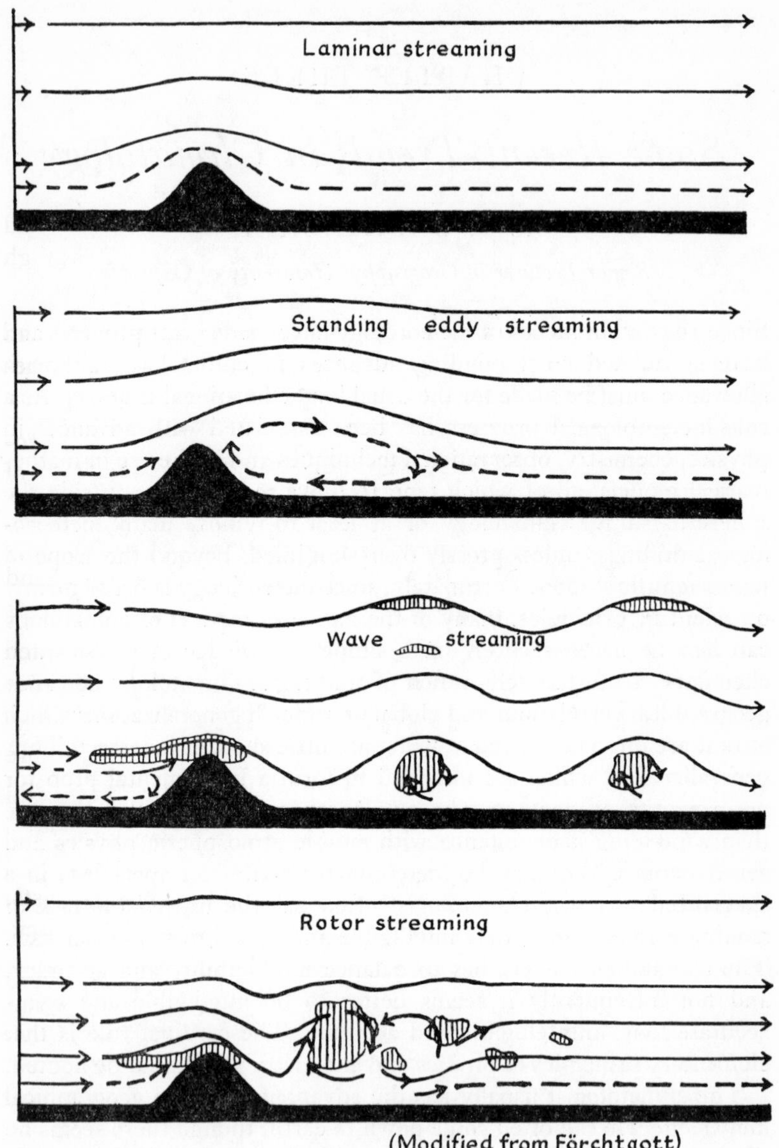

FIG. 3.1. *Types of airflow over a long ridge.*

interrelated aspects of climatology: airflow; precipitation; airmass definition and fronts; the general circulation; and the nature of the lower atmosphere.

AERODYNAMICS AND PLANETARY AIRFLOW

Aerodynamics

The frequent occurrence of orographically-formed cloud-masses above dip-slopes near scarps and of lenticular, banner, and arched clouds over lee-slopes has long drawn attention to airflow across relief-barriers. Today the use of gliders and of powered aircraft has brought great advances in the aerodynamics of relief-influenced airflow. These problems have been studied, for example, by Scorer (1961) and Corby (1954) (Figure 3.1).

With light winds, laminar streaming occurs and a single shallow wave is uplifted symmetrically above the obstacle. With slightly stronger airflow, the crest of the laminar streaming is displaced downwind and a lee eddy forms, often with important climatic effects. With still stronger winds, the lee eddy is replaced by a series of lee-waves which affects all the lower airflow. With very strong winds, especially if the relief barrier is high compared with the airflow, the symmetrical wave streaming breaks down into a complicated turbulence or 'rotor' streaming. An interesting and well-illustrated study of lee waves in the French Alps, published by Gerbier and Bérenger in 1961, shows clearly the appreciable departures from these typical conditions when wind-speeds do not increase regularly towards the tropopause. The characteristic features of wave-streaming airflow across a long mountain range when the wind-speed increases with altitude have been summarized by Wallington (1958; 1960) and are reproduced in a modified form in Figure 3.2.

The aerodynamic effects of smaller obstacles, such as isolated hills, buildings and tree-belts have also been summarized by Caborn (1955 and 1957) and L. P. Smith (1958, pp. 72–77). The extent of the shelter effect depends mainly on the height and permeability of the wind-break. Generally the reduction of wind velocity begins at about 9 times the shelter-belt height to windward and extends to about 30 times the shelter-belt height to leeward. A solid wind-break may in time of strong winds afford practically no shelter to windward and a marked protection (80 per cent or more wind-speed reduction) for only a short distance to leeward before vigorous turbulence (rotor or highly turbulential streaming) occurs. A slightly permeable shelter-belt will develop an eddy ('air-cushion') to windward and a marked lee eddy, which under average conditions affords a shelter of over 20 per cent wind-reduction to distances of twice the height of the windbreak to windward and of 15 to 20 times its height to leeward.

Planetary Airflow

Much progress is being made in the knowledge of the two main surface-wind systems of the globe, the so-called tropical easterlies

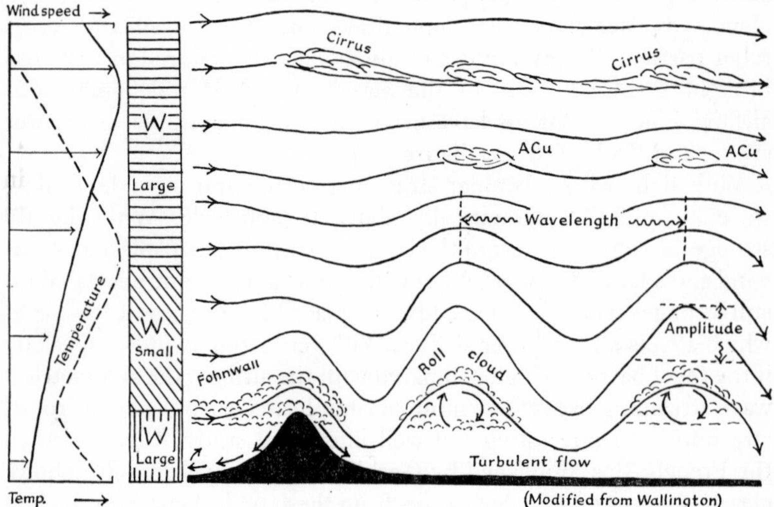

FIG. 3.2. *Features of strong wave-streaming across a long mountain range typically associated with a three-layer troposphere, when an inversion temperature layer in the middle troposphere tends to induce airflow with waves of smaller length and greater amplitude. The maximum streamwave amplitude in the more stable temperature layer is shown. The wavelength is of the order of 2 to 20 miles. ACu denotes altocumulus lenticularis or lenticular cloud. W denotes the general magnitude of the natural or characteristic wavelength determined by airflow and temperature conditions. The decrease of upper wind-speed occurs near the tropopause.*

(Riehl, 1954, pp. 210–34; Koteswaram, 1958) and the sub-tropical and polar westerlies. The former, known as 'Trades' over the oceans, usually play an important or even dominant role in climates between latitudes 32° N and S or upon nearly half the world's area. The two-tier nature of the tropical airspace was recognized a century ago but today the layer-structure and the areal extent of the Trades have been more precisely determined. Crowe (1949 and 1950) has emphasized the great seasonal expansions equatorward and westward of the oceanic Trades from 'root' areas over cold oceanic-upwellings off the west sides of continents. He and Mintz and Dean (1952) show that strong surface Trades occupy 30 million square miles in March and about 40 million square miles in July. Riehl and others (1951 and 1954) have emphasized the vertical stratification of the 'tropical'

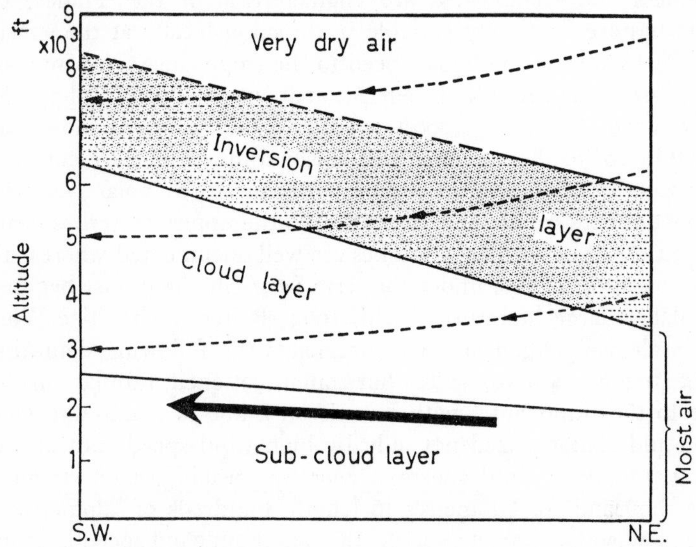

FIG. 3.3. *Schematic cross-section of NE Trades over Pacific between 32° N, 136° W and 21° N, 158° W (Honolulu). Large arrow shows main airflow; arrows on dotted lines indicate general subsidence of airmass through the inversion layer (after Riehl, Yeh, Malkus and La Séur).*

troposphere which, apart from the active portions of disturbances, generally consists of a lower moist and an upper dry layer. The refinement is carried much further. A moist surface-layer, warmed by contact with warm oceans and humidified by evaporation from rough seas, is usually overlain by a strong inversion layer marked by a sharp drop in relative humidity and an appreciable rise in temperature (Figure 3.3). Above is very dry air. The inversion-layer may be near sea-level over sub-tropical littoral deserts but normally in the north Pacific it is at about 4,000 feet in 35° N and rises steadily to

7,000 feet and more inside the tropics where it weakens sufficiently to allow the formation of high cumulo-nimbus clouds and rain. Above this inversion layer the very dry easterly airflow lessens in frequency with height over large areas and, for example, in the northern hemisphere poleward of about 10° N in January and 15° N in July above about 18,000–20,000 feet its mean zonal component changes to westerly or antitrade.

The Westerlies have been described recently by Hare (1960) and Lamb (1959) as two separate vast circumpolar vortices. Whereas steadiness and shallowness are characteristic of the Trades, the Westerlies are extremely variable in direction locally at the surface and usually increase in zonal speed to the tropopause. In winter the polar westerlies prevail at the surface at about 35°–70° N and 35°–62° S but above 16,000 or 18,000 feet (500 mb) westerly flow dominates to within 10° of the equator. The seasonal migrations are considerable; thus in the northern summer the surface or polar westerlies contract northward about 5° latitude and the upper westerlies nearly 10° latitude. The surface westerlies are well documented whereas the upper westerlies have, under the term *jet stream*, only just begun to stimulate a large literature (Riehl, 1962; Reiter, 1963). The World Meteorological Organization recommends the following definition: 'A jet stream is a more or less horizontal, flattened, tubular current, close to the tropopause, with its axis on a line of maximum wind-speed and characterized not only by high wind-speeds but also by strong transverse wind shears. Generally speaking, a jet stream is some thousands of kilometres in length, hundreds of kilometres in width, and some kilometres high; the minimum wind speed is 30 m/s at every point on its axis. . . .'

It is perhaps important to emphasize that although the westerlies generally increase in speed and constancy up to the tropopause, or about 30,000 feet in 'temperate' latitudes and above 45,000 to 50,000 feet or more in the sub-tropics, the broad picture of a single upper westerly belt with a central jet stream in the upper troposphere between latitudes 25°–40° N and S is quite inadequate. Usually there are two or three jet streams or ribbons of higher velocity (Figure 3.4). In January the northern hemisphere sub-tropical jet commonly occurs at about 40,000 feet in 30° N but other ribbons of jet occur between 35° and 80° N. Moreover, the jet streams migrate seasonally with the upper (and polar) Westerlies and their horizontal position advances and retreats irrespective of the seasons. The possible role of these non-seasonal migrations has led to the idea of the *Index Cycle*

(Namias, 1950) which is discussed in many meteorological texts. The jet stream also undergoes vertical oscillations, or long waves, with an amplitude of several thousand miles. These oscillations, together with variations in jet stream velocity and in its motion relative to the

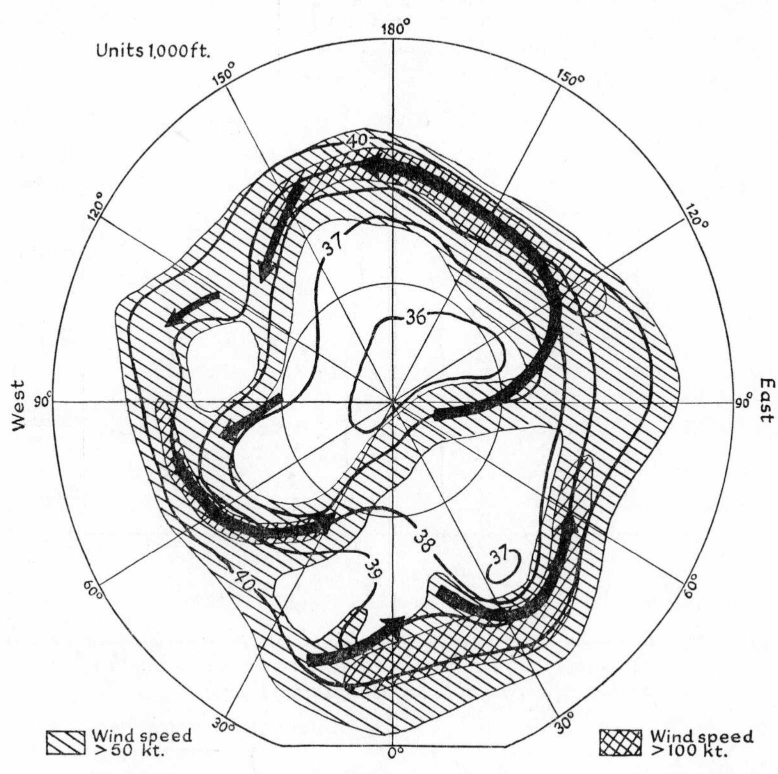

FIG. 3.4. *Upper troposphere airflow at 0300 GMT on 19 December 1953. The 200 mb contours and wind-speed demonstrate the great strength of the sub-tropical jet stream (over 100 knots) at 38,000 to 40,000 feet, whereas at the same time the circumpolar jet reached its greatest development at or near the 300 mb level (28,000 to 30,000 feet) (after Sawyer).*

earth's rotation, are thought to be associated with cyclogenesis and with the position of 'blocking anticyclones' (Rex, 1951; Sanders, 1953; Sumner, 1959). In fact, the jet stream has become a sort of synoptic panacea, which the causal climatologist will find most attractive.

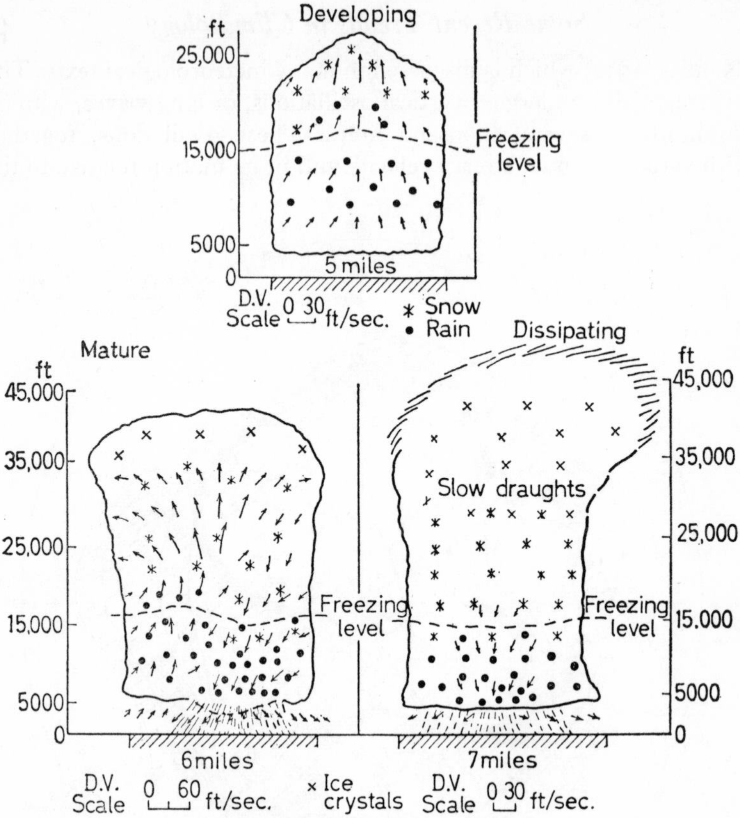

FIG. 3.5. *Idealized cross-section of a thunderstorm cell. Surface rain is shown by dashes; D. V. denotes draught-vector scale, length of arrows being roughly proportional to air-speed (after Byers).*

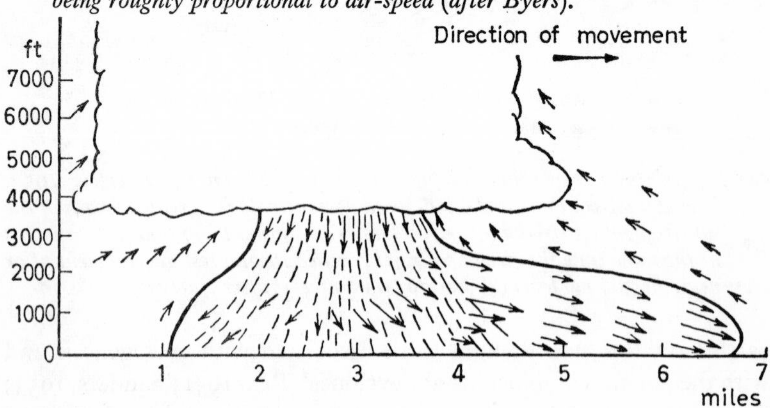

FIG. 3.6. *Schematic section showing the spreading of the downdraught from a thunderstorm cell (after Byers).*

PRECIPITATION

Extraordinary strides have been made recently in the knowledge of precipitation physics, including the processes of formation of cloud-droplets, raindrops and of hail.

Cloud Microphysics: Where the tropopause is high, as in the tropics generally, and in, for example, some warm-air sectors in Britain, the formation of rain at above zero temperatures and without ice-nuclei is firmly established. Here the essential requirement is the presence of some cloud droplets so much above average in size that they fall fast enough to collide and coalesce with smaller droplets. Presumably the larger cloud-drops form on larger or more hygroscopic nuclei.

Outside the tropics, and commonly within, the ice-crystal nuclei mechanism operates. At temperatures below −40° C all cloud droplets freeze automatically or spontaneously but at temperatures between zero and −40° C ice crystals form among supercooled water droplets. These ice crystals grow by accretion at a much faster rate than the water droplets and soon reach small raindrop size. The crystals coalesce as they fall and form snowflakes which nearer earth may melt into raindrops. Details of these cloud microphysical processes, with further references, are given by Mason (1957 and 1959) and Durbin (1961).

Cloud Physics: Modern studies of the thunderstorm and of hail are of great climatological value and prove a real educational asset when such phenomena occur locally. The convectional thunderstorm consists of a collection of cells, each of which may experience a typical life-cycle (Byers and Braham, 1949). The normal cell commonly has a horizontal dimension of about ½ mile up to 6 miles and when triggered off in an unstable airmass may develop at a rapid rate and complete its cycle within 20 to 60 minutes. In the developing stage a slow ascent of air proceeds, usually until ice crystals form (Figure 3.5). Then in the mature stage, precipitation spreads rapidly throughout the cell and the accumulation of water aloft eventually overcomes the ascensional impulses in some lower parts of the cell; a mass of rain-laden air rushes earthward (Figure 3.6) and thrusts before it a violent gust of cold air or a 'line-squall' (Wallington, 1961). Thus the cell is overturned and it now slowly dissipates. However, the sudden down-draught of rain-cooled air from the mature cell usually triggers off the overturning of other cells especially if they are 3 to 5 miles distant. So the thunderstorm rolls across the sky. Although in a

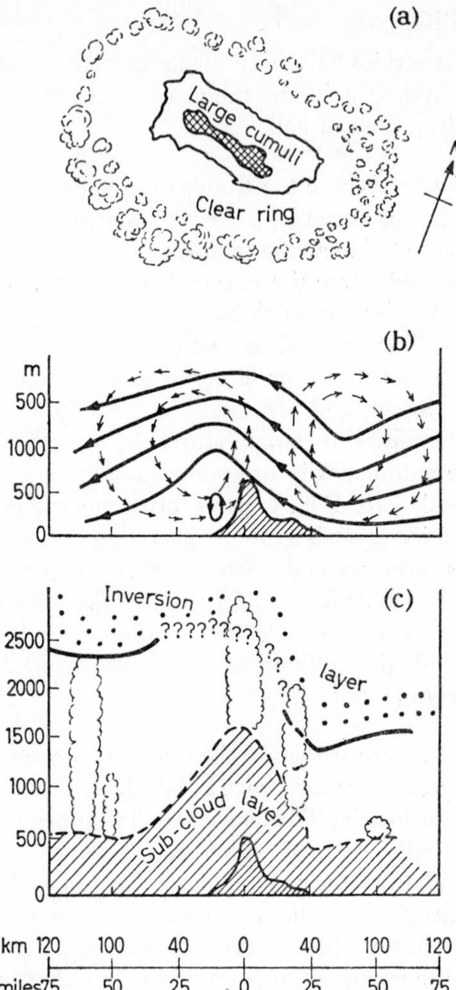

FIG. 3.7. *Schematic reconstruction of day-time orographic-convection cell of tradewind island (Puerto Rico), showing ascent over island and subsidence in surrounding ring on 25 June 1952 (after Malkus). Cross-hatching denotes land over 2000 feet; in (b) streamlines of the main airflow are shown in thick lines and those of the local convection cell in thin lines; in (c) the inversion layer was not measured in parts.*

practically uniform air-mass the cells may have a random or irregular pattern, it seems that they usually group themselves into bands or lines. In some areas these 'progressions' may be due to potential cells passing successively across a more or less stationary trigger-action; in others, there may be 'squall-lines' in the lower atmosphere (Soane and Miles, 1954).

The formation of large hailstones has also attracted much attention and also involves the nature of ascensional impulses in convective or ascensional systems (Ludlam, 1961). Large hail stones, in Europe at least, are often associated with some kind of minor front with a steady but sloping updraught, the slope being opposite to the general direction of the storm's movement. Small hail forms in the usual manner in the upward currents and drops out of the top of the cumulus anvil protruding in front of the storm; in falling, it is caught again in the main ascent and again carried up until it becomes so big that it falls as large hail out of the rear side of the uplift and squall-line. To produce really large hailstones the hail particles that re-enter the storm centre must be the right size to grow as to match closely the increase of speed of the updraught and to stay for a relatively long time in the upward currents.

Recently another form of convective air-movement, the orographic cell, has received revived attention. Although the nature of mountain- and valley-breezes has long been imputed to regional as well as local causes, sufficient credit has not hitherto been given to the potency of orographic cells in the tropics and sub-tropics. The diurnal thermal pulsation of mountainous areas, especially of isolated uplands, greatly encourages local convectional over-turnings. High islands even in Trade-wind areas build up orographic cells which weaken and pierce the inversion layer so that rain-clouds form over higher slopes and, to a lesser extent, a cloud-ring develops some distance offshore (Figure 3.7). This convective influence is often generated in spite of the aerodynamic influences of the relief on the steady Trade airflow (Malkus. 1955).

AIRMASS DEFINITION AND FRONTS

The airmass concept is today accepted generally in climatology and is proving of great geographical value, especially when depicted on equal-area maps instead of on the grossly-misleading Mercator projection. But the airmass is no more rigidly definable than many other 'major regions' used in regional studies. Owing to surface friction and influence the airmass does not move as a strictly rigid unit although it does transpose many of its 'core' characteristics. The idea of airmass depth and of airmass stability or instability aloft seems a necessary addition. The extended shorthand for an airmass then becomes as follows, where Polar maritime airmasses are used as an example:

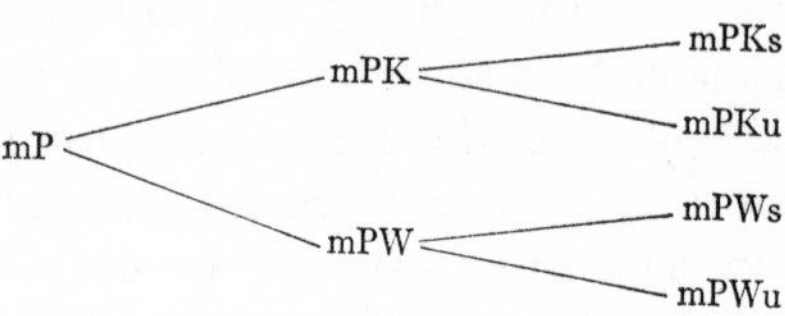

K denotes the airmass is heated from below, W cooled from below, s stable aloft and u unstable aloft. However, the same purpose would also be served if the cross-section of the whole troposphere was always studied in any area.

The airmass concept has already been tied to world climatic classifications, loosely by Flohn (1950 and 1952), and more securely by Miller (1953), Borchert (1953) and Strahler (1960, pp. 189–91). In addition the proportion of airmass and airflow in each month or season is becoming a vital part of regional climatic studies.

It must be admitted that in some areas the airmass which sustains the climate dominates over the fronts which supply the weather. Yet the seasonal change in insolation at the earth's surface and the associated migration of the thermal equator and the fluctuations in areal extent of the main airmasses are so great that the horizontal movement of frontal surfaces can hardly be overemphasized. It seems inevitable that the nature of fronts should provide a constant stimulus to meteorological research, particularly in extra-tropical latitudes where horizontal-mixing of airmasses is prevalent. With an increase of instrumental observations, the nature of temperate-latitude depressions acquires an increasing complexity. As an ideal the simple Abercrombie–Bjerknes–Solberg scheme of a typical 'low' was not likely to be found more frequently than any other ideal, but it was rather oversimplified. To give two examples, cold fronts usually depart from the simple pattern; and correlations with the middle and upper troposphere could not be included scientifically in the earlier patterns as the details were not then available.

Cold fronts were early recognized as being of two main types dependent on whether the warm air was for the most upsliding (anafront) or descending (katafront) at the cold wedge. Subsequently it was increasingly realized that 'there is no *average* cold front', the weather being determined 'by the nature of the warm airmass which is being lifted and by the degree of lift to which the warm air is subjected by the advancing cold air wedge' (*Weather Ways*, 3rd. edn. 1961, p. 80. Met. Branch, Dept. of Transport, Ottawa).

Cold fronts have been studied by Sansom (1961) and Miles (1962). The former showed that the anafront produced greater precipitation and underwent more abrupt changes in temperature and wind, which backed rapidly with height. The katafront revealed only a slight backing of wind with height and brought a rapid clearance followed by fine weather. Miles (1962) has shown that the implicit suggestion that a cold front consists of two distinct airmasses separated by a narrow transition zone is frequently not upheld by observations. Many cold fronts do have a comparatively narrow temperature transition zone aloft but they fail to develop any cloud system sloping up the frontal surface. The type of cold front with moist, cloud-filled air lying above a well-marked cold wedge is, in fact, rare. The commonest type of cold front has a convex air shear zone or temperature transition zone overlain by very dry warm air. Frequently, this warm dry air also protrudes, at 10,000 feet or so, above the warm moist air in the frontal zone which normally extends about fifty miles ahead of the surface cold front. Thus the air above the cold-wedge is very dry and warm while the air ahead of the frontal zone is warm and moist and the change in humidity from very dry to moist commonly occurs in a zone about fifty miles wide almost vertically above the convex snout of the cold-air wedge (Figure 3.8). This humidity transition zone is often the rear edge of the main rain- and cloud-belt associated with the front. Thus the change aloft is in humidity and wind-speed and not in temperature. 'With many well marked surface cold fronts the temperature difference at 700 mb or 500 mb is spread over a distance of at least 200 miles, often without significant horizontal displacement from one level to another' (Miles, 1962, p. 286). It seems that the warm dry air is not necessarily descending in the frontal zone and that it is not very realistic to postulate warm air subsiding down the cold wedge at a katafront.

The search for some causal relationship between surface weather phenomena and upper troposphere conditions in a temperate-latitude depression continues. Frontogenesis or the growth of the depressional wave is being increasingly related to the jet stream which may encourage or discourage, as the case may be, convergence (subsidence) or divergence (ascent) at levels in the middle and upper troposphere. The early concept of frontal depressional models was, in fact, three-dimensional in outlook but its upper air conditions were based largely

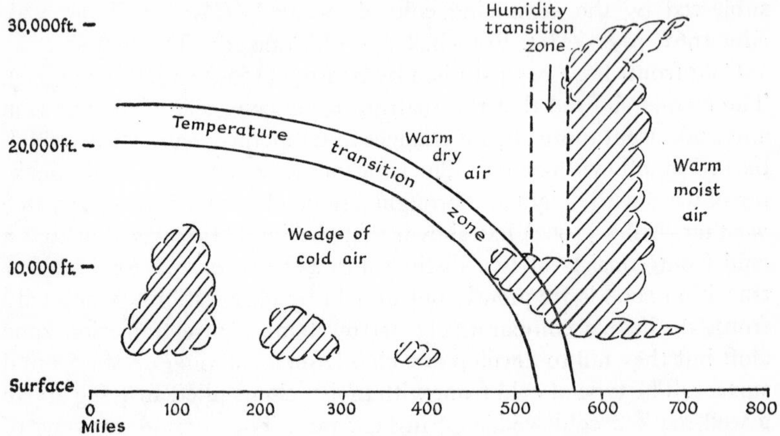

FIG. 3.8. *Schematic model for a common type of cold front* (*after Miles*).

on theory. Today upper-air soundings allow reliable charts to be drawn for selected levels over many continental areas at least; the frontal zones can be studied at various heights and frontal contour charts compiled. The possibility of linking depressions with the circumpolar jet stream is discussed by Murray and Johnson (1952) and by Sawyer (1958). Modern ideas of incorporating the possible influence of upper-air conditions on frontogenesis, are expressed in the three-front model, 'a logical extension of the Bjerknes' model to include upper air data'. Galloway (1958–60) supplies details of this scheme which shows the position of the fronts of Arctic, Polar, and Tropical airmasses at various levels in the troposphere outside the tropics.

The failure of the simple application of the Bjerknes' depression model to weather disturbances in the tropics is not surprising. The dynamic or synoptic meteorology of the tropics remains an academic battlefield. The old term Tropical Front is upheld by some who consider it has 'all the properties of a front' (Bergeron, 1954) while others consider it best to replace the term by the expression 'equatorial air-stream boundaries' when applied to within 10° N and 10° S of the equator (Watts, 1955). However, it is generally agreed that various forms of lows, often non-frontal, and of frontal surfaces operate widely and frequently within the tropics. Perturbations or waves in the easterlies are stressed by Palmer (1951) and Riehl (1954), while most studies of the tropical airspace stress the great distances travelled seasonally by 'troughs' and 'fronts'. The rotating tropical cyclone is today well documented (Malkus, 1958; Dunn and Miller, 1960; Neiburger and Wexler, 1961), and a brief geographical commentary on the complexity of causes of tropical rainfall will be found in Beckinsale (1957).

It happens that the climates of tropical territories demonstrate well the great value and the great difficulties inherent in the airmass concept. As these have been summarized by Trewartha (1961) we need only use the Indian monsoon as an example of modern meteorological trends. The meteorological explanations of the burst of the Indian summer monsoon are highly contradictory but they are strongly unified by insistence on the influence of middle and upper tropospheric conditions, including the jet stream (Yin, 1949; Lockwood, 1965, pp. 2–8; Symposium, 1957–8). Normally the sub-tropical jet stream would migrate northward in summer in Asia, but, according to some theories, it lingers unduly along the southern side of the Himalayan–Tibetan mountain arc. Not until the jet stream migrates over or north of the mountain belt does the monsoon burst in with its typical 'frontal' weather. On the other hand, other suggestions impute the trigger-action to thermal changes in tropopause conditions and consider that the onset of the monsoon slightly precedes the main jet stream migration. Whichever theory proves acceptable, the significance of the new explanation is that a synoptic or upper air concept has replaced an oversimplified climatological generalization. Yet recent studies also reveal that land- and sea-breezes are a scientific reality and differential heating of land and sea may be relegated to a minor status but it cannot be entirely eliminated!

CHANGING MODELS OF THE GENERAL CIRCULATION

A large literature has appeared on aspects of the general circulation since Rossby in his famous essay of 1949 suggested that the observed mean zonal wind profile could be satisfactorily accounted for by complete lateral mixing north of a certain latitude and that the zone of maximum westerly flow (now known as the sub-tropical jet stream) would be near or at the equatorward edge of the lateral-mixing region. The eddies involved in the advection were thought to add energy to the mean zonal airflow.

Subsequently, numerous studies were made of the energizing or 'balance requirements' of the general circulation. These studies emphasized the balance of angular momentum, the balance of energy or the transfer of energy from low to high latitudes, and the water balance. Sheppard (1958) has summarized the balance of angular momentum and Tucker (1960, 1961 and 1962) has discussed in detail various aspects of dynamical climatology which he considers 'the major development of climatology during the last decade'. The general trend has been to raise the importance of dynamic processes at the expense of surface or geographical influences.

However, meteorologists do not agree upon the relative importance of solar heating, with its meridional implications, and of lateral mixing or eddy-action in the energizing of the global circulation as a whole. Starr (1956) suggests that the primary meridional circulation caused by solar radiation is disrupted by the Earth's rotation into large-scale eddies (cyclones and anti-cyclones) and that the rotation acts further in channelling these turbulent motions into prevailing east and west winds. These major circulations are thought to derive their energy of motion from the large eddies, which convert potential energy into kinetic energy. The actual circulation is considered to be almost opposite to that often postulated, as air is presumed to be carried downwards in lower latitudes and upward in higher latitudes. However, many will recall a rather similar three-cell general circulation proposed by Bergeron in 1928, although its latitudinal extent was very different (Figure 3.9).

It must be noticed, however, that no meteorologist has satisfactorily explained the balance of the general circulation without assuming some form of mean meridional circulation, at least in the tropics. Thus, wind observations demonstrate a zonal circulation that consists broadly of tropical easterlies, temperate westerlies, and polar easterlies at the surface; and strong temperate-subtropical westerlies and tropical easterlies in the middle and upper troposphere. At the same time, dynamic analyses, theoretical, observational and experimental, indicate certain broad circulation patterns, of which the chief are:

1. A tropical cell which remains a prime, and to some the prime, feature of the general circulation. It is a convective Hadley cell of

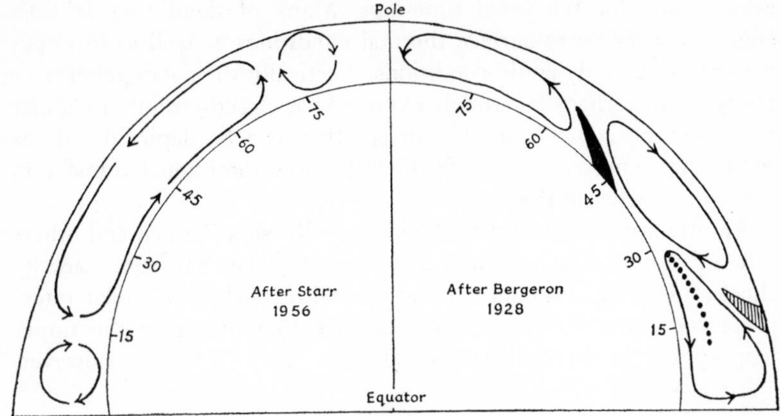

FIG. 3.9. *Old and new ideas on the general meridional circulation of the troposphere.*

remarkable depth for the troposphere, as has been clearly expressed by Palmén (1951 and 1963) (Figure 3.10);

2. In middle latitudes a zone of predominantly lateral or horizontal mixing or in other words a zone where the air-space is usually dominated by large-scale eddies (turbulence) and advection prevails over convection. Here the troposphere is relatively shallow;

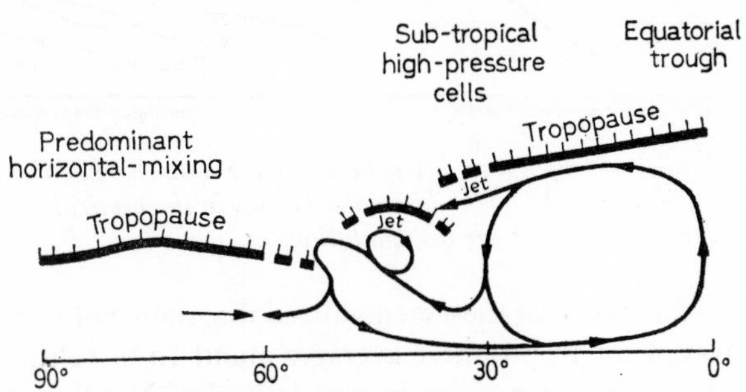

FIG. 3.10. *Scheme for mean meridional circulation in winter (after Palmén). The Polar front, which migrates vast distances, is omitted.*

3. A belt aloft near the junction of (1) and (2) where the upper air westerlies frequently accelerate to form the sub-tropical jet; and

4. Near the poles, some form of air subsidence.

These and other associated features migrate and expand or contract meridionally or latitudinally, or both, with the seasons, allowance being made for the usual time-lags. Many of them vary at least slightly according to surface thermal conditions as well as to upper tropospheric or dynamic conditions. Vertically their components or energy may perhaps be crudely expressed as closed-circuit or cellular zones but horizontally at the surface they can be depicted only as semipermanent cells which frequently allow meridional transfer in the spaces between them.

An attempt to incorporate the ideas of Rossby, Palmén and others into a diagrammatic form useful for climatologists has been made by Birot (1956). Most modern climatological texts show a similar interpretation but a few omit any reference to motions in the upper troposphere in middle latitudes (Figure 3.11). It seems, however,

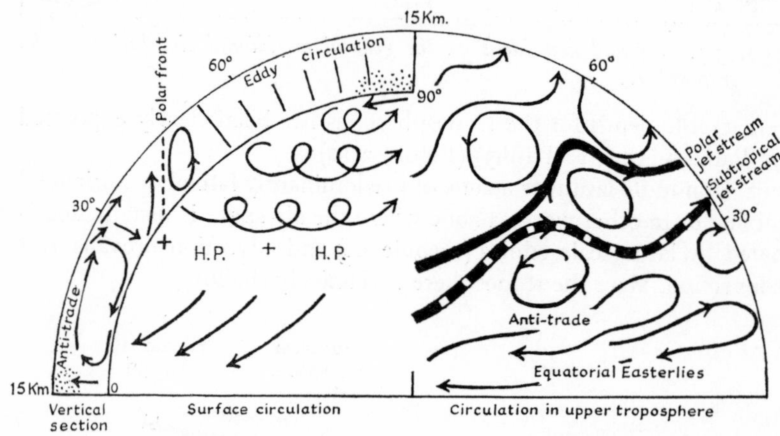

FIG. 3.11. *Model of tropospheric circulation in the northern hemisphere, compiled by Birot after Palmén. The dotted areas in the vertical cross-section denote zones of relative air accumulation.*

unfortunate that some of these generalized diagrams fail to show that the thickness of the troposphere averages about 10 to 12 miles in the tropics and only about 7 miles in mid-latitudes. Needless to say, if the cross-sections of the troposphere shown on global quadrants were drawn true to scale the existence of any general meridional circulation would appear little short of miraculous! Probably the finest exhibition of visual diagrammatic representations of the general atmospheric circulation ever shown was that of the Royal Society (described by Lamb, 1960).

THE NATURE OF THE LOWER ATMOSPHERE

To the progress of the knowledge of the troposphere discussed above must be added the tremendous advances in the knowledge of the lower atmosphere generally. As the ionosphere seems rather outside climatology, reference will be made here only to the ozone region and the stratosphere, or heights between about 60 km and 12 km. The tropopause is now known to be ill-defined and often almost absent above equatorial areas. The lower stratosphere has been shown to have small weather-qualities and to possess many signs of a general meridional circulation. From helium and water-vapour distribution, Brewer (1949) postulated such a circulation sufficient to make 'a significant contribution to the energy of the general circulation'. Subsequently his contentions have been supported by the study of the movement of other tracer substances, such as ozone and radioactive debris (Goldsmith and Brown, 1961; Murgatroyd and Singleton, 1961–2). Newell discusses the problem in detail (1963 and 1964) and, although favouring considerable lateral-mixing, thinks that the effects of 'mean meridional motions cannot be ignored entirely'.

The 'ozonosphere' or region of ozone concentration at about 20 km to 50 km seems to undergo definite seasonal and even diurnal oscillations as well as a general circulation. This layer of atomic oxygen, O_3, absorbs about 5 per cent of the sun's total radiant energy, including nearly all the ultra-violet rays, and forms a warm region overlying the lower stratosphere. Its screening effect is of paramount importance to mankind. Details of its qualities and movements are described fully by Dobson (1963).

There is perhaps no need to warn the climatologist that the effect of changes in the stratosphere and ozone region on the weather of the troposphere is far from being understood. Some of the many secular variations in the circulation pattern of the troposphere are tentatively imputed to stratospheric influences but the extreme complexity of possible correlations are at once evident, as is shown clearly in the relevant chapters of *Changes of Climates* (UNESCO, 1963).

References

GENERAL

BERRY, F. A., BOLLAY, E. AND BEERS, N. R. (eds), 1945, *Handbook of Meteorology* (New York).

BYERS, H. R., 1959, *General Meteorology* (New York).

DOBSON, C. M. B., 1963, *Exploring the Atmosphere* (Oxford).
H.M.S.O., 1961, *A Course in Elementary Meteorology* (London).
MALONE, T. F. (ed.), 1951, *Compendium of Meteorology* (Boston).
PETTERSSEN, S., 1956, *Weather Analysis and Forecasting* (2 vols) (New York).
— 1958, *Introduction to Meteorology* (New York).
SAWYER, J. S., 1957, *The Ways of Weather* (London).
SUTTON, O. G., 1962, *The Challenge of the Atmosphere* (London).
WILLETT, H. C. and SANDERS, F., 1959, *Descriptive Meteorology* (New York).

AERODYNAMICS

CABORN, J. M., 1955, 'The Influence of Shelter-belts on Microclimate', *Quart. Jour. Roy. Met. Soc.*, **81,** 112–15.
— 1957, 'Shelterbelts and Microclimate', *Forestry Commission Bulletin,* **29** (H.M.S.O.).
CORBY, G. C., 1954, 'The Airflow over Mountains', *Quart. Jour. Roy. Met. Soc.*, **80,** 481–521.
GERBIER, N. and BÉRENGER, M., 1961, 'Experimental Studies of Lee Waves in the French Alps', *Quart. Jour. Roy. Met. Soc.*, **87,** 13–23.
SCORER, R. S., 1961, 'Lee Waves in the Atmosphere', *Scientific American,* **204,** 124–34.
SMITH, L. P., 1958, *Farming Weather* (London) (*see especially pp. 72–77*).
WALLINGTON, C. E., 1958, 'Orographic waves . . .', *Met. Mag.*, **87,** 80–87.
— 1960, 'An Introduction to Lee Waves in the Atmosphere', *Weather,* **15,** 269–76.

PLANETARY AIRFLOW

Trades:

CROWE, P. R., 1949, 'The Trade-wind Circulation of the World', *Trans. Inst. Brit. Geog.*, **15,** 37–56.
— 1950, 'The Seasonal Variation in the Strength of the Trades', *Trans. Inst. Brit. Geog.*, **16,** 23–47.
MINTZ, Y. and DEAN, G., 1952, 'The Observed Mean Field of Motion of the Atmosphere', *Geophysical Research Papers.* No. 17, 37–42 (Cambridge, Mass.).
RIEHL, H., 1954, *Tropical Meteorology* (New York).
RIEHL, H. *et al,* 1951, 'The North-east Trade of the Pacific Ocean', *Quart. Jour. Roy. Met. Soc.*, **77,** 598–626.

Westerlies:

HARE, F. K., 1960, 'The Westerlies', *Geog. Rev.*, **50,** 345–67.
LAMB, H. H., 1959, 'The Southern Westerlies', *Quart. Jour. Roy. Met. Soc.*, **85,** 1–23.

Jet Stream:

BERGGREN, R. *et al.*, 1958, 'Observation Characteristics of the Jet Stream', *World Meteor. Organization*, Tech. Note 19, No. 71, T.P. 27 (Geneva).

KOTESWARAM, P., 1958, 'The Easterly Jet Stream in the Tropics', *Tellus*, **10**, 43–57.

NAMIAS, J., 1950, 'The Index Cycle and Its Role in the General Circulation', *Jour. Met.*, **7**, 130–9.

REITER, E. R., 1963, *Jet-Stream Meteorology* (Chicago).

REX, D. F., 1951, 'The Effect of Atlantic Blocking Action Upon European Climate', *Tellus*, **3**, 100–11.

RIEHL, H., 1962, *Jet Streams of the Atmosphere*, Tech. Paper No. 32, Colorado State Univ., 117 pp.

SANDERS, R. A., 1953, 'Blocking Highs Over the Eastern North Atlantic', *Monthly Weather Rev.*, **81**, 67–73.

SAWYER, J. S., 1957, 'Jet Stream Features of the Earth's Atmosphere', *Weather*, **12**, 333–44.

— 1959, 'The Jet Stream', *New Scientist*, **6**, 947–9.

SUMNER, E. J., 1959, 'Blocking Anticyclones in the Atlantic–European Sector of the Northern Hemisphere', *Met. Mag.*, **88**, 300–11.

PRECIPITATION

Cloud Physics:

DURBIN, W. G., 1961, 'An Introduction to Cloud Physics', *Weather*, **16**, 71–82 and 113–25.

MASON, B. J., 1957, *The Physics of Clouds* (Oxford).

— 1959, 'Recent Developments in the Physics of Rain and Rain-making', *Weather*, **14**, 81–97.

Thunderstorm:

BYERS, H. R. and BRAHAM, R. R., 1949, *The Thunderstorm*, U.S. Weather Bureau.

SOANE, C. M. and MILES, V. G., 1955, 'On the Space and Time Distribution of Showers in a Tropical Region', *Quart. Jour. Roy. Met. Soc.*, **81**, 440–8, and **82**, 534–5.

WALLINGTON, C. E., 1961, 'Observations of the Effects of Precipitation Downdraughts', *Weather*, **16**, 35–44.

Hail:

LUDLAM, F. H., 1961, 'The Hailstorm', *Weather*, **16**, 152–62.

Orographic Cells:

MALKUS, J. S., 1955, 'The Effects of a Large Island Upon the Trade-wind Air Stream', *Quart. Jour. Roy. Met. Soc.*, **81**, 538–50, and **82**, 235–8.

AIRMASS DEFINITION AND FRONTS

Airmasses:

BORCHERT, J. R., 1953, 'Regional Differences in World Atmospheric Circulation', *Ann. Assoc. Amer. Geog.*, **43**, 14–26.

FLOHN, H., 1950, 'Neue Anschauungen über die allgemeine zirkulation der Atmosphäre . . .', *Erdkunde*, **4**, 155–9.

— 1952, 'Grundzüge der atmosphärischen zirkulation', *Deutscher Geographentag Frankfurt*, **28**, 105–18.

MILLER, A. A., 1953, 'Air Mass Climatology', *Geog.*, **38**, 55–67.

STRAHLER, A. N., 1960, *Physical Geography* (New York) (*see especially* pp. 189–91).

Fronts:

MILES, M. K., 1962, 'Wind, Temperature and Humidity Distribution at Some Cold Fronts Over S.E. England', *Quart. Jour. Roy. Met. Soc.*, **88**, 286–300.

— 1962, 'Fronts', *Weather*, Schools Suppl. No. 12, 45–48.

SANSOM, H. W., 1951, 'A Study of Cold Fronts Over the British Isles', *Quart. Jour. Roy. Met. Soc.*, **77**, 96–120.

Fronts and Upper Air Conditions:

GALLOWAY J. L., 1958, 'The Three-front Model . . .', *Weather*, **13**, 3–10.

— 1960, 'The Three-front Model, the Developing Depression and the Occluding Process', *Weather*, **15**, 293–309.

MURRAY, R. and JOHNSON, D. H., 1952, 'Structure of the Upper Westerlies', *Quart. Jour. Roy. Met. Soc.*, **78**, 186–99.

SAWYER, J. S., 1958, 'Temperature, Humidity and Cloud Near Fronts in the Middle and Upper Troposphere', *Quart. Jour. Roy. Met. Soc.*, **84**, 375–88.

Tropical Airmass Climatology and Rainfall:

BECKINSALE, R. P., 1957, 'The Nature of Tropical Rainfall', *Tropical Agriculture*, **34**, 76–98.

BERGERON, T., 1954, 'The Problem of Tropical Hurricanes', *Quart. Jour. Roy. Met. Soc.*, **80**, 131–64.

DUNN, G. E. and MILLER, B. I., 1960, *Atlantic Hurricanes* (Louisiana).

LOCKWOOD, J. G., 1965, 'The Indian Monsoon—A Review', *Weather*, **20**, 2–8.

MALKUS, J. S., 1958, 'Tropical Weather Disturbances—Why do so few become hurricanes?', *Weather*, **13**, 75–89.

NEIBURGER, N. and WEXLER, H., 1961, 'Weather Satellites', *Scientific American*, **205**, 80–94.

PALMER, C. E., 1952, 'Tropical Meteorology', *Quart. Jour. Roy. Met. Soc.*, **78**, 126–64.

SYMPOSIUM, 1957–8, 'On the General Circulation over Eastern Asia', *Tellus*, **9**, 432–46 and **10**, 58–75 and 299–312.

TREWARTHA, G. T., 1961, *The Earth's Problem Climates* (New York).

WATTS, I. E. M., 1955, *Equatorial Weather* (London).

YIN, M. T., 1949, 'A Synoptic-Aerologic Study of the Onset of the Summer Monsoon Over India and Burma', *Jour. Met.*, **6**, 393–400.

GENERAL CIRCULATION AND DYNAMICAL CLIMATOLOGY

BIROT, P., 1956, 'Evolution des théories de la circulation atmosphérique générale', *Ann. de Géog.*, **65**, 81–97.

BREWER, A. W., 1949, 'Evidence for a World Circulation provided by the Measurements of Helium and Water Vapour Distribution in the Stratosphere', *Quart. Jour. Roy. Met. Soc.*, **75**, 351–63.

GOLDSMITH, P., 1962, 'Patterns of Fallout', *Discovery*, **23**, 36–42.

GOLDSMITH, P. and BROWN, D., 1961, 'World-wide Circulation of Air Within the Stratosphere', *Nature*, **191**, 1033–7.

LAMB, H. H., 1960, 'Representation of the General Atmospheric Circulation', *Met. Mag.*, **89**, 319–30.

MURGATROYD, R. J. and SINGLETON, F., 1961–2, 'Possible Meridional Circulation in the Stratosphere and Mesosphere', *Quart. Jour. Roy. Met. Soc.*, **87**, 125–35, and **88**, 105–7.

NEWELL, R. E., 1963, 'Transfer Through the Tropopause and Within the Stratosphere', *Quart. Jour. Roy. Met. Soc.*, **89**, 167–204.

— 1964, 'The Circulation of the Upper Atmosphere', *Scientific American*, **210**, 62–74.

PALMÉN, E., 1951, 'The Role of Atmospheric Disturbances in the General Circulation', *Quart. Jour. Roy. Met. Soc.*, **77**, 337–54.

PALMÉN, E. and VUORELA, L. A., 1963, 'On the Mean Meridional Circulation in the Northern Hemisphere During the Winter Season', *Quart. Jour. Roy. Met. Soc.*, **89**, 131–8.

ROSSBY, C-G., 1949, 'On the Nature of the General Circulation of the Lower Atmosphere', *The Atmosphere of the Earth and Planets*, G. P. Kuiper (ed.), Chicago Univ. Press, 16–48.

SHEPPARD, P. A., 1958, 'The General Circulation of the Atmosphere', *Weather*, **13**, 323–36.

STARR, V. P., 1956, 'The General Circulation of the Atmosphere', *Scientific American*, **195**, 40–45.

TUCKER, G. B., 1960, 'The Atmospheric Budget of Angular Momentum', *Tellus*, **12**, 134–44.

— 1961, 'Some Developments in Climatology during the Last Decade', *Weather*, **16**, 391–400.

— 1962, 'The General Circulation of the Atmosphere', *Weather*, **17**, 320–40.

UNESCO, 1963, *Changes of Climate*, 485 pp.

CHAPTER FOUR

Geography and Population

E. A. WRIGLEY

Lecturer in Geography, University of Cambridge

People are the stuff of all social sciences and of history, yet it is remarkable how little population as a general concept has entered into the discussion of social change and function until comparatively recently. There were population discussions of importance and some subtlety before the days of Malthus (1798), but he it was who first elevated the study of population to a central place in the social sciences. His views provoked much discussion in his own lifetime and later, but in the main population has tended to be treated as a variable of secondary importance by social scientists and economists until the post-war surge of interest in underdeveloped countries forced it much nearer to the centre of the stage and underlined the importance of some knowledge of the interplay of nuptiality, fertility and mortality and of these in turn with other sociological and economic features of a society. The same neglect of population has been evident in the writing of history. It is only in very recent years that historians have taken population seriously as an element in general history, in spite of the great importance of such things as the impact of epidemic disease, the average age at marriage, the average size of family, infant mortality, expectation of life, and the proportion of celibates in the adult population, to economic, social and general history. The work of Goubert (1960) and others in France has shown how powerfully an interest in population matters can influence the writing of history. Since the Second World War demography has developed distinctively and has produced important works of general theory as well as a mass of technical literature (e.g. Sauvy, 1956–9).

Population is as important to geography as to history, but much the same criticism can be made of geographers as of historians. Population has not perhaps been so badly neglected, but it has often been treated in a rather perfunctory and wooden way, even though the distribution and density of population, and such things as occupational structure and mobility, have long been basic elements in the study of human geography.[1] In recent years a number of prominent geographers have come to place it much higher in the heirarchy of geographical interests (see, e.g. Trewartha, 1953: also Beaujeu-Garnier, 1956 and Zelinsky, 1962). There has, in short, as so often in the past, been a movement in parallel with those taking place in other subjects. I should like to comment on two aspects of the interest in population which seem to me to make it especially fitting for geography and indeed to hold out the prospect of important new developments for the subject.

CHANGING IDEAS ABOUT THE PLACE OF POPULATION IN GEOGRAPHICAL STUDIES

The first point is connected with the same chain of events discussed above.[2] The whole range of problems involved in the attempt to understand why men come to live where they do, and in what numbers, and how they earn their living, which will perhaps do as a thumbnail definition of human geography, has been very substantially changed by the working out of the Industrial Revolution in western countries and elsewhere. The Industrial Revolution first weakened and then largely destroyed the close traditional ties between a society and the local land, and in so doing made anachronistic the vision of geography of men like Vidal de la Blache. His regional conception had been very well able to deal with the location of population, its density and the manner in which it earned its living under the conditions holding good before the Industrial Revolution, but largely failed when confronted with what came after. But though the old answer no longer satisfied, the question was as interesting and important as ever. It remained the case that human geography was very much taken up with the problem of explaining the location, the size and the economic functioning of populations. The population map remained the point of departure for much else.[3]

The striking thing about the general study of the location of populations and of the industries which afford them employment has been

[1] It is interesting to note that Hettner, having commended Süssmilch's early work on the statistical measurement of social characteristics, goes on to describe the second edition of Malthus' *Essay on Population* as almost a geographical work (Hettner, 1927, p. 70).

[2] See Chapter 1 above.

[3] It is interesting to note that Vidal de la Blache made extensive use of population material in the *France de l'Est* (1917). There are only two maps in this remarkable work. Both are population maps.

the way in which it has underlined the point that populations are nowadays more and more their own justification, to put the matter rather cryptically. In the main (the large retired populations of today are an obvious exception to the rule) men and their families live where there is work for them to do. In pre-industrial societies most people lived on the land and the most important factor in understanding the density and distribution of population was the distribution of land that was of good quality given the agricultural and pastoral techniques of the group and period in question (there were considerable changes in both the density and distribution of population in this country between, say, the late pre-Roman period and the time of the Norman Conquest, but they are intelligible within the terms just proposed in the main: new agricultural techniques turned land which had previously been difficult to use into valuable arable). Given any particular range of agricultural techniques the density and distribution of population could be treated as a dependent variable – something which could be understood in terms of the distribution of land of differing quality.

For a time after the Industrial Revolution it appeared to be possible to understand the great new masses of population growing up in areas previously thinly populated by an extension and modification of the old attitude. The large populations which appeared in places like South Wales, central Scotland, the Pittsburgh area, the Ruhr, central Belgium, the Saar and northern France grew up where they did because coal was necessary in large quantities to most industries, and because the total cost of assembling the necessary raw materials, converting them into finished or semi-finished products and delivering them to an intermediate or final consumer was normally less when the industry was on the coalfield than when it was at any other point (Wrigley, 1961, Chapter 1). Because of this it was possible to add to a treatment of the density and distribution of agricultural populations as a function of the distribution of land of different grades, a treatment of the distribution of manufacturing populations in terms of the distribution of mineral resources, and above all coal.[1] The one treatment paralleled the other and the two in combination accounted for the bulk of productive industry and most of the population. The numbers and distribution of the new manufacturing populations could be made to follow on from the consideration of the poverty or

[1] Making use of the line of reasoning which was given its classic expression in the early part of Alfred Weber's *Uber den Standort der Industrien*, available in translation as *Alfred Weber's Theory of Location of Industries* (Friedrich, 1929).

richness of mineral resources just as the distribution and density of agricultural populations followed from a discussion of soils and natural vegetation. In as much as those employed in tertiary industry (government service, retail and wholesale trading, transport and communications, banking and commercial services, and the professions) were taken into consideration at all they could quite conveniently be treated as dependent upon the populations engaged in agriculture and industry for whom they provided services. Further technological and economic change, however, has rendered this method of analysis largely obsolete. The internal combustion engine, the long-distance transmission of electricity, the use of oil and natural gas, the great gains in economy of fuel use, the triumph of the lorry, the tremendous growth in the durable consumer goods industries, and the associated changes in economic life, have meant that in the twentieth century the best location for more and more industries has become a point close to the largest market. A great city grows still greater like a snowball rolling downhill under its own momentum. Only a few industries continue to be raw material orientated – wood and food processing and sometimes iron and steel. The presence of large tertiary populations may stimulate the growth of secondary, manufacturing industry, rather than vice versa.

The model of the great French social scientist Frédéric Le Play may be invoked at this point to restate this argument in a helpful way. Le Play wrote during the middle decades of the nineteenth century a number of works, of which one, *Les Ouvriers européens* (1855) is of especial interest in this connexion. He formulated three heads under which he felt that information should be brought together and analysis undertaken, popularly known in English as Place, Work and People, or in modern terminology, physical environment, material technology and economic organization, and social characteristics (of which one is the distribution and number of population). He felt, just as the 'classical' geographers felt (see Chapter 1), that it was natural to progress from Place through Work to People. Most geographical textbooks have followed this pattern, beginning with a description of the physical environment, then tackling the characteristics of the economy of the area (agriculture and industry), and finally passing on to deal with population distribution and density and with such things as transport, trade and cities (those industries or aspects of the economy peculiarly bound up with tertiary industry). Grouping material under heads like Place, Work and People and dealing with it in this sequence is the time-honoured method.

The great change which the increasing market orientation of industry has involved is this – that it now makes better sense to invert this three-part sequence and to begin with population distribution in order to explain the distribution of industry rather than vice-versa. It is in this sense that one can justify the remark that populations nowadays are their own justification. It is a variation on the old theme that to them that have shall be given. The presence of ten million people in London constitutes a huge market which is very attractive to most types of manufacturing industry given the economic structure of contemporary society. This causes a large proportion of new industrial plant to be built in the London area and this in turn by affording still more employment reinforces the pre-eminence of London. The tendency to market orientation has long been true of some industries, such as the baking of bread or the garment industry. It is discernible in a very wide range of durable consumer goods industries from motor cars to radios, and is to be seen even in industries which in the past have shown a tendency to raw material orientation because they used a great tonnage of raw materials and wished to minimize total transport costs.[1]

The characteristic modern locational pattern can be seen most clearly in countries which escaped the nineteenth-century Industrial Revolution but have subsequently developed a full range of manufacturing industry. In such countries there are no great coalfield industrial agglomerations of the type which were once so typical to complicate matters. An unusually clear-cut illustration of the modern pattern is afforded by the state capitals of Australia which dominate the Australian manufacturing scene. Melbourne, and Sydney, for example, each account for more than three-quarters of the industrial production by value of their respective states. Each houses more than half the state population but has an even higher share of manufacturing output. A few industries, notably those engaged in processing raw materials, like the N.S.W. dairy factories or the Broken Hill lead and zinc refineries, which reduce the bulk of agricultural or mineral raw materials and so lessen the cost of transporting them, are still located close to raw materials or at

[1] This is true even of the production of primary iron and steel as the history of the Fontana plant near Los Angeles shows (though it is to be remembered, of course, that large markets for steel are also important sources of steel scrap and are therefore major sources of raw material, as it were).

See e.g. Isard and Cumberland (1950).

places determined by the availability of raw materials, but such industries are unusual nowadays. The iron and steel industry is the only major exception to the rule in Australia today and it is perhaps unfortunate that it should attract so much attention in many geographical analyses of Australian industry. The prime fact which any geography of Australia should seek to drive home and explain is the remarkable concentration of industry and population in the state capitals.

The reason for the change in industrial locations over the last two centuries will be familiar. Put rather simply it may be suggested in this way. If transport costs were astronomically high per ton-mile, it would clearly be essential for each manufacturer to seek that location which reduced his transport costs to a minimum since what he would save in this way would far outweigh any additional expenses in other directions to which he might be put. If, on the other hand, transport costs per ton-mile were nil the consideration of transport costs would not enter into any calculation of best location, though other things like differential wage costs in different areas would continue to do so. What has happened in the last two centuries has produced a movement along the spectrum of possibilities between the two extremes, not of course from one end to the other but a good way along the band.[1] Therefore the old constrictions upon industrial location have weakened, those, for example, which a century ago obliged so many industries to seek a coalfield location, and other factors have a relatively greater play. In these circumstances industry has gone towards the great markets, as, other things being equal, it will always tend to do. There the manufacturer is not only aided by close contact with his main consumers and competitors, but is also usually in the largest and most flexible labour market, is close to specialized banking, commercial and professional services, is at the hub of a great transport network, and so on. Some continuing regional differences, like those between London and provincial wage rates, seem rather to reflect the greater attraction of places like London than to counteract it.

An important additional feature of the modern situation is that tertiary employment is of much greater importance than in the past. The tertiary industries between them often nowadays account for more of the working population than secondary, manufacturing

[1] A more economical and effective use of raw materials (e.g. a sharp fall in the amount of coal needed to produce a ton of finished iron and steel products) has, of course, the same effect as a fall in transport costs.

industries; more sometimes than industry and agriculture combined (see Clark, 1940 and Table 1, p. 72). Rising real incomes per head accompanied this change to tertiary employment in the past and are tending to accentuate the importance of tertiary industries today. If the presence of people and the purchasing power of which they dispose is a prime determinant of industrial location it follows that the existence of large tertiary populations instead of being treated simply as a result of the presence of industry and agriculture should now be seen also as a cause of the growth of industry and indeed of the intensification of agriculture. Because Paris is the seat of the government of France, of the chief university, of many important financial and business houses, of some of the major hospitals, of great entertainment industries, and so on, all affording tertiary employment, it is an attractive location for many manufacturing industries, for example the car industry, and for the same reason agricultural land use for many miles round Paris has been greatly changed and intensified (Phlipponeau, 1956). This is as true as the more familiar idea that because there is a prosperous local agriculture there is a shop in the village, a village postman, a schoolmistress, a garage, and so on. One must no longer see tertiary employment as simply the consequence of the presence of the so-called productive industries; it is also an important cause of their presence, and always was, though of less relative importance in the past.

In view of all this, therefore, it makes good sense to begin any discussion of the distribution and density of population and of the economic activities which support it with the population itself. Instead of working from Place through Work to People, it is better to begin with People and proceed through Work to Place, or in some cases from People through Work back to people – that is from the consideration of the presence of sufficient people to constitute an attractive market to the industries which have developed as a result and so back to the size and structure of the population. The treatment this way round is less rigid than the earlier progression in the opposite direction. It leaves it open to anyone attempting to explain population density and distribution to go as deeply into the circumstance of the physical environment as may prove appropriate for the problem in hand. For example in an area like Andean South America where there are still many largely self-sufficient peasant communities it would clearly be important to pay close attention to the characteristics of the local physical environment after having dealt first with the question of the distribution and density of population and the range of material technology at the disposal of the local communities. But it is also possible where the circumstances of the physical environment are unimportant largely to ignore it (as for example in dealing with population and economic activity in London and its vicinity). There is in this procedure no commitment to the priority of the physical environment such as is implied in the 'classical' progression from the treatment of physical environment to the local community, from Place through Work to People. The older procedure was congenial to explanation in terms of cause and effect; the newer to functional analysis and the use of models.

POPULATION AS A CENTRAL FOCUS FOR GEOGRAPHY

The second aspect of an interest in population which may make it especially attractive to geography is that it holds out some promise of being a satisfactory central focus for the subject, a convenient nexus into which all strands can be seen to lead. If it be asked whether there is a common focus for the great mass of published geographical work, ranging from the study of the relationship between wheat yield and precipitation in Kansas to the intensity with which the New Zealand rail network is used or the settlement hierarchy of southern Germany, one possible answer is that they all serve to make clearer either directly or at one remove the circumstances which permit populations of such and such a size, distributed in such and such a fashion, to maintain themselves by such and such activities which afford them a livelihood. Just as in the nineteenth century all the cluster of geographical studies, though very diverse in detail, might be held to point towards the understanding of the ways in which the physical environment influenced social change and function; just as the study of types of farm architecture and of the functional layout of farm buildings or village settlements in the early decades of this century was not only of interest in itself but was related in the minds of the scholars who studied these questions to the more general regional scheme of geography which they had in their mental background (see Chapter 1); so one might suppose that the explanation of the density and distribution of population could serve the same purpose for geography today. This in modern circumstances can be both the starting and the finishing point. A knowledge of the simple facts of population distribution and density is a convenient starting point for the analysis and explanation of these facts, and the course of the argument makes a satisfactory full circle if at its close a fuller understanding of the facts of distribution and density has emerged.

There is nothing very novel or striking about such a suggestion as this. One might say that many of the older types of geography address themselves to the same end, at least in part. The traditional sequence of examination from Place through Work to People, from the physical environment through the material technology of a community to aspects of its social and economic life and organization, did often end with an analysis of population distribution and density among other things. Here the focus lay, at least in the 'classical' treatment of the question, chiefly in the mechanisms by which the physical environment exercised its influence on social change and function, yet there is much common ground between this and the scheme outlined above. Equally the 'regional' view of the subject dominant in the earlier part of this century, with its stress upon the functional interplay between the various elements of landscape, paid much heed to population. Vidal de la Blache, indeed, in the *France de l'Est* (1917) adumbrated a 'population' view of geography in several aspects of his analysis. Any change along the lines suggested can be held to be evolutionary rather than revolutionary; a recombination of some of the traditional elements of geography, the whole to be viewed in a different light; a new focus but much the same set of elements. Not admittedly quite the same set. A line must be drawn somewhere round what can reasonably and fruitfully be included in the subject. It is said that on an average there are only four removes of acquaintance between any two people chosen at random in the population of the United States, between A and B. A knows X who knows Y who knows Z who knows B. In much the same way there is some relationship between all that is taught as geography, between, say, changes in the Cretaceous and the activities of contemporary London, in all the conceptions of geography which have been proposed. There is, after all, connexion between all knowledge, and an endless chain of interconnexion reaching out from any one event to all others. But the temptation to be all-inclusive is one to be resisted. To take *all* that is relevant into account is impossible. It involves too much for the mind to grasp, for any computing programme to handle, for any model to accommodate. One attraction of the view that the chief focus of geographical study should be the attempt to describe and explain the distribution and density of population is that it would enable geographers to be clearer about what should or should not be thought relevant. Things which are relevant only at the second, third, or higher remove from this central question should be scrutinized carefully. A subject develops not only by acquiring new interests but also by dropping old.

The individual scholar is seldom inspired to undertake a piece of research work by his adherence or opposition to a general conception of a subject. For him what is relevant and useful is not determinable in advance. Indeed progress in knowledge has so frequently come by combining interests and techniques from two or more established disciplines that one might be tempted to assert that success comes more often from ignoring disciplinary boundaries than from observing them. Nevertheless, arguments about the general shape and ordering of a subject, about its methodology, are important. They are important because it is natural to ask of any branch of knowledge what its subject matter is and how it is ordered. Upon the answer to this will depend in part the type and quality of people attracted to it. They are important because they help to determine how the subject is taught, how it appears to those beginning their acquaintance with it. And they are important because it is useful constantly to test the compatability and mutual relevance of the components of a field of knowledge, especially in relation to changes in the methodology of cognate subjects. All such arguments form an endless but useful dialogue whose quality both reflects and helps to set the general standard of work and teaching done.

POPULATION AND GEOGRAPHY: AN EXAMPLE

As an illustration of what might be involved in a 'population' view of geography it may be helpful to sketch briefly a study of eastern Australia conceived in this way.

A first look at the density and distribution of population in Australia shows three salient characteristics: that the great bulk of the population is concentrated in the south-east corner of the continent; that very few Australians live more than two hundred miles from the sea; and that the state capitals in four of the five mainland states contain more than half the state population, while there are very few other towns of any size (at the 1961 census Newcastle and Greater Wollongong were the only cities of more than 100,000 people in mainland Australia which were not state capitals). This is the point of departure, and it will also be in a sense the destination. The aim of 'population' geography is to advance the understanding of population distribution and density on the way between these two points.

It is convenient to consider the population as laid down over the surface of Australia in three layers, as it were: a first layer consisting of those engaged in primary industry and their dependants – farmers,

*Table 1. Primary, Secondary and Tertiary Employment (as percentages of total employment)**

A: State
B: Capital city
C: State without capital city

		Primary§	Secondary†	Tertiary†
New South Wales	A	11·2	29·0	54·4
	B	1·1	36·0	56·4
	C	25·9	19·1	51·4
Victoria	A	11·2	32·7	52·3
	B	1·3	40·1	55·4
	C	29·4	18·9	46·6
Queensland	A	20·4	20·6	54·7
	C	2·3	27·8	66·7
	C	31·7	16·1	47·2

* Calculated from *Census of the Commonwealth of Australia*, 30 June 1954, Part I, Table 6.
§ exc. Mining.
† exc. Electricity, Gas and Water.

pastoralists, miners, and the small number of forest workers and fishermen; a second layer composed of those engaged in secondary manufacturing employments and their families; and a third layer made up of those working in tertiary industries and their families. To consider these three groups in this order is not to imply that any one is more basic than any other. They might indeed just as properly be taken in the reverse order. It is, however, convenient to treat them separately in the first instance since the principles relevant to the understanding of the three layers are substantially different from one another. I shall further simplify this sketch by using illustrations drawn only from the three main eastern states, Victoria, New South Wales and Queensland.

The first layer is numerically much the least important. Primary employment in New South Wales, Victoria and Queensland formed only 11·2, 11·2 and 20·4 per cent respectively of the total work force in 1954. Even if the capital city is excluded from the totals in each of the three states, the figures rise only to 25·9, 29·4 and 31·7 per cent. I shall confine my attention for the sake of brevity to employment in the agricultural and pastoral industries. This is, of course, spread quite differently over the face of eastern Australia from the other main branches of employment, reflecting chiefly the availability of well-watered land and ease of access to market. The general picture is well known. There are huge areas given over to sheep or cattle raising with very slight returns per acre and only a light dusting of population. Such areas figure largely in maps of types of land use but contribute relatively little to the total net value of Australian farm output and afford comparatively little employment. The areas of intensive land use are much smaller but very much more productive. The Riverina-Wimmera, the sugar areas of the Queensland coast, the dairying areas of northern New South Wales and the better mixed farming areas of Victoria do not bulk large on a map but they account for a large part of the farm output of the eastern states. Employment in pastoral and agricultural pursuits naturally reflects the differences between the former and latter types of farming. The importance of the small areas of intensive farming is very marked, though often not sufficiently appreciated. For example, the area of good land inside the arc of the Eastern Highlands, the Riverina-Wimmera, contained 66,666 persons engaged in agricultural and pastoral employment in 1954 out of a total in the three eastern states of 363,899 (18·3 per cent). The importance of this area in agricultural employment alone is even more marked – 43,571 in 163,972 (26·6 per cent). Figures of the net value of agricultural production can be used to underline the same point.[1]

[1] In the first post-war decade the gross value of agricultural output in the Riverina-Wimmera was more than 40 per cent of the total for the three eastern states. For statistical convenience the Riverina-Wimmera is here taken to be the divisions of Riverina and South-Western Slope in New South Wales and the divisions of Wimmera, Northern and Mallee in Victoria.

A consideration of the density and distribution of employment in farming is the quantitative starting point for an examination of those aspects of the material culture and economic structure of the community with which it is associated, and also of the characteristics of the environment which have complemented them in determining the distribution and density of primary employment. In connexion with the Riverina-Wimmera this might involve, for example, the discussion of such things as the structure of agricultural prices within which the Australian farmer operates; the advantages of this part of Australia because of its soil, rainfall and evaporation characteristics; the changes brought about in recent years by the use of subterranean clovers and superphosphates in raising production per acre; the amount and the variability of flow in the Murray and Murrumbidgee rivers compared with those elsewhere in Australia; the significance to the irrigated districts of the progress of the Snowy River Scheme; and so on. A comparison of the trends of agricultural employment and population in the Riverina-Wimmera with those in, say, the Western Districts of New South Wales might also be helpful in bringing to light those factors which in recent decades have tended to lead to a more and more intensive use of the best land rather than the taking in of new land in order to secure an increase of production. The history of the soldier settlement schemes in the Mallee after the First World War makes a very interesting contrast in this connexion with the history of the contemporary settlement in the irrigation districts.

A similar treatment of the sugar growing areas of Queensland would embrace other matters, including, for example, the discussion of the sense in which the policy of the Australian government is the foundation of the sugar economy of the area. Again the treatment of the very scantily populated sheep and cattle grazing areas inland must include the discussion of the peculiar sensitivity of the ecological balance in marginal areas and the great difficulties experienced in securing the provision of social overhead capital for these areas. In dealing with dairying areas of northern New South Wales the significance of access to the Sydney and Brisbane markets needs stressing; and so on.

At the conclusion of the section dealing with the first layer of the population map the reader will have been led as far into related questions as the explanation of the density and distribution of population dependent on primary industry requires. The level at which the discussion is pitched can be varied to suit the needs of the audience. It might not be the same, for example, for those who could be expected to be familiar with writings in the tradition of von Thünen as for those who could not. It is not, in short, geared to a particular technique of analysis but is flexible enough to be used at all levels.

Next in this brief adumbration comes secondary, manufacturing employment. This comprises a much larger share of the total work force than primary employment. Table 1 shows that manufactures in 1954 employed 29·0 per cent of the work force in New South Wales, 32·7 per cent in Victoria, and 20·6 per cent in Queensland. In the capital cities of these three states the figures were 36·0, 40·1, and 27·8 per cent; while in the states without the capital cities the figures were 19·1, 18·9 and 16·1 per cent. The outstanding feature of the distribution of manufacturing employment in Australia is its concentration in a few large cities (more than half the total of persons in secondary employment in the whole of Australia are in Melbourne and Sydney alone). This section might well begin with a discussion of the general circumstances of economic life today which make it more profitable for most manufacturers to establish their plants close to the largest market than at any other point, and also of the conditions in which this generalization does not hold true. The dominance of Sydney within New South Wales may be used as an example of the range of issues which will arise at this stage of the analysis. In 1954 78·0 per cent of the total industrial labour force of New South Wales was to be found in greater Sydney. Table 2 shows the distribution of employment in eight main divisions. In four of the eight divisions more than 70 per cent of employment was concentrated in Sydney. They conform to the rule that the largest market is the best point of manufacture. In the other four divisions the percentages were lower. The lowest, 30·4 per cent, was in Sawmilling and Wood Products where much employment occurs near the source of supply because so much is saved in transport costs by processing the raw material at this point. For this reason many of the workers in this industry were found to be in the Hunter and Manning and North Coast divisions of the state where there are substantial local timber reserves. A similar explanation accounts for the comparatively high percentage of employment outside Sydney in the Food, Drink and Tobacco industry. It is convenient to erect freezing and canning plants, butter and cheese factories, slaughterhouses, etc., in agricultural areas near the point of production because the raw material is perishable or because there is a large loss of weight in the process of manufacture and so a useful saving in transport costs may be had. In the Electricity, Gas and Water group much of the employment is of the service type and is therefore distributed roughly in proportion to the spread of total population with knots of concentrated employment where there are large power stations or gas works.

*Table 2. Industrial Employment in New South Wales**

	Founding, Engineering, etc.	*Ships, Vehicles, etc.*	*Textiles*	*Clothing, Boots, etc.*	*Food, Drink & Tobacco*	*Sawmilling and Wood Products*	*Paper and Printing*	*Electricity, Gas & Water*	*TOTAL*
Sydney	82763	38881	13662	38682	31498	5150	25264	18598	317551
Rest of Cumberland	1968	1040	408	920	1735	194	233	500	7984
North Coast	548	846	30	349	2935	3005	384	767	9552
Hunter and Manning	20899	5100	2826	2723	4045	3459	1139	3087	46944
South Coast	13320	626	171	1619	1250	967	310	1194	21933
Tablelands	3418	1393	837	1622	2252	1534	662	1336	15141
Slopes	1228	1464	200	994	2692	1462	563	1013	10654
Plains	136	258	4	73	284	581	64	148	1666
Riverina	242	348	7	155	1211	395	126	236	3093
Western	123	174	2	74	382	197	108	556	1709
Other	36	11	1	18	18	3	14	6	134
Total	124641	50141	18148	47229	48302	16947	28867	27441	436361
Percentage in Sydney	66·4	77·5	75·3	81·9	65·2	30·4	87·5	67·8	72·8

* Calculated from the *Census of the Commonwealth of Australia*, 30 June 1954, Part I, Table 6.

There remain the metal and engineering industries. In many branches of engineering the dominance of Sydney is almost complete as a further breakdown of this group would show, but the manufacture of primary iron and steel takes place not in Sydney but at Newcastle to the north and Greater Wollongong to the south of Sydney where good coking coal outcrops to the sea and the chief raw materials necessary for iron and steel making can be brought together at low cost. Hence the large employment figures in this category in the Hunter and Manning and South Coast divisions. The employment figures make a convenient point of departure for a discussion of the location of iron and steel manufacture. This has been less affected than most other large industries by the tendency to market orientation of the last half century because

*Table 3. Tertiary Employment (expressed as percentages of total employment)**

A: Capital city
B: State without capital city

		Building	Transport	Communication	Finance	Commerce	Public Authority and Professional	Amusement Hotels, etc.
New South Wales	A	7·2	7·5	2·3	3·7	17·7	13·5	6·3
	B	9·8	6·5	2·1	1·6	13·1	10·1	6·0
Victoria	A	7·3	6·2	2·3	3·4	17·2	13·0	6·1
	B	9·9	5·7	2·1	1·5	12·4	9·7	5·4
Queensland	A	9·6	8·1	2·8	3·6	19·7	16·3	6·6
	B	9·4	7·2	1·9	1·7	12·2	9·0	6·0

* Calculated from the *Census of the Commonwealth of Australia*, 30 June 1954, Part I, Table 6.

of the very large weight of raw materials involved and the importance of keeping down the final cost of the product by seeking the point of least cost assembly of the major raw materials involved. As with the food and wood industries it is important in order to understand the distribution of the iron and steel industry to know of the distribution of the raw materials which are used in these industries since the economic and technological circumstances of their production make it profitable to locate manufacture in many cases close to the point at which the raw materials are produced. In other industries where this is not the case there is little point in examining the sources of raw material supply when discussing location. Details of the size and value of production in each industry or industrial area can, of course, be introduced within a framework of analysis of this sort as easily as within a more conventional framework.

Tertiary employment is quite different in its distribution from either primary or secondary. Primary employment is almost entirely outside the great cities. Secondary employment is predominantly within them. Tertiary employment is much more evenly distributed. In 1954, for example, it constituted 56·4 per cent of the employment of Sydney itself, 54·4 per cent in the state as a whole, and 51·4 per cent in the state without its capital city. Table 3 shows the chief differences between Sydney and the rest of the state. In Building employment was relatively greater outside Sydney; in Transport, Communication, and Amusement and Hotels the pattern was similar in the capital city and outside; but in Finance, Commerce and Public Authorities the greater relative importance of tertiary industry in Sydney was marked since these three together employed 34·9 per cent of the total Sydney work force but only 24·8 per cent were similarly employed in the rest of New South Wales. The reason for the difference is very simply understood, at least in outline. As many primary school teachers are needed per thousand children of primary school age in a small country town or in an irrigation district as in Sydney itself; but the employment of men and women to teach in universities can be much more highly concentrated. Branch banks are needed everywhere roughly in proportion to the totals of population but central banking functions can be carried on in one great centre. Hence the distribution of tertiary employment in part reflects the distribution of primary and secondary employment in combination (the primary school teacher or postman type of tertiary employment), but an important element in tertiary employment is not tied in this way but occurs largely in state capitals. Since the existence of this second type of tertiary employment significantly increases the population and purchasing power of the cities in which it occurs it plays an important independent role in stimulating industrial growth. No analysis of the concentration of secondary employment in the Australian state capitals would be complete without reference to the importance of this point. The phenomenon can be observed in its purest form in the case of Canberra where there would be no secondary employment at all were it not for the decision of the Commonwealth in its early days to seek out a new site for a federal capital, but it has also operated powerfully in all the state capitals.

Some considerations necessary to the understanding of the distribution and density of Australian population can be introduced when dealing separately with each layer of the population cake, but some can best be taken after this stage when the population can be treated as a whole. Any treatment of the functional relationship between city and region can then take place. For example, there is scope for an interesting discussion of the differences between Brisbane on the one hand and Sydney or Melbourne on the other within their respective states. Brisbane is very eccentrically placed within Queensland close to the state boundary and contains a much smaller fraction of the population of the state than either Sydney or Melbourne. It is unable because of its position to dominate the industrial, service and administrative activities of Queensland as Sydney and Melbourne do in New South Wales and Victoria. Maryborough, Rockhampton, Mackay, Townsville and other coastal towns are better placed than Brisbane to serve parts of the Queensland hinterland and restrict Brisbane's dominance. The effect of Brisbane's position can be traced in Queensland's economy, transport system and social and political life. The pattern of Queensland road and rail communication is quite different from that of the more southerly states. Jealousy of the southern half of the state has long been a political force to reckon with in the north and gave rise in the past to strong separatist feelings; and so on. An analysis of this sort can be much more illuminating than the bald and misleading assertion that Sydney and Melbourne dominate their respective states to an unhealthy degree. Other general questions of resource utilization, population trends, the water problem in Australia, the transport network, and so on, might also be taken most conveniently at this stage.

CONCLUSIONS

This sketch is too brief to allow an extended discussion, but the prime virtue of the idea of a 'population' geography should be apparent. It is that it provides a nexus in which each line of inquiry can be seen to be anchored; that, to change the metaphor, it makes available a touchstone of relevance which can be applied to any body of material and which will ensure its *connexité* – to use the word to which Brunhes gave currency. If the geography of an area is treated in this way few things which are conventionally present in geographical works need necessarily be excluded, but inclusion would depend upon relevance to the central question of the distribution and density of population. Nor would a 'population' geography imply just a retreat from the periphery to a central citadel. It would involve the acquisition of new interests. This may be seen by comparing it with landscape geography. Whatever the virtues of landscape geography, its weaknesses are very grave. In the circumstances of modern industrial society a great part of the people and the activities by which they make a living escape the net. Unless this is thought an acceptable price something else must replace it. A concept of the subject suitable for a pre-industrial world will not serve for the world today. Geographers interested in modern industrial communities have in recent years materially changed the subject, devising in the process new models and statistical tools to assist them. If it is objected to this that a means should not be confused with an end, that statistical ingenuity needs to be harnessed to some more general conception of what geographers would be at, then perhaps the answer may lie in a 'population' view of the subject.

References

BEAUJEU-GARNIER, J., 1956–8, *Géographie de la population*; 2 vols (Paris).

CLARK, C., 1940, *The Conditions of Economic Progress* (London).

FRIEDRICH, C. J. (ed.), 1929, *Alfred Weber's Theory of the Location of Industry* (Chicago).

GOUBERT, P., 1960, *Beauvais et le Beauvaisis de 1600 à 1730.*

HETTNER, A., 1927, *Die Geographie: Ihre Geschichte, Ihr Wesen, und Ihre Methoden* (Breslau).

ISARD, W. and CUMBERLAND, J. H., 1950, 'New England as a Possible Location for an Integrated Iron and Steel Works', *Econ. Geog.*, **26,** 245–59.

LE PLAY, P. G. F., 1855, *Les Ouvriers européens* (Paris).

MALTHUS, T. R., 1798, *An Essay on the Principle of Population as it affects the Future Improvement of Society* (London).

PHLIPPONEAU, M., 1956, *La Vie rurale de la banlieue parisienne* (Paris).

SAUVY, A., 1956–9, *Théorie génerale de la population*; 2 vols (Paris).

TREWARTHA, G. T., 1953. 'A Case for Population Geography', *Ann. Assn. Amer. Geog.*, **43,** 71–97.

VIDAL DE LA BLACHE, P., 1917, *La France de l'Est* (Paris).

WRIGLEY, E. A., 1961, *Industrial Growth and Population Change* (Cambridge).

ZELINSKY, W., 1962, *A Bibliographic Guide to Population Geography*, Univ. of Chicago, Dept. of Geog., Research Paper 80.

CHAPTER FIVE

Trends in Social Geography

R. E. PAHL

Lecturer in Sociology, University of Kent at Canterbury

> The earth's covering of human dwellings is a phenomenon more geographical, more closely bound to natural conditions, than the earth's covering of human beings itself. . . . Truly geographical demography is above all the demography of the habitation. [Jean Brunhes, 1920]

> No geography can properly be regarded as 'Social' unless it draws its material from active study of men and women in their work and homes. [T. W. Freeman, 1961]

Since the idea that 'geographers start from soil, not from society' (Febvre, 1932, p. 37) was until recently widely held by most geographers, and is indeed still held by some, it is easy to understand why social geography has been slow to develop. We may define the field of the subject as the *processes and patterns involved in an understanding of socially defined populations in their spatial setting*. The social geographer is thus concerned with 'the lesser divisions of cities, town and country', which, as Visher noted as early as 1932, are likely to become more and more sharply differentiated according to *social* criteria. Such a definition immediately involves us in a wide range of problems. By taking socially defined groups and considering the processes acting upon them the geographer is involved with data common to other social sciences. However, in the same way that a geomorphologist has to acquire some knowledge of geology and physics, so the social geographer has of necessity to have some competence in human ecology and sociology in order to understand the processes at work in his field of study. Again, the patterns or models which the social geographer may discern or erect, however closely based on rigorous quantitative methods, must never be allowed to dominate thought so that they be made to fit all societies and all places at all periods of time. The social geographer, as much as the geomorphologist dancing on the coffin of the Davisian cycle, must constantly test by empirical analysis whatever middle order generalizations may be currently held at a theoretical level.

Social geography must be seen as being more than a loose agglomeration of such sub-divisions as medical geography (Howe, 1963), religious geography (Boulard, 1960; Zelinsky, 1961), population geography, linguistic geography (Jones and Griffiths, 1963), and now, maybe, electoral geography or the geography of education opportunity. This is no more than saying that economic geography is not limited to the study of the distributions of everything from tea production to atomic power stations. This is not to say that the mapping of distributions is not important; much cartographic work requires great ingenuity in choice of data and construction of indices. It is simply that distributions, however well presented, are not enough. Description does not necessarily imply comprehension and understanding: explanation may involve a wider range of variables than has hitherto been considered in such work. It is debatable whether the best map is that which answers or that which provokes most questions. Just as economic geography is now more concerned with the theories of the location of economic activity, so social geography has become concerned with the theoretical location of social groups and social characteristics, often within an urban setting. It is here that the links within the allied field of social ecology are closest.

THE ORIGINS OF SOCIAL GEOGRAPHY IN BRITAIN

However important it is to emphasize the development of social geography in an urban setting, before discussing the development of this branch of the subject from its origins in the Chicago school of human ecology, it might be useful to develop a parallel theme – the development of a school of social geography out of the more traditional human geography in Britain. In the inter-war period sociology was still suspect in many British universities, but it was becoming increasingly accepted that anthropologists, who at the time were producing some of the great functional studies of primitive communities, had something useful to contribute to human geography. The reaction by Barrows (1962, p. 6) in the U.S.A. in 1933 against the determinism of Semple and Huntington – 'How can an inanimate thing like soil or topography influence man? It would be as foolish to expect it to send me an invitation to a birthday party.' – was typical of the school of geographers who approved of the theory of adjustment, so that different human groups adjust in different ways in different places. Unfortunately, attempts 'to deal with the broad features of economic pattern and to consider their relation to physical environment, to social organization and to major factors in the growth of civilization' (Daryll Forde, 1934, p. vi) were entirely concerned with primitive agricultural communities. Whilst it is natural that distance should add enchantment it is curious that only in the last decade or so have local rural communities in Britain been given the same attention that those in Nigeria or Malaya appear to have received. Much of the drive towards an understanding of the social geography of rural communities has come from the University of Wales, where the link between anthropology and geography has been most fruitful, and led, in the 1940's, to the start of the most important work on Llanfihangel Yng Ngwynfa in northern Montgomeryshire (Rees, 1950) which has done so much to influence later work. A selection of further studies of Welsh rural communities, emanating from the same school (Jenkins *et al.*, 1960), helped to deepen the understanding of the cultural basis of community adjustments. A more recent work still, based on a Devon parish (Williams, 1963), emphasizes the spatial relationships of social and economic change and analyses the 'enduring relationship between society and the physical environment' (p. xx). Family farming in this instance is the main manifestation of this relationship. The pattern of the spatial and structural elements of land holding in this area depends partly on the quality of the farmland itself and partly on 'an attitude towards the relationship between family and land' (p. 80). Williams considers that an attempt to create a model of the dynamic family-land relationships is premature, and certainly any model of a stable and simple structure of family-land relationships would be unsatisfactory, when change is part of the system. Further work away from Highland Britain is essential if a meaningful picture of the social geography of rural communities in Britain is to emerge. A study by a social anthropologist of a Cheviot parish (Littlejohn, 1963) gives some indication of the sort of differences that may emerge in different socio-geographic settings, but studies of society-land relationships under the dominance of a major urban centre, say in south-east England, are badly needed.

In a sense the anthropo- or socio-geographic study of a modern rural community was not a great advance on studies of the traditional peasant communities in the 'Man and his Conquest of Nature' school. Certainly kinship analysis may have been substituted for housetype analysis as a meaningful clue to the 'adjustment' of the social group to the land, but in an *urban* environment the relationship with the physical environment may be completely without significance. Initially the study of the origins and morphology of towns provided interest. However, there was a real danger of arid classifications according to origin or present function, and work on the historic structure of towns could easily degenerate into antiquarianism. The importance of the physical environment was implied in much of the early work, with little regard for the social and economic reality of the community, in the manner in which rural communities were studied. So many towns, for example, with an area in the centre designated by urban geographers as the 'medieval core', were held to have much in common on that account, whereas the *use* to which this central area was put might range from slum dwellings, through a tourist centre, to palaces for the rich. Too often towns were discussed solely in terms of their generalized present economic functions – railway towns, spinning towns, political centres and so on with no real attempt to come to grips with the significant sub-divisions based on meaningful criteria.

GEOGRAPHY IN AN URBAN ENVIRONMENT

The tacit determinism in much of human geography, which led workers on the subject to search so readily for the (implicit) influence of the physical environment in patterns of settlement and economic functions of society, held up developments in an urban environment. Once the point had been made that the marshy patch in an eighteenth-century map accounted for a recreation ground in the present city, or that present street patterns followed old field boundaries, there seemed little more to say. In order to find the roots of urban social geography, therefore, we are obliged to go to the United States and consider pioneer work in human ecology, which later proved to be a fertile source of ideas. Thus one of the first serious attempts to formalize the diffuse information on cities was made by R. E. Park whose paper 'The City: Suggestions for the Investigation of Human Behaviour in the Urban Environment' first appeared in 1916.

> There are forces at work within the limits of the urban community – within the limits of any natural area of human habitation, in fact – which tend to bring about an orderly and typical grouping of its population and institutions. [Park, 1952, p. 14]

Park went on to try to discover and explain the regularities which appear in man's adaptation to space in an urban area. In ecological terms a high degree of interdependence and division of labour results in competitive co-operation for space use. As a result *natural areas* of the city emerge. 'They are the products of forces that are constantly at work to effect an orderly distribution of populations and functions within the urban complex' (1952, p. 196). Park has been criticized for seeing in the rapidly developing city of Chicago too much that was not typical of cities in other parts of the world. However, he himself wrote 'The ecological organization of the community becomes a frame of reference only when, like the natural areas of which it is composed, it can be regarded as the product of factors that are general and typical' (1952, pp. 198–9). He even went so far as to claim that he had 'covered more ground, tramping about in cities in different parts of the world, than any other living man' (Park, 1950, p. viii).

The 'forces' which Park mentions are implicitly ecological, and organize at a biotic or sub-social level of society, and it is for this tacit determinism that Park and his followers, the early classical school, have been criticized. For an account of the various models and theories produced by human ecologists the textbooks by Hawley (1950) and Theodorson (1961) are particularly useful. Hawley and others have argued that economic data, being readily available, are often the best indices of social phenomena. Activities will tend towards a central location depending on whether their need for accessibility is such that they can withstand the high cost associated with the land values of central locations. Such central areas – the so-called Central Business Districts (C.B.D.s) – may extend along major lines of communication, which themselves extend by time-cost distance the factor of centrality. Models, whether based on concentric rings round a central point in the C.B.D. or modified by sector development along lines of communication, make physical space, through land values, the final determinant. Now it is quite clear that individuals are not scattered randomly through space, but on the other hand the fact that segregation, according to social and economic criteria, exists does not thus destroy individual will or volition. In this connexion Firey's work (1947) is important, since his empirical analysis of Boston emphasized socio-cultural values as basic to the understanding of 'socially defined populations in their spatial setting'. He argues that *symbolic values* become linked with a spatial area and then social groups seek identification with such an area as an end in itself. There may indeed be a conflict betweeen economic interests and social symbolism as determinants of land use. Firey is concerned to show that 'social values are real and self-sufficient ecological forces' (Firey, 1947, p. 87) and this is demonstrated in his urban study of Boston. Beacon Hill, a district five minutes' walking distance from the city centre, has been an upper class residential district for a century and a half. Three minutes away on the north slope of the same hill is a decayed area of transient roomers, Jewish and Italian immigrants, where prostitution and other such activities flourish. The question is why the South slope has remained fashionable when other such hill sites similarly placed have changed? On the basis of a strictly rational or economic theory this central site should have been developed by the business establishments and exclusive apartment houses, which have repeatedly tried to locate themselves on Beacon Hill.

It is not necessary to describe here the way in which Firey builds up his thesis, based on considerable historical and sociological insight. In short, the relationship of the upper class families, as a social system, to this particular social environment is by no means 'biotic' or subcultural as the space determinists would presumably suggest. Emrys Jones in his more recent work on Belfast again shows that 'human motivation . . . itself tends to conform to a pattern reflecting current social values' (Jones, 1960, p. 268). 'What the geographer must avoid at all costs is the direct and simple correlation between the land and modern urban land use which is suggested by concomitant distributions' (p. 279). In 1840 the Malone ridge, on which the University of Belfast is built, was invested with social values, thus attracting residential building, overcoming the physical differences as between one part of the ridge and another.

Whatever the town or city, it is becoming increasingly accepted that geographers have much to learn from social ecologists and urban sociologists. There is a fashionable trend at the moment to believe that 'only by the complete rejection of uniqueness can geography resolve its contradictions' (Bunge, 1962, p. 13). A certain school of theoretical geographers would agree with Bunge that 'If only we could rake our lawns optimally we could be close to being able to arrange our cities optimally' (p. 27). This is nice to know; but surely the central place theory, which Bunge attempts to illuminate by his metaphor of raking leaves into piles, should be considered in the light of changing technology. There are other ways of clearing lawns than by creating Bunge piles. Unfortunately, so long as geographers sit puzzling over their lawn-raking type problems then the necessary empirical research on urban areas is not going to get done, and workers in other disciplines will feel that geographers can contribute little more than pedantic area delimitations, or the maps and diagrams to illustrate the data gathered by others.

A further point has to be considered. Most of the work on urban social geography has evolved in capitalist societies, where the so-called free play of market forces gives rise to differentially valued areas and the differences in land use related to economic forces. However, the situation is different in a centrally planned economy. How does this affect the theories of urban growth so far propounded? The social geography of Prague gives some interesting indications. The pre-war development of the city showed a concentric differentiation in social areas, which, in the terminology of a sociologist in a country with a centrally planned economy, were due to 'a society which embodied its social differences and injustices in its very territorial structure' (Musil, 1960, p. 237). During the last fifteen years, however, the trend to social area differentiation has halted and Prague is becoming sociologically more homogenous. In this connexion it is interesting to quote from the guide to the Prague development plan.

> The structure of a town has always been reflection of the social order . . . to-day our people are building a classless society . . . without the conflicts between town and country, manual and clerical workers . . . a town in this sort of society needs a functional structure that differs radically from the one it had under capitalism. The problem of reconstructing a city must be viewed as a complex whole . . . the inequality between the favourable living conditions of people in the newer districts of Prague and the dismal living conditions in the working class suburbs and the city centre must be eliminated . . . it is very important to reconstruct the industrial zone on the same lines as the residential area. [*Rebuilding Prague*, 1962, p. 18]

Now simply because the political or cultural factor changes the situation to be expected under the conditions which give rise to American cities' growth and morphology, this does not mean that segregation ceases to exist. There are strong indications that certain socio-economic groupings segregate themselves in certain districts of Prague, as in the British new towns certain areas appear to attract certain types of people, without such areas being very different in terms of rent or architectural qualities. That segregation continues to persist suggests that the social geographer working in urban areas is likely to become more social in his orientation, as the 'free play of market forces' becomes less free and as the impact of social mobility on the British class structure emphasizes what Goldthorpe and Lockwood describe as the *normative* and *relational* aspects of class rather than the economic aspect alone. A greater concern with total life style, for status differentiation and status enhancement, is likely to lead to the emphasizing of small differences between one residential area and another. Already the indications are strong that new, privately-built estates in the towns and cities of England are fairly homogenous internally, in terms of the social characteristics of their populations, but are socially sharply differentiated from neighbouring estates. Segregation appears to be on the increase.

URBAN GEOGRAPHY IN DEVELOPING COUNTRIES

A most significant and important trend which has developed in the past twenty years is the growing interest in the urban areas of the economically less developed countries.

> The rapid growth of cities, especially of large cities, is an outstanding feature of the modern age. Between 1800 and 1950, the population of the world living in cities with 20,000 or more inhabitants increased from 21·7 to 502·2 million, expanding about 2·6 times in the same period; 2·4 per cent of the world's population lived in urban centres of 20,000 or more in 1800, 20·9 per cent in 1950. [U.N., 1957, p. 113]

Between 1900 and 1950 the population living in cities of 100,000 or more increased by 444 per cent in Asia and 629 per cent in Africa. The phenomenal urban growth of Asia in the first half of the twentieth century has meant that now one-third of the world's large city (100,000 +) population is Asian. Certain countries in Latin America have higher proportions of their populations in localities of 20,000 or more than European countries, such as France or Switzerland. More than one-fourth of all Latin Americans live in cities of 20,000 or more and about one-fifth live in cities of 100,000 or more.

In the same way that a human geographer of an earlier period might visit Latin America or India to do field work on, say, peasant cultivation so, now, social geographers are visiting the booming, choking cities of these areas. Evidence of this interest is given in a recent volume entitled *India's Urban Future*, which is a selection of papers, some of which are contributed by geographers. The separation of residential land use from business or industry is by no means clear-cut in Indian cities. They have not grown in an orderly fashion but rather by addition and agglomeration. Large areas of cities are not very different from villages, so that the life-styles and attitudes of such 'urban' populations are much the same as those of villagers. The analysis and delimitation of the C.B.D. is irrelevant in the Indian situation where 'they can scarcely be said to exist except in the Indo-British seaports' (Turner, 1962, p. 67). Binuclear or polynuclear patterns have been described in various Asian and African cities, being mainly the result of the historical and cultural background of such cities. The social factor is crucial. Indian society is highly differentiated into mutually exclusive groups, and, although castes may lose some of their exclusiveness in urban areas, they continue to exercise a divisive effect in segregating the population.

Singh's study of Banaras (1955), although only very slightly orientated towards social geography, provides a useful case study to compare with Boston and Belfast. The *Inner Zone* of the city is a closely developed labyrinth of lanes connecting the temples, and the main activity is the pursuit of religion. It is mainly a Hindu area with a great concentration both of 'religious minded people' and of the high status homes of the rich. The religious status of the area has led to vertical expansion for residential (not office!) use. Next to the inner zone is the *Middle* or transitional belt of Muslim settlement. Again almost entirely built up, mosques are characteristic and the outer part of this zone contains the poor homes of moslem weavers. The *Outer Zone* is a curious mixture of slums and of the spacious houses of those rich people who were unable to find a house at the centre. In addition, the British administration added the *Civil Lines*, an area of wide, metalled roads and officers' bungalows, and the *Cantonment*, which was designed as a military garrison (Banaras was an important military garrison in the Second World War). There is also the isolated University quarter. In between these areas are formless masses of one- to three-roomed mud houses, typically with a verandah in front, which cannot be described as anything but suburban slums.

Thus Banaras shows considerable social and cultural segregation in its urban geography. There seems to be an almost random scatter of cottage industries throughout the city and the business centres, similarly scattered, provide evidence of what Singh calls 'haphazard' growth, although he does suggest that the concentration of 'feminine commodities' in one lane is on account of the number of women who pass that way to an important temple!

Singh is much less convincing when he attempts to describe the

'Umland' or sphere of influence, of Banaras, taking as his criteria the supply zones of vegetables, milk and grain and agricultural products, and also the circulation zone of newspapers, he derives Umlands of 72, 180, 4,000 and 20,000 square miles respectively. The latter area he claims to be the 'culturally integrated area of the Umland'. Such an approach was thought to be entirely inappropriate by Lambert, who felt that what should be considered was 'not the radiation of influence from cities into rural areas, but rather the presumed radical change in life style which confronts the villager when he moves to the city or the city moves out to encompass the village' (Turner, 1962, p. 131). Urban influences on rural areas appear to depend, in India, not so much on the actual *position* of the village, but on the receptiveness to the absorbtion of new ideas of social groups within the village and the degree of potential flexibility present in the social structure, so that, curiously, 'many of the villages whose unity and isolation have been emphasized lie very close to towns and cities' (p. 126).

Useful teaching material in the social geography of urban areas in economically underdeveloped countries may be found in the three UNESCO reports and also the 1957 U.N. *Report on the World Social Situation*. One might also mention some studies in Africa – Kuper (1958) on Durban, Southall and Gutkind (1957) on Kampala, the Sofers (1955) on Jinja and Marris (1961) on Lagos, which, although not primarily geographical works, provide useful insights into urban ecology and are more easily obtainable than works in French on African towns and in Spanish and Portuguese on some in Latin America. Some further ecological studies are listed in Theodorson's selection of papers (1961, pp. 438–9). The empirical studies need to be related to some sort of theoretical framework; an attempt to provide a typology of urbanization has been recently put forward by Riessman (1964, pp. 198–235) and is likely to be a useful teaching aid.

THE INCREASING DOMINANCE OF SOCIAL FACTORS

We have discussed the study of British towns and rural communities and have made some mention of the social geography of the rapidly growing cities in Asia. Singh's work on the Umland of Banaras was strongly influenced (far too strongly in fact) by some work on urban hinterlands in Britain and the United States carried out some 10–20 years ago. C. G. Galpin's pioneer study, *The*

Social Anatomy of a Rural Community, was published as early as 1915, but it was not until some twenty years later in this country that R. E. Dickinson started to lead the breakaway from the geographers' preoccupation with land use, and in *City, Region and Regionalism* challenged the rigid town/country dichotomy by arguing that 'an area of common living can be designed only in the key trait of that common living, that is, in terms of *social* considerations, not of a particular set of physical factors which condition that pattern of living in part' (Dickinson, 1947, p. 9; my emphasis). However, Dickinson went on to argue that the measurement of the 'service factor' would provide an accurate method of determining the hinterland of urban centres. It was indeed satisfying to see the way the work of Green (1950) or Bracey (1952) could be used to grade service centres, and it was easy to believe that one was bringing order and insight to settlement geography. This was especially so when work based on the analysis of bus time-tables, before the rapid spread of the ownership of motor-cars, could provide a guide to the number of journeys made from a place. However, much of such work became historical geography almost before it was published and it is open to the criticism that it applies a rigid static framework to a developing and dynamic system. It is becoming increasingly understood that the social area of a community may not bear much relationship to the service factor and the detailed studies of rural communities by Rees (1950), Williams (1963) and Littlejohn (1963) are useful correctives here. Further, the superimposition of one pattern over another is becoming accepted and analysed (Pahl, 1965). Farmers may have a different pattern of activities and linkages from farm workers, and middle-class commuters from other villagers. The daily journey to work is increasing in the rural as well as in the urban areas, as Lawton (1963) has shown nationally, and as some interesting work conducted by certain Planning Departments illustrates in detail. More and more research workers in this field are obliged to sub-divide their populations, generally according to social criteria, and to describe the movement of such sub-groups. It is quite inappropriate to consider the population of any settlement, however small, as homogeneous. In many parts of the country an analysis of bus journeys simply reflects the journeys made by those without cars – the working class wives, the old and the poor.

Perhaps the most fundamental trend in the work of social geographers in the last twenty years has been the developing interest in the growth and structure of metropolitan regions. An early description of the growth of the London metropolitan region was given in 1911:

> In many districts, urban and rural, outside the boundary, both the volume of population and its abnormal rate of increase must be partly attributed to their situation with respect to the Metropolis, and although the distance to which the Metropolitan influence extends cannot be defined with accuracy, it can hardly be put at less than thirty miles from the centre. . . . The rates of increase in many seaside towns, which, though outside the thirty-mile limit, are within easy reach of London, have also been far above the average. Men whose daily work is in London often reside at a distance corresponding to a railway journey of not less than an hour's duration. [Report of the London Traffic Branch of the Board of Trade (1911), Cmd. 5972, p. 4]

Parallels could easily be found in America and elsewhere, but certainly the growth of the London Region has been outstanding. The inter-war expansion was described in the Barlow Report (Cmd. 6153, 1940) and Abercrombie later described the 'unbridled rush of building (which) was proceeding in the form of a scamper over the home counties' (1945, p. 2). The post-war trend continued, 788,012 people adding to the region's total from 1951–61, with a redistribution of population such that Central London declined by 11·2 per cent and the inner county ring increased by 46·5 per cent (only half the increase being due to the creation of six New Towns in the area). An American who came to study the situation felt that 'We are more ignorant concerning the physical manifestations of the interlocking character of the Metropolitan community than we are on other seemingly remote questions, such as the internal heat of stars' (Foley, 1961).

For the situation in the United States there is an interesting summary of the social geography of metropolitan regions in the work of Wissink (1962). His argument is that 'the cultural system of a nation – in particular its space-related values, objectives, instruments and institutions – gives rise to a national type of city, a basic theme on which the individual cities are variations' (p. 287). Thus the booming rural-urban fringe round American cities (see also Dobriner, 1963) is paralleled in Britain, with a different cultural and political tradition, by, for example, the planned New Towns of the London Region.

Wissink draws on a broad range of American sociological and ecological literature to provide a valuable summary of the various approaches to the study of the city in relation to its hinterland and also to the history, definition and description of the rural-urban fringe. This book is extremely valuable as a teaching aid and would seem to be an important addition to university textbooks in social geography. In particular it is a useful corrective to the rather arid ecological classifications of Bogue and his associates.

It is estimated that in the United States by 1975, 57 per cent of the total population of metropolitan areas will be in the suburban and fringe area (Dobriner, 1963, pp. 146–7). Over three-fifths of the United States' population is now living within Standard Metropolitan Areas (formally defined in 1950 by the U.S. Bureau of the Census) so that the centrifugal movement out to their peripheries involves a considerable movement of population.

Gottman, in his useful survey of the urbanized north-eastern seaboard of the United States, claimed that all previous patterns of urban regions based on central cities and hierarchies of suburbs and satellite towns were totally inadequate tools with which to analyse his 'megalopolis'. A 'totally new order in the organization of inhabited space' is emerging (Gottman, 1961, p. 9) (thus Dickinson's work published in 1947 is now quite outmoded (ibid., p. 736)). Gottman can find no orderly pattern for this but only a 'nebulous structure'; this structure is perhaps held together by the motor-car.

Returning to this country, it has often been argued that the centrifugal movement can be explained in terms of an anti-urban reaction to the industrial city, with its roots in the nineteenth century and earlier. Ebenezer Howard's vision of 'Garden cities' and its realization at Letchworth and Welwyn Garden City has been maintained in the pressure for more new towns. Vigorous urban renewal or the creation of a new *city* does not seem to meet with the same degree of public support in this country. The green belt idea is a good example of the social basis of land use policy; thus the London green belt was to be 'where organized large-scale games can be played, wide areas of parks and woodlands enjoyed and footpaths used through the farmland' (Abercrombie, 1944, p. 8). The actual use of the green belt has recently been discussed by Thomas (1963), and it is clear that to sterilize land for development, in areas where the economic pressures to build are high, is to put social values before economic ones. In whatever ways the green belt is used in practice and however much it is a 'moat' which commuters have to pay extra to cross, it does help to prevent the continuous spread of urban building, although of course it helps to intensify pressures on its outer side.

It is the interpretation of the national Town and Country Planning legislation by the local authorities which does so much now to determine the distribution of social groups in space. Hence, when a geographer attempts to grapple with the practical problems involved in the megalopolitan London Region, as Peter Hall does in his *London 2000* (1963), this should be welcomed, rather than superciliously rejected as not being 'academic geography'. Clearly there is much in such an unashamedly polemical work which one might want to disagree with or modify. However, as a teaching aid and stimulant, it is first class; would that more present-day geographers were less afraid of their so-called reputations and would follow Hall's lead! The Report on *Traffic in Towns* (H.M.S.O., 1963) estimates an increase of 16½ million vehicles between now and 1980, a trebling of traffic in a little under twenty years. The Registrar General has estimated that the population of England and Wales will rise by 7,318,000 to 54,086,000 between 1962 and 1982. Thus by 1980 there is likely to be in the order of 450 vehicles for every 1,000 of the population. It is clear that, whatever may have been the trends of social geography in the past, if social geography is going to be taught adequately in universities, training colleges and schools, then professional geographers cannot but concern themselves more with the problems that surround them. It may well be that geography will acquire an increasing relevance in future years.

We have moved some way from the changing relationships of farmers to the land at Ashworthy, through the accounts of Boston, Belfast and Banaras, to an analysis of the development of the vast and complex metropolitan regions. Whether one is documenting the characteristics of the rapid urbanization of the economically underdeveloped countries, or the complexity of commuting patterns in the city-regions of advanced countries, it is clear that change is ever-present. The patterns are constantly in a state of flux. Individual families are frequently on the move and whole communities are in the process of changing their characters. It is almost as if there were some sort of cycle at work. At an early stage of the cycle the social factors are important; for example, the Jewish ghettoes are strictly segregated. Then comes the economic stage when the maximum profit determines the use of land and the models of theoretical geographers are useful and illuminating. Thirdly, the social factors reassert themselves and people move to certain areas, not because it is cheaper to do so, but because a form of segregation is part of the way people adapt to their position in society. This is not necessarily an economically determined position. Social geography is the discipline which is fundamentally concerned with the spatial manifestations of social change. In the same way that economic geographers are concerned with the *processes* involved in the study of industrial activity, so too is the social geographer interested in the social processes involved in the spatial location of social groups. To answer questions concerning the distribution of West Indians in London or Catholics in Belfast, the social geographer is obliged to go beyond the simple stage of mapping distributions.

Emrys Jones has expressed some central issues of social geography as follows:

> If the factor being studied has an effect on the residence of people then its distribution will not be a random distribution but irregularities will occur which will exhibit segregation. . . . Human motivation . . . itself tends to conform to a *pattern* reflecting current social values. [Jones, 1960, pp. 199–200 and p. 268; my emphasis]

Hence it is impossible for the social geographer to create a model which would be suited to all societies at all periods of time. Take, for example, the segregation of a country's *élite*. This may congregate round the coast or at a religious centre; it may select a fashionable area of the centre of a town or city or it may choose a particular suburb or area at the periphery. At one period of time the *élite* may be scattered throughout the countryside; at others it may concentrate in a particular place. A particular part of a town may be fashionable at one period, may decay in time, and then become revalued and renovated as fashion changes. Houses built for the artisan at one period may be used by the *élite* of another period, even though these houses may require great expense to make them comfortable and even though they may be economically poorly sited in relation to the workplaces of the earners. Hence it is impossible to limit study to the actual dwelling, since of much greater importance are the socio-economic characteristics of those who live in the dwellings. Under the forces promoting urbanization and in industrially highly developed societies, the field of social geography is concerned less with the relationships of social groups to the physical environment than with the patterns and processes involved in the segregation of social groups and settlements in space. The prime geographical factor is *distance*, whether actual physical distance or economic distance measured in time-cost terms. One might also add that 'social distance' is also of interest to the social geographer. It appears that there is some indication that social mobility and geographical mobility are related and thus, in a society in which certain sections are able to move up a promotional ladder with differential economic rewards, there are strong pressures for these economic differences to be reflected in social differences and hence in the segregated estates mentioned above. The social geographer is thus interested in the broad changes of population structure and distribution as the very necessary first stage in his analysis. From the basic demographic aspect, the analysis might proceed to such aspects as the geography of mortality rates, religion or occupational groups. The binding framework of settlement pattern, whether of villages, towns, cities or metropolitan regions, can be analysed within a particular socio-cultural background. The mobility of socially defined groups is of fundamental interest, as is the changing function and nature of communities as their physical and social space relationships change. Naturally the social geographer is concerned with the *social* aspects of changing space use and is thus likely to have some training in and understanding of sociology. The necessity for some economics to be taught in school geography is already understood, the equal necessity for some sociology to be included is not so readily accepted.

THE TEACHING OF SOCIAL GEOGRAPHY

The preceding part of this chapter has been intentionally discursive. It is an essential part of the argument that social geography does not have a grand structure or theory which may be described and that it would lose much of its point if such a structure or theory were built and then rigidly accepted. The social geographer is obliged to read widely in the field of sociology and the selected essays of Louis Wirth—*On Cities and Social Life* (1964)—provide a useful starting point. A more controversial, but highly stimulating, introduction to sociological insight is *The Sociological Imagination* (1958) by C. Wright Mills. Wirth and Mills would together provide a good introduction for the non-sociologically-trained geographer to a new world of ideas. General textbooks of social geography are useful to provide some general background – for example, those by Pierre George (1961) and R. E. Dickinson (1964). However, in order to teach an imaginative and constructive approach to social geography, it might be helpful to use some of the studies of modern British communities which, fortunately, have become more available in the last fifteen years.

If we take, for example, the English Midlands and we want to introduce a class to certain aspects of social geography, then there are several interesting studies which do much to illuminate the situation, although these are not very appropriate for use by pupils at school. There is an account of Coseley in the Birmingham conurbation by Doris Rich (1953), which relates mobility for leisure time activities to different social and physical factors. About one-third of the leisure time spent outside the home was spent outside Coseley. Using similar methods of investigation, modified according to the local situation, the results of the Coseley survey could be tested and compared with the pattern in the area of the school or college. That is to say, rather than simply measuring what the facilities of a place are, the investigation would be concerned with where people *actually* go in their leisure time. Journeys may be quantified and expressed diagrammatically and an attempt made to account for the pattern. Social factors may then be balanced with economic ones as determinants of mobility patterns.

Studies of Banbury by Margaret Stacey (1960) and of two neighbourhoods in Oxford by Mogey (1956) provide further valuable social insights into the complexities of modern communities. A discussion of these studies in the context of aspects of the modern geography of the area could usefully lead to a consideration of the local area once more and the way it is changing. Quite obviously with an expanding economy, the growth of new industries and the enormous number of houses that need to be built annually, there are few communities that will not be changing in some way in the near future. This emphasis on change presents a challenge to the social geographer to analyse the processes making for the decline of one area and the growth of another, with the consequent development of new patterns of mobility for economic and social reasons. Towns are changing in relation to each other and villages are losing one function to take on another.

It is often possible for schools and colleges to do useful work which may have practical value in the future planning of the communities as well as providing an understanding of the social geography of the area. Local planning authorities may often have basic maps of changes in housing, population and journeys to work and would welcome detailed traffic surveys or mobility patterns of the inhabitants of particular streets or neighbourhoods to provide checks to their more general work. Similarly a class could do practical work on the census material, using either the county volume or more detailed information which could perhaps be made available by the local Planning Department.

Clearly if the study of the local community is taken as the means of gaining an imaginative insight into social geography, there is the danger of assuming that the patterns and processes which emerge and operate in one particular society will necessarily hold true for another. In order to guard against this sort of parochialism there is much to be said for the study in depth, possibly with field work, of another quite different area. The guide produced by Birou, Burdet and Lapraz (1957) to the study of French rural communities provides a mass of information and ideas which could serve as the basis for much useful teaching in social geography. Indeed this is so useful that it is difficult to suggest a better introduction for teachers to practical work in the subject.

It has been the intention of this essay to show that the social geographer, concerned with such things as the commuting patterns of the chief earner, the differential mobility of his family and the new estates on which such people live in the exploding cities of the economically advanced countries, is simply the modern counterpart of the human geographer, interested in, for example, the movement of Swiss peasants up to the upland pastures in spring and down to the vine harvest in the lower valley in the autumn. This is not to say that commuting is more important than seasonal transhumance; it is simply often more relevant to the lives of those in predominantly urban countries. Again, the urban geography of Brisbane is not more or less important than the geography of the Queensland sugar industry; it is simply becoming more relevant to an urban society. The social geographer, armed with sociological insight and conscious of the importance of space, can approach the local community with techniques which may provide a most intellectually satisfying analysis.

References

ABERCROMBIE, P., 1945, *Greater London Plan 1944* (H.M.S.O.).

'BARLOW REPORT', 1940, *The Royal Commission on the Distribution of the Industrial Population, Report* (H.M.S.O.).

BARROWS, H. H., 1962, *Lectures on the Historical Geography of the United States as given in 1933* (Chicago).

BIROU, A., BURDET, R., and LAPRAZ, Y., 1957, *Connaitre une Population Rurale* (Economie et Humanisme, 262 Rue Saint-Honoré, Paris 1er).

BOGUE, D. J., 1950 A, *Metropolitan Decentralisation. A Study of Differential Growth* (Oxford, Ohio).
— 1950 B, *Structure of the Metropolitan Community. A Study of Dominance and Sub-Dominance* (Ann Arbor, Michigan).
— 1955, 'Urbanism in the United States', *American Jour. of Sociology*, **60** (1955), 471–86.
BOULARD, F., 1960, *An Introduction to Religious Sociology* (London).
BRACEY, H. E., 1952, *Social Provision in Rural Wiltshire* (London).
'BUCHANAN REPORT', 1963, *Traffic in Towns* (H.M.S.O.).
BUNGE, W., 1962, *Theoretical Geography* (Lund).
DICKINSON, R. E., 1947, *City, Region and Regionalism* (London).
— 1964, *City and Region* (London).
DOBRINER, W. M., 1963, *Class in Suburbia* (Englewood Cliffs, New Jersey).
FEBVRE, L., 1932, *A Geographical Introduction to History* (London).
FIREY, W., 1947, *Land Use in Central Boston* (Cambridge, Mass.).
FOLEY, D., 1961, 'Some Notes on Planning for Greater London', *Town Planning Review*, **32**, 53–65.
FORDE, DARYLL, 1934, *Habitat, Economy and Society* (London).
GEORGE, P., 1961, *Précis de Geographie Urbaine* (Paris).
GOLDTHORPE, J. H. and LOCKWOOD, D., 1963, 'Affluence and the British Class Structure', *Sociological Rev.* (New Series), **11**, No. 2, 133–63.
GOTTMAN, J., 1961, *Megalopolis* (New York).
GREEN, F. H. W., 1950, 'Urban Hinterlands in England and Wales: An Analysis of Bus Service', *Geog. Jour.*, **116**, 64–81.
HALL, PETER, 1963, *London 2000* (London).
HATT, P. K. and REISS, A. S., 1957, *Cities and Society* (Glencoe, Illinois).
HAWLEY, AMOS H., 1950, *Human Ecology* (New York).
HOWE, G. M., 1963, *National Atlas of Disease Mortality* (London).
JENKINS, D. *et al.*, 1960, *Welsh Rural Communities* (Cardiff).
JONES, E. and GRIFFITHS, I. L., 1963, 'A Linguistic Map of Wales, 1961', *Geog. Jour.*, **129**, 192–6.
JONES, EMRYS, 1960, *A Social Geography of Belfast* (London).
KUPER, L. *et al.*, 1958, *Durban, a Study in Racial Ecology* (London).
LAWTON, R., 1963, 'The Journey to Work in England and Wales: Forty Years of Change', *Tijdschrift voor Economische en Sociale Geografie*, **54**, 61–69.
LITTLEJOHN, J., 1963, *Westrigg – the Sociology of a Cheviot Parish* (London).
MARRIS, P., 1961, *Family and Social Change in an African City* (London).
MILLS, C. WRIGHT, 1958, *The Sociological Imagination* (New York).
MOGEY, J. M., 1956, *Family and Neighbourhood* (Oxford).

MUSIL, J., 1960, 'The Demographic Structure of Prague' (in Czech), *Demografie* (Prague), **2,** No. 3, 234–48.

PAHL, R. E., 1965, *Urbs in Rure: The Metropolitan Fringe in Hertfordshire* (London School of Economics Monograph).

PARK, R. E., 1950, *Race and Culture* (Glencoe, Illinois).

— 1952, *Human Communities – The City and Human Ecology* (Glencoe, Illinois).

REBUILDING PRAGUE, 1962, *K. Problemum Vystavhy Prahy* (Prague).

REES, A., 1950, *Life in Welsh Countryside* (Cardiff).

REISSMAN, L. 1964 *The Urban Process – Cities in Industrial Societies* (Glencoe, Illinois).

RICH, D., 1953, 'Spare Time in the Black Country', *Living in Towns*, Leo Kuper *et al.* (London).

SINGH, R. L., 1955, *Banaras: A Study in Urban Geography* (Banaras).

SOFER, C. and R., 1955, *Jinja Transformed* (Kampala, Uganda).

SOUTHALL, A. W. and GUTKIND, P. C. W., 1957, *Townsmen in the Making* (Kampala, Uganda).

STACEY, M., 1960, *Tradition and Change* (Oxford).

THEODORSON, G. A., 1961, *Studies in Human Ecology* (Evanston, Illinois).

THOMAS, D., 1963, 'London's Green Belt: The Evolution of an Idea', *Geog. Jour.*, **129,** 14–24.

TURNER, ROY (Ed.), 1962, *India's Urban Future* (Berkeley and Los Angeles). (See: Brush, J. E., *The Morphology of India's Cities*, pp. 57–70. Ellefson, R. A., *City-Hinterland Relationships in India*, pp. 94–116. Lambert, R. D., *The Impact of Urban Society Upon Village Life*, pp. 117–140. Hoselitz, B. F., *A Survey of the Literature of Urbanisation in India*, pp. 425–43).

UNESCO, 1956, *Social Implications of Industrialisation and Urbanisation in Africa South of the Sahara* (Paris).

— 1957, *Urbanisation in Asia and the Far East* (Calcutta).

— 1961, *Urbanisation in Latin America* (Paris).

UNITED NATIONS, 1957, *Report on the World Social Situation* (New York).

VISHER, S. S., 1932, 'Social Geography', *Social Forces*, **10,** 351–4.

WILLIAMS, W. M., 1963, *A West Country Village* (London).

WIRTH, L., 1964, *On Cities and Social Life* (Chicago and London).

WISSINK, G. A., 1962, *American Cities in Perspective* (Assen, The Netherlands).

ZELINSKY, W., 1961, 'Religious Geography of the United States', *Ann. Assn. Amer. Geog.*, **51,** 139–93.

CHAPTER SIX

Changing Concepts in Economic Geography

P. HAGGETT

Professor of Geography, University of Bristol

It has been a recurring criticism of our military command that it was well prepared to fight the Boer War by 1914 and adept at trench warfare by 1939. How far such criticism may be unjust of military preparedness there were disturbing signs in the academic world that geographers were in the post-war period girding their loins to fight pre-war battles. The long debate over possibilism and determinism was carried wearily on long after it had been dismissed in other subjects as '. . . simply a misunderstanding of history' (Bronowski, 1960, p. 93). Broadsides continued to be launched at quantitative methods which had been tested, tried and assimilated decades before in subjects no better suited to their use. Textbooks in regional geography continued to be fashioned around the same conceptual constructs which, as Wrigley (Chapter 1) points out, Vidal de la Blache had seen collapsing about him two generations before.

Ackerman (1963) has seen the origins of this frustration in the extreme separatism of geographers as a group; a separatism traced in Hartshorne's *Nature of Geography* (1939). He suggests that in neglecting the course of science as a whole, we neglected the axiom that the general progress of science determines the progress of its parts. How far Ackerman's case may be true, it is certain that the last decade has seen strenuous efforts to close the gap. More external concepts, notably from the fields of *mathematical statistics* and *systems analysis*, have been introduced into geography in the last decade than in any comparable period. These imports have been of critical importance in the new geography, but since they are wholly general in application rather than specific to economic geography I have not treated them at length here. Burton (1963) has provided an extensive review of quantification in geography and Chorley (1962) has demonstrated the impressive potential of systems analysis for geographical research.

I propose then to discuss some of the other aspects of contemporary change in economic geography and leave the quantitative and systems-analysis revolution as a self-evident truth. I have in mind three lines of more distinctive development: convergence of economic geography with other branches of human geography; the extension of models in teaching and research; and the growing interest in chance or stochastic processes. This is not an exhaustive list of possible topics, but it does include three that I regard as important in my own approach to economic geography.

THE IDEA OF CONVERGENCE

Convergence with Other Parts of Human Geography

In a penetrating study of the relations between history and geography, Darby (1954) has suggested that contacts between disciplines can be both *active* and *passive*. Using this analogy we can think of an active relationship between geography and economics in which the one is actively influencing the other or a passive juxtaposition. McCarty's *Geographic Basis of American Economic Life* (1940) is concerned with the active role of regional geography in shaping economic activity; conversely Chisholm's *Rural Settlement and Land Use* (1962) is concerned with the active role of the economics of transport in structuring familiar features of regional geography. Other contributions, like that of Ginsburg's *Atlas of Economic Development* (1961), are harder to classify and may well represent a more passive juxtaposition of economics and geography in equal partnership. Certainly the swing of the pendulum has now set in strongly towards the second sort of active relationship with a growing emphasis on the role of economics in determining the form of geographic patterns. To be sure this emphasis is currently on a rather specialized part of economics, transport economics, with studies of highway economics and urban change in the United States to the fore (Garrison, Berry, Marble, Nystuen and Morrill, 1959; Berry, 1959), but there are signs that this is being extended to the wider range of economic analysis.

It is of course as convenient as it is misleading to regard economics as a fixed point and plot the course of geography in relation to it. Economics itself has been undergoing developments at least as fundamental (Kendall, 1960). Robbins (1935, p. 17) in his *Nature and Significance of Economic Science* drew attention thirty years ago to the swing in economics from a classificatory to an analytical position: 'We do not (now) say that the production of potatoes is economic activity and the production of philosophy is not . . . in so far as either kind of activity involves the relinquishment of other desired alternatives, it has its economic aspect.' But this move away from a classificatory view of economics had little apparent effect on textbooks of classical 'economic' geography (Smith, Phillips and Smith, 1955; Jones and Darkenwald, 1954) where rubber gathering and steel production continued to be regarded as the quintessence of economic activity. In so far as this remains true, economic geographers may be guilty of teaching a view of economics that has dwindling acceptance among economists.

Even if geographers were unfamiliar with the changing functions of economics they were unable to ignore the changes in the landscape they described. And here they saw, as Vidal de la Blache had seen, the fundamental trends towards an urban world; a world where the number of large cities with over one million inhabitants had doubled since 1930 to over a hundred. Although accurate measurement has been blurred by the problems of city boundaries and the varying thresholds adopted for 'urban population' (from 250 in Denmark to 10,000 in Spain (Alexander, 1963, p. 528)) the growing dominance of the metropolis in the organization of world economic activity has become self-evident.

This trend created both problems and opportunities in economic geography. It meant a newer and more complex world in which the traditional stand-bys of climate and soil provided a less sure guide to its puzzling land-use patterns; it meant that traditional locational theories like those of Weber (1909) lost their immediate relevance in a welter of 'footloose' industrial expansion; it meant that the study of processes, processes of urban growth, processes of diffusion, processes of system adjustment, emerged as one way of making sense of a rapidly changing world.

It was this fundamental concern with the centres of organization, the cities and the city hierarchy, that underscores the idea of convergence. For in concentrating on the cities, on viewing the economic landscape as an urban-centred system, the economic geographers came back towards those workers in settlement geography, historical geography, and social geography who had been continuing to work on their own separate aspects of urbanization. Indeed so complete has been the fusion that it is often difficult to distinguish old divisions in new studies. Historians, such as Russell (1964), have adopted concepts from economic geography like basic–non-basic activities (Isard, 1960) and applied them to Domesday population. Workers in the field of settlement geography (Berry and Pred, 1961) are then welding an integrated 'human', 'social' or 'behavioural' geography in which economics is playing an important but not overbearing role. In the agricultural field Spencer and Horvath (1963) have reminded us that at least five main groups of factors need to be considered before the regions such as the 'Corn belt' of the United States can begin to be satisfactorily explained. They identify sets of psychological factors (e.g. farming attitudes), political factors (e.g. farm subsidies), historical factors (e.g. 'lags' in the spread of technological knowledge), and agronomic factors (e.g. improvement of hybrid corns) besides the conventional economic factors.

Convergence with Regional Geography

Regional studies have been booming in North America since World War II. A recent survey (Perloff, 1957) reported about 140 U.S. universities had established programmes in regional studies, while two new institutions, the Regional Science Association and Resources for the Future, have polarized regional research on a new scale. In Britain the Hailey and Parry Committees on Afro-Asian and Latin American studies in British universities have seen the founding of new regional research centres; at Cambridge, a new South Asian Studies Centre has been created with a geographer, B. H. Farmer, as its first director.

While such regional studies tend to deal with many features and involve the use of several academic disciplines, the strongest development has come from economics or, more specifically, from econometrics. Thus it is that the first major textbook on regional science, Isard's *Methods of Regional Analysis* (1960) is essentially concerned with economic regions. The problems he sees as paramount is the economic 'performance' of a region (p. 413); what industries does it need to smooth out employment irregularities?; how can it optimize the use of its often niggardly resource endowment? Questions of this kind throw the weight of interest solidly towards economic development and Fisher (1955, p. 6) has summed up this view: '. . . the most helpful region . . . is what might be called the *economic development region*'.

The approach of economists to regions has been strongly mathematical. Two of the most important tools used have been those of *input–output analysis* and *linear programming*. In input–output analysis an attempt is made to trace the flow of goods between industrial sectors and express these as a matrix of input–output coefficients. The difficulty lies in the lack of data on many movements and in the sheer size of the matrix; indeed Meyer (1963, p. 33) has shown that if we are interested in five regions, each of which has fifty industrial sectors, then the matrix would contain 62,500 coefficients. Leontief (1963) in a very readable summary has shown, however, that when such matrices are complete they form a very powerful tool both in showing how regions work and pay their way, and equally important, in diagnosing defects in regional structures. The second major technique, linear programming (Isard, 1960, p. 413–92), is used where the problem demands that some function is maximized or minimized. A typical case might be to relocate hospital, school, or commercial 'collecting areas' so as to minimize the cost of transportation and optimize the size of units; such a study was carried out by Yeates (1962) for schools in Grant county in Wisconsin and is being used at least by one rural English county, Somerset. Perhaps the most ambitious study using mathematical programming in locational programmes is the Penn-Jersey study which relates housing development to a very complex and complete set of environmental conditions (Herbert and Stevens, 1960).

Whether the present interests of economists in regions is a major departure or whether in the future '. . . regional economics may increasingly be indistinguishable from the rest of economics' as argued by Meyer (1963, p. 48) remains to be seen. Whatever the long run importance for economics the impact on geography has been catalytic. Both economic geographers and regional geographers have been either exposed to the literature or drawn in to participate on inter-disciplinary regional research of a rigorously high standard. As Garrison's review (1959–60) indicates, the boundary line work has been immensely productive both of new ideas and new techniques and this is already being translated into action in a few geography schools. No one who follows through the research theses published by the University of Chicago's geography department since 1948 can be unaware of both the nature and pace of the revolution.

Convergence with Physical Geography

This trend is more speculative. Hartshorne (1959) has summarized the views of orthodox geographers in regarding the divisions between physical and human geography as a fundamental dichotomy. Whether this was regarded as a source of division and weakness or as a fundamental educational advantage in a pedagogic world of arts and science, the existence of the difference was never in dispute. It is one of the curious by-products of an apparently disrupting influence, quantification, that it brought geographers to a realization of certain common problems of relating form and process. Bunge in his *Theoretical Geography* (1962, p. 196) has memorably found many things in common:

> Davis's streams move the earth material to the sea and leave the earth etched with valleys; Thünen's agricultural products are moved to the market and leave their mark on the earth with rings of agriculture; . . . farmers scattered on plains move to their hamlets and form Christaller's hexagonal network on their landscape; . . . agricultural innovations creep across Europe, as do glacial fronts, to yield Hägerstrand's regions of agricultural progress and terminal moraines.

Whether Bunge's refreshingly unified view of a discipline organized around the duals of 'geometry and movement' is a pipe-dream or prophecy, the immediate situation is that there is now more co-operation and borrowing across apparently immutable boundaries than for many decades. Geography has, in Ackerman's idiom, dropped both its internal and external separatism.

THE IDEA OF MODELS

Types of Models

In everyday language the term model has at least three different usages. As a noun, model implies a representation; as an adjective, model implies ideal; as a verb, to model means to demonstrate. We are aware that when we refer to a model railway or a model husband we use the term in different senses. In scientific usage Ackoff (Ackoff, Gupta and Minas, 1962) has suggested that we incorporate part of all three meanings; in model building we create an idealized representation of reality in order to demonstrate certain of its properties.

Models are made necessary by the complexity of reality. They are a conceptual prop to our understanding and as such provide for the teacher a simplified and apparently rational picture for the classroom, and for the researcher a source of working hypotheses to test against reality. They convey not the whole truth but a useful and apparently comprehensible part of it.

It is clear then that classifications are very simple models. How vitally important such classifications are in the development of understanding may be seen from the order which Linnaeus and others brought to botany, an order which was to be the starting point for models of evolution and genetics. Why the Linnaean classification was to be so important was difficult to foresee. Other classifications had been tried before and we may find the answer simply in the fact that it worked. Our own working classifications need to be extended and continued despite early failures for I can see no *a priori* grounds for thinking our own periodic tables or fundamental indices are not waiting to be found.

There are indeed several typologies of models, of which that by Chorley (1964) is the most fully developed for the earth sciences. A simple three-stage breakdown has been suggested by Ackoff (Ackoff *et al.*, 1962) into *iconic*, *analogue* and *symbolic* models, in which each stage represents a higher degree of abstraction than the last. Iconic models represent properties at a different scale; analogue models represent one property by another; symbolic models represent properties by symbols. A very simple analogy is with the road system of a region where air photographs might represent the first stage of abstraction (iconic); maps, with roads on the ground represented by lines of different width and colour on the map, represent the second stage of abstraction (analogue); a mathematical expression, road density (Taaffe, Morrill and Gould, 1963), represents the third stage of abstraction (symbolic). At each stage information is lost and the model becomes more abstract but more general.

It will be clear that the examples cited so far are 'static' models, little more than rarefied descriptions. Most of the models used in economic geography to date tend to be of this kind – Christaller (1933) and Lösch (1954) describe the pattern of settlement with its basic hexagonal structure as a static economic landscape, Auerbach (1913) describes the structure of cities by a rank-size rule, Von Thünen (1826) describes the structure of land use as a rent-distance function, Weber (1909) describes the location of industry by a weight-loss rule, Zipf (1949) describes movement between centres in terms of Newtonian physics, and so on. Each represents a rule-of-thumb model of how part of the economic landscape behaves at a given point in time rather than providing a predictive model into which changes and forecasts can be built.

This concentration on static rather than dynamic models is rather a serious shortcoming of economic geography at this stage. Berman (1961, p. 300; cited by Meyer, 1963, p. 39) has put it: 'It may be argued that dynamic models are harder to construct than static, or that we cannot begin to fashion dynamic models until we have a static model of some believability. But for practical purposes . . . a crude dynamic model may be better than a highly tooled, multi-jeweled static creation.' Economic geography lacks the (dynamic) iconic models of the geomorphologist (e.g. the wave tanks of the coastal geomorphologist) or dynamic analogue models (e.g. the hydraulic models of sector flow of the economist). The contrast between static and dynamic models is nowhere more clearly seen than in biology: Charles Darwin's great contribution was to take a well-known static model, evolution, and inject it with a dynamic mechanism, the mechanism of natural selection. A dynamic infusion for geography's static models is being pursued through the study of stochastic processes, the subject of the final section.

Approaches to Model Building

In economic geography, model building has proceeded along two distinct and complementary paths. In the first, the builder has 'sneaked up' on a problem by beginning with very simple postulates and gradually introducing more complexity, all the time getting recognizably nearer to real life. This was the approach of Von Thünen in his first 1826 model of land use in his *Isolierte Staat* (Chisholm, 1962). In this 'isolated state' he begins by assuming a single city, a flat uniform plan, a single transport medium, and like simplicities and in this simple situation is able to derive simple rent gradients which yield a satisfying alteration of land-use 'rings'. But Von Thünen then disturbs this picture by reintroducing the very things that he originally assumed inert and brings back soil differences, alternative markets and different transport media. With their introduction the annular symmetry of the original pattern gives way to an irregular mosaic far more like the pattern we observe in our land-use surveys. Nevertheless Von Thünen's model has served its point; in Ackoff's terminology it has 'demonstrated certain properties' of the economic landscape.

The second method is to 'move down' from reality by making a series of simplifying generalizations. This is the approach of Taaffe (Taaffe *et al.*, 1963) in his model of route development. The study begins with a detailed empirical account of the development of routes in Ghana over the period of colonial exploitation. From the Ghanaian pattern a series of successive stages is recognized. In the first, a scatter of unconnected coastal trading posts; in the last, an inter-connected phase with both high-priority and general links established. This Ghanaian sequence is finally formalized as a four-stage sequence common to other developing countries like Nigeria, East Africa, Malaya and Brazil.

Not all such models have developed inductively from observations within geography. Some of the most successful have come from borrowing ideas from related fields, especially the field of physics. Thus Zipf (1949) attempted to extend Newton's 'divine elastic' of gravitation to social phenomena and his $P_i \, P_j/d_{ij}$ formula for the interaction between two cities of 'mass' P_i and P_j at a distance d_{ij} is a direct extension of Newtonian physics. When modified by Isard's refined concept of distance (Isard, 1960) and Stouffer's addition of intervening opportunity (Stouffer, 1962) it has proved a very powerful predictive tool in the study of traffic generation between points. A less widely known borrowing was used by Lösch (1954, p. 184). He related the 'bending' of transport routes across landscapes of varying resistance and profitability to the sine formula for the refraction of light and sound. While such borrowing may have its dangers, it is a most fruitful source of hypotheses that can be sobelry tested for their relevance to the problems of economic geography. A book like D'Arcy Thompson's *On Growth and Form* (1917) illustrates how many subjects find common ground in the study of morphology; there is inspiration still to find in his treatment of crystal structures or honeycomb formation, as Bunge (1964) has illustrated.

Perhaps the biggest barrier that model builders in economic geography will have to face in the immediate future is an emotional one. It is difficult to accept without some justifiable scepticism that the complexities of a mobile, infinitely variable landscape system will ever be reduced to the most sophisticated model, but still more difficult to accept that as individuals we suffer the indignity of following mathematical patterns in our behaviour.

THE IDEA OF CHANCE PROCESSES

Indeterminacy and Game-theory

In the spirit of optimism that seized science after Newton's triumphant demonstration of his laws of gravitation there was much nonsense dreamed about scientific prediction. It was the French mathematician Laplace who suggested it was conceptually possible to forecast the fate of every atom of the universe both forwards and backwards through time. Although all doubted that the technical possibility lay remotely far in the future it served as a grail towards which science might slowly progress. The break-up of such physical determinism with the rise of quantum physics and the enunciation of Heisenberg's 'uncertainty principle' in 1927 is a part of scientific history of unrecognized importance to geography.

The realization that even physical laws were statistical approximations of very high probability based on immense uniform populations seeped rather slowly through to the social sciences. Economics itself recovered slowly from what Bronowski (1960, p. 67) has called '. . . the fatal reasonableness of Adam Smith's *Wealth of Nations*' and remained largely wedded to a causal system of growing complexity. This attempt to reduce economics to a set of principles has come under increasing attack from both within and without economics departments. Kendall regarded the system '. . . as mistaken as the attempts of the early physicists to explain everything in terms of four elements, or of the early physicians to explain temperament in terms of four humours' (Kendall, 1960, p. 7). The breakthrough here waited until 1944 with Von Neumann and Morgenstern's *Theory of Games and Economic Behaviour* (1944). In this the uncertainty principle was introduced through a mathematical treatment of games, substituting the formalism of demand, supply and perfect assumptions for the half guesses and probabilities of an uncertain market. This is the world which we know intuitively as individuals making our own economic decisions; a world which is neither wholly rational nor wholly chaotic but a probabilistic amalgam of choice, calculation and chance.

Gould (1963) has made direct use of Von Neumann and Morgenstern's ideas in a study of land-use patterns. He illustrates the problem of farmers in a small west Ghanaian village who have five major crops which yield very differently in 'wet years' and 'dry years'. Yams for example yield nearly eight times as heavily in wet as in dry years, while another crop, millet, has much lower yields but is little affected by weather fluctuations. In this situation Gould uses game-theory to derive the crop combination that optimizes yield under these unpredictable conditions. His answer, to specialize in maize (77 per cent of the area), and rice (23 per cent of the area) accords roughly with the land use actually adopted in the area. Since the 'natural' solution was evolved only through trial and error, error which meant starvation, the practical implications of such problem-solving techniques is clear.

Origins of Indeterminacy

How does such randomness arise? Morrill (1963) in his study of town location in Sweden suggests that randomness enters the locational process in three ways. First, randomness arises from the imperfectness of human decisions. We are not always able to distinguish between equally good choices and we cannot always recognize optimum locations even should these exist. There are, Morrill contends, basic uncertainties in the pattern of human behaviour that we cannot wish away. Secondly, randomness arises from the multiplicity of equal choices. There are far more potential routeways than there are routes, more town sites than towns. Thirdly, randomness arises from our inability to take into account the effect of many small sources in any reasonably comprehensible view of reality. Following Newton's view that an alighting butterfly disturbs the earth it is clear that each locational decision stems from an infinity of cause and counter-cause which leads back only into an unending labyrinth. The net effect of all these small causes may be considered as random even though each may be, in the last analysis, rational; we can in practice only hope to disentangle some of the major threads, the rest we can regard as 'noise', a background Brownian motion.

These problems were recognized by medieval scholars like Aquinas with his search for the First Cause, and are reflected in contemporary geography by Meinig's (1962) study of the routes chosen by the railways in the Pacific north-west. He argues that not single routes but sets of possible routes were normally available to the developing company from the purely engineering viewpoint. To explain the location of the route actually selected would need detailed archival research into boardroom decisions and '. . . leave one stranded in the thickets of the decision-making process' (p. 413). Meinig argues that what graft, ignorance, whimsy or good sense leads to a given locational decision we may never know.

Development of Stochastic Models

One of the ways in which such indeterminacy is being built into a number of locational models is through the use of probability matrices. Fig. 6.1 shows a hypothetical example of part of such a matrix where it is postulated that eight centres are to be established within an area. Clearly there are more possible locations than centres and we need some method of assigning them to the area. If the area is graded into three different levels of attractiveness, I, II, III (Fig. 6.1–A), so that the probability of each area being chosen is respectively 3:2:1 (Fig. 6.1–B), then we can assign a sequence of numbers between zero and ninety-nine to the areas (Fig. 6.1–C) which represent the 'chances' of each cell in the area receiving a centre. The actual assignment process is random in that eight numbers are drawn between zero and ninety-nine from a random numbers table to determine the location of the centres. With numbers 05, 42, 59, 61, 67, 78, and 80 the pattern is as shown on Fig. 6.1–D. The case is of course a trivial one and actual matrices used (e.g. by Morrill, 1962) contain far more centres over a far wider area with additional growth and spacing constraints 'built in' to make the model a better approximation of reality. Nevertheless the basic principles of simulation by setting determinate behaviour in a probability framework are apparent even in this simple case; we clearly cannot predict the outcome of *individual* events, but we can simulate the *general* locational pattern.

Using techniques of this kind with rules related to distance from original points of growth, migration proportional to size of settlement, the availability of the transport net, and so on, Morrill (1962) was able to build up the pattern of settlement around hypothetical centres and check this with the evolution of settlement in a specific area, the Värnamo area, of southern Sweden (Morrill, 1963). The advantages of running this type of model is that where data is available, it is possible to adjust the 'rules of the game' to make the model simulate what we know actually to have taken place. Morrill was able to do this for twenty-year periods between 1860 and 1960. It is tempting to allow the model to run on into the future; not to predict what *will* happen but to predict what could, *might* happen if conditions remained unchanged. Such a model might have very important consequences for testing alternative legislative schemes for land-use development in an area of rapid growth, say the south-east of England or the north-eastern seaboard of the United States. One disadvantage of the method is the very great amount of computation needed. In Morrill's small area the migration phase of the model alone required the computation of probabilities for nearly 150,000 migration paths.

Neyman and Scott (1957) have carried this idea a stage further with a general stochastic theory which they suggest may be applicable to phenomena as unlike as star galaxies and animal populations. Basically their scheme hinges on chance, distribution of population centres, chance variations in population increase, chance mechanisms

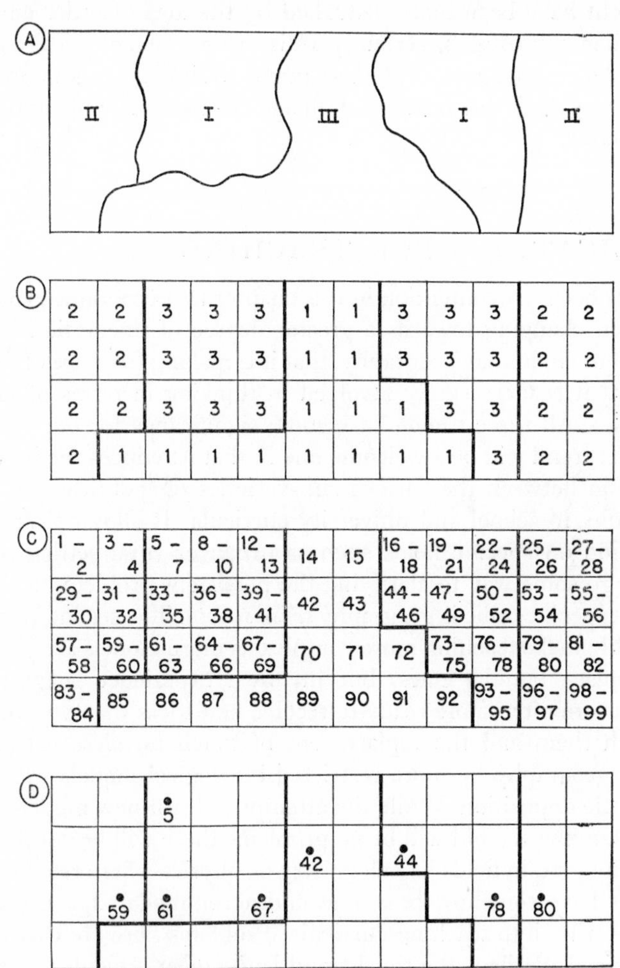

FIG. 6.1. *Hypothetical probability matrix for the allocation for eight settlements in relation to three environmental classes, I, II, III.*

of dispersal, and chance mechanisms of survival. The curious result of these random processes is to build up regular hierarchies and patterns not unlike those of the Christaller–Lösch landscape. The idea has recently been carried much further by a geographer (Curry, 1964) and there is every indication that if Lösch had survived he would have taken up this idea with enthusiasm. He wrote in 1940: 'I doubt that the fundamental principles of zoological, botanical, and economic location theory differ very greatly' (Lösch, 1954, p. 185). He might have been more disturbed by the idea of order emerging from chance processes, certainly a disturbing metaphysical concept, but evidence from geomorphology points to similar random processes creating recognizable order in drainage systems (Leopold and Langbein, 1962).

IMPLICATIONS FOR TEACHING

It has been my contention in this chapter that economic geography has been changing towards a greater degree of integration with its neighbours in human geography, that it is making freer use of models, and that it is increasingly involved in thinking in terms of random processes. All three trends have some significance for our teaching. The first trend will be a welcome one in that it reduces the inevitable confusion between the various sub-varieties of 'political' or 'social' geography in school and university curricula. It allows clearer concentration on specific fields such as urban or rural settlement and should go some way to clarifying the persistent problem of whether or not a specific topic is geography *sensu strictu*. The second trend will inevitably make teaching more complex as more models are produced and replace familiar ones, but my own experience suggests that students are often more ready to receive new ideas than we are ready to teach them and the replacement of much formless teaching of human geography by more restricted but logical models will surely meet little opposition. While the attempt to learn new models before understanding the old will bring problems these will be considerably less acute than in fields like chemistry or physics where revolutionary changes have somehow been met and accommodated, more or less satisfactorily, into teaching curricula. Problems here are easily overemphasized. Perhaps the third trend is the most difficult to comprehend and to teach in that it represents a major change in thinking patterns. The concept of stochastic processes is not an immediately assimilable one and our best hope here may lie in the spread of elementary statistical teaching into school mathematics with a basic concern with probability and chance. The encouraging work by the Association of Teachers of Mathematics promises revolutionary change in the 'numeracy' of the children who will come to read geography in the next decade; an increasing number of schools' VIth-form time-tables are being freed to allow mathematics and geography to be taken together.

My main impression as one who teaches as well as researches in this field is that human geography is entering a most exciting phase. Problems that puzzled Vidal de la Blache and Ellsworth Huntington are beginning to yield to an attack that owes as much to changed thinking as to the hardware of the statistical revolution. It is of vital importance in this field as in others that the thread linking research, university teaching and school teaching, a thread already pulled taut, should not be allowed to part.

References

ACKERMAN, E. A., 1963, 'Where is a Research Frontier?', *Ann. Assn. Amer. Geog.*, **53**, 429–40.

ACKOFF, R. L., GUPTA, S. K. and MINAS, J. S., 1962, *Scientific Method: Optimising Applied Research Decisions* (New York).

ALEXANDER, J. W., 1963, *Economic Geography* (New York).

AUERBACH, F., 1913, 'Das Gesetz der Bevölkerungskonzentration', *Petermann's Mitteilungen*, **59**, 74–76.

BERRY, B. J. L., 1959, 'Recent Studies Concerning the Role of Transportation in the Space Economy', *Ann. Assn. Amer. Geog.*, **49**, 328–42.

BERRY, B. J. L. and PRED, A., 1961, 'Central Place Studies: a Bibliography of Theory and Applications', *Regional Science Research Institute Bibliographic Series*, **1**, 1–153.

BRONOWSKI, J., 1960, *The Common Sense of Science* (London).

BUNGE, W., 1962, *Theoretical Geography* (Lund).

— 1964, 'Patterns of Location', *Michigan Inter-University Community of Mathematical Geographers, Discussion Papers*, **3**, 1–39.

BURTON, I., 1963, 'The Quantitative Revolution and Theoretical Geography', *Canadian Geographer*, **7**, 151–62.

CHISHOLM, M. D. I., 1962, *Rural Settlement and Land Use: an Essay in Location* (London).

CHORLEY, R. J., 1962, 'Geomorphology and General Systems Theory', *U.S. Geol. Survey, Prof. Paper, 500–B*, 10 pp.

CHORLEY, R. J., 1964, 'Geography and Analogue Theory', *Ann. Assn. Amer. Geog.*, **54**, 127–37.

CHRISTALLER, W., 1933, *Die zentralen Orte in Süddeutschland* (Jena).

CURRY, L., 1964, 'The Random Spatial Economy: an Exploration in Settlement Theory', *Ann. Assn. Amer. Geog.*, **54**, 138–46.

DARBY, H. C., 1954, 'On the Relations of Geography and History', *Inst. Brit. Geog. Pub.*, **19**, 1–11.

FISHER, J. L., 1955, 'Concepts in Regional Economic Development Programmes', *Regional Science Association, Papers*, **1**, W1–W20.

GARRISON, W. L., 1959–60, 'Spatial Structure of the Economy', *Ann. Assn. Amer. Geog.*, **49**, 232–39 and 471–8; **50**, 357–73.

GARRISON, W. L., BERRY, B. J. L., MARBLE, D. F., NYSTUEN, J. D. and MORRILL, R. L., 1959, *Studies of Highway Development and Geographic Change* (Seattle).

GINSBURG, N., 1961, *Atlas of Economic Development* (Chicago).

GOULD, P. R., 1963, 'Man against His Environment: a Game Theoretic Framework', *Ann. Assn. Amer. Geog.*, **53**, 290–97.

HARTSHORNE, R., 1939, *The Nature of Geography* (Lancaster, Pa.).

— 1959, *Perspective on the Nature of Geography* (London).

HERBERT, J. D. and STEVENS, B. H., 1960, 'A Model for the Distribution of Residential Activity in Urban Areas', *Journal of Regional Science*, **2**, 21–36.

ISARD, W., 1960, *Methods of Regional Analysis* (New York).

JONES, C. F. and DARKENWALD, G. C., 1954, *Economic Geography* (New York).

KENDALL, M. G., 1960, 'New Prospects in Economic Analysis', *Stamp Memorial Lecture*.

LEONTIEF, W., 1963, 'The Structure of Development', *Scientific American*, **209**, 148–66.

LEOPOLD, L. B. and LANGBEIN, W. B., 1962, 'The Concept of Entropy in Landscape Evolution', *U.S., Geol. Survey, Prof. Paper, 500–A*, 20 pp.

LÖSCH, A., 1954, *The Economics of Location* (New Haven).

MCCARTY, H. H., 1940, *The Geographic Basis of American Economic Life* (New York).

MEINIG, D. W., 1962, 'A Comparative Historical Geography of Two Railnets: Columbia Basin and South Australia', *Ann. Assn. Amer. Geog.*, **52**, 394–413.

MEYER, J., 1963, 'Regional Economics: a Survey', *Amer. Economic Rev.*, **53**, 19–54.

MORRILL, R. L., 1962, 'Simulation of Central Place Patterns Over Time', *Lund Studies in Geography, Series B, Human Geography*, **24**, 109–20.

— 1963, 'The Development and Spatial Distribution of Towns in Sweden', *Ann. Assn. Amer. Geog.*, **53**, 1–14.

NEUMANN, J. VON and MORGENSTERN, O., 1944, *Theory of Games and Economic Behaviour* (New York).

NEYMAN, J. and SCOTT, E. L., 1957, 'On a Mathematical Theory of Population conceived as a Conglomeration of Clusters', *Cold Spring Harbor Symposia on Qualitative Biology*, **22,** 109–20.

PERLOFF, H. S., 1957, *Regional Studies at U.S. Universities* (Washington).

ROBBINS, L., 1935, *An Essay on the Nature and Significance of Economic Science* (London).

RUSSELL, J. C., 1964, 'A Quantitative Approach to Medieval Population Change', *Jour. Econ. Hist.*, **24,** 1–21.

SMITH, J. R., PHILLIPS, M. O. and SMITH, T. R., 1955, *Industrial and Commercial Geography* (New York).

SPENCER, J. E. and HORVATH, R. J., 1953, 'How Does an Agricultural Region Originate?', *Ann. Assn. Amer. Geog.*, **53,** 74–92.

STOUFFER, S. A., 1962, *Social Research to Test Ideas* (New York).

TAAFE, E. J., MORRILL, R. L. and GOULD, P. R., 1963, 'Transport Expansion in Underdeveloped Countries: a Comparative Analysis', *Geog. Review*, **53,** 503–29.

THOMPSON, D'ARCY, 1917, *On Growth and Form* (Cambridge).

THÜNEN, J. H. VON, 1826, *Der Isolierte Staat in Beziehung auf Landwirtschaft und Nationalökonomie* (Hamburg).

WEBER, A., 1909, *Ueber der Standort der Industrien* (Tübingen).

YEATES, M., 1962, 'The "Transportation Problem" in Geographical Research', *Northwestern Univ., Dept. Geogr. Disc. Paper*, 2: 1–9.

ZIPF, G. K., 1949, *Human Behaviour and the Principle of Least Effort* (Cambridge).

CHAPTER SEVEN

Historical Geography: Current Trends and Prospects

C. T. SMITH

Lecturer in Geography, University of Cambridge

As long as geographers are concerned with the study of places and what they are like, how they differ from each other, and how their parts are interrelated, they will want to know how these places came to be what they are, and what they were like in the past. It is obvious that such studies may be pursued with varying relevance to the geography of the present day, and with varying interest and importance to geographers, according to their predilections. But historical studies have a contribution to make which lies very much closer to the heart of the subject, for they are frequently essential to the understanding of why things are where they are. This problem of location is, indeed, seen by some as the central theme of the subject of geography as a whole.

Now it is apparent that the problem of the location of phenomena of interest to the geographer may be treated either in a genetic way or a functional way or both. Thus, the location of an industry, a town, or a political frontier, for example, may be discussed in terms of the ways in which it is related to the conditions of its present environment and how it fits into or functions in its present context. For some types of study this may be enough. Studies of crop distribution, for example, may sometimes satisfactorily concentrate on the context of a physical, social, economic and political environment in which little reference need be directly made to the past. The same is true, in general, of studies of urban spheres of influence, migrations of populations, the movement of trade and traffic, or the economic operation of an industry on a given site. But it is equally obvious that many geographical features require historical study for a satisfactory explanation of how they come to be where they are. The location of farms, villages, towns, industries and communications, for example, can only be understood in terms of a sustained consideration of the conditions under which they were founded, grew, and survived to the present.

Experience of examining the work of candidates for Advanced Level and other examinations reveals many misconceptions and very many quite appalling gaps in the knowledge of even good candidates about the importance of the historical past to an understanding of the facts of human and regional geography. Towns and industries seem to appear fully grown overnight as an anonymous 'response' to bridging points, heads of navigation, proximity of raw materials, fuel or water. It is rare indeed for any candidate to give an indication of a real historical perspective or to hint at the differences which separate economies and societies of the past from those of the present. It is, on the other hand, far too common for considerations widely different in their chronological relevance to be lumped together in a cavalier manner. Some of the commonly used textbooks are, unfortunately, greatly deficient in explaining the circumstances under which, for example, cities, industrial regions or even certain zones of highly specialized agriculture have developed and survived. The over-simplifications and distortions which have been so frequently perpetrated may in part be the result of undue emphasis on 'the influences of the physical environment' and a misguided neglect of 'non-geographical' factors, but they also arise from the failure of historical geographers to fill the many gaps which occur between what is available in historical writings and what is needed by the modern geographer.

It is clearly necessary to be much more precise and rigorous about the problems and topics which emerge out of these general and preliminary comments. First, the aims and purpose of historical geography should be clarified, and in particular, what function it should play in relation to the mainstream of geography. Very different answers have been given to this question in the past, as will be seen. Secondly, it may be profitable to sum up briefly the current trends which have been shown by research in the subject and to attempt to chart the broad directions of future activities in the light of new developments in mainstream geography. Finally, presentation often produces difficult literary problems of organization. Place, time and topic present three major headings under which a study must be organized and it is the ordering and relative placing of these, together with their integration into a modern geographical study, which create a challenge to the ingenuity of authors. But this is an issue about which little need be said here in view of Darby's excellent presentation of the problem and discussion of the various means which have been taken solve it (Darby, 1963).

THE AIMS AND METHODS OF HISTORICAL GEOGRAPHY

Every practitioner may have his own views about the meaning of historical geography, and these will almost certainly be very strongly coloured by the nature of his own research work in the field, but one could, perhaps, reduce to some six or seven the definitions which have been given to the subject in the past. Of these, three or four are archaic and should be of no more than historical importance. Historical geography is a term which has been used as a synonym for the history of geography as a discipline, and although Sauer seems to have included it within the boundaries of historical geography in 1941 (Sauer, 1941), it is not normally now so used. In the nineteenth century it was commonly used to mean the history of exploration and discovery and also the history of the mapping of the earth (Baker, 1936; Gilbert, 1932). Keith Johnston's study of the development of knowledge about the surface of the earth was called *A Sketch of Historical Geography* in 1872, and the term still survives in continental literature in these senses, as in *Petermanns Geographische Mitteilungen* or the *Bibliographie Géographique Internationale*. As long as geography could be seen as primarily concerned with the description and naming of geographical features and with survey and discovery, historical geography could clearly and legitimately retain these meanings, but they are now also archaic. Similarly, the preoccupation of historians with political and national history was reflected in the emergence of an historical geography which dealt with the history of changes in the boundaries of political units, with or without lists of the battles, conquests, marriages confiscations and the like by which new pieces of territory were won or lost. Freeman's *Historical Geography of Europe*, was of this type. Mirot's *Géographie Historique de la France* (1929) is also in this genre, and Kretschmer (1904) wrote a German counterpart, dealing with the historical geography of Central Europe. Yet it should be noted that both Freeman and Kretschmer held much wider views about the nature of historical geography than appears in the contents of these two books. Kretschmer in particular put much greater stress on a more modern concept of historical geography as the reconstruction of past human and regional geographies.

Other views about the nature of historical geography cannot be so rapidly dismissed, however. They may be listed briefly as: the operation of the geographical factor in history, the evolution of the cultural landscape, the reconstruction of past geographies, and the study of geographical change through time. As will be seen, all are subject to a number of limitations, and discussion about them must be qualified by a few remarks. Firstly, the value of all of these portmanteau definitions is limited by the difficulty of comprehending in a single slogan attitudes or points of view which have many subtle shades of meaning. As will be seen, each of the four concepts quoted above shade one into the other; many studies would be difficult to classify under no more than one of the headings given above; yet the advantages of classifying them in this way are probably greater than the danger of creating that kind of spurious opposition between imaginary viewpoints which is not uncommon in methodological writings.

Secondly, it is also apparent that each of these headings is closely associated with a particular view about the nature of geography as a whole. Each of them appears to stand in a symmetrical and orderly relationship to geography as a whole in a way which is logical, but which rarely fits at all exactly the somewhat rambling and haphazard way in which knowledge in fact grows. The tendency for profitable research to be done at the margins between disciplines is well known and almost axiomatic, but the new fields of study which are thus created may tend to gravitate towards the discipline from which most of the research has been conducted even when it may logically belong elsewhere. Whether particular topics have been approached from history or geography may be largely a matter of accident, personality, the structure and traditions of institutions concerned with teaching and research, or it may even be partly a result of the geography and history of the area concerned. A single example must suffice.

Until fairly recently much of the work on field systems, agrarian structures and the history of rural landscapes has been done in England by historians, but by geographers in France. It may be hazarded that a number of factors have contributed to this state of affairs: the structures of open-field agriculture are still an integral part of the rural geography of France and therefore a suitable topic for geographical study, whereas in England the enclosure movements have removed the visible evidence of open fields from the landscape. In

France the close association of history and geography, coupled with an early preoccupation from the time of Brunhes with the facts of the 'cultural landscape', may also have worked towards a greater geographical interest in this direction, whereas in England a more precocious development of economic and social history and the lack of any obvious and outstanding relevance of the problem for geography tended, perhaps, to swing the topic another way. And finally the interests of men such as Bloch, Dion, Seebohm, Gray, Hoskins and Darby have clearly had important consequences.

THE HISTORICAL FACTOR IN GEOGRAPHY

In the 1870's Wimmer considered that the primary functions of historical geography were two-fold: it should be concerned with the operation of the geographical factor in history and also with the interrelationship of phenomena in space at a particular period, or the geography of past periods (Wimmer, 1885). It is the first of these views that must be examined at this stage. Many of the writings on this theme of the geographical factor in history, and there is quite a considerable bulk, use the term geographical in an ambiguous way. Where, as in the hands of more recent writers such as Whittlesey or East, the term 'geographical' in this phrase is given its full value, the study of the operation of the geographical factor in history may in practice involve a study not very different from reconstructing the geography of a past period as a part of the necessary context within which the flow of historical events may be the better understood. But where, as is often the case, the adjective 'geographical' indicates simply the facts of physical geography, this study of the geographical factor in history is seen as an old friend in a very thin disguise, for it may simply be rephrased as 'the control/effect/influence of the physical environment on man's activity in the past, or on history and historical events'. To some extent it carries with it the dusty haze which surrounds the acrid discussions of determinism and possibilism; and it suggests a view of causation by which it was thought that phenomena could best be understood by studying, listing and classifying the operation of particular groups of factors and their effects, each group producing its own particular pattern of determinism: social, economic, technological determinisms thus take their place side by side with the geographical determinism.

Both historians and geographers converged on this theme from different directions, however. In the last quarter of the nineteenth century growing interest in economic and social history meant a new concentration on the everyday activities of past societies. Historians were therefore becoming interested in many of the features which also interested geographers: agriculture, settlement, field systems, towns, industries, trade and communications; and they also became interested therefore in the simple facts of physical geography to which these features were in some degree related. Thus, J. R. Green in the *Making of England* (1882) used the evidence of geology, relief, soils and vegetation together with knowledge of navigable river systems in order to throw a flood of new light on the history of England in the Dark Ages. G. A. Smith's *Historical Geography of the Holy Land* sought to discover what geography had to contribute to questions of biblical criticism, venturing boldly into speculative realms on the effects of environment on religious thought. It is interesting and revealing that H. B. George in his *Relations Between History and Geography*, 1901, should still find it worth making the point, now so much a part of the historian's stock in trade, that 'geographical knowledge affords much valuable data for solving historical problems'.

As long as geography was seen as essentially concerned with the relationship of man with his environment geographers could clearly search for material in the records of the past just as logically as in those of the present for examples of 'responses' to physical environment which could be catalogued and classified as a stage *en route* to the formation of geographical laws.

The work of E. C. Semple is full of this type of selection from history, sometimes cavalier in wrenching information from a proper and necessary historical context (e.g. Semple, 1903, 1932). Possibilists, on the other hand, have used the past as a source from which to collect information about the ways in which man has used his physical environment through many generations and are concerned to demonstrate the varied uses to which it might be put. From here it is but a very short step to a kind of study in which the same place is studied at different times and a kind of comparative regional geography built up dealing with the same place at different times instead of similar places in their modern setting. Conclusions might thus be expected to emerge about the use made of position and resources under very different social, economic and technological environments. Something like this has indeed been the task of many writers in the French regional school, and it is an approach which has yielded much of interest and significance. But by the time the study of the geographical factor in history has reached this level of interpretation, particularly with the emphasis on social, economic and political matters that is the logical consequence of the possibilists' position, there is very little difference at all from a viewpoint which is defined as a reconstruction of past geographies.

Two charges in particular were levelled against this view of historical geography by Hartshorne (1939). It seems right to object that the elucidation of the geographical factor in history is a task for the historian rather than the geographer, since he may be the better equipped to set the geographical factor in its full context and to put the operation of the geographical, as opposed to other factors, into its proper context, but it is worth noting that this is a criticism which flows from a changed attitude to causation and explanation, for it implies a modern attitude towards the study of complex interrelationships rather than the now largely outdated attempt to study causes. The other serious objection to this type of historical geography, that it sets out to clarify history, not geography, seems to be largely valid. And one might also concur with Sauer that at its worst, this type of historical geography may be nothing more impressive than 'adding the missing environmental notations to the work of historians' (Sauer, 1941).

THE CHANGING 'CULTURAL LANDSCAPE'

Whatever its methodological inadequacies, the idea of geography as the study of the landscape, with its corollary that historical geography should be the study of the changing cultural landscape, has been extremely productive of good work and stimulating ideas. This is not the place to explore the roots of the concept in the ambiguities of German *Landschaft* or French *paysage*, the irritating discussions over what may be called problems of visibility, or the attraction of an approach to the subject which rids it of the duality and question-begging of the man–environment approach, yet which offers the geographer a moderately reliable touchstone whereby to distinguish what is relevant to him and what is not. Historical geography as the study of the changing landscape is a very obvious extension of this idea, and one which was followed even before 1914 by Brunhes, in for example the first volume of the Hanotaux's *History of the French Nation* which is a general introduction to the human geography of

France, or by Kretschmer in his historical geography of Central Europe or by Wimmer long before him. Many French and German geographers have followed this line of attack, which has been taken up in England notably by R. E. Dickinson and Darby in geography and by some historians, chiefly Hoskins (1955) and Beresford (1957). Whittlesey has sometimes followed this approach in the U.S.A. and Sauer's emphasis on culture-history has been very closely associated with studies which concentrate upon the changes wrought by man upon the land.

Studies of the changing landscape lead naturally to the extension that historical geography should be concerned particularly with the transformation of natural landscapes by man, and this theme is given form and substance by Darby's systematization of the major items: the clearing of the woodland, the drainage of marsh, the reclamation of heath, the changing arable, the landscape garden and towns and industry. These are indeed the headings under which 'the changing landscape' is considered in a paper of 1951 which parallels a discussion by S. W. Wooldridge of the physical landscape of Britain (Darby, 1951). With a few additional headings, which might include irrigation, soil erosion and conservation, perhaps, or a greater emphasis on the features associated with population and settlement this approach can be seen to yield a set of more or less standardized categories under which historical geography may be written. And what is more important, these are categories which are so clearly relevant to and peculiar to geography that there is little possibility of that kind of confusion with economic history that has so bedevilled the subject in the past. Yet in some respects it is a system which is fore-shadowed already in the writings of Brunhes and his insistence on the classification of landscape features as the essential facts of geography. It may be noted, however, that part of the price which is paid for this rediscovery of unity in geography by way of landscape studies is the danger that man may be deposed from a central position in the study of geography and regarded as no more than an anonymous automaton whose task it is to produce the visible features of the cultural landscape, as impersonally as the processes responsible for soil creep. By this token, man plays the role of geomorphological agent, and while this is a casting which has been very productive of new ideas in human geography and in geomorphology, it should surely be regarded as only one of several roles.

One of the most appealing virtues of the approach to historical geography through the changing cultural landscape is the apparent symmetry with genetic geomorphology. Both are seen to be concerned with the evolution of landscape features; both, in the words of H. C. Darby (1953), are concerned with laying the foundations of geography, though one must stress, as he does, that these are no more than foundations, on which better things are ultimately to be built by social and economic geographers. Consideration of this analogy between historical geography and geomorphology may draw attention, in particular, to some of the problems of evidence and method which beset both of these subjects. Genetic geomorphology sets out to understand landscape features by arranging them according to the manner and chronology of their development. Past circumstances which relate to the development of landforms must be reconstructed in the light of knowledge about the processes which act to produce the observed landforms themselves. Evidence external to the landscape features themselves comes only from the composition of deposits of various kinds. Now it is also clear that in historical geography progress may be made in reconstructing past geographies by a careful analysis of the manner and chronology by which features of the landscape have taken the form they now possess. Emphasis on the study of such features and their interpretation is undoubtedly the greatest single contribution of this school of thought. Studies of the morphology of towns, rural settlement patterns, agrarian structures and field systems arise directly out of this approach, and have leaned very heavily on reconstructions of chronological sequences out of the pattern of landscape features (the arrangement of field systems, for example). Indeed this method of approach has been dignified by such terms as 'the morphogenesis of the cultural landscape', and the analogy with genetic geomorphology is even carried to extremes by attempts to characterize the development of settlement structures in terms of 'structure, process and stage' and by attempts to identify 'cycles' of development, as for example, the attempt to characterize a cycle of development in French *bocage* landscapes.

In breaking the type of circular argument which results from interpreting landscape elements in terms of themselves, historical geography has always so obviously had recourse to documentary evidence (analogous perhaps to the evidence of the deposits for the geomorphologist) that his normal task has more frequently been seen as the reconstruction of past geographies than the understanding of landscape features which derive from the past. Indeed, as techniques of dating and interpreting the ecology of relatively recent deposits are improved and refined in precision, and as knowledge of geomorphological processes becomes more elaborate, so attention in geomorphology has itself begun to shift beyond an emphasis on the reconstruction of the narrative of events towards a much more informed reconstruction of the circumstances of particular periods of the past in the light of more elaborate methods of analysis.

The next stage in the discussion should clearly be concerned with the approach to historical geography as the reconstruction of past geographies, but before proceeding to it, it seems worth making two further points about the study of landscape. Firstly, it sometimes appears that disproportionate attention is given to landscape elements which are either minor in themselves or which have a significance which is marginal to what is normally considered to be part of geography. Roof forms, house types, the distribution of marl pits, for example, may seem to fall into the first category and the study of landscape gardening, so obviously relevant in itself to landscape studies, seems to fall in the second category in so far as it seems to lead us away from what is normally comprehended in geography towards social history and even towards the history of the fine arts. The same may be said to be true, for example, of the study of baroque town plans and it is also quite obviously true of the study of architectural forms.

A second question is thus raised. To identify and put into chronological order the elements of a landscape is an interesting exercise, and may well be a difficult one to carry out (Yates, 1960), but though Yates succeeded admirably in his selected area, it is obvious that the understanding of the distributions of relict forms depends very heavily on knowledge of the historical circumstances of their origin and survival. Dating alone is not enough. It is perhaps more frequently profitable to regard the landscape elements as a source of invaluable evidence for the reconstruction of past geographies than as the phenomena which are to be explained by historical study.

The magnificence of medieval churches may thus be seen as possible evidence for the distribution of medieval prosperity or piety; the distributions of roof and house types fall into place as items, together with field systems, settlement patterns and documentary evidence, which throw light on, for example, cultural contrasts between northern and southern France, as in the excellent study of Limagne (Derrau, 1949). Marl pits in Norfolk are revealed as a hitherto neglected source of evidence for the distribution of activity by eighteenth-century improvers. The distribution of 'ridge and furrow' is important as a potential source of information about early farming activities in the open fields. Study of the origin of the Norfolk Broads began as an exercise in the origin of a landscape feature first thought to be of natural origin and only at a much later date considered to be of historical origin. But in the historical part of the study, the most interesting problems which emerged were those which had to do with the economic, social and physical circumstances of the period when the turf pits were abandoned and were subsequently flooded. It is helpful and interesting to consider the landscape or the topographical map as a palimpsest, but it is not enough to identify and put into chronological order the fragmentary inscriptions which are legible. The ultimate aim of the study should surely be to read and interpret the inscriptions themselves.

THE RECONSTRUCTION OF PAST GEOGRAPHIES

By far the most orthodox and, indeed, unexceptionable view of historical geography is that it should be concerned with the reconstruction of the geographies of past times. Now it is clear that these can be as varied as the adjectival geographies of modern times. There can be not only regional geographies of the past but there can also be urban and rural or agricultural and industrial geographies. Many of the most important historical geographies to have been published employ this method. A series of cross-sections is presented in H. C. Darby's classic work on the historical geography of England before 1800, and this is the method which is used so systematically and ingeniously by Ralph Brown in 'Mirror for Americans' – a reconstruction of the geography of the Eastern Seaboard as it was in 1811, using contemporary sources. It is the method used by Fernand Braudel in his equally classic work on the history and geography of the Mediterranean world in the second half of the sixteenth century, though he calls it *geohistory* rather than historical geography. These are but fairly recent examples of a theme which goes back at least as far as the last quarter of the nineteenth century. Kretschmer considered that the task of historical geography was to discover the changing relationships of land and people at particular periods according to their causal interdependence, and like Whittlesey at a much later date advised that 'It is necessary to study periods of peace and stability before or after great changes.

It must not be concerned with processes of development. It describes and explains the geographical interrelationships for a fixed period' (Whittlesey, 1929). Kretschmer's became the standard view, with a few variations in emphasis. Thus, Mackinder wrote that historical geography is a study of the historical present: 'the geographer has to try and put himself back into the present that existed, let us say one thousand or two thousand years ago; he has got to try and restore it'. And finally the view that historical geography was concerned with the reconstruction of the geography of a past period or periods was the only view of the subject which Hartshorne would allow in 1939 though it is important to recognize that his opinions seem to have changed since then (Hartshorne, 1959).

There appear to be two reasons why consideration of the geography of past periods should be of significance to the geographer. First, if successive cross-sections are made of a given area, a kind of comparative geography may be built up in which it may be possible to view over a whole period of time the way in which such factors as resource and position, soils and climate have been used under varying conditions of technology, social structure, population trends and so on. Useful conclusions may be drawn about the relationships of man and environment from such comparative studies, and since both physical position and physical environment are substantially the same, the historical method should make it possible, at least in terms of theory, to isolate some of the extremely complex variables of any geographical situation. It was probably this which Wimmer had in mind when he wrote as early as 1885, 'The individual aim of historical geography is to compare the geographies of different periods in the same area.' East (1935) has written in the same vein that: 'The significance of historical geography is the more readily grasped when it is possible to review side by side the geography of a whole series of historical periods.'

Secondly, the reconstruction of the geography of past periods is necessary before the relationship between past and present geography can be fully understood. J. B. Mitchell (1954) has written in this context that the value of the work of the historical geographer, *qua* geographer, 'lies in the fact that some elements of the geographical design that develop in response to passing conditions are extremely stable in their form or long lasting in their effects, and the understanding of the present demands the study of the geography of the period of their establishment and development'. Little further comment seems necessary, however, since this is clearly very similar to the view already expressed above in connexion with the study of relict forms in a cultural landscape. Such forms can only be properly understood when placed in the general geographical or cultural context of their origin or even, in certain cases, the context of the conditions which allowed them to persist. Thus, like the case of the dog that did not bark in the night, the survival of so much Georgian and Regency architecture in the market towns of East Anglia is in itself a negative comment on the development of these areas in the nineteenth century.

The concept of historical geography as a reconstruction of past periods is orthodox and it is also clear and distinctive. But this clarity and distinctiveness need further examination, for they rest on assumptions about the nature of explanation in geography and history which were never wholly valid. To put the matter very briefly indeed, there is general agreement that geography is essentially concerned with the functional interrelationship of the phenomena it studies. Nineteenth-century German writers stressing *causal interdependence* and the interrelatedness of the *zusammenhang* of phenomena, French writers such as Brunhes stressing the idea of *connexité*, Mackinder, Hettner and subsequent authors, notably Hartshorne, are all agreed on this point. The area of disagreement had been largely about whether geography should be concerned with genetic studies, that is, studies of growth and development, or the processes which have operated to produce change. Wimmer, Kretschmer, Mackinder, and Hartshorne in 1939 felt that genetic studies belonged properly to historical method and should be excluded from geography as much as possible. It is from this position, in fact, that historical geography can be most clearly seen as a series of period pictures or cross-sections, using its own methods to produce a geography which would be methodologically quite distinct from any kind of history of the same place during the same period. Geography should be concerned with functional interrelationships; genetic studies were essentially historical.

Unfortunately, it is quite clear that this position leads to the empirically untenable conclusion that although geographers may legitimately consider how an industry or town may function at the present time, it would not fall within their task to study how that industry or town came to be where it is. Most geographers have rejected this extreme position, including Hartshorne in his more recent methodological work.

There is, however, another and more strictly logical reason for rejecting this excessively simple view. The opposition between functional and genetic studies is more apparent than real. Genetic studies may, in fact, involve one of two different components. Many so-called genetic studies are, indeed, no more than an attempt at reconstructing the functional relationships of some period in the past. Thus, a study of industrial location frequently proceeds by showing how successive patterns of industrial location were adjusted to the physical, social and economic circumstances of particular periods. The second component is, however, rather different in character and concerned with the evolution of situations continuously through time, and may be conveniently labelled the dialectic method. This has involved '. . . a pre-occupation with linked historical sequences. A dialectic explanation of an historical situation will demonstrate how it arose from the situation which preceded it; a dialectic prognosis will show how a certain future is fashioned by forces operating in the past. Past, present and future are but steps in the "ascending ladder of necessity": a ladder in which every step is so fully supported by the step below and so fully supports the step above that its relative position between the two will tell us all we need to know about itself as well as about the ladder as a whole' (Postan, 1962, p. 399). Dialectical materialism is one example in the historical field of a theoretical approach using this type of method; and in the geographical field it would seem that the idea of *stage* in the Davisian trinity of structure, process and stage owes a great deal to this kind of dialectical thought.

Now it is apparent that many historians have retreated from the ambitious claims of a dialectic method such as this; so much so that Postan quotes a historian's fears 'lest by playing down historical changes we remove history itself from the work of historians' (Postan, 1962, p. 399). Instead, historians have increasingly pursued a more limited aim of understanding the relationships within societies in a much more precise and detailed way than was thought possible by nineteenth-century historians. Studies of change through time of a single element or of a complex of related elements offer the possibility of evaluating the part which it may play in a society, but this is by no means the same idea as the study of changing phenomena in their totality through a continuous flow of events. From the publication of Namier's classic study historians, indeed, have tended more and more to concentrate on the functional interrelationships of phenomena at a given period, and although the emphasis is different from that of the historical geographer, it is quite obvious that neither history nor geography can claim exclusive rights to methods of analysis.

It is thus practically undesirable and methodologically unnecessary to exclude genetic study from geography. But if genetic studies of particular elements in the complex situations of modern geography are desirable, this must also be logically true of geographies of past times. And so the conceptual clarity of a series of relatively static period pictures is greatly obscured. It does, indeed, then become highly desirable to link successive period pictures by studies of intervening changes and the social and economic determinants of change (Broek, 1932; Darby, 1960).

There are, moreover, other difficulties involved in restricting historical geography to the large scale period picture, and the fact that these are practical difficulties and can be quite briefly discussed does not make them any less important than the methodological problems which have been raised at great length above. Whittlesey and others have followed Kretschmer in considering that periods of stability should be chosen as suitable for the reconstruction of period pictures and it is true that the geography of certain periods may be particularly significant in contributing to a geographical understanding of areas such as the Great Plains. But stability is by no means contemporaneous over a whole area or over a whole sector of an economy or society. Moreover, in a series of geographical period pictures even the simple elimination of repetition by omitting those sectors in which change has not taken place will clearly result in a concentration of some sort of change rather than stability. The axiom itself that periods of stability should be chosen needs careful examination, for in studies of settlement, for example, or of clearing and reclamation, agriculture, town growth and communications, it is precisely on the moment of change and on the process of change that most geographers concentrate their attention, for it is frequently this which is the most interesting and profitable study. The kind of questions which are currently being asked often demand consideration of change as well as static relationships – how did settlement take place? Why did a town foundation succeed here and fail there? In what circumstances and with what result did the clearing of forest take place in this or that area?

GEOGRAPHICAL CHANGE THROUGH TIME

Indeed, in many cases the position of the historical geographer may be quite indistinguishable from that of the historian in so far as he concentrates on what has been described above as 'studies in the changes of a single element or of a complex of related elements in order to evaluate the part it plays' – but, of course, in a geographical situation related to place rather than an historical situation related to society. A. H. Clark stresses this point also in asking for a greater emphasis on the description of the processes by which 'selected elements . . . that are believed to contribute largely to regional character have changed through time' (Clark, 1960). There is much to be said for an approach to historical geography which would see it as a study in 'geographical change through time' as Clark suggests (Clark, 1954). To accept this view is, in a sense, to do no more than recognize what historical geographers are already doing in their studies of settlement, field systems, changing industrial locations, urban growth and so on. There are few methodological stumbling blocks, and although it begs questions about the nature of geography like any other of the viewpoints mentioned above, it is admirably loose and permissive rather than precise and restrictive. But it offers only a partial solution and extends the field without adequately defining the whole. For it is unsatisfactory to *insist* on the study of change. Changing situations may enable a clearer picture to emerge about the function of position or resource (for example in the successive uses of river, sea and land routes in the history of the Low Countries), but it is clearly a prerequisite to examine, say, the trading structure of seventeenth-century Holland in a relatively static context. And it is still basic for geographers to study how the things which interest them about periods and places are interrelated.

What conclusions, then, emerge from this analysis of viewpoints that have been taken in the past about historical geography? The first and most important comment to make is that no doctrinaire solution to the problems of relevance which beset historical geographers is possible, and that there is no formula which will decide for them what is history and what is geography. The problem itself is often artificial and arid, since what matters is the contribution being made to knowledge, but it is frequently helpful to bear in mind the rule of thumb that geographers are essentially concerned with places and what they are like, whether in the past or the present. No precision instrument can be fashioned for this purpose out of the most carefully constructed definition. Secondly, each of the attitudes which have held the field at one time or another have served a useful function in connexion with prevailing views about the nature of geography and also about the nature of explanation. Thirdly, each of these viewpoints has helped to contribute to, or has facilitated, or has (perhaps most often) simply accepted and recognized a real contribution to knowledge which was being made to geography or history or both.

Finally, the abandonment of doctrinaire attitudes towards the nature and content of historical geography opens the way for a much more flexible approach to the organization of work and its presentation. This is a theme which has quite recently been discussed at some length by Darby (1962). There is little that needs here to be added except that the extent of variation in the methods of presenting historical elements in geography is not only a measure of the difficulty of the problem, but it is also a measure of the degree to which historical geography has felt itself emancipated from the restrictions of outworn methodologies. It is obvious that the nature of the material itself may dictate the form of organization, particularly in research topics. The aim and purpose of an author may perhaps lead him to adopt a retrospective approach, or to adopt a topical approach by which changes in settlement or agriculture, for example, are followed through time. This has now become so common as to be orthodox.

CURRENT TRENDS AND PROBLEMS IN HISTORICAL GEOGRAPHY

Trends in the literature of historical geography in recent decades illustrate these themes. The relationship of man to his environment was from the nineteenth century onwards a theme which stimulated and was often associated with studies of primitive societies, among which adjustment to physical environment played a part much more obvious and direct than among Western industrialized societies. Prehistory was the corresponding field in which appreciation of position, relief, soils, drainage and water supply could throw new light on human geography and could also contribute something to archaeology. But there has been a shift away, in recent years, from the kind of study in prehistory made by Fleure, Daryll Forde, Cyril Fox or Wooldridge and Linton on Anglo-Saxon settlement. Prehistorians have learnt their geographical lessons, but it is also true that the links which connect prehistory (in Europe, but not in America and Asia) with modern geography tend to be long and tenuous, debatable and often of little apparent importance in the modern world. They have left relict field patterns and traces of settlement patterns, and these now form a starting point for studies in rural history. They have affected distributions of race and language, but geographical interest in these topics has waned with the decline of political geography, having reached its peak, perhaps, in the period before 1939.

The study of period pictures has long been standard in the repertoire of historical geographers and will undoubtedly continue to be so, for much remains to be done that is profitable and that still fills a gap between the parochialism of much local history and the larger interests of the economic historian, who still occasionally tends to ignore regional differences from one area to another and who is not usually *primarily* interested in areal differentiation. The success of *Historical Geography before 1800* (Darby, 1936) and the value of the Domesday Geography make it clear that there are considerable possibilities and that there would be great value in the further extension of this principle by detailed examination and mapping, wherever possible, of the data relating to population, tax returns and land use which are available in many forms from 1250 to the sixteenth- and seventeenth-century population censuses and hearth taxes. Much has already been done on a local scale and there is a wide field of opportunity still for pilot studies of small areas. But the evaluation of the data, the compilation of maps and the interpretation and comparison from one period to another of conclusions drawn from such studies on a larger scale represents a major task for the future. Topographical accounts, travellers' writings and early maps are other sources of peculiar interest to historical geographers of the early modern period, and these too are being exploited at greater depth than was possible only a few years ago. And only a beginning has been made on the vast resources of the nineteenth century for the reconstruction of the geography of an industrializing country. This is not the place to list such sources in any detail, and a broad outline is available in Darby's paper on 'Historical Geography, Twenty Years After' (Darby, 1960). But two points are worth making. The first is that the analysis and mapping of such quantitative data as is available for the early periods seems to be one of the fields in which there are considerable possibilities for judicious use of elementary statistical techniques to compare distributions at successive periods, to help in throwing light on the factors involved in particular distributions, and perhaps occasionally to help in checking the validity of samples. The second is that it may be possible to provide a regional framework for specific purposes which could profitably replace unsatisfactory political units or a modern regional framework which is often equally unsatisfactory. Such regions, sometimes provided ready-made like the agricultural regions identified by authors of the *General View of Agriculture* are more often the product of detailed labours on, for example, land-use statistics (Mitchell's land-use divisions of medieval Suffolk based on studies of the *Inquisitiones Nonarum* of 1341 is a case in point).

Studies in the genesis of landscape features and in geographical change through time have both led to considerably more emphasis on tracing the evolution of agriculture, settlement, trade and industry in what have been called 'vertical' or topical studies through time. It would be tedious to make a survey of the contributions which have been made to the historical aspects of systematic studies in agricultural and industrial geography, etc., but it may be helpful to pick out the areas of considerable growth in recent years.

Changes in physical landscapes in historical time have been studied more intensively with new standards of criticism and new sources of material drawn from local and national archives. Studies of climatic change before the appearance of instrumentation have been made with the help of meticulous attention to year-by-year weather conditions and cropping in monastic accounts, or by systematic study of the dates of vine harvests in France, and these add new criteria to the very many others which have been used to throw light on climatic conditions in historical time and their relationship to economic conditions. The view of man as an agent shaping the face of the land is apparent in the title of the work edited by Thomas on *Man's Role in Changing the Face of the Earth*, and it is in this light that one might also view a considerable recent output on the clearing of woodland, reclamation of heath, drainage and irrigation. Clearing, soil erosion and sedimentation have been treated from a historical point of view in work on the French Alps and by Haggett in work on Brazil. Many studies of this kind have combined techniques of research in physical and historical geography in a new and profitable way which seems often to demonstrate the absence of any great gulf between historical and 'scientific' methods.

The study of the Norfolk Broads (Lambert *et al.*, 1960) led to the accumulation of evidence both from stratigraphy and historical documents in order to explain how they originated. And in the Low Countries the history of reclamation and flooding have been greatly informed by the examination of recent sedimentation, and here, as in Western Germany and France, steps have been taken to establish the medieval history of clearing and even the major outlines of the development of land use by studies of pollen and of other plant remains.

Studies of rural settlement, field systems and agrarian structures have undoubtedly been given a great stimulus by the emphasis which has been placed on the understanding of elements in the landscape. In Western Europe, pioneer works in settlement and field systems were written between the end of the nineteenth century and the late 1920's. But after the publication of such standard works as that of Meitzen in Germany, Marc Bloch in France, the works of Seebohm and Gray in England, and after the publication of the proceedings of the Commission on Rural Habitat by the International Union in 1928, work in this direction appeared to dry up. During and since the Second World War there has been a great expansion of such studies on a more local and often on a more intensive scale than the pioneer works of early writers. In France, the Low Countries, Germany and Sweden a great deal of progress has been accomplished, not only towards the understanding of present rural landscapes and their problems, but also towards a much more profound knowledge of the historical circumstances under which field systems and settlement patterns developed. In England historians, notably Hoskins, Beresford and Allison, have continued the traditions established by social historians or by the long line of local topographers, but geographers too have made important contributions, particularly in recent years. Jones, Sylvester, Vollans, Thorpe and Bowen may be mentioned, but there are many others who are or have been researching in this field with considerable profit and mutual benefit for historian and geographer. This type of study has been one of the growing points of geography as a whole, and it has become a mainstay of agrarian history. Articles on this kind of theme have indeed served to fill the pages of new or relatively new periodicals such as *Études rurales* and the *Agricultural History Review*.

The organized study of settlement, both urban and rural, and the study of field systems clearly fills one of the gaps that was left between history and geography in the late nineteenth century, and which was only partially filled by the pioneer studies of that time and later. In the inter-war years it was largely neglected, and its revival has come mainly after the Second World War. One is inclined to speculate why this should be so and to list some of the qualities about this type of study which seem to have made it peculiarly susceptible to attack at this period and by people equipped with geographical training.

Firstly, there is usually a relationship of settlement and field system to the characteristics of the physical environment. It was indeed this fact which attracted geographical interest in the early phase of settlement studies. Relief, soils and water supply were the factors quoted by geographers to counter the ethnic views of Meitzen and his followers. This interest in the facets of physical environment is of enduring interest. Field systems and rural settlement patterns are prominent landscape features in areas of continental Europe where enclosure movements have not taken place, and the relevance of medieval and early modern rural history is all the more apparent. Indeed, one is inclined to wonder how far the early reticence of English geographers on this theme is associated with the extent to which enclosures obscured earlier arrangements of fields. Thirdly, the sources of evidence are heavily weighted towards the kind of sources geographers may feel most competent to tackle – early maps and plans, air photographs, and, of course, field observation and measurement.

Fourthly, in the absence of the abundance of documentation such as one often has in English studies, the most appropriate methods of study have often been those of genetic geomorphology. It is in connexion with studies of rural settlement and field systems that the ugly phrase 'morphogenesis of the cultural landscape' has been forged (Vadstena Symposium, 1960, 1961). In more general terms, it might be added that this is also a field in which interest is frequently centred on social and technological factors rather than strictly economic matters, and its revival thus coincides with renewed interest in social aspects of geography.

Rather similar comments apply also to urban studies. For here, too, there was a wave of interest in the late nineteenth century which was concerned with questions about the origin of towns and the nature of their institutions. This wave of interest also receded before 1939 and it is only since the war that there has been a revival of interest in urban history by historians, geographers, archaeologists, architects and town-planners. Many of the qualities which made rural settlement an attractive field of settlement for geographers apply also to the study of urban history, particularly when it is centred on problems connected with physical growth, economic function and with the changing functions of particular areas within the town. But it seems unnecessary to elaborate on this theme, essentially similar to that of rural settlement in certain ways.

It may be useful, though hazardous, to conclude by suggesting the ways in which historical geography may develop in the near future.

The application of statistical methods is a development which geography shares with economic history. It may be in the more refined analysis of the quantitative data relating to distributions of prosperity, population and land use that statistical methods have most to contribute. If so, the study of distributions at particular periods may be provided with more reliable tools for the comparison of successive distributions and the checking of data. Studies such as those of Buckatzsch (1950) provided a pointer in this direction, perhaps. Secondly, one may expect historical studies of vegetation and sedimentation to make much greater progress as it becomes possible to set conclusions drawn from pollen analysis and the organic content of recent sedimentary deposits dated by radio carbon methods against the fragmentary record of the documents or place-names. The clearing of forest, soil erosion, the reclamation of heath and the drainage of marsh may be the better understood, and so also may the history of land use in some areas.

In the study of rural settlement there is also need for more systematic co-operation in the use of evidence from different disciplines: local or architectural historian, geographer, soil scientist, and particularly archaeologist may profitably work together, as in the village survey project of Norfolk. A beginning has already been made towards the elaboration of a theoretical model for the study of the expansion of settlement through time making certain assumptions about the nature of new colonization (Bylund, 1960). But this is no more than a beginning and the application of models of this type seems to be greatly limited by the difficulty and sometimes the impossibility of gauging the approximation of real systems to the theoretical. Nevertheless, in matters of rural settlement, both in the sense of settlement and colonization and in the sense of the interpretation of settlement forms and patterns there is much scope, it would seem, for a more systematic approach through generally acceptable classifications and terminologies.

It is perhaps in urban geography and in theories of location that many of the recent advances have been made in the subject as a whole. This is no place to enter into a discussion of the utility and validity of the theoretical concepts which these involve: but it is not impossible that new approaches to problems of location, 'spatial analysis' or the *Standortsproblem* may provide a conceptual framework of reference which may greatly help in the organization and interpretation of past geographies.

An illustration may clarify this approach. In modern highly

industrialized countries the multiplication of consumer industries and the emancipation of industry from the coalfields, the universal availability of power, and the market-orientation of many 'footloose' industries have all tended to mean that an industrial geography based on the study of individual commodites has become increasingly impracticable, or at best, divorced from the realities of modern industrial structures. Geographers have been faced for some time by the need to examine industrial development either in the general terms of the economist or in terms of the study of industrial regions or of individual cities and conurbations. This must necessarily complement the study of a few selected industries which are, for one reason or another, considered suitable for analysis: iron and steel, heavy chemicals, textiles. The concentration of industry on coalfields or on other favoured sites, e.g., ports and capitals, which formerly gave coherence to industrial geography is now lost.

The situation which has emerged with emancipation from the coalfields has something in common with the 'industrial' geography of pre-industrial society. Before the rise of coal as a dominant locative factor, there were relatively few industries – iron, cloth, ship-building, etc. – which warranted detailed studies of location. Then, as in recent years, perhaps the greater part of industry took place in very scattered locations in the small workshops and in the homes of both rural and urban populations. Market forces seem often to have dominated the localization of crafts and even of rural industry to some degree. It is perhaps relevant therefore for the historical geographer of pre-industrial societies to use and modify the ideas and principles of organization developed in the literature on central place theories, urban hierarchies and spheres of urban influence. It seems likely, for example, that ideas such as these may help to make coherent and intelligible the varied occupational structure of small towns and regional centres, and they may help to interpret some aspects of the distribution of rural domestic industry round the towns which supplied the raw material and in which the finished product was marketed, e.g. in the framework knitting industry of the East Midlands.

It is also evident that new trends will transgress the boundaries of the various viewpoints about the nature of historical geography which have been outlined above and it is right that they should. Topical studies in historical branches of systematic geography seem to offer possibilities of fundamental development, particularly in terms of a better understanding of the processes by which geographical change through time may take place. Some of the developments in urban geography, notably those concerned with central place theory and spheres of influence, statistical techniques of regional analysis and the further development of distributional studies would seem to give a new and considerable stimulus to studies of the geography of particular periods. So also will the transference to historical cases of geographical studies of 'underdevelopment' and progress therefrom. It is perhaps not too ambitious to hope that these may all be fields in which the geographer has something to contribute to historical understanding as well as to the study of place, whether in the past, present or future.

What have recent trends in historical geography to offer to the teaching of geography in schools, and in what ways may they help to correct some of the shortcomings pointed out in the beginning of this chapter? Perhaps one of the most important opportunities it opens up is towards a much deeper understanding of local environments through an ability to follow up the wider context of writings on urban and rural settlement, field systems or enclosure history. There is still abundant opportunity for field studies and practical work which may overlap with local history (an opportunity for co-operation and mutual profit rather than an occasion for competition), but which can usually avoid the barrenness of many land-use studies. The ideas which come with the study of local urban and rural settlement may help students to make a profitable link between textbook abstractions about other and larger regions, and the landscapes of these areas as they are seen in film or filmstrip, or increasingly in the modern affluent world, on holidays abroad.

It is also clear that recent emphasis in historical geography towards clarifying the ways in which the past has contributed to present geographies is of much greater relevance to the teacher than academic exercises in the geography of past periods. Textbooks are rightly introducing more and more material by which the importance of a cultural heritage or of the early settlement of an area may be gauged by the student. Study of Latin America, for example, gains immeasurably by knowledge of the cultural contribution of the Indian cultures, the Spanish colonial period and nineteenth-century independence. Appreciation of the use of position, routes and resources gains by some knowledge of their use in the past. Use of the Rhine can as well be linked with the growth and vicissitudes of the towns and ports in its neighbourhood as with the movement of oil and coal. And it is also clear that crudely deterministic 'explanations' of the location of industry or the siting and growth of cities can be judiciously avoided by the introduction of a historical summary of how they came to be where they are. But it is for the historical geographer no less than the historian to provide the textbook writer with the data, to summarize as accurately as possible, and to provide methods by which problems and local studies may be the more readily understood.

References

BAKER, J. N. L., 1936, 'The Last Hundred Years of Historical Geography', *History*, New Series, **21**, 193–207.

BERESFORD, M. W., 1957, *History on the Ground* (London).

BROEK, J. O. M., 1932, *The Santa Clara Valley, California* (Utrecht).

— 1943, 'Relations between History and Geography', *Pacific Historical Rev.*, **10**, 321–5.

BUCKATZSCH, E. J., 1950, 'The Geographical Distribution of Wealth in England, 1086–1843', *Econ. Hist. Rev.*, 2nd series, **3**, 180–202.

BYLUND, E., 1960, 'Theoretical Consideration regarding the Distribution of Settlement in Inner N. Sweden', *Geog. Annaler*, **42**, 225–32.

CLARK, A. H., 1954, 'Historical Geography', Chapter 3 in *American Geography: Inventory and Prospect*, ed. P. E. James and C. F. Jones (Syracuse).

— 1960, 'Geographical Change as a Theme for Economic History', *Jour. Econ. Hist.*, **20**, 607–17.

DARBY, H. C. (ed.), 1936, *An Historical Geography of England Before 1800* (Cambridge).

— 1951, 'The Changing English Landscape', *Geog. Jour.*, **117**, 377–98.

— 1953, 'On the Relations of Geography and History', *Trans. Inst. Brit. Geog., Pub. No.* **19**, 1–13.

— 1960, 'Historical Geography, Twenty Years After', *Geog. Jour.*, **126**, 147–59.

— 1962, 'The Problem of Geographical Description', *Trans. Inst. Brit. Geog., Pub. No.* **30**, 1–13.

DERRUAU, M., 1949, *La Grande Limagne* (Clermont Ferrand).

EAST, W. G., 1935, *Historical Geography of Europe* (London).

FREEMAN, E. A., 1881, *Historical Geography of Europe* (London).

GILBERT, E. W., 1932, 'What is Historical Geography?', *Scot. Geog. Mag.*, **48**, 129–36.

— 1951, 'The Seven Lamps of Geography: an Appreciation of the Work of Sir Halford Mackinder', *Geog.*, **36**, 21–40.

HARTSHORNE, R., 1939, *The Nature of Geography* (Lancaster, Pa.).

— 1959, *Perspective on the Nature of Geography* (Chicago).

HOSKINS, W. G., 1955, *The Making of the English Landscape* (London).
KRETSCHMER, K., 1904, *Historische Geographie von Mitteleuropa* (Munich).
LAMBERT, J. M. *et al.*, 1960, 'The Making of the Broads', *Roy. Geog. Soc.*, Research Series No. 3.
MIROT, A., 1929, *Manuel de la Géographie Historique de la France*, 2 vols (Paris).
MITCHELL, J. B., 1954, *Historical Geography* (London).
POSTAN, M., 1962, 'Function and Dialetic in History', *Econ. Hist. Rev.*, 2nd series, **14,** 397–407.
SAUER, C. O., 1941, 'Foreword to Historical Geography', *Ann. Assn. Amer. Geog.*, **31,** 1–20.
SEMPLE, E. C., 1903, *American History and Its Geographic Conditions* (Boston).
— 1932, *The Geography of the Mediterranean Region: Its Relation to Ancient History* (New York).
SMITH, D. M., 1962, 'The British Hosiery Industry at the Middle of the Nineteenth Century', *Trans. Inst. Brit. Geog., Pub. No.* **32**, 125–42.
VADSTENA SYMPOSIUM, 1961, 'Morphogenesis of the Agrarian Cultural Landscape', *Vadstena Symposium*, 1960, *I.G.U. Congress, Geog. Annaler*, **43,** 328 pp.
WHITTLESEY, D., 1929, 'Sequent Occupance', *Ann. Assn. Amer. Geog.*, **19,** 162–5.
WIMMER, J., 1885, *Historische Landschaftskunde* (Innsbruck).
YATES, E. M., 1960, 'History in a Map', *Geog. Jour.*, **126,** 32–51.

PART TWO

TECHNIQUES

CHAPTER EIGHT

The Application of Quantitative Methods to Geomorphology

R. J. CHORLEY

Lecturer in Geography, University of Cambridge

It would be wrong to assume that the changes which have taken place in geomorphology during the past quarter of a century or so have resulted from the application of radically new techniques to the study of landforms. The new approach to the subject drives its roots much deeper than this, and involves a series of responses to questions which are fundamentally different from those posed either by Davis or by the denudation chronologists. In fact these questions were being asked before Davis formulated his cyclical approach to the subject, but it is only recently that even a small body of geomorphologists has thought it worth while to try to answer them. Partly this attempt has been promoted negatively, in that the shortcomings of both the cyclic and denudation chronology approaches to the subject had by the Second World War so restrictively formalized geomorphology as to reduce the virile and proliferating study of the late nineteenth century to a series of narrow scholastic exercises. There was, however, a much more positive reason for the recent change of geomorphic emphasis, in that modern quantitative methods both of the study of geomorphic processes and forms (the latter largely through improved mapping and aerial photography) began to yield data which could be processed by simple statistical techniques. It is the purpose of this chapter to start from the 'ecological niche' created in the western world by the decline of the Davisian system and denudation chronology, to balance the factors which have impeded and promoted quantification in geomorphology, and to proceed to a general outline of the quantitative treatment of geomorphic data.

THE WEAKENING OF DENUDATION CHRONOLOGY

The decline of the Davisian geomorphic system was treated in Chapter 2 and there is no need to return to this matter here. However, although some indication has been given regarding the failure of denudation chronology to form an adequate geomorphic basis for *geographical* work, it is necessary here to show in what ways its former dominance over *geomorphic* work in Western Europe and the United States has diminished. It should be made clear at the outset that studies of the sequential development of landforms having changes of baselevel as a fundamental focus of interest still form a respected and fruitful branch of geomorphology, and that many of the criticisms which follow apply most strikingly to later, more derivative, work in localities poorly suited both to the aims and methods of denudation chronology. The point which I wish to make in this respect is that much of the former dominance claimed for denudation chronology in the field of geomorphology has now vanished, and that both a cause and an effect of this is the development of what has been termed, rather inaccurately, quantitative geomorphology.

One of the major limitations on the study of denudation chronology is that it cannot exist satisfactorily on a purely morphological plane. Many of the most impressive studies of this type rely heavily on the known origin and date of associated *deposits* (commonly terrace deposits or those resting on erosion surfaces), and it is characteristic that the higher the elevation or the longer the uninterrupted period of erosional history the more difficult it is to produce an unambiguous denudation chronology for the area. For a few classic areas significant historical reconstructions have been made, but for most regions attempts at a denudation chronology commonly end with the presentation of a cleverly-integrated body of ambiguous circumstantial evidence the interpretation of which is strongly coloured by the previous findings of similar workers in other areas. The ambiguity of purely *morphological* evidence relating to valley features was pointed out in Chapter 2, and Rich (1938) has given a summary of many of the arguments against purely morphological denudation chronology. It is instructive to take the best-known work of the latter sort, Johnson's book on the Appalachians (1931), and read it in the light of more modern work (e.g. Flint, 1963) which is stressing lithological control over elevation, rather than that of baselevel. Denudation chronology did provide, however, the only important pre-Second World War stimulus for quantitative work in morphometry (i.e. the study of the *geometry* of landscape), which mainly took the form of altitude/frequency analysis (e.g. Hollingworth, 1938). Such studies were, of course, based upon certain articles of faith, chiefly that areas of low slope are probably erosional surfaces related to appropriate baselevels, and that elevation above sea-level is a measure of relative age for such features. Irrespective of such commitments, however, this quantitative data was collected in such a subjective manner and treated by means of such coarse and uncritical techniques as to destroy the real significance of quantification and to reduce these studies to a quasi-scientific veneer embellishing essentially qualitative work.

Another aspect of denudation chronology which has tended to bring it into disrespect is the very prevalent use of '*ad hoc* postulates'. The abandoned eustatic swings of sea-level or the cavalier casting of Cretaceous covers which are required by some ambitious reconstructions are examples of such *ad hoc* postulates which are employed to 'explain' certain observed features without any thought as to the other concomitant effects of such occurrences. Long ago Chamberlin (1897) formalized the method of 'multiple working hypotheses' for use in the earth sciences where evidence is ambiguous and sparse. The inability of some workers in denudation chronology to follow up the other associated effects which might be logically deduced from their hypotheses has done little to create respect either for their aims or their methods.

It is, however, on a much more fundamental plane that the study of denudation chronology is being shown to provide an inadequate *general* basis for the study of geomorphology. The historical bias inherent in the denudation chronology approach, together with the assumption that rates of change involving landforms are commonly very slow, has meant that landforms have been viewed in much the same manner as the light from a distant star, in which what is perceived is merely a reflection of happenings in past history. Denudation chronologists have therefore almost universally directed their attention to those (often minute) elements of landscape having supposed evolutionary significance, in such a manner as to exclude from geomorphology the study of the basic dynamics which are controlling the continuous development of the major part of all landscapes. This dichotomy between the so-called *historical hangover* and *dynamic equilibrium* approaches has been presented elsewhere (Chorley, 1962 and 1964) as highlighting the fundamentally different methodological approaches to geomorphology of W. M. Davis and G. K. Gilbert (1877). Gilbert was primarily concerned with the manner in which equilibrium landforms become adjusted to geomorphic processes, and an interest in the progress towards such adjustment and the changes to which such adjustment is susceptible through time replaced for him a simple cyclical basis such as that which preoccupied Davis. Obviously, however, most present landscapes possess in highly variable proportions evidence of past history and of the reasonably contemporaneous processes which are gaining ascendency, blurring and destroying them at very different rates. These rates are related to the rates of operation of geomorphic processes which may be very slow (as in the case of some ancient pediment surfaces in Africa) or practically instantaneous (as for the features of hydraulic geometry). Most landform assemblages lie somewhere between these extremes, and one of the most pressing needs, the existence of which has done so much to give modern work its distinctive character, is for this fact to be recognized and for the relative importance of past and present processes to be evaluated. The key to such investigations lies obviously in the proper understanding of the current rates of operation of geomorphic processes, and it is precisely in this aspect of geomorphology that most ignorance exists. If James Hutton has anything to teach the modern geomorphologist it is that we should not assume the inadequacy of present 'causes' without establishing the surest quantitative grounds for so doing. We can, after all, directly observe and measure present conditions (which is more than we can for those in the past!) and before we sweep many of our geomorphological problems under the mat of historical speculation we should try to learn in what ways they can be treated by attention to presently-observable conditions.

THE DISTRUST OF QUANTIFICATION

The influences which have tended to inhibit quantification in geomorphology stem largely from two sources – the past historical and geographical affinities of the subject. Many of the historical influences follow naturally from the previous discussion of denudation chronology. Chief among these is the blanket belief that most topographic forms are essentially 'fossil' and relate to antique processes operative in climatic, tectonic or historical circumstances different from those of the present. From this point of view studies of the relationships between present forms and processes are irrelevant. As has been shown above, the truth of the above assertion varies widely with the environmental circumstances, and a recent example indicates the dangers of its uncritical application. In a work on cliff forms in the Colorado Plateau Ahnert (1960) identified rounded cliff tops with processes occurring during past pluvial conditions, and from this assumption went on to deduce a fundamental disharmony between these conditions and those obtaining at present. Shortly afterwards Bradley (1963) demonstrated that the existence of such rounding was widely due to the occurrence of pressure-release joints in the sandstone – from which it may be assumed that the existence of cliff rounding is a structural matter and not in the least indicative of the nature of past processes. Most purely *morphological* evidence is so ambiguous that theory feeds readily on preconception.

Closely associated with these historical inhibitions are those relating to the measurement of form and process and their affinities. Precise measurement and quantitative expression of the geometry of landforms were not only irrelevant to the denudation chronologist but absolutely unnecessary for the Davisian synthesis, and, for example, Strahler (1950 A) has noted the lack of precisely surveyed slope profiles in Davis' work. Davis did, in passing, point to the parabolic form of the typical longitudinal stream profile, but his reasoning was that of the intuitive artist rather than that of the quantitative scientist. It is also commonly believed that most landforms are too complex to treat satisfactorily in a quantitative manner, although one of the important features of recent work is that significant and diagnostic aspects of landscape geometry have been isolated and so treated. Thus, for example, Horton (1945) analysed areal properties of the erosional drainage basin, Strahler (1950 B) the straight middle segments of valley side slopes, and the same author (1952) the hypsometric aspects of drainage basins (in which a 3-dimensional problem was reduced to a 2-dimensional one).

Allied with the above belief in the irrelevancy and complexity of form is the traditional attitude to geomorphic processes, which have commonly been regarded as of little consequence in the analysis of existing landforms; . . . 'I regard it as quite fundamental that Geomorphology is primarily concerned with the interpretation of forms, not the study of processes' (Wooldridge, 1958, p. 31). Although this extreme attitude is less fashionable now than hitherto, there still remains the view that process (even more than form) is usually of such a complex character that significant measurement is well-nigh impossible. Processes commonly seem to operate either too rapidly, too slowly, too infrequently, too capriciously, or too variably to make for ease of observation and measurement. In part these objections have been met in a similar way to those relating to the complexity of form, in that significant *aspects* of process have been isolated and treated – e.g. *bankfull* discharge (Leopold and Wolman, 1957; Dury, 1961), soil moisture *changes* on slopes (Young, 1960; Kirkby, 1963), the '*significant* wave' (Shepard, 1963, pp. 54–55), and the '*weighted resultant* wind' (Bagnold, 1951, pp. 80–82; Rosenan, 1953). Such considerations have naturally given rise to a recent interest in the magnitude and frequency of geomorphic processes (Wolman and Miller, 1960). However, the 'traditional' view of such processes commonly held by geomorphologists in the humid temperate regions is that rates of operation are too slow to be susceptible of useful measurement. Even excepting the increased erosional rates observed in other environments (see Langbein and Schumm, 1958, in Chapter 2), the few measurements which have yet been made of humid temperate processes (e.g. Young, 1960; Kirkby, 1963) by no means support this view. Likewise, measurable rates of diastrophic movement have been reported (Gilluly, 1949; Chorley, 1963; Schumm, 1963). In short, we can no longer proceed on the assumption that current rates of operation of earth processes can be ignored in an assessment of existing landforms. Probably in some instances they ultimately can, but even such elimination must only be made on the basis of reliable measurement.

The second source of inhibition to quantification in geomorphology derives from its many geographical affinities. Bunge (1962) has pointed to the geographer's traditional preoccupation with the unique and distrust of generalization (i.e. 'idiographic' attitude) during the present century, and there is no doubt that the attraction which geographers have commonly felt towards the intuitive and artistic aspects of natural and social science has militated against the quantification of geography in general and of geomorphology in particular. It is characteristic for such scholars to hold the erroneous belief that 'you can prove anything with statistics'; for them to dwell with satisfaction on the fact that on Charles Darwin's death his rulers were found to be of inexact length and his conversion tables incorrect; and for them to believe that the important, interesting and worthwhile aspects of natural phenomena are somehow bound up with departures from predictable regularity (Chorley, in press). However, even in geography quantification is proving an increasingly valuable research stimulus (Ackerman, 1963), so much so that it can be legitimately held that the quantitative 'revolution' is already upon us and that it is profitless to pursue methodological discussions which are based upon a disregard of this fact (Burton, 1963). Such a view is even more applicable to geomorphology.

THE PROMOTION OF QUANTIFICATION

It is less easy to isolate the factors which have prompted quantification specifically in geomorphology from those operative in the fields of natural and historical science as a whole. Indeed, in common with students of these subjects, geomorphologists are becoming increasingly impressed with the weight of quantitative arguments and are developing a healthy distrust of their purely visual sense and of the basic preconceptions which seem to have restricted their science in the past.

The employment of reasoning based on measurement is not new in geomorphology (e.g. Geikie, 1868), but since the 1930's the increasing opportunity for such work in terms of facility of measurement, availability of data, and the imaginative and technical advances in treatment has given post-war studies much of their distinctive flavour. The desire to relate form to process, inspired by the pioneer work of Gilbert, has given rise to the important fluvial studies of the hydrologist R. E. Horton (1945), as well as the associated work of Strahler (1950 B) and his students; to the work on hydraulic geometry mainly associated with the name of L. B. Leopold (with Maddock, 1953; with Wolman, 1957; with Wolman and Miller, 1964); to research on dune formation (e.g. Bagnold, 1941) and periglacial forms (Peltier, 1950; Jahn, 1961); and to investigations into beach forms and dynamics (e.g. Krumbein, 1944). It is one of the features of this work that much of its stimulus has come from the investigation of practical engineering problems and from recent improvements in topographical maps and air photographs. The quantitative analysis of landscape geometry (i.e. 'morphometry') has been especially revivified by the latter improvements.

The developments of quantification in geomorphology can be viewed as indicating a general progress of the discipline towards a more secure scientific footing through its concern with measurement and the analysis of data (Strahler, 1954; Melton, 1957), through its attempt to break out of the narrow scholasticism which characterized much of its pre-war development, and through its increasing association with kindred sciences. It is becoming recognized that sensible quantification of those aspects of the subject which call for it translates many geomorphic questions into a common language which enables them to be attacked in the light of the experience of the other sciences, such that geomorphology can both draw on and contribute to the wealth of scientific experience which is the most distinctive intellectual achievement of our age. There is no remedy for bad or irrelevant measurement, or excuse for unnecessarily elaborate treatment and analysis, but once it is recognized that useful properties and associations can be expressed in quantitative terms then one is *automatically* committed to a programme (no matter how simple in conception and execution) of operational definitions, choice of scales of measurement, sampling and collection of data, descriptive statistical methods and analytical methods. It is with a brief treatment of the features of such a programme that the remainder of this chapter is concerned.

ASPECTS OF QUANTIFICATION

The appropriateness of quantification in geomorphology depends basically upon the character of the science which one wishes to exploit. The methodological features of an historical natural science have been recently set forth in some detail (Albritton, 1963), and, for those workers concerned with geomorphic phenomena as part of the latter phases of historical geology, the main preoccupations continue to be the elucidation of *what has occurred* and *in what order*. Quantification, of course, often plays an important part in both these aims, but it is not indispensable in setting up a sequence of recognizable events. One has only to recall Davis' reticence on the subject of the precise dating of cyclic events to be impressed with the fact that, even in an historically-oriented science, quantification of *time* is not immediately basic to the main thesis, which is one of relativity. Where a conflict of views on the value of quantification in geomorphology occurs, it commonly resolves itself into a dispute between those who view the whole subject as part of historical geology and those who do not.

Those who find the quantitative approach to the study of landforms valuable can recognize four major classes into which geomorphic parameters may be grouped. Those relating to *force* or *energy*, of which discharge and wave energy are examples; to *strength* or *resistance*, by far the most neglected of the groups, which includes infiltration capacity and shearing resistance; to *time*, involving absolute dating of events, rates of erosion, etc; and to *form*. This latter class, including slope angles, drainage density, relief and height, is the one which has obviously been most subjected to quantification in the past, and measurements of stream profiles and hypsometric attributes have even become part of the standard techniques in denudation chronology.

Whichever of these classes is the subject of interest, it is necessary for the geomorphologist to adopt a formalized procedure for the collection of his quantitative data. It is necessary at an early stage, having decided on the nature of the problem and the type of questions to be asked about it (i.e. the *design of the experiment*; Krumbein, 1955), to lay down some clear *operational definition* involving a precise statement of the attribute concerned, such that there shall be no confusion regarding what parameter has actually been measured. Thus 'valley side slope angle' or 'stream discharge' are manifestly inadequate as operational definitions. It is surprising how the establishment of a meaningful operational definition assists in clarifying the character of the investigation in the mind of the investigator. Next one has to decide on an appropriate *scale of measurement* (Stevens, 1946) with reference to which the above parameter may be stated and the units in which it shall be expressed. Space does not permit an extended treatment of the theory of scales of measurement here, and it must suffice to point to the ratio scale as the most versatile, and to suggest that the expression of much data in geomorphology (but much more so in geography as a whole) falls short of this ideal. Another group of highly versatile numbers are those which are derived in such a manner as to make them 'dimensionless' (e.g. Strahler's (1952) 'hypsometric integral'). *Errors of measurement* must be a prime consideration at this stage and a conscious attempt made to reduce 'operator variation' in measurement – i.e. involving important differences in the recognition and measurement of identical phenomena either by the same 'operator' or by different operators, should more than one be involved in the investigation. Many of these errors can be eliminated by a meaningful operational definition and a logical experimental design. Finally, before the actual collection of data can take place a *sampling design* must be established. This topic, too, is a very large one, for obviously geomorphic sampling can be carried on both in space and time. However, two concise guides to areal sampling have been given by Strahler (1954) and Krumbein (1960), although it is largely a matter of trial and error at this stage in determining the size of sample necessary to give an adequate representation of a given phenomenon.

The first step in the treatment of the data collected as the result of the procedure outlined above is to organize and describe it in some convenient manner. Such *descriptive statistical methods* are described fully in standard texts (e.g. Croxton and Cowden, 1939; Croxton, 1959), as well as for geographical (Gregory, 1963), geological (Miller and Kahn, 1962) and geomorphic data (Strahler, 1954). These commonly involve the plotting of frequency diagrams (e.g. histograms), the recognition of the 'characteristics of the population' (i.e. whether it is normally distributed, skewed, etc.), the definition of the 'measures of central tendency' (the mean, mode or median), and the definition of the 'dispersion' (e.g. the standard deviation) of the data. Descriptive statistical methods involve both the description of the sample data and the drawing of inferences regarding the general characteristics of the total 'population' from which the sample was drawn. This latter inference enables one to introduce the concept of *probability* upon which the whole of statistical analysis rests.

The most simple analytical techniques which have proved useful in geomorphic research involve the testing of the significance of difference between data grouped into classes. The *'chi-square' test* (Croxton and Cowden, 1939, pp. 282–7; Strahler, 1954, pp. 9–10) is used to investigate the significance of difference between frequencies of 'variates' (i.e. occurrences) within classes, and in this respect is useful in testing the normality of distributions. The *'t' test* (Croxton and Cowden, 1939, Chapter 12; Strahler, 1954, pp. 12–14) enables one to test the significance of difference between the arithmetic mean values of two samples. This type of analysis has proved most important in quantitative geomorphic reasoning, and Strahler (1950 B) demonstrated a significant difference between maximum angles of valley-side slopes which were being basally corraded and slopes where the basal accumulation of debris was taking place (Figure 8.1A). This significance of difference between sample means cannot always be objectively analysed by visual inspection, and in figure 8.1C, for example, there is no reason to assume (at a probability of 5 per cent) that the difference in mean valley-side slope angles shown by the two samples from the Athens sandstone and the Pennington sandstone and shale in part of the folded Appalachians reflects a real difference between mean angles on the two formations (i.e. the observed sample

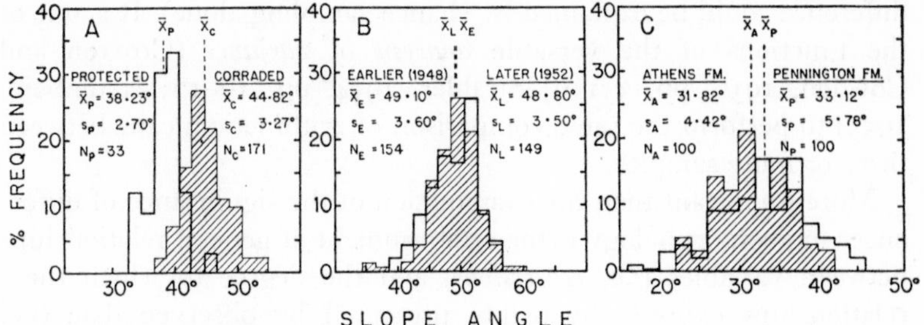

FIG. 8.1. *Paired histograms showing:*

A. Maximum slope angles of valley-sides in the Verdugo Hills, California, the bases of which are being protected and corraded, respectively (after Strahler, 1950 B).

B. Maximum slope angles of valley-sides in badlands at Perth Amboy, New Jersey, measured at an interval of four years (after Schumm, from Strahler, 1954).

C. Valley-side slope angles measured on the Athens sandstone and the Pennington sandstone and shale of western Virginia (after Miller, from Strahler, 1954).

(In all instances $\bar{X}$ = the arithmetic mean of the sample, s = the standard deviation and N = the number of variates in the sample.)

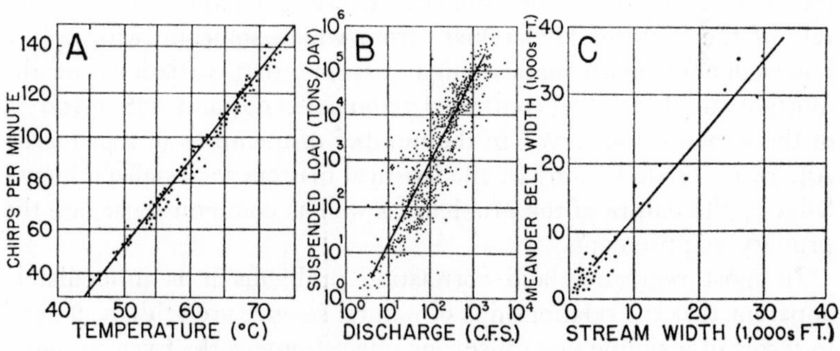

FIG. 8.2. *A. Linear relationship between the number of cricket chirps per minute and the temperature (after Croxton and Cowden, 1939).*

B. Non-linear (logarithmic) relationship between the amount of suspended load and the discharge of the Powder River at Arvada, Wyoming (after Leopold and Maddock, 1953).

C. Linear relationship between meander belt width and stream width for some Wisconsin rivers (after Bates).

difference could be explained by chance sampling alone). It is one of the functions of the versatile *analysis of variance* (Croxton and Cowden, 1939, pp. 351–9; Strahler, 1954, pp. 16–17; Krumbein, 1955) to perform the same comparison of arithmetic means between three or more samples.

More important than the examination of the significance of differences in geomorphology is the establishment of general relationships between variables (i.e. *regressions*) and the degree to which these relationships express the reality presented by observed data (i.e. *correlation*). Some regression lines are shown in figure 8.2, with a high degree of correlation indicated in figure 8.2A and poorer correlation in figure 8.2B. It will be readily appreciated that reasoning from such associations commonly involves the assumption that one variable (the *independent*) is exercising some control over the other (the *dependent*), but especial care must be taken in this respect. Sometimes a causal relationship can be readily supported (as in figures 8.2A and 8.2B), but on other occasions the observed relationships may result from the yoked behaviour of two dependent variables responding in harmony to variations in some undisclosed independent variable or variables – in figure 8.2C, meander belt and stream widths are responding to differences in other variables, notably stream discharge and the calibre of bed debris. In other instances a high degree of correlation may result from purely *chance* co-variation, as when a correlation coefficient of 0·87 was obtained between the membership of the International Machinists Union and the death rate of the State of Hyderabad – from which someone with only a knowledge of mathematics might assume that variations in the mortality of Hyderabad controlled 75·69 per cent (i.e. $0{\cdot}87^2 \times 100\%$) of the variations observed in the number of members of a primarily American Trade Union! In all statistical matters a specialized knowledge of the nature of the problem, as well as common sense, are the primary requirements.

In most regression and correlation problems it is immediately apparent that the relationships of natural science are seldom, if ever, in terms of a simple one cause–one effect framework. Even in figure 8.2A, where there is a very high degree of correlation, such deviations which occur from the general regression trend indicate that temperature, although by far the most important, is not the only factor controlling the rate of cricket chirping (others might be the age of the cricket, the absolute humidity, etc.). The poorer correlation exhibited by figure 8.2B indicates similarly that discharge is not the only factor controlling the rate of suspended sediment transport in a river, and that other factors are involved (i.e. calibre of the sediment, specific gravity of the sediment, temperature of the water, etc.). The recognition of this multi-factor or *multivariate* character of most problems in the earth sciences has long existed, but until comparatively recently it was possible to do little more than to pay lip service to the complexity of most problems by the use of such phrases as 'X seems to be the most important control' and 'other things being equal, Y is a function of X'. This difficulty often resulted in the past in the assumption of gross oversimplifications in cause and effect, and, in the most extreme instances, in an extremely narrow determinism. Refined statistical techniques, facilitated in recent years by the application of electronic computers, now enable one to face up more realistically to the challenge of a multivariate reality, such that the following kinds of questions can be asked – and largely answered:

1. How many significant controlling factors are involved in determining the effect with which we are concerned?
2. Are these factors interrelated in some manner?
3. Can these controls be ranked in order of importance, both relatively and absolutely?
4. Do the factors retain this importance under all circumstances? – i.e. Does the intervention of one factor reinforce or damp-down the effect of another?

Of these questions, the first can only be approached through experience and intuition, although the relevance of a choice made in this manner can be later checked statistically, and the last three are amenable to solution by the techniques of multivariate statistics. These techniques can be divided into two broad groups – eliminative investigations and more sophisticated multivariate techniques.

Eliminative investigations are those involving the careful selection of a number of instances (either situations simulated and controlled in the laboratory, or 'real world' situations) such that the assumed controlling variables are successively eliminated or included in various associations. This has been the standard method of testing the factors which, for example, control crop yields by the use of experimental plots. Examples in geomorphology include the eliminative laboratory investigations into the factors controlling tractive capacity (i.e. the mass movement of stream bed load) (Gilbert, 1914) and the angle of repose of fragmented material (Van Burkalow, 1945). The number of factors involved in most natural phenomena, together with the restricted number of locations in which their effects can be examined in detail, has tended to inhibit eliminative field work in geomorphology, but an example of this technique has been provided in the examination of factors influencing the geometry of erosional landscapes (Chorley, 1957).

An obvious limitation facing eliminative investigators is that questions 2 and 4 (above), involving the mutual effects of independent variables, cannot be readily answered. This is because one can never realistically hold one set of factors constant while the changing effect of others is examined (i.e. other things are *never* equal!) for, as Sir Ronald Fisher pointed out, nature 'will best respond to a logical and carefully thought out questionnaire; indeed, if we ask her a single question she will often refuse to answer until some other topic has been discussed'. Such a 'questionnaire', in which all the assumed independent variables together with the dependent variable are examined and evaluated simultaneously in a large number of different situations, combinations and magnitudes, forms the basis of *multivariate analysis*. The development of statistical methods assisted by the use of high-speed electronic computers now enables the data from such a questionnaire to be processed and evaluated in such a manner that all four of the above questions can be answered. Two such, closely-allied, methods are those of *multiple correlation* and *multiple regression*. Melton (1957) examined by means of the first method the control exercised by five factors over drainage density (i.e. the total length of drainage lines per unit area) and found that all five operating together could account for more than 93 per cent of the observed variation in drainage density between different localities. Using a multiple regression technique Krumbein (1959) evaluated the influence of four factors in controlling beach firmness, concluding that together they contributed over 76 per cent of the observed firmness variations.

Quantitative techniques, supported by statistical analysis, provide a standardized, rigorous, conservative and objective framework for the investigation of many of the problems of earth science – although those of an historical character are less obviously susceptible to such treatment at present. However, these techniques and analyses are only an adjunct to, and not a substitute for, the initial qualitative stage of any investigation. This stage is entirely a matter for the exercise of experience, controlled intuition, imagination and creativity, in which quantitative methods are of no help – although they may subsequently be used to test the efficiency of this qualitative design. Quantitative methods and statistical analyses are merely tools, but tools with which one may sharpen the imagination and, like Galileo's telescope, which enable this imagination to operate on higher planes than ever before.

References

ACKERMAN, E. A., 1963, 'Where is a Research Frontier?', *Ann. Assn. Amer. Geog.*, **53**, 429–40.

AHNERT, F., 1960, 'The Influence of Pleistocene Climates upon the Morphology of Cuesta Scarps on the Colorado Plateau', *Ann. Assn. Amer. Geog.*, **50**, 139–56.

ALBRITTON, C. C., 1963, *The Fabric of Geology* (Reading, Mass.), 372 pp.

BAGNOLD, R. A., 1941, *The Physics of Blown Sand and Desert Dunes* (London), 265 pp.

— 1951, 'Sand Formations in Southern Arabia', *Geog. Jour.*, **117**, 78–85.

BRADLEY, W. C., 1963, 'Large Scale Exfoliation in Massive Sandstones of the Colorado Plateau', *Bull. Geol. Soc. Amer.*, **74**, 519–28.

BUNGE, W., 1962, *Theoretical Geography* (Lund), 210 pp.

BURTON, I., 1963, 'The Quantitative Revolution and Theoretical Geography', *Canadian Geog.*, **7**, 151–62.

CHAMBERLIN, T. C., 1897, 'The Method of Multiple Working Hypotheses', *Jour. Geol.*, **5**, 837–48.

CHORLEY, R. J., 1957, 'Climate and Morphometry', *Jour. Geol.*, **65**, 628–38.

— 1962, 'Geomorphology and General Systems Theory', *U.S. Geol. Survey, Prof. Paper 500-B*, 10 pp.

— 1963, 'Diastrophic Background to Twentieth-century Geomorphological Thought', *Bull. Geol. Soc. Amer.*, **74**, 953–70.

— 1964, 'The Nodal Position and Anomalous Character of Slope Studies in Geomorphological Research', *Geog. Jour.*, **130**, 70–73.

— (In press), 'The Application of Statistical Methods to Geomorphology', in *Essays in Geomorphology*, ed. by G. H. Dury (London).

CROXTON, F. E., 1959, *Elementary Statistics with Applications in the Medical and Biological Sciences* (Dover Paperbacks), 376 pp.

CROXTON, F. E. and COWDEN, R., 1939, *Applied General Statistics* (New York), 944 pp.

DUNCAN, O. D., CUZZORT, R. and DUNCAN, B., 1961, *Statistical Geography: Problems in Analysing Areal Data* (The Free Press of Glencoe, Illinois), 191 pp.

DURY, G. H., 1961, 'Bankfull Discharge: an Example of Its Statistical Relations', *Int. Assn. Scientific Hydrology*, 6th Year, No. 3, 48–55.

FLINT, R. F., 1963, 'Altitude, Lithology, and the Fall Zone in Connecticut', *Jour. Geol.*, **71**, 683–97.

GEIKIE, A., 1868, 'On Denudation now in Progress', *Geol. Mag.*, **5**, 249–54.

GILBERT, G. K., 1877, *The Geology of the Henry Mountains*, U.S. Dept. of the Interior (Washington) (Chapter 5, 'Land sculpture').

— 1914, 'The Transportation of Debris by Running Water', *U.S. Geol. Survey, Prof. Paper 86*, 263 pp.

GILLULY, J., 1949, 'Distribution of Mountain Building in Geologic Time', *Bull. Geol. Soc. Amer.*, **60**, 561–90.

GREGORY, S., 1962, *Statistical Methods and the Geographer* (London), 240 pp.

HOLLINGWORTH, S. E., 1938, 'The Recognition and Correlation of High-level Erosion Surfaces in Britain: a Statistical Study', *Quart. Jour. Geol. Soc.*, **94**, 55–84.

HORTON, R. E., 1945, 'Erosional Development of Streams and Their Drainage Basins: Hydrophysical Approach to Quantitative Morphology', *Bull. Geol. Soc. Amer.*, **56**, 275–370.

JAHN, A., 1961, *Quantitative Analysis of Some Periglacial Processes in Spitzbergen* (University of Warsaw, Poland), 54 pp.

JOHNSON, D. W., 1931, *Stream Sculpture on the Atlantic Slope* (New York), 142 pp.

KIRKBY, M. J., 1963, 'A Study of the Rates of Erosion and Mass Movement on Slopes with Special Reference to Galloway' (*Unpublished Ph.D. Thesis, Cambridge University*).

KRUMBEIN, W. C., 1944, 'Shore Processes and Beach Characteristics', *U.S. Beach Erosion Board, Tech. Memo.* 3.

— 1955, 'Experimental Design in the Earth Sciences', *Trans. Amer. Geophys. Union*, **36**, 1–11.

— 1959, 'The "Sorting Out" of Geological Variables illustrated by Regression Analysis of factors Controlling Beach Firmness', *Jour. Sedimentary Petrology*, **29**, 575–87.

— 1960, 'The "Geological Population" as a framework for Analysing Numerical Data in Geology', *Liv. and Man. Geol. Jour.*, **2**, 341–68.

LEOPOLD, L. B. and MADDOCK, T., 1953, 'The Hydraulic Geometry of Stream Channels and Some Physiographic Implications', *U.S. Geol. Survey, Prof. Paper 252*, 57 pp.

LEOPOLD, L. B. and WOLMAN, M. G., 1957, 'River Channel Patterns: Braided, Meandering and Straight', *U.S. Geol. Survey, Prof. Paper 282-B*, 39–85.

LEOPOLD, L. B., WOLMAN, M. G. and MILLER, J. P., 1964, *Fluvial Processes in Geomorphology* (San Francisco), 522 pp.

MELTON, M. A., 1957, 'An Analysis of the Relations among Elements of Climate, Surface Properties and Geomorphology', *Office of Naval Research Project NR 389–042*, Tech. Rept. 11, Dept. of Geol.. Columbia Univ., New York, 102 pp.

MILLER, R. L. and KAHN, S. J., 1962, *Statistical Analysis in the Geological Sciences* (New York), 357 pp.

PELTIER, L. C., 1950, 'The Geographic Cycle in Periglacial Regions as it is related to Climatic Geomorphology', *Ann. Assn. Amer. Geog.*, **40,** 214–36.

RICH, J. L., 1938, 'Recognition and Significance of Multiple Erosion Surfaces', *Bull. Geol. Soc. Amer.*, **49,** 1695–722.

ROSENAN, E., 1953, 'Comments on the Paper by R. A. Bagnold; In "Desert Research"', *Research Council of Israel, Spec. Pub. 2* (Jerusalem), 94.

SCHUMM, S. A., 1963, 'The Disparity between Present Rates of Denudation and Orogeny', *U.S. Geol. Survey, Prof. Paper 454-H*, 13 pp.

SHEPARD, F. P., 1963, *Submarine Geology*, 2nd Edn. (New York), 557 pp.

STEVENS, S. S., 1946, 'On the Theory of the Scales of Measurement', *Science*, **103,** 677–80.

STRAHLER, A. N., 1950 A, 'Davis' Concepts of Slope Development Viewed in the Light of Recent Quantitative Investigations', *Ann. Assn. Amer. Geog.*, **40,** 209–13.

— 1950 B, 'Equilibrium Theory of Erosional Slopes, approached by Frequency Distribution Analysis', *Amer. Jour. Sci.*, **248,** 673–96 and 800–14.

— 1952, 'Hypsometric (area-altitude) Analysis of Erosional Topography', *Bull. Geol. Soc. Amer.*, **63,** 1117–42.

— 1954, 'Statistical Analysis in Geomorphic Research', *Jour. Geol.*, **62,** 1–25.

VAN BURKALOW, A., 1945, 'Angle of Repose and Angle of Sliding Friction; an Experimental Study', *Bull. Geol. Soc. Amer.*, **56,** 669–708.

WOLMAN, M. G. and MILLER, J. P., 1960, 'Magnitude and Frequency of Forces in Geomorphic Processes', *Jour. Geol.*, **68,** 54–74.

WOOLDRIDGE, S. W., 1958, 'The Trend of Geomorphology', *Trans. Inst. Brit. Geog.*, **25,** 29–35.

YOUNG, A., 1960, 'Soil Movement by Denudational Processes on Slopes', *Nature*, **188,** 120–22.

CHAPTER NINE

Scale Components in Geographical Problems

P. HAGGETT

Professor of Geography, University of Bristol

One of the characteristic features of geographical research is its concern with a particular scale of reality. If we conceive this scale as a continuum running from the reality of the electron microscope, up to the one-to-one reality of our everyday life, through to the astrophysical reality of the galaxies, and on finally to the dimensionless reality of mathematics, then geographical research occupies a rather well-defined position in this succession (Haggett, Chorley and Stoddart, 1965). It ranges from highly localized studies of individual villages or river basins at magnitudes of 10^{-1} square miles through to world wide studies of the order of 10^7 square miles. Unlike the microscopic sciences where results have to be brought up to a one-to-one world for our understanding, geography is macroscopic in that it has to shrink reality to make it comprehensible.

The concern of this chapter is to draw attention to this shrinkage problem. It suggests that scale obtrudes into geographical research in three main ways: in the problem of covering the earth's surface; in the problem of linking results obtained at one scale to those obtained at another; and in standardizing information that is available only on a mixed series of scales. For convenience these are simply referred to here as the 'scale-coverage problem', the 'scale-linkage problem', and the 'scale-standardization problem'.

THE SCALE COVERAGE PROBLEM

Nature of the Problem

The scale coverage problem is simple and immediate. The earth's surface is so staggeringly large that, even if we omit the sea-covered areas, each of the profession's 3,000 nominal practitioners (Meynen,

1960) has an area of about 5,000 square miles to account for! These gross ratios caricature rather than characterize the problem. But if we agree with Hartshorne that the purpose of geography is '... to provide accurate, orderly, and rational description and interpretation of the variable character of the earth surface' (Hartshorne, 1959, p. 21) or follow Sauer in regarding it as a 'focused curiosity' (1952, p. 1) then we need to be aware of the magnitude of the task we set ourselves, or alternately the size of the object we are trying to get into focus.

This can hardly be regarded as a new problem. From at least the time of Eratosthenes the size of the problem has been apparent and it may well be that our predecessors were more keenly aware of its importance. Many a doubtful isopleth now strays self-importantly across areas that our more honest forbears might have filled with heraldic doodles or labelled 'Terra Incognita'. There are more reassuring signs, however, that the scale-coverage problem is partly being solved at this time by changes both inside and outside the discipline.

Internal Solutions: Sampling

Sample studies have long been used in both research and teaching. Platt (1942, 1959) was acutely aware of the '. . . old and stubborn dilemma of trying to comprehend large regions while seeing at once only a small area' (1942, p. 3) and he skilfully used sample field studies to build up an outstandingly clear series of pictures of the regions of Latin America. Similarly Highsmith (Highsmith, Heintzelman, Jensen, Rudd and Tschirley, 1961) has used a world-wide selection of sample studies as the basis for an extremely useful teaching manual in economic geography.

There is an important difference, however, between these attempts to use sampling to circumvent the scale problem, and the way in which sampling is now being used in research. This essential difference is between *purposive* and *probability* sampling. While it is outside the scope of this chapter to discuss sampling theory [there are excellent general summaries by Cochran (1953) and Yates (1960)] it is important to note that in probability sampling it is possible to estimate how accurate the survey is likely to be from the information actually collected during the sample survey. This means that given a limited budget the accuracy of any sample survey can be determined; or, vice versa, given a fixed limit of accuracy, the necessary size of sample and time-cost estimates can be made. The simplest case of

Table 4. Model of Sampling Systems

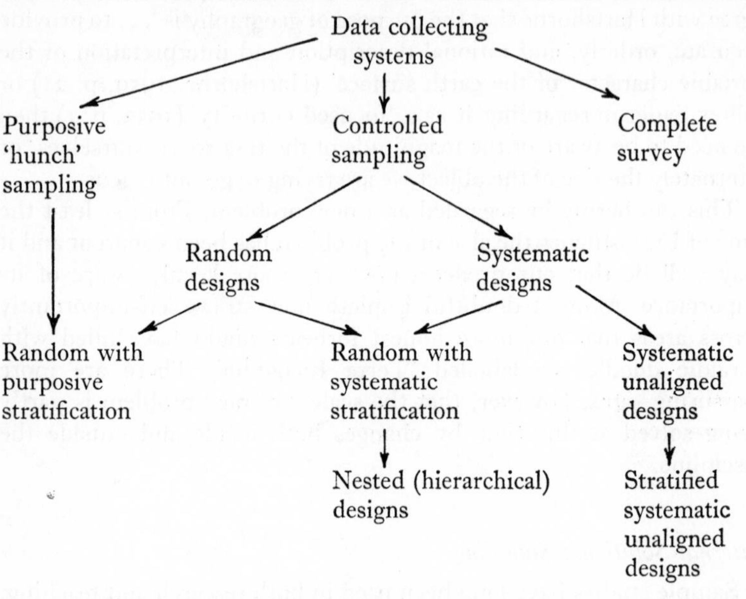

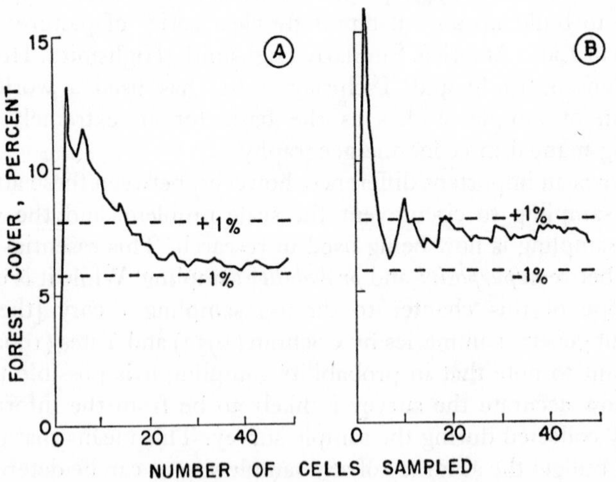

FIG. 9.1. *Changes in sampling error with size of sample (source: Haggett, 1963).*

this type of relationship is in the simple random design where the random sampling error (accuracy) is proportional to the square root of the number of observations (effort). An empirical illustration of this relationship is shown in Figure 9.1 where as sample size is steadily increased the values settle down around the known true value.

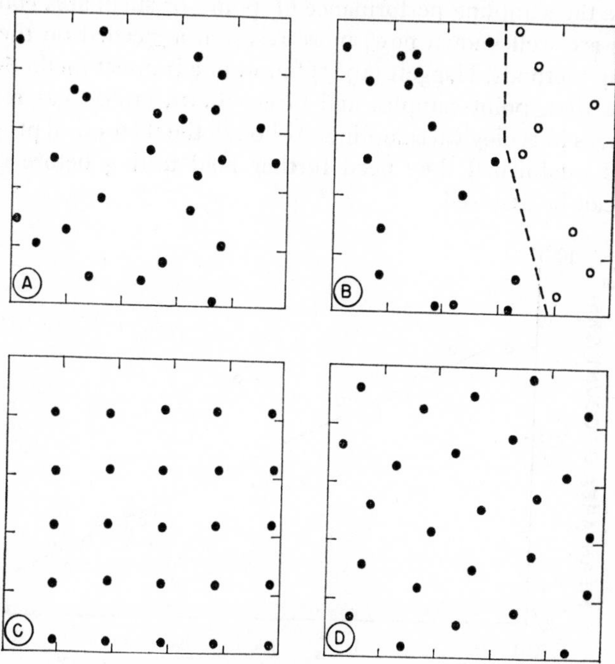

FIG. 9.2. *Alternative types of sampling design (source: Krumbein, 1960; Berry, 1962).*

Various types of sampling design have been used in recent years for specific problems (Table 4). Four of these designs are shown in Figure 9.2. Where the problem is exploratory and little is known about the characteristics of the 'population' being studied then a simple random design may be adopted, *A*. Wood (1955) introduced stratification, *B*, into the random design to allow certain parts of his study areas, eastern Wisconsin, to be more heavily sampled than others. The disadvantage of random designs is that they are more difficult to use as control points in mapping the results and systematic samples, *C*, may be substituted in their place where mapping is a prime consideration. Berry (1962) in an extensive study of the application of sampling design to land-use surveys of flood plains has found that a compromise design, the 'stratified systematic unaligned sample', *D*, gave the most accurate results with the additional advantages of facilitating both punched card storage and machine mapping.

While the sampling performance of 'point' or small area collecting systems are well known now, more research is needed on their line sampling methods. Haggett (1963) found line transect methods more accurate than point samples and Greig-Smith (1964) has reported advantages in ecological sampling. Although the theoretical properties are well established they need further field testing before relative merits can be assessed.

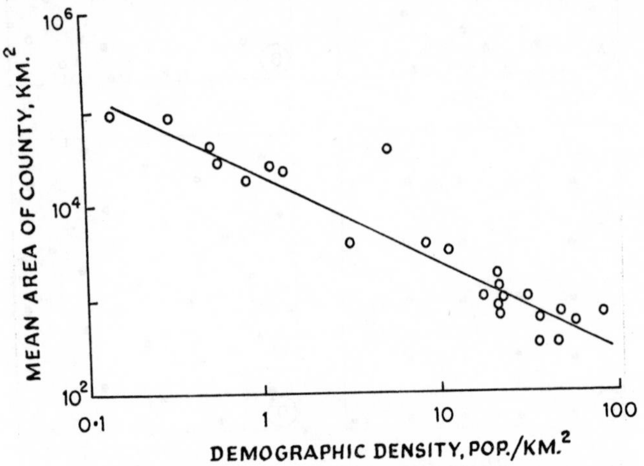

FIG. 9.3. *Relationship of the area of Brazilian county divisions to state population densities.*

External Solutions: Improving Sources

More information is available about the earth's surface today than at any previous time. The trickle of maps and census reports from government agencies a century ago has now risen to a torrent and there are indications that it is increasing logarithmically over time. This vast increase has not, however, been evenly spread so that the information contrasts between one part of the earth's surface and another are rather acute. Commonly the level of information is related to the development of the area. Figure 9.3 shows the relationship between the fineness of the data-collecting grid and the population density for one country: Brazil. Similarly, Berry (Ginsburg, 1961, p. 110) has pointed to the inverse relationship between the economic development of an area and the amount of information available about that development.

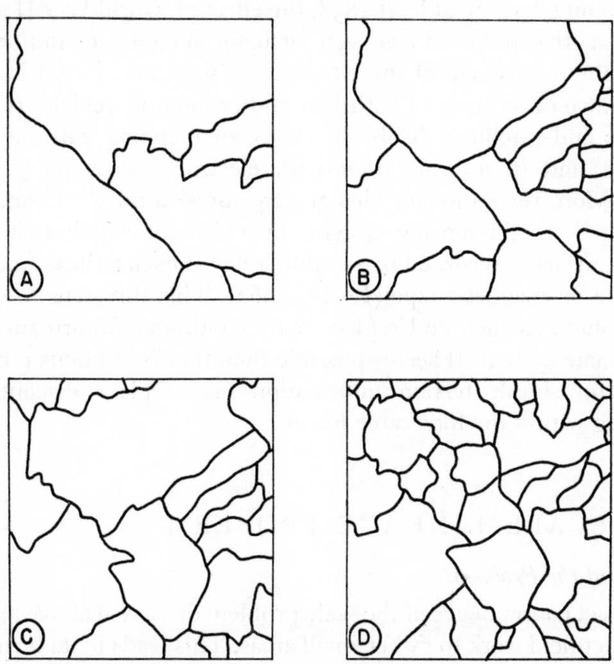

FIG. 9.4. *Increasing dissection of administrative areas in an area of rapid population increase; Santa Catarina state, Brazil, 1870–1960.*

Comparisons over time also run into difficulties. The very fact of improving information may make comparisons with earlier periods invalid. Figure 9.4 shows the successive subdivision of an area of rapid population increase over successive thirty-year time periods. Although far more is known in detail about the area in the final period, 1950, than in the initial period, 1870, the degree of comparable detail is controlled by the coarsest grid or the largest areal denominator. Dickinson (1963) has illustrated similar problems of subdivision and boundary change in England and Wales, while Hall (1963) has noted the problems in tracing the industrial growth of London from census data. For map coverage, Langbein and Hoyt (1959) have shown that for the United States there are some curious lacunae in both coverage and age with the poorly mapped areas being revised rather less frequently. Again the gap in coverage is tending to grow.

A vitally important supplement to such 'archival' data in maps and censuses is the growth of airphoto coverage. Although this has a history going back to at least 1858, the effect of World War II and the 'cold war' that followed has been virtually to complete and/or revise the airphoto coverage of the whole earth's surface. Rapid improvements have been made in both lens and camera, in vehicles (through to U2's and satellites) (Colwell, 1960), in mapping with electronic plotters, and in interpretation with electronic scanners (Latham, 1963). More revolutionary changes are foreshadowed in completely automated terrain sensing systems in which information about the earth's surface is recorded by satellite, relayed back to base, and made available on magnetic tape (Lopik, 1962). This threatens to cut out the airphoto as such and replace it by continuous information on a co-ordinate system. It seems possible then that continuous recording of certain simple terrain information may replace discontinuous mapping within the forseeable future.

THE SCALE LINKAGE PROBLEM

Nature of the Problem

A direct consequence of the scale problem discussed above has been to restrict field work to rather small areas. This leads in its turn to the second problem of researchers, that of '. . . seeing the link between their own local field work and the standard regional courses on the continents of the world' (Bird, 1956, p. 25). This scale linkage problem was brought home forcibly to the writer in comparing the inferences on forest distribution based on a small 100 square kilometre survey area made in 1959 with a later survey of the same features over a wider surrounding area (Haggett, 1964). These dangers have been very neatly summarized by McCarty: 'In geographic investigation it is apparent that conclusions derived from studies made at one scale should not be expected to apply to problems whose data are expressed at other scales. Every change in scale will bring about the statement of a new problem, and there is no basis for assuming that associations existing at one scale will also exist at another' (McCarty, Hook and Knos, 1956, p. 16). Such difficulties are not of course confined to geographical research. Duncan, Cuzzort and Duncan (1961) have pointed out the difficulty of linking individual with mass economic behaviour, while Bronowski (1960, p. 93) has raised the whole problem of individual action in the 'stream of history'.

For the more particular problems of scale linkage two sets of solutions are discussed here. First the qualitative attempt to accommodate scale in regional systems and secondly, the quantitative attempt to isolate and measure the impact of scale at series of levels.

Qualitative Solutions: Scale in Regional Systems

The fact that scale realizations have long troubled geographers is rather plainly shown in the series of attempts that have been made to define regions on scale terms. With formal regions, the early system applied by Fennemann (1916) to the landform divisions of the United States with his recognition of major divisions, provinces, and sections had a major effect on other writers (Table 5). Unstead (1933)

Table 5. Comparative Scales and Terminology of Regional Systems

Approx. size (sq. mls.)	*Fennemann (1916)*	*Unstead (1933)*	*Linton (1949)*	*Whittlesey (1954)*	*Map scales for study**
10^{-1}			Site		
10		Stow	Stow	Locality	1/10,000
10^2	District	Tract	Tract		1/50,000
				District	
10^3	Section		Section		
		Sub-Regn			
				Province	
10^4	Province		Province		1/1,000,000
		Minor Regn			
10^5	Major Divn		Major Divn	Realm	1/5,000,000
10^6		Major Regn	Continent		

* Whittlesey, 1954.

in an interesting paper on 'systems of regions' put forward the scheme which filled in at the smaller levels the system Fennemann has begun at the larger. Linton (1949) integrated both preceding systems in a seven-stage system which ran through the whole range from the smallest unit, the site, to the largest, the continent. More recently Whittlesey (In James, Jones and Wright, 1954, pp. 47–51) presented a 'hierarchy for compages' with details of the appropriate map scales for study and presentation and followed this with a model study on Southern Rhodesia to illustrate his method (Whittlesey, 1956). The decade since the Whittlesey scheme was put into operation and the call was made to '. . . fill this lacuna in geographic thinking' (James *et al.*, 1954, p. 47) has not seen any rush to adopt it. Of the few significant papers published in this field, only one, that by Bird (1956) subjected Whittlesey's scheme to field testing. Bird's two-scale comparison of the western peninsulas of Brittany and Cornwall suggested that, while a general (or small-scale) approach showed the two areas to be similar, the intensive (or large-scale) study showed that the two peninsulas were quite dissimilar in most details. Bird's deft illustration of a fundamental and very common geographic problem passed scarcely without comment.

The second major move in the period since Whittlesey's papers came from Philbrick (1957) who published a very full scheme based on the concept of a sevenfold hierarchy of functions. Corresponding to each function is a nodal point with its functional region. Here scale is introduced through the 'nesting' concept with each order of the hierarchy fitting within the next highest order. As a theoretical model Philbrick illustrates the case where each central place of a given order is defined to include four central places of the next lower order. This gives a succession for a seventh-order region of 4 sixth-order places, 16 fifth-order places, and so on down to the final level of 4,096 first-order places. His attempt to apply this scheme to the eastern United States with New York and Chicago in the role of seventh and sixth-order centres was only partly successful but the attempt to introduce a scale component into a system of nodal regions has given an important lead.

Quantitative Solutions

Quantitative attempts to isolate and measure scale components in geographical patterns have not been widely attempted and there is, to the writer's knowledge, no extensive literature in this field. Hence the sections which follow two possible methods of attack – those of filter mapping and of nested sampling – are illustrated from work in hand at Cambridge on patterns of forest distribution in central Portugal.

(1) *Filter Mapping:* The basic ideas of filter mapping can be seen from a fairly simple example. Figure 9.5 shows the stages by which a given distribution may be broken down into regional and local components. Map *A* shows the original distribution in a section of central Portugal. For statistical purposes this pattern can either be expressed as a ratio of (forested/non-forested area) or as a fraction

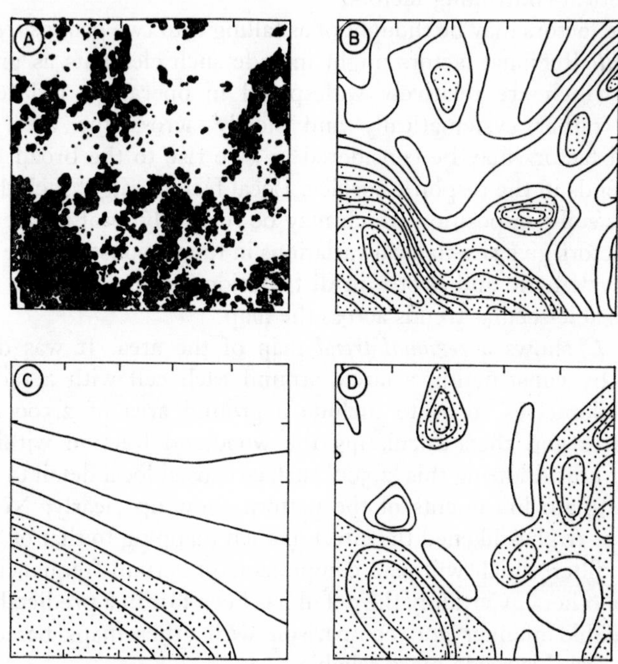

FIG. 9.5. *Mapping of regional and local components in a land-use pattern: Tagus–Sado basin, central Portugal. Shaded areas show densities above average.*

(percentage forest in total land area), i.e. either 0·352 or 26·30. By covering the area with a rectangular grid these ratio values can be collected for small areas (in this case square cells of forty square kilometres) and contoured. The resulting map, *B*, completely describes the area in two-dimensional form. Like contour values for terrain it could be converted to a three-dimensional plaster model but in any case can be regarded conceptually as a three-dimensional trend surface.

This surface may be thought of statistically as a *response surface* (Box, 1954). That is the height (i.e. degree of forest cover) at any one point may be regarded as a response to the operation of that complex of '. . . geology, topography, climatic peculiarities, natural composition, economic disparities, and local and regional history' (Köstler, 1956, p. 82) which together determine forest distribution. Variation in the form of the surface may be regarded as responses to corresponding areal variations in the strength and balance of these hypothetical controlling factors.

These factors may be thought of as falling into two groups, *regional* and *local.* Regional factors might include such elements as growing season which are relatively widespread in operation and tend to change rather systematically and slowly across the area. Such regional factors may be considered to give rise to the broad larger-scale trends in the response surface. Local factors might include such items as soil composition which may be relatively local in operation. Such factors give rise to local variations in the response surface which are unsystematic and spotty in distribution and do not give rise to recognizable secular trends across the map.

Map *C* shows a *regional trend* map of the area. It was derived simply by constructing a circle around each cell with a radius of 28·20 kilometres so as to include a ground area of 2,500 square kilometres and then calculating the woodland fraction within this circular unit. Plotting this '2,500' surface caused local detail to be lost but the main lineaments of the pattern show up clearly. Nettleton (1954, p. 10) has likened the effect of such mapping to that of '. . . an electric filter which will pass components of certain frequencies and exclude others'. Certainly the detail has been lost in a predictable and controllable manner and comparison with other maps based on a similar 'grid' is made more reliable.

Separation of the local anomalies can be very rapidly derived from the regional map. For each cell the values of the original 40 square kilometre cell are subtracted from those of the 2,500 square kilometre cell. Positive values (i.e. where local values exceed regional values) are shaded and negative values (i.e. regional values exceed local values) are unshaded. Map *D* therefore shows the operation of local factors as a pattern of positive and negative residuals.

Clearly an infinite number of trend-surface maps can be drawn and the nature of the resultant maps will vary with the grid interval chosen. To this extent the trend map is a quantitative expression of a qualitative choice. However, by including details of the generating grid with the map (in the same way that scale and orientation are conventionally included on a map) and by standardizing mapping around multiples of conventional levels – the 100 square kilometre unit would seem a useful basis for both aggregation and subdivision – this disadvantage can be nullified.

An alternative approach to filter mapping which is not dependent on a regular grid has been proposed by Oldham and Sutherland (1955). By using orthogonal polynomial equations they were able to compute a 'best fit' quadratic surface. This approach which has been further developed by Krumbein (1959) is taken up again in another context later in the chapter, but it is worth noting at this stage that more rigorous bases for determining trend surfaces are being explored and applied. These alternatives are discussed at length by Chorley and Haggett (In Press).

(2) *Nested Sampling:* One approach to the problem of local and regional variation that cuts out the need for complete information on all the area considered is that of nested sampling. (Olson and Potter, 1954; Krumbein, 1960). It is particularly valuable in exploratory studies where there is a need to cover as large a region as possible but at the same time pay attention to local variations. The basic idea of the nested approach (also termed the 'multilevel' or 'hierarchical' approach) is to divide the region into a few major areas of equal size. Several of these major regions are then chosen at random and broken down into a number of smaller sub-regions. Several of these sub-regions are chosen at random and broken down again, the process being continued until the smallest meaningful unit is reached or data cease to be available. Figure 9.6 illustrates this process by breaking down a 150 by 100 kilometre quadrangle into six 'regional units' each 50 by 50 kilometres square, and then subdividing each square a further four times until the smallest units, squares 3·125 by 3·125 kilometres, are reached. By selecting randomly at each level only two of the four available squares, only 96 of these smaller units are selected for study out of a possible total of over 1,500 such units within the original area, i.e. a sampling fraction of $\frac{1}{16}$. Their location is shown in Map *B*. Sampling on this hierarchical framework ensures not only that every part of the region is represented but that the field work time in visiting each point is reduced well below that of a simple $\frac{1}{16}$ random sample.

The main value of collecting data in this frame comes at the analysis stage. Here any 'local' value (X) can be regarded as being generated by the sum of independent deviations at each level of variability; i.e.:

X – Overall mean value (150 × 100 km)
+Region (50 × 50 km) deviation from overall mean value.
+Sub-region (25 × 25 km) deviation from region mean.
+District (12·5 × 12·5 km) deviation from sub-regional mean.
+Sub-District (6·25 × 6·25 km) deviation from district mean.
+Locality (3·125 × 3·125 km) deviation from sub-district mean.

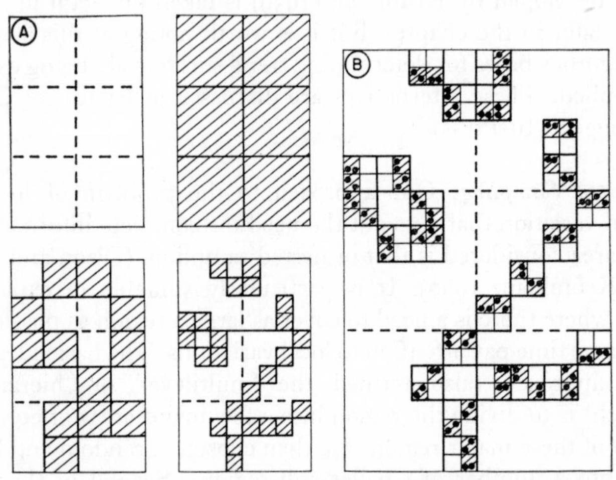

FIG. 9.6. *Five-stage nested sampling procedure for extracting scale components: Tagus–Sado basin, central Portugal.*

Values for each level can be determined using an appropriate type of variance analysis so that it is possible to specify the contribution each level makes to the total variability of the pattern. Table 6 shows the

Table 6. Tagus-Sado Basin, Central Portugal: Contribution of Five Areal Levels to Variability in a Forest Pattern

Level	*Areal unit (with area in square kilometres)*	*Variance component*	*Percentage contribution*
I	Region (2,500)	208	32
II	Sub-region (625)	0	0
III	District (156)	258	41
IV	Sub-district (39)	51	7
V	Locality (10)	124	19

results of an analysis made by the writer into the pattern of variability shown by the forest distribution in central Portugal using this approach. The original pattern is shown in Figure 9.5A. It indicates the great contrast between the increment in variability at the third level, the district, compared to the negligible effect of the next higher level. In this case it proved possible to link the high variability at levels I, III and V with the operation of specific factors at those levels.

It is clear that either by mapping techniques applied over the whole area (filter mapping) or by carefully selected sampling designs (nested sampling) the known variability of areal patterns can be broken down and examined. Techniques developing outside geography, notably in geophysics and in plant ecology (Greig-Smith, 1964) are being increasingly applied and the final forging of traditional cartography and applied statistics is proving a most powerful tool in dissecting areal problems.

THE SCALE STANDARDIZATION PROBLEM

Nature of the Problem

While the two previous scale problems apply with equal force to work in all sides of geographical research the third scale problem applies with particular force to work in human geography. Here two almost intractable problems are faced. Firstly, much of the data is released for areas rather than for points; secondly, these areas vary wildly in size and shape both between countries and within countries. This variation had little more than nuisance value when the limit of sophistication was the choropleth map but as more refined statistical analysis is carried out the problem has grown in importance. Weighting for area as proposed by Robinson (1956) can overcome some of the problems but there remains a range for error and misinterpretation so great that Duncan found it necessary to devote the greater part of his pioneer book on statistical geography to just such problems in analysing areal data (Duncan, Cuzzort and Duncan, 1961).

The problem can be seen at its simplest in city comparisons. If we ask an apparently simple question: 'Is City X bigger in population than City Y?' then the answer will often hinge on our areal definition of the city. Dickinson (1963, p. 68) has shown that Liverpool may be either smaller or larger than Manchester depending on how each city is demarcated. Similarly, Duncan has shown that Chicago may be more densely or less densely populated than Detroit depending on which of the available definitions – that of the 'city', the 'urbanized area', or the 'standard metropolitan area' – is taken (Duncan *et al.*, 1961, pp. 35–36). Table 7 shows how this problem extends to inter-

Table 7. Belgium and Netherlands, 1947: Comparison of Apparent Commuting Differences

Region	*Out-commuters*	*Mean size of sub-divisions**
Belgium	40·0%	1,880 hectares
Netherlands	15·2%	6,670 hectares

* Weighted according to population resident in each sub-division.
Source: Chisholm, 1960, p. 187.

national comparisons of commuting movements. Chisholm (1960) has demonstrated how the apparent differences in commuting described by Dickinson (1957) between two neighbouring states probably owes more to differences in the size of administrative sub-divisions than to any inherent differences in social behaviour. This arises from the definition of a commuter as 'moving to work outside his residence area': clearly *ceteris paribus* the smaller the residence areas the greater the chance of a worker being recorded as a commuter.

Unless great care is taken such 'mirage effects' are likely to become more common as more refined indices are derived. Kendall as a statistician has warned that with certain coefficients of geographical association we can get any coefficient we choose by juggling with the collecting boundaries! (Florence, 1944, p. 113). It is an open question whether detailed medical maps (e.g. Murray, 1962) in which mortality indices are most carefully standardized for age and sex should not equally well be standardized for the areas for which they are collected. Certainly we need to be reassured that the apparently unhealthy small pockets of disease in Lancashire and Yorkshire owe nothing to the fragmented system of local government areas.

These problems are central and critical and in suggesting a series of part-solutions below, the writer is keenly aware that these are really ways of making the best of a bad job. The eventual answer lies in (*a*) collecting more field data ourselves on a controlled sampling basis; and (*b*) persuading local and central government to follow the lead of the Swedish census in collecting and publishing data for specific

x, y co-ordinates rather than for irregular, inconvenient and anachronistic administrative areas.

Aggregation Solutions

Where the collecting units are many and irregular a fairly simple counter-measure is to group them into fewer but more regular areas. Such a technique was adopted by Coppock (1960) in a study of parish records in the Chilterns. Here not only are parishes irregular in shape and size and running orthogonally across the major geological boundaries but the farms on which the data was collected themselves had land outside the parishes for which their acreages were recorded. Grouped parishes in this case allowed both more regular units and reduced the farm 'overlap' problem since with larger units the farm area outside the combined parish boundaries was proportionally a rather insignificant part of the larger total area.

It is important to remember in using the aggregation method that it is possible to throw the baby out with the bathwater in the sense of throwing away detail to gain uniformity. Haggett (1964) has suggested that the coefficient of variation may be used as an indication of both loss of detail and of gains in uniformity, and that only when the latter exceeds the former is the detail loss justified. Table 8 shows a

Table 8. South-east Brazil, 1950. Comparison of Original and Grouped Counties

Characteristics	*County (município)*	*Super-county*
Number	126	24
Mean area	133 square miles	699 square miles
Coefficient of variation	74·2	7·91

Source: Haggett, 1964, p. 371.

case in point in a regression analysis of county-data in south-eastern Brazil. Here the original 126 counties were grouped into only 24 'super counties' but the 82 per cent loss in detail was less than the percentage gain in uniformity as measured by comparisons in the coefficient of variation. Where units are in any case very regular, as in the American Middle West (Weaver, 1956), the method is hardly justified.

Aggregation and testing will be very much speeded when computer programmes are developed for rapidly checking all possible number of ways in which contiguous units can be combined and re-combined. In view of the enormously large number of possible combinations it is uncertain that the combinations used so far are the optimum ones in terms of uniformity, of size, shape, and number.

Grid-Type Solutions

The difficulties of dealing with aggregate areas is that they are themselves highly irregular in shape if not in size. Attempts have therefore been made to collect information not in areal units but in regular frames or grids.

An outstanding example of this type of work is the *Atlas of the British Flora* (Perring and Walters, 1962) where field data on the occurrence of British vascular plants was collected for the 100 square kilometre grid-squares of the British National Grid System. This grid system was also used by Johnson (1958) in a study of the location of factory population in the West Midlands. Grid systems have the great advantage of ease of mapping and the flora maps were all directly plotted by computer. This has very considerable merit in an era where more maps are being produced directly from punched-tape data (Tobler, 1959) and allows very ready comparison between the original data and controlling factors. It also allows micro-analysis by breaking down the original squares into smaller ones or macro-analysis by combining such squares into larger units, and on the lines suggested by filter sampling above.

In both examples cited, data was either collected on a regular grid pattern or was precisely located and could therefore be assigned to a grid. Where data only exists for irregular administrative areas the transfer to a grid is more complex. Robinson, Lindberg and Brinkman (1961, p. 214) used a regular hexagonal grid in a study of population trends in the Great Plains of the United States. County data was transferred to the grid by measuring how much of each hexagon was made up by any one county and multiplying this value by the population density of the county. The sum of all the county parts gave the average value for the hexagonal unit. This principle was first used by Thiessen in 1911 for calculating the average rainfall over watersheds and its accuracy clearly hinges on two principles. First, the degree to which population density (or any similar measure) can logically be regarded as uniform over the areal sub-division and, second, the number of such sub-divisions which make up the regular unit. Where each hexagon contains a number of undivided counties the assumptions under the first principle become less limiting as the 'split' counties contribute less to the total value. Again the problem is one of optimizing through linear programming both the reliability of each grid-unit (by increasing its size) and the number of such grid-units (by decreasing their size).

Grid-free Solutions

Clearly the two foregoing methods, aggregation and grid-type solutions must involve some loss in detail in that the revised units are fewer than the original. Attention has recently been directed to the

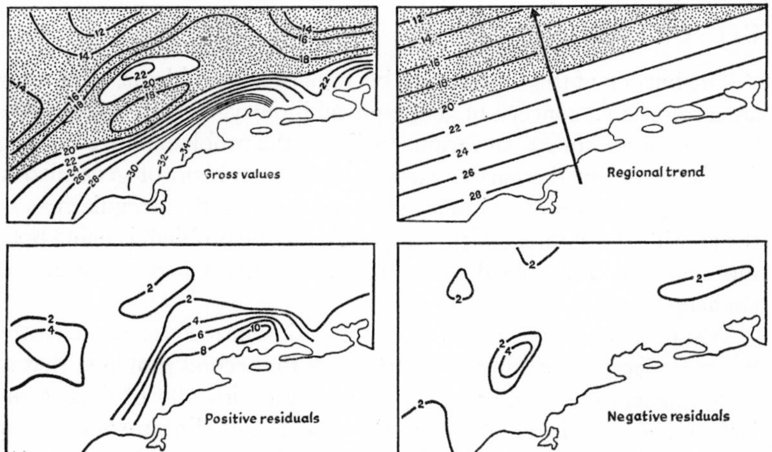

FIG. 9.7. *Regional trend surface with positive and negative residuals: south-east Brazil. (source: Haggett, 1964).*

problem of how generalized maps can be made which retain all the original data. This problem first came to light through geophysical problems, e.g. through meteorology where general weather patterns have to be mapped from irregular and often highly localized weather recording stations and through petroleum prospecting where basin and facies characteristics may have to be mapped from irregular well and bore records. Krumbein (1959) has illustrated how computers may be used to derive an algebraic formula which gives the average surface which 'fits' the irregular control points best. This 'best fit' polynomial surface uses all the available records and builds them into a generalized picture. It is of particular importance where gaps in the areal spread of records are found and has been used by Whitten (1959) to fill in the 'ghost stratigraphy' of crystalline areas.

Use of these methods in landform analysis has been undertaken by Chorley (1964) in a study of the topography of the Lower Greensand, while simpler first-order trend surfaces, i.e. simple planes (Figure 9.7) have been used by Robinson and Caroe (Garrison, 1964) in a study of population in Nebraska, and by Haggett (1964) in a study of forest distribution in south-east Brazil. There is every indication that they will be more widely used in the future (Chorley and Haggett, In Press).

CONCLUSION

The burden of this chapter has been that scale problems, while they are a traditional concern of geographical inquiry, have in recent years become more explicit. In various forms, in the problem of overcoming the sheer size of the earth's surface, in the problem of generalizing over various levels of inquiry and in overcoming the irregularities of the administrative areas with which we deal, the problem is being both more acutely recognized and steps are being taken to meet the challenge.

While each of the three problems has significance for research the major impact on teaching comes through the second problem, that of scale-linkage. Here we are at the root of the difficulty of linking the large-scale sample study with the small-scale lineaments of the continental pattern. Some textbooks, notably Mead and Brown (1962) in their study of North America, go some way towards a solution by reproducing air photographs and portions of 1/62,500 maps alongside regional transects. Ideas that factors may change with scale may be introduced for familiar phenomena. The location of a light manufacturing plant in north-west London might be explained in terms of regional factors (location within industrial north-west Europe), local factors (access to the London and Midlands market), and site factors (land characteristics such as bearing capacity of soils for foundations and heavy machinery). Each explanation nests within the other and operates within the general restraints set by the next highest factor. This has the advantage of resolving some of the apparently conflicting hypotheses by restricting each to a particular level of generalization. It also accords with the known practice of locational decision-making (McLaughlin and Robock, 1949).

It is clear that open recognition of the problem of working within a scale continuum clarifies some problems and raises others. While it is undoubtedly more comfortable to work on in happy or perverse oblivion of the problems its recognition brings, there are encouraging signs that with recognition comes part-solutions; from such part-solutions we hope some final answers may emerge.

References

BERRY, B. J. L., 1962, 'Sampling, Coding and Storing Flood Plain Data', *U.S. Dept. Agric. Farm Econ. Div., Agric. Hdbk.*, **237**, 1–27.

BIRD, J., 1956, 'Scale in Regional Study: Illustrated by Brief Comparisons between the Western Peninsulas of England and France', *Geog.*, **41**, 25–38.

BOX, G. E. P., 'The Exploration and Exploitation of Response Surfaces', *Biometrics*, **10**, 16–30.

BRONOWSKI, J., 1960, *The Common Sense of Science* (London).

CHISHOLM, M. D. I., 1960, 'The Geography of Commuting', *Ann. Assn. Amer. Geog.*, **50**, 187–8 and 491–2.

CHORLEY, R. J., 1964, 'An Analysis of the Areal Distribution of Soil Size Facies on the Lower Greensand Rocks of East-central England by the use of Trend Surface Analysis', *Geol. Mag.*, **101**, 314–21.

CHORLEY, R. J. and HAGGETT, P. In Press. 'Trend-surface Mapping in Geographical Research', *Trans. Inst. Brit. Geog.*

COCHRAN, W. G., 1953, *Sample Survey Techniques* (New York).

COLWELL, R. L. (Ed.), 1960, *Manual of Photographic Interpretation* (New York).

COPPOCK, J. T., 1960, 'The Parish as a Geographical-Statistical Unit', *Tijds. v. econ. soc. Geogr.*, **51**, 317–26.

DICKINSON, G. C., 1963, *Statistical Mapping and the Presentation of Statistics* (London).

DICKINSON, R. E., 1957, 'The Geography of Commuting: the Netherlands and Belgium', *Geog. Rev.*, **47**, 521–38.

DUNCAN, O., CUZZORT, R. P. and DUNCAN, B., 1961, *Statistical Geography: Problems in Analysing Areal Data* (Glencoe, Ill.).

FENNEMAN, N. M., 1916, 'Physiographic Divisions of the United States', *Ann. Assn. Amer. Geog.*, **6**, 19–98.

FLORENCE, P. S., 1944, 'The Selection of Industries Suitable for Dispersal into Rural Areas', *Jour. Roy. Stat. Soc.*, **107**, 93–116.

GARRISON, W. L. (ed.), 1964, *Quantitative Geography* (New York).

GINSBURG, N., 1961, *Atlas of Economic Development* (Chicago).

GREIG-SMITH, P., 1964, *Quantitative Plant Ecology* (London).

HAGGETT, P., 1963, 'Regional and Local Components in Land-use Sampling: a Case Study from the Brazilian Triangulo', *Erdkunde*, **17**, 108–14.

— 1964. 'Regional and Local Components in the Distribution of forested areas in South-East Brazil: a Multivariate approach', *Geog. Jour.*, 130, 365-80.

HAGGETT, P., CHORLEY, R. J. and STODDART, D. R., 1965, 'Scale Standards in Geographical Research: A New Measure of Areal Magnitude'; *Nature*, **205**, 844–7.

HALL, P., 1962, *The Industries of London since 1861* (London).

HARTSHORNE, R., 1959, *Perspective on the Nature of Geography* (London).

HIGHSMITH, R. M., HEINTZELMAN, O. H., JENSEN, J. G., RUDD, R. D. and TSCHIRLEY, P. R., 1961, *Case Studies in World Geography* (New York).

JAMES, P. E., JONES, C. F. and WRIGHT, J. K. (eds.), 1954, *American Geography: Inventory and Prospect* (Syracuse).

JOHNSON, B. L. C., 1958, 'The Distribution of Factory Population in the West Midlands Conurbations', *Trans. Inst. Brit. Geog.* Pub. **25**, 209–23.

KRUMBEIN, W. C., 1959, 'Trend Surface Analysis of Contour-type Maps with Irregular Control-point Spacing', *Jour. Geophys. Res.*, **64**, 823–34.

— 1960, 'The "Geological Population" as a Framework for Analysing Numerical Data in Geology', *Liv. and Man. Geol. Jour.*, **2**, 341–68.

KRUMBEIN, W. C. and SLACK, H. A., 1956, 'Statistical Analysis of Low Level Radioactivity of Pennsylvanian Black Fissile Shale in Illinois', *Bull. Geol. Soc. Amer.*, **67**, 739–62.

KÖSTLER, J., 1956, *Silviculture* (Edinburgh).

LANGBEIN, W. B. and HOYT, W. G., 1959, *Water Facts for the Nation's Future: Uses and Benefits of Hydrological Data Programmes* (New York).

LATHAM, J. P., 1963, 'Methodology for an Instrumented Geographic Analysis', *Ann. Assn. Amer. Geog.*, **53**, 194–209.

LINTON, D. L., 1949, 'The Delimitation of Morphological Regions', *Trans. Inst. Brit. Geog. Pub.* **14**, 86–87.

LOPIK, J. R. VAN, 1962, 'Optimum Utilization of Airborne Sensors in Military Geography', *Photogramm. Eng.*, **28**, 773–8.

MCCARTY, H. H., HOOK, J. C. and KNOS, D. S., 1956, 'The Measurement of Association in Industrial Geography', *Univ. Iowa. Dept. Geogr. Rept.*, **1**, 1–143.

MCLAUGHLIN, G. E. and ROBOCK, S., 1949, *Why Industry moves South* (Kingsport).

MEAD, W. R. and BROWN, E. H., 1962, *The United States and Canada* (London).

MEYNEN, E., 1960, *Orbis geographicus 1960* (Wiesbaden).

MURRAY, M., 1962, 'The Geography of Death in England and Wales', *Ann. Assn. Amer. Geog.*, **52,** 130–49.

NETTLETON, L. L., 1954, 'Regions, Residuals and Structures', *Geophysics*, **19,** 1–22.

OLDHAM, C. H. G. and SUTHERLAND, D. B., 1955, 'Orthogonal Polynomials: Their Use in Estimating the Regional Effect', *Geophysics*, **20,** 295–306.

OLSON, J. S. and POTTER, P. E., 1954, 'Variance Components of Cross-bedding Direction in Some Basal Pennsylvanian Sandstones of the Eastern Interior Basin: Statistical Methods', *Jour. Geol.*, **62,** 26–49.

PERRING, F. H. and WALTERS, S. M., 1962, *Atlas of the British Flora* (London).

PHILBRICK, A. K., 1957, 'Principles of Areal Functional Organization in Regional Human Geography', *Econ. Geog.*, **33,** 299–336.

PLATT, R. S., 1942, *Latin America: Countrysides and United Regions* (New York).

— 1959, 'Field Study in American Geography: the Development of Theory and Method exemplified by Selections', *Univ. Chicago. Dept. Geogr. Res. Pap.*, **61,** 1–405.

ROBINSON, A. H., 1956, 'The Necessity of Weighting Values in Correlation of Areal Data', *Ann. Assn. Amer. Geog.*, **46,** 233–6.

ROBINSON, A. H., LINDBERG, J. B. and BRINKMAN, L. W., 1961, 'A Correlation and Regression Analysis applied to Rural Farm Population Densities in the Great Plains', *Ann. Assn. Amer. Geog.*, **51,** 211–21.

SAUER, C. O., 1952, *Agricultural Origins and Dispersals* (New York).

TOBLER, W., 1959, 'Automation and Cartography', *Geog. Rev.*, **44,** 534–44.

UNSTEAD, J. F., 1933, 'A System of Regional Geography', *Geog.*, **18,** 175–87.

WEAVER, J. C., 1956, 'The County as a Spatial Average in Agricultural Geography', *Geog. Rev.*, **46,** 536–65.

WHITTEN, E. H. T., 1959, 'Composition Trends in a Granite: Modal Variation and Ghost Stratigraphy in Part of the Donegal Granite', *Jour. Geophys. Res.*, **64,** 835–48.

WHITTLESEY, D., 1956, 'Southern Rhodesia: an African Compage', *Ann. Assn. Amer. Geog.*, **46,** 1–97.

WOOD, W. F., 1955, 'Use of Stratified Random Samples in a Land-use Survey', *Ann. Assn. Amer. Geog.*, **45,** 350–67.

YATES, F., 1960, *Sampling Methods for Census and Surveys* (London).

CHAPTER TEN

Field Work in Geography, with Particular Emphasis on the Role of Land-Use Survey

C. BOARD

Lecturer in Geography, London School of Economics

In the space of a year geography has lost two of its greatest advocates of field work. Mainly as a result of the activities of Professor Wooldridge and Geoffrey Hutchings field work has progressed far in the last half century, so that it is a universally respected approach to the study of geography. The generation of geographers trained by Wooldridge is today playing a major part in training yet more geographers in the same well tried methods. It is the wish of the pioneers in this approach that 'other teachers will carry on the method' (Wooldridge and Hutchings, 1957, p. xi). This suggests that the time is not inopportune for an appraisal of the position of field work. In looking back to the development of the methods characteristic of this approach and by comparing its influence in British geography with its role in North American geography, it should be possible to arrive at an evaluation of the distinctive part that this approach has played in the progress of geography.

WOOLDRIDGE AND 'REAL FIELD WORK'

The kind of field work as practised by Wooldridge is essentially British, and represents the trade-mark of much British geographical writing. It has its roots in the observation and records of naturalists such as Gilbert White of Selborne (1720–93), an uncompromising protagonist of first-hand observation. The careful collection of specimens and painstaking recording of occurrences from Nature are characteristic approaches of the field sciences, botany, geology, and zoology. Geographical field methods owe much to those of geology;

in fact Wooldridge's training as a field man was as a geologist. It is in physical geography that the influence of field geologists like Archibald Geikie is felt most of all. In an essay on 'Science in Education' (1905), Geikie pointed out that deficiencies of 'literary methods' could be overcome by cultivating the faculty of observation. He realized that everyone was not equally endowed with this faculty and that training was required so that the student could see much more in the world around him 'than is visible to the uninstructed man' (Geikie, 1905, p. 296). There is little doubt that views such as these inspired Wooldridge, whose pronouncements on field work are in sympathy with those of Geikie and the great field geologists. Indeed, it may be said that because much of the early field work in geography was done by geologists, who had become interested in landforms, the currently accepted form of geographical field work concentrates on the natural landscape. Anxious that geography was becoming town centred and too concerned with man rather than land, Wooldridge (1949 A, p. 14) enters an impassioned plea for the teaching of 'natural history geography'. For Wooldridge (1949 B, p. 3) the Weald was one of the true field laboratories, used by field scientists of all breeds. Although most commentary on field work in geography has been made by Wooldridge and Hutchings, it has also become fashionable to include references to the necessity for field work in the inaugural addresses by new occupants of chairs of geography (Linton, 1946; Balchin, 1955; Edwards, 1950; Monkhouse, 1955; Pye, 1955). Such comments stressing the value of field work in geographical education point out other praiseworthy advantages ranging from an open-air life to informality between instructor and instructed. There is clearly little need now to defend the position of field work in geography.

Lest it be supposed that all geographical field work is of this didactic kind, it must be asserted here and now that American geography usually reflects a different brand of field study. This will be distinguished by the name field research. At the same time, it must be said that field research has also been characteristic of British geography. Nevertheless, it will be contended here that field teaching, as distinct from field research, has had a peculiarly powerful influence upon British geographical work. Wooldridge and East (1958, p. 163) in fact recognize this distinction, insisting on the need for training before undertaking research in the field. They give as an example Platt's specimen studies of landscapes in Latin America (1942), whose 'objective is to provide an *interpretative description* of the countrysides and regions of a vast, diversified and on the whole little known continent'. *Real field work* as Wooldridge and East (1958, p. 161) regard it is 'the close examination and analysis in the field of an accessible piece of country, showing one or more aspects of areal differentiation'. Their exposition of its methods demonstrates that by this is meant training in field work. The danger is that these should be equated with field work as a whole.

It would be equally misleading to suggest that all field work in British geography was influenced by the principles of field teaching. Morphological mapping as developed by Linton and Waters (1958) has all the hallmarks of field research, in that 'it is capable of universal application' (Waters, 1958, p. 15). By careful observation and plotting, the indivisible morphological units of the earth's surface can be mapped empirically. No prior knowledge of geology or geomorphology is presupposed and little room is left for subjective interpretation (Waters, 1958, p. 16). Such detailed morphological mapping owes much to similar but scattered work done by other geomorphologists, whose intention was not primarily didactic (Sparks, 1949, pp. 167–8; Hare, 1948, pp. 301–7).

In order to see the extent to which field teaching has affected field research, or is likely to affect it in the future, it is necessary to pay more detailed attention to the aims and methods of field teaching in Britain. In the nineteenth century, it must have seemed to those who looked at the writings of geographers that they shared a strange long-sightedness with Mrs Jellyby whose eyes appeared to 'see nothing nearer than Africa' (Dickens, 1890 edn., p. 30). This cult of 'otherwhereitis' as it was termed by Wooldridge (1950, p. 9), was gradually broken down by geographers and others interested in education. Their pleas for making geography more realistic (Branford, 1915–16, p. 97; Fairgrieve, 1937, p. 16; Hunt, 1953, p. 277; Hutchings, 1962; Wooldridge, 1955, pp. 78–9) are amply recorded in the pages of the *Geographical Teacher* or *Geography*. 'The only true geographical laboratory is the world outside the classroom' (Incorporated Association of Assistant Masters, 1954, p. 180). This statement, and others like it, owe much to the notion that by making use of one's immediate surroundings, the raw material for teaching geography could be made convincing. Fairgrieve's advocacy of the study of the 'home region' as 'the only criterion by which the rest of the world may be judged' (1937 B, p. 251) stressed the second-hand nature of much geographical knowledge. He went on, 'the standards which are known to the children themselves are the only ones which may be used to measure other places'. This theme is taken up by Wooldridge (1955, p. 80) when he makes one of his frequent references to G. K. Chesterton's belief that 'to make a thing real you *must* make it local'. Apart from the possibility offered by local field work, the only way of making far-off lands more real was by using material published in the form of detailed first-hand accounts of 'sample areas' (Roberson and Long, 1956). These, however, are still no substitute for the pupils' own first-hand study of the ground.

After several decades of field teaching, Wooldridge and Hutchings in their guide to field excursions around London (1957, pp. xi and xii) state their view that the aim of geographical fieldwork 'is regional synthesis and though this depends largely upon labours in the library and map room, it cannot dispense with sensitive field observation'. This places field work in true context as but one way of collecting information.

THE METHOD OF FIELD TEACHING

Both Wooldridge and Hutchings have recently provided a fairly comprehensive account of the aims and methods of field teaching. Wooldridge (Wooldridge and East, 1958, p. 161) dismisses as not being real field work, surveying, visits to farms and factories, and censuses. For him the starting point is the comparison of the map with the ground (Wooldridge, 1948 B, p. 2). This is an essential process in map reading and is a way by which the student may gain an appreciation of the scale of phenomena (Hutchings, 1962, p. 6). Since 'the ground, not the map, is the primary document' (Wooldridge and East, 1958, p. 162), the student should work from the ground to the map. Wooldridge (1948 B, p. 4) also points out '. . . the fact that, over a great range of studies, reality is in the field'. Since the map is frequently deficient, because of the exigencies of convention and scale, a second principle of geographical field work is to make 'significant additions' to it, from observations in the field (Wooldridge and East, 1958, p. 165). Hutchings (1962) points out the need for making sketch maps and annotated field sketches of landscapes. Both of these are considered as adjuncts to the description of landscape which forms a major part of work in the field. In reviewing current practice in one university geography department, Wise (1957, p. 20) maintains that students are introduced to areas of progressive complexity in their three-year training to develop an eye for country. Some areas are in fact so 'subtle' or difficult to interpret, that Wooldridge and Hutchings (1957, pp. xiii and xiv) consider them 'not really suited to those beginning to learn the craft of field observation'. Much of the drift covered country of low relief in Eastern England falls into this category. In still another way, it can be seen that the difficulties of field work are not minimized when Wooldridge (1948 B, p. 3) pointed out to laboratory-bound scientists:

'I wish sometimes they could come with us and practise thinking on their feet in all weathers when rain or perspiration drips from one's person and the bar-parlour with its insidious temptation to spirituous theorizing insistently beckons.'

When one turns to the interpretation of field evidence, relatively little guidance is offered. 'The high Art is to teach the learner to use his eyes and draw his own conclusions' (Wooldridge, 1960, p. 3). Hutchings (1960, p. 6), in insisting that a landscape sketch is in in itself an interpretation, writes, 'The geographical draughtsman weaves into his drawing some of his *knowledge* of the things he is depicting.' There is no doubt that one of the main advantages deriving from field study lies in the appreciation of interrelationships of things in space (Wooldridge, 1960, p. 3). This, however, has a corollary in which lie pitfalls for unwary, lesser minds. Field teaching quite properly concentrates on 'observable field data' (Wooldridge, 1955, p. 79), but the explanations for visible patterns are not always, nor ever entirely, to be found by visual inquiry. It is difficult to imagine undertaking a survey of a market town in order to explain its present character, without resorting to some form of interviewing, or the close inspection of the interior of buildings. It is Hutchings' view that, although 'the story of a village, and especially of its social and economic working, is by no means legible in visible signs' (1962, p. 10), not much of it can be considered geography, and therefore in field work it is better not to interrogate the local inhabitants. This would confine geography to a study of the visible landscape. Wooldridge and Goldring (1953, p. 4) are perfectly well aware that the facts of modern economic and social geography are not amenable to the methods of field teaching. They justify the exclusion of the patterns of modern times from their study of the Weald because 'their study involves distinct methods and ways of thought which are not those of the "field man".'

THE NATURE OF FIELD RESEARCH

Before turning to examine the effect field teaching has had on the progress of geographical research, it is instructive to consider the characteristic methods of field research. Many non-descriptive works base their initial hypotheses upon chance remarks and marginal observations of other workers, not necessarily geographers. In other instances the field itself is held to be 'the primary source of inspiration and ideas' (Wooldridge, 1948 B, p. 2). In many cases both field evidence and previous literature have combined to suggest new topics of research. Wooldridge's own work on the Pliocene history of the London Basin (1927) clearly indicates his debt to Barrow (1919, and Barrow and Green, 1921) for its point of departure. Another work finding inspiration from the landscape but the point of departure in previous literature is McCann's study of raised beaches in Western Scotland (1963). Stevens (1959) attempts to illustrate Wooldridge and Linton's contention that geomorphology is of geographical significance by showing that morphological differences which have their roots in geological history have a profound effect on the land-use pattern in north-east Hampshire. Linton (1955), Palmer (1956) and Palmer and Neilson (1962) provide studies of tors, which are good examples of features of the physical landscape that have given rise to speculation and thorough investigation. In the field of historical geography Mead's account (1954) of ridge and furrow in Buckinghamshire shows his indebtedness to a vision of the countryside as well as to the stimulus of a remark by an economic historian. An analysis of the distribution of strip-lynchets by Whittington (1962) is largely inspired by the evidence of so-called cultivation terraces in various parts of Britain. There is no doubt that 'a large part of the evidence in our various subjects and specialisms (in field sciences) is evidence obtained in the field' though supplemented by work in the laboratory (Wooldridge, 1948 B, p. 2). It is less certain that the field 'inspires a great part of both the matter and the *method* [my italics] of our subjects' (Wooldridge, 1948 B, p. 2). Although in economic geography the field evidence may provide the initial problem, as in the case of Hodder's account of rural markets in Nigeria, the methods of field investigation are not those usually associated with the physical field sciences. They demand 'techniques more commonly associated with the social anthropologist' (Hodder, 1961, p. 158). Undoubtedly, in any profound studies of contemporary human geography visual observation has to be supplemented by inquiry. The frequent use of the specimen farm in the reports of the Land Utilisation Survey of Britain is a case in point (Stamp, 1948, pp. 335–50). In some cases, geographical hypotheses may be suggested by the contemplation of patterns on maps, such patterns not being visible in the field. Much of the interpretative work of the Land Utilisation Survey follows this pattern. Still other land-use studies have stemmed from the visual recognition of areal correspondence such as that between pasture land and flood plain (Berry, 1962).

THE ROLE OF THE VISIBLE LANDSCAPE

Visible elements in the landscape form a large part of the subject matter for both geomorphologists and historical geographers who make use of the genetic method. We are reminded that 'geomorphology is the historical geography of the physical landscape' (Wooldridge, 1948 A, p. 28). Some of the dangers that befall historical geographers, who rely too heavily on legacies in the present landscape as a guide to the interpretation of its evolution are expressed by Gulley (1961, p. 308). Can one be sure that conditions, both physical and social, in the past would have the same value as today? (Kirk, 1952, pp. 156–7, 159–60). Would the identification of a series of relicts from former landscapes be an adequate basis for an explanation of its evolution? Some features survive better than others. Strip-lynchets for instance are more likely to be preserved on more resistant rock formations. Ridge and furrow patterns clearly survive best when undisturbed by the plough (Mead, 1954, p. 36), whose long use may have destroyed them in the chalk belt of Buckinghamshire. Such considerations threaten the value of the palimpsest concept. In 1893, Geikie compared the surface of the country to a palimpsest bearing the writing of earlier centuries (Geikie, 1905, p. 56) and was followed by Maitland (1897, p. 15) who called the Ordnance Survey 'One Inch' map 'that marvellous palimpsest' from which it was possible to distinguish at least two distinctive forms of village settlement pattern. Wooldridge and Hutchings (1957, p. 44) urge 'we must need attempt the same *genetic* method if we are to understand our country in any real sense' and warn of missing the essential and interesting content of geography if one allows the dominance of London in south-east England to obscure legacies from the past or the fact 'that much of the pattern seen on the ground or the map is of ancient establishment'. Such an attitude plainly will not serve contemporary economic and social geography. Even if the historical geographer does attempt to interpret the legacies of former landscapes in the present landscape, the historian rarely substitutes for his body of material the fragmentary evidence provided by the faded writing beneath the latest script on the palimpsest. The archaeologist in the field has, perforce, to employ the technique of piecing together evidence of a previous culture from its material remains. By extending those methods to the study of the present landscape, with only the evidence of that landscape on which to base one's conclusions, some geographers have encouraged the neglect of evidence not visible in the field. Wooldridge and East (1958, pp. 30–31) have commented on attempts to treat of cultural elements of landscape which are 'unbearably trite' chiefly because they ignore the invisible aspects of the American farm or the city. Unfortunately the morphological approach to economic and social geography has been encouraged by the powerful influence of Sauer (1925, p. 32) who saw 'the cultural landscape as the culminating expression of the organic area' and 'the geographic area in the final meaning' (p. 46). Sauer interpreted Vidal de la Blache and Brunhes to American geographers, representing that they were important for their studies of the application of morphology to the works of man in the landscape. The morphological approach adopted by Brunhes is seen stripped to its bare essentials in his *Human Geography* (1952 edn., p. 36) – 'ridding our mind of knowledge of man, let us try to see and note the essential facts of human geography with the same eyes and in the same way as we discover and disentangle the morphological, topographical and hydrographical features of the earth's surface'. The restrictive view of the landscape morphologist has frequently been adopted by the geographical field teacher, with the result that an imperfect picture and explanation of the region under study is inevitable.

Field work in American geography seems to have been identified much more with research than with teaching (Davis, 1954). Jones and Sauer (1915, p. 522), insist that 'field work raises many questions which must be solved, if at all, after leaving the field'. They see field work as a method of testing preconceived theories. For Sauer (1924, p. 20) field work was to be equated with the systematic survey method. Platt, another great exponent of field methods, said of geography: 'In the discipline a major approach has been through field study, in which geographers go directly to the source of all geographical knowledge and confront the raw and undisturbed phenomena with which they have to deal' (1959, p. 1). Not all

American geography was so systematic, as Strahler points out. W. M. Davis' qualitative approach to landscape 'appealed then, as it does now, to persons who have had little training in basic physical sciences, but who like scenery and outdoor life' (Strahler, 1950, p. 209). Sauer (1956, p. 295), reviewing his own professional life, bewailed the abandonment of geomorphological work by geographers, as this provided a strong incentive to field observation and training the eye. On the need for such a training he is in no doubt, saying, 'there are those who never see anything until it is pointed out to them' (Sauer, 1956, p. 290).

REGIONAL SURVEYING AND FIELD WORK

Geographical field work in Britain came in the 1920's and 1930's to be dominated by the regional survey movement. Although its practitioners claimed to be inspired by Frédéric le Play, they felt his influence largely through the indefatigable and versatile Patrick Geddes who interpreted Le Play, adding some of his own characteristic ideas and techniques. In spite of the fact that much of Le Play's original work (1855, p. 22) relied heavily upon interviewing, the methods of the Le Play Society, judging from their published reports of field work (e.g. Fleure and Evans (eds.) 1939) only rarely included systematic interviewing. Le Play's succinct recipe for survey method, *Lieu*, *Travail*, *Famille*, was transformed by Patrick Geddes into *Place*, *Work*, *People*, or *Folk*. Although Geddes provided a framework for social investigation, in the hands of geographers in foreign lands the intimate contact of the investigator with the family on the land was forgotten.

An earlier manifestation of the regional survey movement sprang from summer meetings organized by Geddes in Edinburgh from 1887. Allied to these discussions which drew such scholars as Herbertson, Fleure, Fawcett, George Chisholm and Elisée Reclus (Fagg, 1928–9), was the 'Outlook Tower' which provided through its exhibits, stained-glass windows and *camera obscura* a visual synthesis of the surrounding region (Boardman, 1944, pp. 177–92). The regional survey method emanating from this source bore some similarities to the field techniques of the naturalists. They both relied essentially on visual observation, training novices to perceive features and relationships not seen by ordinary men. The regional survey of the 1920's which was inspired by Geddes (Fagg and

Hutchings, 1930, p. 35) was held by Fagg (1915, p. 24) to be 'the organized study of a region and its inhabitants . . . from every conceivable aspect, and the correlation of all aspects so as to give a complete picture of the region'. A key step in collecting information about a region is the undertaking of a surface utilization survey (Fagg and Hutchings, 1930, Ch. V). Even the collection of more data than we actually require was justified on the grounds that specialized workers may have missed their significance at first (Fagg and Hutchings, 1930, p. 64).

After recording the facts of the use to which the land is put the next step is interpretation. This is difficult without instruction in the methods of field geography, such as geographical landscape drawing, where the very act of drawing a landscape is an exercise in interpretation (Hutchings, 1960, p. 6). Nevertheless, there still remains the danger of ignoring factors and interpretations invisible even to the trained eye. Buchanan (1952, p. 4) warns against the risk of thinking along with the landscape purist 'that the landscape carries with itself the answers to the questions it asks'. It would be a mistake to explain a land-use pattern simply in terms of altitude, relief, soil texture and drainage, important though they be. Just as important are the size and scale of enterprise of the farms whose land-use pattern is being examined. Furthermore, the characteristics of the farmer himself may well affect his decisions to plough, to sow a straw crop or a cash root such as potatoes (Butler, 1960, p. 36). Comparing the map with the ground will hardly lead us to the right answers in this case. The map is a generalization of the landscape, with features such as names and boundaries added. Even the landscape is only a selection of reality, the universe with which we must deal. Field teaching which trains us to squeeze more out of the landscape and to use the map to help synthesize features of a wider area than we can see at any one time, is an essential preliminary to field research. It should not be mistaken for it. In field research the map is merely a tool, one of several useful for solving problems.

FIELD WORK AND LAND-USE SURVEYS

The effects of the influence of field teaching on geographical research can be seen in much British work on land use. In contrast, much American land-use work had a definite purpose from the outset, whereas the origins of British land-use studies may be traced to the regional survey, which had a purely general academic interest until the town planners discovered its value (Pepler, 1925, p. 80). Although some of the earliest land-use surveys in the United States were academic, in origin they sprang from purposeful attempts to test hypotheses on the nature of the relationship between the environment and the cultural landscape (Whittlesey, 1925, p. 187). In an early paper on land-use survey, Sauer (1919, p. 51) points out that such work can provide the geographer with a definite purpose and that 'it eliminates facile generalizations and the plausible and insecure reconnaissances that constitute too large a part of geographical literature'. In Britain on the other hand, the surface utilization survey provided a systematic method of accumulating a mountain of information about various aspects of a region, and could be undertaken by schoolchildren and students. Thus it fitted very neatly into field teaching as an exercise that could be done with a minimum of supervision. It concentrated for obvious reasons on the visible aspects of the landscape. It followed that with the influence of field teaching so strong and with the prevalent enthusiasm for man/land relationships, much but not all of the interpretation was in terms of the distribution of elements of the natural environment, and occasionally socio-economic or what were vaguely termed historical factors. The chief method of interpretation seems to have been the comparison of the land-use map with maps of other distributions (Stamp, 1960, p. 56).

It is instructive to trace the development of land-use studies in Britain from the days of pilot surveys often done by individual members of the Geographical Association to the flowering and final fruition of the Land Utilisation Survey of Britain. When Stamp became the Chairman of the Geographical Association's Regional Survey Committee he was able to channel a good deal of enthusiasm for regional survey into the systematic survey of land use of the whole of Britain (Stamp, 1948, pp. 3–4). What is often overlooked is the growing interest in the countryside during the inter-war era. Whether the townsman sought out bucolic pleasures through hiking, cycling, the Youth Hostel movement, or through the increasingly popular small family car, the effect was the same. Some like Sir George Stapledon were urging that this invasion of the countryside should furnish an opportunity for instruction. Conditions called for a national policy for land use (Stapledon, 1935, p. 7). The exigencies of planning for post-war reconstruction did more than anything else to confer upon land-use survey the status of a primary and fundamental source of information for physical planning of future development (Stamp, 1951, p. 5). The motives for the second national survey of land use are stated to be mainly intellectual curiosity, a desire to know and understand the face of our country and especially its regional contrasts (Coleman, 1960, p. 347). Furthermore, the first survey is now out of date and needs revision if it is to be of value for planning.

However, it would be quite misleading to suggest that American work in land-use survey did not have an academic purpose or did not spring from a curiosity in the countryside. An active band of young geographers was led by Sauer, who pioneered a land-use survey in Michigan before 1920 (Sauer, 1917). He saw in intensive field work, or geographic survey, a way of making 'the sprightly sketchiness of observation of the geographer-traveler' more profound and objective (Sauer, 1924, p. 25). The major task of this method was to represent the way in which land was utilized. The keenness to study the 'superficially familiar scene' (Sauer, 1924, p. 32) was characteristic of other Mid-Western geographers and is reflected in the pages of the *Annals of the Association of American Geographers* in the 1920's. Indeed, Jones and Finch (1925) record the activities of a group of like-minded geographers who met to study these questions in the field, in 1924 and 1925. They agreed that the synthesis of economic life and natural environment which was achieved in land-use survey and recorded on a map of land use and physical features 'compels the observer to group together in the field phenomena which occur together and thus is much superior to synthesis in the office of related facts' (Jones and Finch, 1925, p. 151). A number of other experimental studies were undertaken, for example by Whittlesey (1925). But, in the United States, it did not become possible for even these enthusiastic geographers to command sufficiently large resources for field mapping to complete a countrywide land-use survey. In any case, coverage of topographic maps of suitable scale was not available. Nor is there any evidence that such a complete survey was ever contemplated. In fact there are very early signs that it was realized that intensive field study was only possible in restricted areas 'of strong geographic individuality and high interest' (Sauer, 1924, p. 31). This necessitated field work at two levels, the less intensive for large areas.

In Britain, on the other hand, Stamp saw that the completion of a standardized survey of land use was perfectly within the bounds of possibility. The good topographic map coverage and goodwill of local education authorities, followed by the generosity of local patrons were not insignificant in allowing the survey and its publication to go forward. Thus, the very different conditions under which land-use survey was being pioneered in America and Britain, were almost bound to affect its later development as a geographical field technique. This situation had two main effects. American work was the first to develop methods to save time, effort and money by the increasing use of sampling techniques and similar statistical devices. But, in Britain, there was no such incentive to use these economical techniques, so that the major effort went into completing the mapping. In this lay the great danger that the use of new methods to interpret the patterns thus produced were completely overshadowed by the sheer exhaustion of completing and checking the map. Interpretation was thus relegated to second place. Although American work on land use was not to find newer methods of explaining patterns until the 1950's, sampling techniques were no new thing. Sauer (1956, p. 298), when reflecting on the training of geographers, is impelled to point out the disadvantages of wasting valuable field time. 'Mapping soon runs into diminishing returns. . . . Routine may bring the euphoria of daily accomplishment as filling in blank areas. . . . Time consuming precision of location, limit, and area is rarely needed.' He placed the 'unit area' scheme of mapping 'below almost any other expenditure of effort'. These trends of thought exemplify the willingness with which American workers have sought new attitudes to and methods for land-use survey. As a result their field and mapping techniques have been equally applicable to research and teaching. This is far from true of the conventional attitudes to land-use survey among British geographers.

THE EFFECT OF REGIONAL SURVEY ON LAND-USE STUDIES

From our mid-century vantage point we can now see that British work on land-use survey came out of the regional survey stable and was primarily concerned with the production of an inventory of the facts of land occupance. In this way the British effort was spread out over a relatively wide front, the effort being spent much as the power of storm waves is dissipated on a wide, shallow sandy beach. Such an attack on the problems of agricultural geography failed to bring to light any methods suited to the interpretation of some of the fundamental aspects of agricultural patterns. It actually held back research in this part of geography. That land use and agriculture were regarded as interchangeable terms is clear from an examination of the reports of the Land Utilisation Survey and other work, for instance in New Zealand (Fox, 1956, p. 9). In particular, land-use regions were being differentiated on a basis of agricultural criteria of wider application than the form that use of land took. A notable exception to this tendency is found in Wooldridge's report (1945) on land use in the North Riding of Yorkshire, where the land-use regions are clearly delineated on a basis of land-use patterns.

The author's own work in South Africa (Board, 1962, p. 170) although paying lip service to the need for such a procedure goes on to point out that some boundaries of land-use regions were drawn so as to 'coincide with the transition from one type of farming to another'. The names of several of the land-use regions (e.g. Chalumna Native Trust Agricultural Area) are patently more closely related to the interpretation of the land-use pattern than to the pattern itself. This illustrates the essential difficulty – the impossibility of interpreting land-use patterns in any terms other than farming type areas, or agricultural regions. Only where the type of farming is homogeneous over the whole area of study (a very rare occurrence) is there a chance of the interpretation being made principally in terms of, say, the physical environment. In effect in the Border Region (Board, 1962) the distribution of different types of farming was the subject of interpretation. The land-use pattern was merely one manifestation of that distribution; perhaps the best single guide, but certainly not the only one. At the time of the completion of that survey the author, like many others, was unaware of the importance of work in America which was to make possible the more satisfactory, rigorous interpretation of land-use patterns.

The two reasons why land-use survey held back progress in research into the geography of agriculture were, first, this difficulty of finding objective methods by which to interpret the pattern of land use, and secondly, the excessive concentration on morphological aspects of land use, to the exclusion of the processes which moulded the agricultural landscape. This latter was particularly the inheritance of the regional survey movement.

These trends are combined in the difficulty which faces all workers intending to explain the pattern of land use in a given area. Standard practice was to analyse elements of the pattern, one by one, mapping each separately. Major factors thought to be responsible for the distribution were then isolated and compared with the land-use pattern. Stamp (1960, p. 56) points out that this was achieved by mapping these factors and comparing the succession of maps. 'It is often difficult to do this with a series side by side, and hence a very common device is to use a series of transparencies, where certain facts can be printed on transparent paper and two or more maps placed one above the other.' This procedure was a very useful by-product of the preparation of maps to illustrate the county reports (Wilson, 1964), the separation drawings of individual elements being drawn at the scale of 1:63,360 on tracing paper. The method of interpretation, however, remained subjective.

A further problem was introduced by the fact that interpretation had to be made of the static distribution of a constantly changing pattern. Changing patterns could be studied only by re-mapping (which could scarcely be contemplated), or by case-studies using maps of land use at different times. This could easily lead to short-term responses to changing economic conditions being ignored. Such changes were studied chiefly through the medium of agricultural statistics, as published for whole counties, or more rarely for individual parishes. It was naturally quite difficult to integrate such figures into a pattern of land use at a particular time. Coppock's studies of agricultural changes in the Chilterns (1957, 1960) include parish maps showing changes over comparatively short periods. But these can be related to land-use changes only in the text and by implication. Maps of changes in land use appear separately (Coppock, 1954, p. 130). The period of change is, however, a relatively long period of time and is entirely conditioned by dates when surveys happened to be done. Such information is of course very useful, but it tends to favour the identification of long-term trends. Classic examples of the use of historical material on land use are seen in Willatts' report (1937) on Middlesex and the London Region. Nevertheless, the difficulty remains that the explanation of changes, or indeed of the pattern at any one time, depends upon a subjective process of visual comparison, even though it is simplified by the close juxtaposition of maps of geology, relief and land-use elements. Thus, it is difficult to show how the land-use pattern is a reflection of the interaction of the farmer's resources, choice, ability to adapt to changing technology and price levels. This problem has usually been overcome by the parallel study of the specimen farm (Stamp, 1948, p. 348).

THE IMPORTANCE OF VISIBLE FEATURES IN LAND-USE STUDIES

Allied to these disadvantages of traditional methods is over-concentration on visible relationships between land use at the time of survey and elements of the physical environment. It is contended that this derives in part from the methods adopted by the regional survey school of geographers. It is also related to this difficulty of relating land use to the less obvious factors affecting it. If complete mapping is the objective of a land-use survey, there is rarely time for a thoroughly adequate inquiry into the practices and motives of in-individual farmers. By concentrating on the factors which are relatively easy to map and see in the field, the land-use surveyor runs the risk of underestimating the importance of social and economic factors, which generally operate in a more subtle way. This was well appreciated by the forester Bourne (1931, p. 52), whose classic work on regional survey recognizes the need for 'farm survey' as well as 'regional survey' because 'from the point of view of the farm manager this simple Regional Survey of a developed area does not bring to light all the combinations of conditions which constitute existing farms'. It cannot even be accepted that every farmer is motivated by a desire to maximize his profit (Butler, 1960, p. 45). Even if he is, his knowledge of improved techniques and his ability to introduce them into his existing farming system, or into a modified farming system, will vary from farm to farm independently of the natural environment. Similarly, changes in relative price levels, as for instance between wheat and barley, may lead to the cultivation of barley on soils normally regarded as too heavy and more suited to wheat. Buchanan (1959, p. 10) has reminded us that: 'The crop that best fits the physical conditions will be the preferred crop if, and only if, its advantage in physical yield over its competitors is reflected in an advantage in financial yield.' The ploughing up of much heavy land west of Cambridge which carried the ridge and furrow indicative of an earlier phase of cultivation, was made possible on the one hand by the system of deficiency payments for cereals and, secondly, by the employment of heavier, crawler tractors financed by forward-looking farmers. The wheat and barley acreages are adjusted from season to season, according to prevailing prices. A final illustration will suffice to emphasize that the study of land use cannot be made without taking into account the traditional restraints exercised in some areas by a land tenure system. It has been argued that the legislation to protect the crofting system, conferring a large measure of security of tenure, has been responsible for the failure to make use of the holdings of absentee tenants. In some townships land is left uncultivated for this reason when in general the size of the holdings of crofters is too small for them to be viable (Scotland, Cmd. 9091, 1954, pp. 40, 41). Such an explanation for uncultivated arable land would be obtainable only by inquiry, and generalizations only by survey or commission procedure. It is heartening to see that some modern surveys, under the auspices of the Geographical Field Group (Moisley, 1961, p. 30), are carrying out surveys of land use supplemented by surveys of farmers (crofters in this case) and their families. What is perhaps also interesting about these surveys is that the Group responsible for their organization is a lineal descendant of the Le Play House student group which had the blessing and encouragement of Geddes (Geographical Field Group, 1962).

One further problem facing those who seek to interpret land-use patterns is not peculiar to British studies. In the attempt to analyse that aspect of pattern which concerns the shape of individual elements of the pattern, there has been little advance beyond the subjective approach indicated by Stamp (1948, pp. 88–89), where complementary patterns of arable land and permanent pasture are examined side by side. The use of words for description puts a strain on the interpreter because of their limited range and the subjective way in which they can be applied. The search for a more consistent method of describing shape in land-use patterns has already begun in America (Latham, 1958 and 1963).

So far British work on land-use survey and mapping has reached a cul-de-sac. Although complete surveys are feasible, we are not much nearer to being able to interpret the patterns thus obtained, so that the relative importance of different causal factors can be weighed.

THE SEARCH FOR SPEED AND EFFICIENCY IN AMERICAN LAND-USE STUDIES

In contrast with British work, it is contended that conditions in America were conducive to the early development of new methods, devised to overcome some of these obstacles and to take land-use studies out of the purely descriptive, or speculative, phase into a phase of greater certainty in the field of explanation. Because it was not tempted by the possibility of producing complete surveys,

American work turned to sampling as a way of accumulating more information about wider areas. Even Finch in his classic work on the Montfort region (1933, p. 9) pointed out that the value of statistics of land use determined from a field-by-field study 'seemed hardly to warrant the expenditure of the time necessary for their preparation. A more ambitious project might recommend the use of Hollworth [*sic*; for Hollerith?] machines or other mechanical devices in a more facile accomplishment of this tedious statistical analysis.' The geologist Trefethen (1936) quickly took this matter up and, using the method developed by Rosiwal (1898) by line traverses in two directions at right angles to each other, was able to produce land-use statistics for 50 to 100 square miles for a week's work. This he contrasted with Finch's 47 square miles in 120 man-days. Thus a technique first used for petrographic analysis was put to use in geography. It was not long before Proudfoot (1942) and Osborne (1942) continued this work. Proudfoot compared the accuracy of traverse line sampling as against area measurement of land-use type by planimeter. Osborne's attention was drawn to the relative efficiency of types of line traverse sample – the stratified random and the systematic, concluding that the latter were more so. Some years later, Steiner (1958) also compared the efficiency of systematically and randomly located area samples, by comparing land-use statistics thus obtained with quantities known from other sources. Wood (1955) has made use of a stratified random sample of areas so that aspects of land use may be mapped in the field over a limited area representative of Eastern Wisconsin. The data for these areas indicated that another complete source of land-use statistics could be used to make a map of the whole area with an ascertainable reliability. Dot maps of different land-use types were thus produced by a combination of sample field mapping and correlation with available statistical data. Anderson (1961) has pointed out that the United States National Inventory of Soil and Water Conservation Needs (1962) has made use of a 2 per cent area sampling rate to speed the field collection of data on soil and land use. In this way, the whole country may be covered by a uniform survey. Arising from the programme of investigation into flood-plain problems being carried out by Chicago University, Berry (1962) has compared the relative efficiency of different sampling systems for accumulating land-use statistics. He concludes that the stratified, systematic, unaligned sample of points at which a record of land use is obtained is the most efficient. Berry, however, makes the interesting comment (1962, p. 14) that improved, speedier methods of field work are not provided by sampling. Nevertheless, this method does provide a quick and unambiguous way of estimating the area of a particular category of land use, it side-steps the problem of the detailed delimitation of the actual area of separate categories of land use, and most important of all, it is versatile in that it facilitates the interpretation of the land-use pattern through a study of spatial associations.

BRITISH STUDIES TAKE UP THE SEARCH

Work in Britain has been principally concerned with the assessment of woodland area by sampling. The Forestry Commission have been carrying out some experiments with different types of random samples in order to establish to within ± 2 per cent the area of woodland in different regions of Britain (Locke, 1963). Haggett (1963) sampled woodland cover shown on 1:63,360 Ordnance Survey maps to test the relative efficiency, both from the point of view of accuracy and time taken for obtaining area measurements. Furthermore, the tests were designed to indicate the most appropriate sampling methods for both regional and local levels of sampling. Haggett found that at the regional level, the area of woodland cover could be estimated to within ± 1 per cent by using sixteen cells of 5 square kilometres per one 100 km by 100 km square, selected by stratified random sampling. At the local level, 40-line traverse samples were preferred, since they gave a mean error of $< \pm 1$ per cent forest cover.

THE INTERPRETATION OF LAND-USE PATTERNS

It has been emphasized that the use of sampling in many cases enables more rapid accumulation of land-use statistics. Over wider areas, where land-use patterns are of coarser grain this may be more important still. Furthermore, as Berry (1962) noticed, sampling designs help in the progression to the next stage, that of explaining the patterns of land use. It is always possible to sample the maps which are the product of a complete survey, in order to test the relationship between aspects of that pattern and factors thought to have influenced it. But unless there is an actual need for the map of land use, for whatever purpose, there seems little point in laboriously collecting and plotting the distribution of land use, if sufficient data can be collected in another way. As yet, there have been comparatively few attempts to explain land-use patterns in a consistent and non-subjective fashion. Wood's study of land use in Wisconsin refers to the inverse relationship between crop land and farm land in trees, which enabled him to predict the area of the first from the second. He comments significantly: 'neither of these two ingredients of the landscape may be expected to cause the other; nevertheless they are closely related' (1955, p. 364). Zobler appears to be one of the first to have suggested a method of relating natural factors to land use, 'the key to the understanding of the geographic adjustment to agricultural resources' (1957, p. 89). Zobler obtained land-use acreages from air photographs in part of New Jersey. The area was divided into physiographic regions and a number of other factors likely to have influenced land use were also noted. Included in the series of χ^2 tests, is one to see whether the acreages of different types of land use differ significantly from purely chance variations as between all possible pairs of physiographic regions. It was possible to see that in some cases an apparent difference in physiography was not reflected in differences in land use. Another series of tests were made to see whether land-use variations occurred with reference to type of soil. Berry (1962) goes further than Zobler by expressing the degree of the association between one land-use category and an aspect of the natural environment. He makes use of frequencies (presence or absence of permanent pasture) obtained by a point sampling design. Relating these frequencies to the presence or absence of flood plain, and employing the coefficient ϕ to express the degree of association, he applies the χ^2 test to check the significance of the relationship. In this case the value of ϕ ranges between 0 and 1. If the coefficient Q (Hagood and Price, 1960, p. 358–70) were to be used, by extending the range of values of the cofficient to -1 through to $+1$, one is able to allow for unexpected relationships (negative values) as well as expected relationships (positive values). Although Berry does not map the occurrences of pasture which do not fall on flood-plain, or vice versa, it would be quite feasible and indeed illuminating to map such so-called anomalies. The field investigator then may well feel that concentrations of exceptions such as these warrant further inspection. In Britain, only Haggett (1964) and Reid (1963) have so far used statistical techniques to interpret land-use patterns. Both used multiple regression in an attempt to assess the relative influence that different factors had on proportions of different kinds of land use. Reid obtained his data by sampling land-use maps. Haggett was examining the factors responsible for the extent of deforestation in part of Brazil. With the results of a second land-use survey now being published in Britain, it is important that the methods used by Haggett and particularly Reid should be widely appreciated. Since high-speed electronic computers are becoming part of the equipment of modern geographical investigations (Coppock, 1962 and 1964) even in Britain, it is not too much to hope that complicated exercises such as these will become commonplace methods of dealing with the multiple explanations of land-use patterns.

FARM SURVEYS AS AN ALTERNATIVE TO LAND-USE SURVEYS

Sampling and statistical analysis have long been employed in another field of agricultural investigation. The economic surveys of farming, normally based upon a sample of farms, is well known both in America and in Britain. It is only relatively recently that such techniques have been applied to geographical investigations of patterns of farm-type. Birch's study of the Isle of Man (1954) is almost the only British example. There are a number of similar studies which have been carried out by agricultural economists in recent years. These include Bennett Jones' study of the pattern of farming in the East Midlands (1954), Mitchell's work on the dairy farms of the Somerset Levels (1962) and Jackson, Barnard and Sturrock's study of the distribution of types of farm in Eastern England (1963). These studies are not fundamentally interested in land use, but in the pattern of the functional units of the agricultural landscape (Birch, 1954, p. 144). Sample surveys of the type developed by agricultural economists may well be used to supplement land-use surveys. They certainly provide a depth of understanding rarely achieved in the more subjective interpretation of land-use patterns and in the use of specimen farms. At least it is possible to say that the data from most sample surveys is representative of a certain area, or class of farms; the selection of case studies, whether specimen parishes, or farms, is frequently conditioned by the availability of data, or the willingness of the farmer to provide information. At the same time, it is difficult to see how the two approaches can be fused into one method of describing and interpreting agricultural patterns. The land-use survey, with its emphasis on area *qua* area is difficult to marry with the farm survey approach, whose emphasis lies on the operating unit. The one is more characteristic of regional geography, the other, of economic geography. Both seem equally valid in their particular fields.

The problem may be illustrated by examining what would happen if a stratified, systematic, unaligned point sample of land use were made of a certain area. Each occurrence of a category of land use could be associated with a continuous distribution of, say, a natural phenomenon, such as relief. It would not be possible to associate these measures of land use with size of farm, sampled on the same basis, because of the strong likelihood of double counting. Some way would have to be found to generalize the information on farm size, and perhaps land use too, so that the two patterns may be associated by virtue of their distributions. Reid (1963) suggests that mapping of sample data of land use by point sampling is less successful than line sampling, and this is also Dahlberg's impression (1963). Line sampling of both land uses would yield a map of proportions of different uses, but line sampling of farms to relate to that distribution would produce some very curious results because of the fragmented character of many farms, quite apart from their irregular shape. What would be considered the area of a farm could best be obtained by inquiry from the operator himself. It would appear that line sampling would not be an efficient way of collecting data on farm size. The most obvious method would be to generalize from a systematic point sample of farms, with the object of producing a map of farm size of a certain reliability; but it would have the disadvantage of over-representing the large farms. In view of these difficulties, one is forced back to the traditional method of sampling farms, from lists of farms or farm operators. Farm size, if mapped on a basis of this sample, would be valid only for the area from which the sample was drawn, but it would be representative of the whole farm population.

There is ample scope here for pilot investigations into the problems of mapping and interpreting sample data over area. Whatever the future has in store in the way of new techniques of mapping and interpretation, there will still be a place for the traditional, complete mapping of land use. The difficulties encountered by Latham (1958, 1963) in an attempt to work out satisfactory measures of shape and orientation for elements in land-use patterns reinforce this view. Although Latham demonstrates that it is possible to express such aspects of land-use patterns, by means of 'uninterrupted distance' measurements along rotated sample traverse lines, such measures are useful only for regional summaries (Latham, 1958, p. 254). There is, as yet, at the detailed local scale, no substitute for the field by field land-use map. But the analysis of such maps has been considerably aided by the development of modern methods.

CONCLUSION

Although it may be fairly claimed by Hunt that land-use survey is 'a means of carrying economic geography into the field' (1953, p. 284), it is more extravagant to insist that, apart from the land-use map, 'there is no other way of showing the distribution of *all* significant activities in their true spatial context' (Hunt, 1953, p. 285). Herein exists the confusion that has arisen between field teaching and field research. Hunt is in fact arguing that land-use survey is of great value in field teaching, but this is not to say that land-use survey has no place in field research. Similarly, it is contended that field teaching is an essential preliminary to field research. No one can expect to embark on field work for a research project unless he has first taken the trouble to understand the significance of the features which are visible in the landscape. As Marsh put it 'Sight is a faculty; seeing, an art' (1864, p. 10). One cannot map river terraces or temporary pasture until one can recognize them. If field research is to be regarded as the preserve of past masters in the art, instruction through field teaching is the means by which one may aspire to membership of the brotherhood. It is only by the exercise to the fullest extent of our powers of intuition, observation and discrimination that we shall achieve a deeper understanding of patterns on the surface of our earth.

References

ANDERSON, J. R., 1961, 'Toward More Effective Methods of obtaining Land Use Data in Geographic Research', *Prof. Geog.*, **13** (6), 15–18.

BALCHIN, W. G. V., 1955, 'Research in Geography' (Swansea), 23 pp.

BARROW, G., 1919, 'Some Future Work for the Geologists' Association', *Proc. Geol. Assn., Lond.*, **30**, 1–48.

BARROW, G. and GREEN, J. F. N., 1921, 'Excursion to Wendover and Buckland Common Near Cholesbury', *Proc. Geol. Assn., Lond.*, **32**, 32–46.

BEAVER, S. H., 1962, 'The Le Play Society and Field Work', *Geog.*, **47**, 226–39.

BERRY, B. J. L., 1962, *Sampling, Coding and Storing Flood Plain Data*, U.S. Department of Agriculture Handbook No. 237 (Washington, D.C.), 27 pp.

BIRCH, J. W., 1954, 'Observations on the Delimitation of Farming Type Regions with Special Reference to the Isle of Man', *Trans. Inst. Brit. Geog.*, Pub. No. **20,** 141–58.

— 1960, 'A Note on the Sample-Farm Survey and Its Use as a Basis for Generalized Mapping', *Econ. Geog.*, **36,** 254–9.

BOARD, C., 1962, *The Border Region: Natural Environment and Land Use in the Eastern Cape* (Cape Town), 238 pp.

BOARDMAN, P., 1944, *Patrick Geddes, Maker of the Future* (North Carolina), 504 pp.

BOURNE, R., 1931, 'Regional Survey and Its Relation to the Stocktaking of the Agricultural and Forest Resources of the British Empire', *Oxf. For. Mem.*, **13,** 169 pp.

BRANFORD, V. V., 1915–16, 'The Regional Survey as a Method of Social Study', *Geog. Teach.*, **8,** 97–102.

BRUNHES, J., 1952, *Human Geography* (translated from the French abridged edition of 1947) (London), 256 pp.

BUCHANAN, R. O., 1952, 'Approach to Economic Geography', *Ind. Geogr. Jour.* (Madras), Silver Jubilee Souvenir Volume, 1–8.

— 1959, 'Some Reflections on Agricultural Geography', *Geog.*, **44,** 1–13.

BUTLER, J. B., 1960, *Profit and Purpose in Farming. A Study of Farms and Smallholdings in Part of the North Riding*, Leeds, Economic Section, Department of Agriculture, Leeds University.

COLEMAN, A. M., 1960, 'A New Land-use Survey of Britain', *Geog. Mag., Lond.*, **33,** 347–54.

COPPOCK, J. T., 1954, 'Land Use Changes in the Chilterns, 1931–51', *Trans. Inst. Brit. Geogr.*, Pub. No. **20,** 113–40.

— 1957, 'The Changing Arable in the Chilterns, 1875–1951', *Geog.*, **42,** 217–29.

— 1960, 'Crop and Livestock Changes in the Chilterns, 1931–51', *Trans. Inst. Brit. Geog.*, Pub. No. **28,** 179–96.

— 1962, 'Electronic Data Processing in Geographical Research', *Prof. Geog.*, **14**(4), 1–14.

— 1964, 'Crop, Livestock, and Enterprise Combinations in England and Wales', *Econ. Geog.*, **40,** 64–81.

DAHLBERG, R. E., 1963, Personal Communication.

DAVIS, C. M., 1954, 'Field Techniques', Ch. 24 in James, P. E., Jones, C. F. and Wright, J. K. (eds), *American Geography: Inventory and Prospect* (Syracuse).

DICKENS, CHARLES, 1890 edn, *Bleak House* (London).

EDWARDS, K. C., 1950, *Land, Area and Region* (Nottingham), 20 pp.

FAGG, C. C., 1915, 'The Regional Survey and the Local History Societies', *South East Nat.*, **20**, 21–30.

— 1928–9, 'The History of the Regional Survey Movement', *South East Nat.*, **33**, 71–94.

FAGG, C. C. and HUTCHINGS, G. E., 1930, *An Introduction to Regional Surveying* (Cambridge), 150 pp.

FAIRGRIEVE, J., 1937 A, 'Can We Teach Geography Better?', *Geog.*, **21**, 1–17.

— 1937 B, *Geography in School* (London), 417 pp.

FINCH, V. C., 1933, 'Geographic Surveying', in Colby, C. C. (ed.), Geographic Surveys, *Bull. Geog. Soc.* Chicago, **9**, xiii, 75 pp.

FLEURE, H. J. and EVANS, E. E. (eds), 1939, *South Carpathian Studies, Roumania*, II, Le Play Society, 60 pp.

FOX, J. W., 1956, *Land Use Survey. General Principles and a New Zealand Example* (Auckland), 46 pp.

GEIKIE, A., 1905, *Landscape in History and Other Essays* (London), 352 pp.

GEOGRAPHICAL FIELD GROUP, 1962, *A Short History of the Geographical Field Group* (Nottingham), 2 pp. (Roneoed).

GULLEY, J. L. M., 1961, 'The Retrospective Approach in Historical Geography', *Erdkunde*, **15**, 306–9.

HAGGETT, P., 1963, 'Regional and Local Components in Land-use Sampling: A Case-study from the Brazilian Triangulo', *Erdkunde*, **17**, 108–14.

— 1964, 'Regional and Local Components in the Distribution of Forested Areas in South-East Brazil: A Multivariate Approach', *Geog. Jour.*, 130, 365–380.

HAGOOD, M. J. and PRICE, D. O., 1960, *Statistics for Sociologists* (New York), 575 pp.

HARE, F. K., 1947, 'The Geomorphology of a Part of the Middle Thames', *Proc. Geol. Assn., Lond.*, **58**, 294–339.

HODDER, B. W., 1961, 'Rural Periodic Day Markets in Part of Yorubaland, Western Nigeria', *Trans. Inst. Brit. Geog.*, Pub. No. **29**, 149–59.

HUNT, A. J., 1953, 'Land-use Survey as a Training Project', *Geog.*, **38**, 277–86.

HUTCHINGS, G. E., 1960, *Landscape Drawing* (London), 134 pp.

— 1962, 'Geographical Field Teaching', *Geog.*, **47**, 1–14.

INCORPORATED ASSOCIATION OF ASSISTANT MASTERS IN SECONDARY SCHOOLS, 1954, *The Teaching of Geography in Secondary Schools* (London), 512 pp.

JACKSON, B. G., BARNARD, C. S. and STURROCK, F. G., 1963, *The Pattern of Farming in the Eastern Counties. A report on Classification of Farms in eastern England.* Occasional Papers, Farm Economics Branch, School of Agriculture, Cambridge University, No. 8, 60 pp.

JONES, R. BENNETT, 1954, *The Pattern of Farming in the East Midlands* (Sutton Bonington), University of Nottingham School of Agriculture, Department of Agricultural Economics, 176 pp.

JONES, W. D. and FINCH, V. C., 1925, 'Detailed Field Mapping in the Study of the Economic Geography of an Agricultural Area', *Ann. Assn. Amer. Geog.*, **15**, 148–57.

JONES, W. D. and SAUER, C. O., 1915, 'Outline for Field Work in Geography', *Bull. Amer. Geog. Soc.*, **47**, 520–5.

KIRK, W., 1952, 'Historical Geography and the Concept of the Behavioural Environment', *Ind. Geogr. Jour.* (Madras), Silver Jubilee Souvenir Volume, 152–60.

LATHAM, J. P., 1958, *The Distance Relations and Some other Characteristics of Cropland Areas in Pennyslvania: An Experiment in Methodology for Empirically Analyzing, Regionalizing and Describing Complexly-Distributed Areal Phenomena*, Philadelphia University of Pennsylvania, Wharton School of Finance and Commerce. Technical Report No. 4, NR No. 389–055, Office of Naval Research.

— 1963, 'Methodology for an Instrumented Geographic Analysis', *Ann. Assn. Amer. Geog.*, **53**, 194–211.

LE PLAY, P. G. F., 1855, *Les Ourvriers Européens* (Paris), 301 pp.

LINTON, D. L., 1946, *Discovery, Education and Research* (Sheffield), 17 pp.

1955, 'The Problem of Tors', *Geog. Jour.*, **121**, 470–87.

OCKE, G. M. L., 1963, Personal Communication.

CCANN, S. B., 1963, 'The Late Glacial Raised Beaches and Re-Advance Moraines of the Loch Carron Area, Ross-shire', *Scot. Geog. Mag.*, **79**, 164–9.

MAITLAND, F. W., 1897, *Domesday Book and Beyond* (Cambridge), 527 pp.

MARSH, G. P., 1864, *Man and Nature; or Physical Geography as modified by Human Action* (London), 560 pp.

MEAD, W. M., 1954, 'Ridge and Furrow in Buckinghamshire', *Geog. Jour.*, **120**, 34–42.

MITCHELL, G. F. C., 1962, 'The Central Somerset Lowlands: the Importance and Availability of Alternative Enterprises in a Predominantly Dairying District', *Selected Papers in Agricultural Economics* (Bristol), **7**, 295–453.

MOISLEY, H. A., 1961, *Uig, a Hebridean Parish*; Parts I and II (Nottingham), Geographical Field Group, iv, 55 pp.

MONKHOUSE, F. J., 1955, *The Concept and Content of Modern Geography* (Southampton), 31 pp.

OSBORNE, J. G., 1942, 'Sampling Errors of Systematic and Random Surveys of Cover-type Areas', *Jour. Amer. Statist. Assn.*, **37**, 256–64.

PALMER, J., 1956, 'Tor Formation at the Bridestones in North-east Yorkshire, and Its Significance in Relation to Problems of Valley-side Development and Regional Glaciation', *Trans. Inst. Brit. Geogr.*, Pub. No. **22,** 55–71.

PALMER, J. and NEILSON, R. A., 1962, 'The Origin of Granite Tors on Dartmoor, Devonshire', *Proc. Yorks. Geol.* (*Polyt.*) *Soc.*, **33,** 315–40.

PEPLER, G., 1925, 'Regional Survey as a Preliminary to Town Planning', *South East Nat.*, **30,** 81–89.

PLATT, R. S., 1942, *Latin America: Countrysides and United Regions* (New York), 564 pp.

— 1959, *Field Study in American Geography, The Development of Theory and Method exemplified by Selections*, Chicago, University of Chicago, Department of Geography, Research Papers No. 61, 405 pp.

PROUDFOOT, M. J., 1942, 'Sampling with Transverse Traverse Lines', *Jour. Amer. Statist. Assn.*, **26,** 265–70.

PYE, N., 1955, *Object and Method in Geographical Studies* (Leicester), 19 pp.

REID, I. D., 1963, *The Application of Statistical Sampling in Geographical Studies* (Liverpool University, M.A. thesis), 112 pp.

ROBERSON, B. S. and LONG, M., 1956, 'Sample Studies: The Development of a Method', *Geog.*, **41,** 248–59.

ROSIWAL, A., 1898, 'Ueber geometrische Gesteinen analysen. Ein einfacher Weg zur ziffermassigen Feststellung des Quantitatsverhaltnisses der Mineralbestandtheile gemegter Gesteine', *Verh. geol. Reichs Anst. Wien*, 1898 (5 & 6), 143–175.

RUSSELL, E. J., 1925–6 (1926), 'Regional Surveys and Scientific Societies', *Geographical Teacher*, **13,** 439–47.

SAUER, C. O., 1917, 'Proposal of an Agricultural Survey on a Geographical Basis', *19th Annual Report of the Michigan Academy of Science*, pp. 79–86.

— 1919, 'Mapping the Utilization of the Land', *Geog. Rev.*, **8,** 47–54.

— 1924, 'The Survey Method in Geography and Its Objectives', *Ann. Assn. Amer. Geog.*, **14,** 17–33.

— 1925, 'The Morphology of Landscape', *Univ. Calif. Pub. Geog.*, **2** (2), 19–53.

— 1941, 'Foreword to Historical Geography', *Ann. Assn. Amer. Geog.*, **31,** 1–24.

— 1956, 'The Education of a Geographer', *Ann. Assn. Amer. Geog.*, **46,** 287–99.

SCOTLAND, 1954, *Report of a Commission of Enquiry into Crofting Conditions* (Cmd. 9091), 100 pp.

SPARKS, B. W., 1949, 'The Denudation Chronology of the Dip-slope of the South Downs', *Proc. Geol. Assn., Lond.*, **60,** 165–215.

STAMP, L. D., 1948, *The Land of Britain: Its Use and Misuse* (London), 507 pp.

— 1960, *Applied Geography* (Harmondsworth, Penguin Books, Ltd.), 208 pp.

STAPLEDON, R. G., 1935, *The Land Now and Tomorrow* (London), 323 pp.

STEINER, R., 1958, 'Some Sampling of Rural Land Uses' (abstract), *Ann. Assn. Geog.*, **48,** 290.

STEVENS, A. J., 1959, 'Surfaces, Soils and Land Use in North-east Hampshire', *Trans. Inst. Brit. Geogr.*, Pub. No. **26,** 51–66.

STRAHLER, A. N., 1950, 'Davis' Concepts of Slope Development Viewed in the Light of Recent Quantitative Investigations', *Ann. Assn. Amer. Geog.*, **40,** 209–13.

TREFETHEN, J. M., 1936, 'A Method for Geographic Surveying', *Amer. Jour. Sci.*, **32,** 454–64.

UNITED STATES OF AMERICA, DEPARTMENT OF AGRICULTURE, 1962, *Agricultural Land Resources: Capabilities; Uses; Conservation Needs.* Agricultural Information Bulletin, No. 263 (Washington, D.C.), 30 pp.

WATERS, R. S., 1958, 'Morphological Mapping', *Geog.*, **43,** 10–17.

WHITTINGTON, G., 1962, 'The Distribution of Strip Lynchets', *Trans. Inst. Brit. Geog.*, Pub. No. **31,** 115–30.

WHITTLESEY, D. S., 1925, 'Field Maps for the Geography of an Agricultural Area', *Ann. Assn. Amer. Geog.*, **15,** 187–91.

WILLATTS, E. C., 1937, *Middlesex and the London Region*, The Report of the Land Utilisation Survey of Britain, Part 79, 117–304.

— 1951, 'Some Principles of Land Use Planning', in *London Essays in Geography* (Rodwell Jones Memorial Volume), eds: Stamp, L. D. and Wooldridge, S. W., 289–302.

WILSON, E., 1964, Personal Communication.

WISE, M. J., 1957, 'The Role of Field Work in the University Teaching of Geography', *Journal for Geography* (Stellenbosch), **1**(1), 17–23.

WOOD, W. F., 1955, 'Use of Stratified Random Samples in a Land-Use Survey', *Ann. Assn. Amer. Geog.*, **45**, 350–67.

WOOLDRIDGE, S. W., 1927, 'The Pliocene History of the London Basin', *Proc. Geol. Assn., Lond.*, **38,** 49–132.

— 1945, *The North Riding of Yorkshire*; The Report of the Land Utilisation Survey of Britain, Part 51, 351–417.

— 1948 A, 'The Role and Relations of Geomorphology' (Inaugural lecture at King's College, London), reprinted in Stamp, L. D. and Wooldridge, S.W. (eds), 1951, *London Essays in Geography*, 19–31.

— 1948 B, *The Spirit and Significance of Field Work*; Address at the Annual Meeting of the Council for the Promotion of Field Studies, 8 pp.

— 1949 A, 'On Taking the Ge- out of Geography', *Geog.*, **34,** 9–18.

— 1949 B, 'The Weald and the Field Sciences', *Adv. Sci.*, **6**(21), 3–11.

WOOLDRIDGE, S. W., 1950, 'Reflections on Regional Geography in Teaching and Research', *Trans. Inst. Brit. Geog.*, Pub. No. **16**, 1–11.

— 1955, 'The Status of Geography and the Role of Field Work', *Geog.*, **40**, 73–83.

— 1960, *Field Studies Council: Retrospect and Prospect*; Address at the Annual meeting of the Field Studies Council, 4 pp.

WOOLDRIDGE, S. W. and EAST, W. G., 1958, *The Spirit and Purpose of Geography*, 2nd Edn (London), 186 pp.

WOOLDRIDGE, S. W. and GOLDRING, F., 1953, *The Weald* (London), 276 pp.

WOOLDRIDGE, S. W. and HUTCHINGS, G. E., 1957, *London's Countryside Geographical Field Work for Students and Teachers of Geography* (London), 223 pp.

ZOBLER, L., 1957, 'Statistical Testing of Regional Boundaries', *Ann. Assn. Amer. Geog.*, **47**, 83–95.

— 1958, 'The Distinction between Relative and Absolute Frequencies in using Chi Square for Regional Analysis', *Ann. Assn. Amer. Geog.*, **48**, 456–7.

CHAPTER ELEVEN

Field Work in Urban Areas

M. P. COLLINS

Lecturer in Town Planning, University College London

The existence of urban settlements can be traced back to the fourth millenium B.C. when they played an important part in facilitating advances in learning, the arts, technology and social organization by providing a relatively secure and economically stable environment. The emergent cultures became localized in the towns, and in time nations became known by the splendour and achievements of their cities. By virtue of their function cities have seldom remained static in either size or form for very long and now they have grown to such an extent that the contiguous built-up area constitutes a distinct region in its own right. The advent of the city region has been accompanied, however, by a host of problems which have still to be resolved. London, for example, continues to expand and attract both industry and population from other parts of the country despite the timely warning contained in the Barlow Report, and in this context it is very disturbing to learn that the metropolis is still committed to developments which will provide employment for an additional 400,000 office workers (Hookway *et al.*, 1963). Whilst the regional and national consequences of this trend have still to be evaluated, it is probable that over two-thirds of the country's population will be living in the South of England by the year A.D. 2000 (Childs, 1962).

TOWN CLASSIFICATION

Despite the urgency and complexity of these problems very little more is known about the forces which control the present-day composition, morphology and functions of our cities than is known about such ancient cities as Mohenjo-Daro and Babylon. In the field of urban studies little progress has been made in establishing definitive techniques, let alone laws, which stand the acid test of universal application. Most of the present day studies of towns and cities tend to be descriptive rather than interpretative and fail to explain the dynamics of urban evolution. Even the more sophisticated approach of the *Centre for Urban Studies* fails to come to grips with the subject when examining the character and composition of 157 urban areas (Moser and Scott, 1961). This study examines the urban areas with a population of over 50,000 inhabitants in 1951 and employs fifty-seven separate indices to establish the basic character of each urban area; such as the size and changing structure of the population, the number of households, the type and condition of housing, the social class, health, education and voting habits. This material is compared, analysed, and classified mathematically, using correlation coefficients to measure the closeness of the relationship between any given pair of variables (there being 1,596 possible pairings). Many of the relationships seem to be straightforward and predictable, whilst others are apparently irrational. Having completed this basic correlation, the results are summarized by means of four new indices which describe the characteristics of each urban area almost as well as the original fifty-seven factors. This condensation involves the use of component analysis to calculate the four components which account for at least 60 per cent of the basic differences between the 157 urban areas. Component one reflects differences in social class; component two the differences in the amount of intercensal population change; component three the differences in post-1951 developments; and component four the housing conditions. The end-product is a classification which distinguishes three main categories of urban areas: viz.

Group A: Comprised of resorts, administrative and commercial towns;
Group B: Comprised mainly of industrial towns; and
Group C: Comprised of suburbs and suburban type towns.

The use of these statistical techniques to measure the degree of association between observable phenomena removes the subjective element from the task of analysis and synthesis, but it is important to remember that they only establish the intensity of the association without attempting to define or explain it (Chadwick, 1961). In other words, a causal relationship can only be inferred with varying degrees of probability depending on the calculated value of the correlation coefficient. Viewed in this light it is difficult to subscribe to the present enthusiasm for techniques which regard 'association' as synonymous with 'causality'.

LAND USE IN URBAN AREAS

Although geographers have made laudable attempts to examine the use of land within the United Kingdom (Stamp, 1950; Best and Coppock, 1962) there has been no similar effort to examine the detailed pattern of land use within the built-up areas of towns and cities. Most of the present detailed land-use studies of urban areas derive from the Town and Country Planning Act of 1947 (Section 5, sub-section 1), which required local planning authorities to undertake a detailed survey of their respective administrative areas and keep it up to date. The Ministry of Town and Country Planning advised the local planning authorities as to the best method of conducting the survey, classifying the land uses and of presenting the results in map form (Ministry of Town and Country Planning, 1949 A). The land-use surveys which have finally emerged, however, vary considerably and do not submit readily to comparison due to differences in presentation, content and even sometimes in the basic classification. As the land-use survey constitutes a vital part in any urban study it is essential that the classification should be sufficiently broad to give a valid picture of the distribution of functional elements within the built-up area. This information should be plotted on a series of large scale maps, preferably at a scale of 1/1250, with an accompanying schedule of explanatory data, noting the date at which the survey was conducted. The following suggested classification indicates the main categories of land use which should be distinguished, together with points of detail, whilst conducting the actual survey.

1. Residential Use: This includes hotels, boarding houses, residential clubs, hostels, welfare homes and nursing homes. It is important to note the type of dwelling (i.e. house, maisonette or flat), the number of storeys, whether the building is occupied or vacant, and to make a subjective assessment of the physical condition of the building. When this material has been analysed it can be amplified further by examining the information contained in the electoral rolls, which will confirm whether the dwellings are in multi-occupation and also provide the basis for establishing the probable population of any particular area or district. These population estimates are arrived at by multiplying the figures obtained from the electoral rolls by a conversion factor which represents the ratio between the total population and the total registered electors for the same administrative area. By checking the yearly incidence of surnames it is possible to assess the

'neighbourhood potential' of the area in so far as a stable population infers the existence of a defined sense of place and strong local allegiances.

2. Open Space: This use can be classified further as follows:

(*a*) Public open space, i.e. parks, recreation grounds, commons and heathlands which are freely available for the enjoyment of the public at large. The number of games pitches, the number and capacity of the play apparatus, together with an estimate of the number of persons actually using these facilities should also be noted. The provision of public open space can be measured in terms of acres per 1,000 persons as is the case in London, for example, where the London County Council has prepared an open space deficiency map which indicates both the lack and maldistribution of public open spaces (London County Council, 1951 and 1961).

(*b*) Private open space, i.e. allotments, cemeteries, woodlands, golf-courses, sports-grounds, school playing fields and the grounds of private institutions.

3. Public Buildings: These can be classified further as follows:

(*a*) Places of assembly, i.e. museums, art galleries, churches, exhibition halls, chapels, missions and other halls, sunday schools, cinemas, theatres, concert halls, music halls and dance halls, non-residential clubs, gymnasia, skating rinks, small exhibition and amusement arcades, and fun fairs.

(*b*) Other public buildings, i.e. libraries, baths, clinics, dispensaries, police stations, courts of law, and fire brigade stations.

(*c*) Institutions, i.e. hospitals, health and social centres, and schools.

4. Industry: This primary use can be sub-divided into the twenty-four standard categories which form the basis of the Industrial Census Tables (Central Statistical Office, 1958). Later these industries can be grouped into three major categories which reflect their probable impact upon the environment of the surrounding residential areas, e.g.

(*a*) Light Industry, viz., buildings in which the machinery and processes carried on are such as could continue in any residential area without detriment to amenity, having regard to noise, vibration, fumes, smoke, ash, dust or grit.

(*b*) General Industry, viz. buildings in which the machinery and processes carried on are considered to be detrimental to residential amenity having regard to noise, vibration, fumes, etc.

(*c*) Special Industry, viz. buildings in which the machinery and processes carried on are considered to be obnoxious because of poisonous fumes, unpleasant odours, the need to treat products in an offensive condition with consequent health hazards to neighbouring dwellings (Ministry of Housing and Local Government, 1963 A).

Wherever possible an attempt should be made to interview the factory manager in order to ascertain the following additional information:

(*a*) When was the factory founded?

(*b*) If the factory was not founded on this site why did it move here?

(*c*) What are the assets and drawbacks of the present site?

(*d*) Where do the raw materials come from and how are they transported?

(*e*) What is the nature of the finished product, where is it marketed and how is it transported there?

(*f*) How many people does the factory employ, where do they live and how do they travel to work?

(*g*) Is the factory tied to the local area or region?

5. Commerce: This use includes buildings used as warehouses, wharves, depositaries, stores, garages, builders' yards, timber stores, contractors' yards (including local authority and Ministry of Works Depots), post office sorting and parcels offices, grain silos, etc. As in the case of the industrial uses, additional information should be solicited by interviews, in an attempt to establish the relationship between site and function with its consequential impact upon the surrounding region.

6. Offices: This use includes buildings used primarily for administrative purposes including banks, post offices, central and local government offices, buildings occupied by registrars and other public and quasi-public servants. Local offices providing professional services but still employing general clerical workers, e.g. solicitors, accountants, estate agents, travel agents, etc., should all be included in this category. Information should be solicited by interviews to establish the number of persons employed in the building, their place of residence and method of travelling to work.

7. Shops: In order that a proper study can be made of the pattern of retail trading it will be necessary to study this particular use in detail, and the following points should be noted:

(*a*) The total number of shops.

(*b*) The type of shop, e.g. foodstuffs, household goods, luxury goods, clothing, and establishments providing personal services, noting any significant variations in the provision of these facilities in centres of varying sizes.

(*c*) The number of department stores, multiple and chain stores, etc.

8. Statutory Undertakers: Including the basic utilities, viz. Gas, Electricity and Water, the railways, and the land under the jurisdiction of various government departments.

9. Vacant and derelict buildings, cleared land, etc.

When this survey material has been collected and presented in a series of analytical maps which highlight the more salient patterns of incidence and distribution, it is worth remembering that there is an accepted colour notation for these main land-use categories, e.g. residential use is indicated by a light red-brown, public open space by green, private open space by yellow-green, public buildings by red, industry by purple, commerce by grey, offices by light blue, shops by dark blue, and vacant land by yellow. There is no standard notation for statutory undertakers, although Gas and Electricity Undertakings are usually coloured purple, railways grey, and the service departments by a red verge around the boundary of the land in question. This basic land-use survey can be carried out in greater detail depending upon the size of the built-up area and the time and number of persons available. In the Central Business Area of larger towns it will be found that the buildings commonly encompass a wide variety of these land-use categories, necessitating the use of a letter subnotation to augment the basic colour notation. Various combinations have been tried by local authorities with only a limited degree of success and this problem presents a challenge for all who are interested in cartography.

This land-use survey will establish the geographical complexities of the urban way of life, by indicating the daily journeys which have to be made before the inhabitants can find employment, shop, and utilize the educational, social, recreational, cultural and commercial facilities. Further investigation will be needed, however, before it is possible to attempt to delimit the neighbourhood and communal structure of the built-up area. Although the geographer has demonstrated his ability to analyse the constituent elements which comprise a particular area, he has not evidenced a similar proficiency in the more complex art of synthesis with a view to establishing the criteria for delimiting urban, let alone geographical, regions. Whether this failure is due to the untenability of the concept of the region or the paucity of techniques is open to debate, but it cannot be denied that even in the more restricted field of urban studies only limited progress has been made in defining the catchment areas of neighbouring service centres.

COMMUNITIES AND URBAN FIELDS

Reference has been made to the existence of a neighbourhood and community structure within the build-up area of urban settlements, and in this context the term 'community' refers to a primary 'face to face collectivity' which is comprised of an admixture of formal and informal social groups. Occasionally it may possess a civic identity which is distinct from that of its component social groupings, but in essence it describes a mass of individuals who live in a conscious though undefined physical association (Anderson, 1960). The precise nature of these human relationships is difficult if not impossible to define. It varies from town to town, but there is little doubt that its intensity is inversely proportional to the actual size of the town in question. The inhabitants of a large town are interdependent in all sorts of specialized and external ways, yet often they lack the personal intimacy and emotional security which springs from a central unifying urban value or sense of place. The focal point of the urban community is usually the town centre with its concentration of shops, entertainment facilities, restaurants, cultural facilities, and professional and commercial services. Nearly all towns possess only one such centre and its sphere of influence encompasses the whole of the built-up area, which constitutes the urban community in place. The urban field concept rightly stresses the importance of the town as the seat for centralized services, and attempts to delimit its sphere of influence by superimposing a series of generic regions which reflect particular functions of the town (Smailes, 1944). Needless to say the results are cartographically disastrous and provide only an indication of the 'zones of divided loyalty' where the inhabitants can choose between equally accessible and hence rival town centres. This procedure cannot readily be applied in the larger cities such as London where there are numerous urban communities with their respective regional service centres. In these larger cities the distribution of specialist institutions such as secondary and grammar schools, hospitals, technical college libraries, etc., is sporadic in appearance, for it is dictated by the differing tributary populations that are needed to sustain each type of institution. Inevitably these institutions serve metropolitan rather than regional needs and their catchment areas tend to cut right across the urban community structure as expressed in terms of tribute to the regional service centre. In other words the generic or functional regions cannot be superimposed as a means of delimiting the metropolitan community structure, for they are not related to a common point of urban nodality. But perhaps the greatest defect of the urban field concept is its failure to take account of the quantitative use of central area facilities, despite the fact that this is, perhaps, a more sensitive index for assessing the regional importance of rival town centres (Herbert, 1961).

THE NEIGHBOURHOOD STRUCTURE

As soon as an attempt is made to define the urban community in spatial terms it usually becomes necessary to do so on a neighbourhood basis, for the 'neighbourhood' usually has more precise physical associations (Forshaw and Abercrombie, 1943). By definition the neighbourhood concept implies an intensity of social communion which results from the close juxtaposition of a group of families or individuals within a prescribed physical setting. To some extent it reflects the historical evolution of the urban area, for the resultant morphology tended to create relatively self-contained residential enclaves. Attempts to define the neighbourhood in terms of residential proximity are useful for ensuring that residential services and amenities are properly distributed, as is usually the case in the 'New Towns'. However, it does not always fit the geographical realities of urban friendships now that people tend to be drawn together by common interests rather than by the fact that they live in the same street or district.

In most towns the neighbourhood structure can be delimited in terms of the existing land-use pattern, for it highlights the potential physical barriers within the built-up area. Railway viaducts and cuttings, canals with their associated belts of industry and commerce, main roads and the grounds of large institutions all tend to inhibit movement and create inward-looking and relatively self-contained residential enclaves. Usually these 'enclaves' possess a local shopping complex comprised of half a dozen food shops, a chemist, a draper, an ironmonger, a hairdresser, a shoe repairer, a branch post office, and sometimes a bank and a branch library. These local sub-centres are an essential complement to the regional service centre and the acquisition and further intensification of commercial services in no way enables them to compete with the main centre during the week-end period.

A cursory examination of the residential areas in any large town often fails to reveal the identity of any particular locality, and this is most certainly the case in London where the residential areas have tended to assume the identity of the nearest service centre. In those areas where this process of 'urban adoption' has proved impracticable, the task was accomplished by the Postmaster General who used district numbers. Needless to say, these areas possess a diversity of character and often a very real sense of place for the inhabitants who were born there, but the uniformity of housing types and layouts (often the result of bye-law control) present a formidable challenge to newcomers who attempt to establish a local 'sense of place'. This term has been used frequently and it implies something which is experienced, almost subconsciously, a feeling of security which is engendered by the familiarity of everyday surroundings. It usually implies a degree of kinship which springs from having shared the same experiences – perhaps from childhood to old age. These experiences are set against the back-cloth of buildings, streets, trees, parks and, most important of all, of friends and relations. Each building and element in the urban scene acquires a subjective character which cannot be related to its aesthetic qualities, until gradually the whole area is seen in this light by its inhabitants. This process marks the emergence of the neighbourhood in place, and its formal name invokes memories, nostalgia and a sense of kinship and belonging. Sometimes, however, an area can assume a character or atmosphere which is not obvious to its actual inhabitants, as in Soho where the residents are more conscious of the squalor and poor housing conditions than its 'cosmopolitan or continental' character (*The Times*, 1963).

SERVICE CENTRES

As stated previously, the residential areas have often tended to establish their identity in terms of the nearest service centre, which is the most significant place of assembly. The service centre is very susceptible to changes in the economic potential of the surrounding area, and redevelopment at higher net residential densities has led to a renewal of prosperity in centres which were previously deemed to be decaying. This is especially true in London despite the fact that there is no causal relationship between many of the suburbs and the service centres which supply their daily needs. Indeed, most of the suburban areas are dependent upon the concentration of economic opportunity within the central area, and have only utilized the historic pattern of settlement as a convenient framework for siting additional commercial services and urban amenities. Consequently the historic townships in Greater London have acquired large populations, within a relatively short period of time, without developing the economic facilities to support them. For five days in each week these historic townships play a minor role in the economic life of the community, and it is only on Saturdays that they assume their rightful role as regional centres.

The regional function of the service centres presents an absorbing field for investigation and their geographical distribution, size, composition, functions and probable catchment areas are already being studied in some detail (Carruthers, 1957 and 1962; Collins, 1960 A and B; Lomas, 1964; Parker, 1962; Smailes and Hartley, 1961; Manchester University, 1964). It has been established that a hierarchy of service centres exists within most regions, but the criteria which have been used to rank these centres, viz. the provision of specified facilities within the central area together with an assessment of their utilization, as reflected by rateable values and the nodality of bus services, is inadequate and the actual rankings which result are questionable (Carruthers, 1962). The subtleties of this particular classification were not supported by a series of survey visits which indicated that there is a marked degree of similarity in the composition of service centres in South London, and that the differences were quantitative rather than qualitative (Collins, 1960 A and B). The distribution of such stores as Woolworth's, Marks and Spencer's, Littlewood's, British Home Stores, Times Furnishings, Dolcis, Saxone's, Sainsbury's, Boots the Chemists, together with that of cinemas, shows a remarkable consistency of incidence within the service centres in South London.

The land-use survey will provide much useful information about the composition of such service centres and the variations in the basic morphology will provide some indication of the probable regional status of any particular centre. This material can be considered under the following headings:

1. The number of shops which comprise the service centre.
2. The number of shops which specialize in the five major categories, viz. foodstuffs, clothing, household goods, luxury goods and commercial services such as hairdressing, cleaning, etc.
3. The number of department, variety, chain and multiple stores.
4. The number of establishments providing professional services such as solicitors, accountants, estate agents, etc.
5. The number of banks.
6. The number of local and national government offices.
7. The number of commercial offices.
8. The number of places of entertainment.

More detailed information can be obtained by measuring the actual allocation of floor space between competing land uses. These figures can be ascertained by taking measurements from the 1/1250 ordnance survey sheets which are quite adequate for this purpose, providing that a fairly detailed field survey has been carried out, and they will serve as a guide to the probable employment potential of the service centre. The rating lists also provide some indication of the regional importance of the centre in as much as the current rateable value reflects the amount of capital which has been invested in the centre, and by inference its importance as seen through the eyes of entrepreneurs.

I have attempted to carry out this type of study for South London, and it revealed the existence of a hierarchy of service centres which is essentially complementary despite the fact that the larger centres have attracted most of the major redevelopment schemes. So far four categories of service centres have been distinguished for the convenience of presentation, and this somewhat arbitrary classification is based solely on the provision of central area facilities with no assessment of their use. The first category is comprised of 9 centres with over 300 shops (Figure 11.1A), ranging from 301 shops in Bromley to 425 shops in Croydon. Even in this category over 70 per cent of the total shops are concerned with satisfying the basic needs of the resident population. The accessibility of these 9 centres has been assessed in terms of bus services and it would appear that most residents of South London can reach one of these centres after completing a twenty-minute journey; although this criteria is of less significance within the County of London due to the closer spacing of the centres. The second category is comprised of 12 centres with between 200 to 300 shops (Fig. 11.1B), ranging from 215 shops in Bexleyheath to 270 shops in Deptford. The centres in the outer

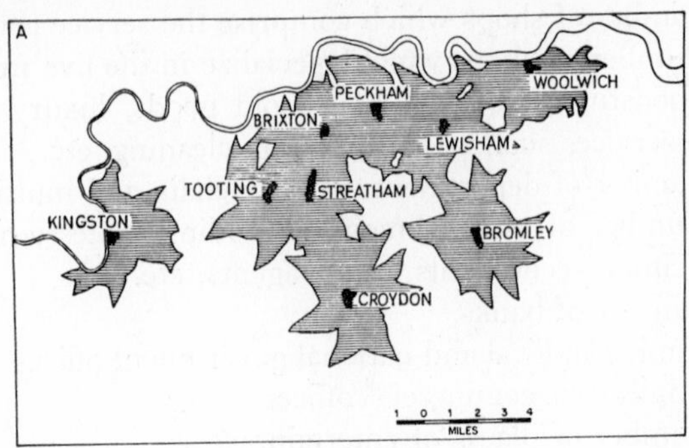

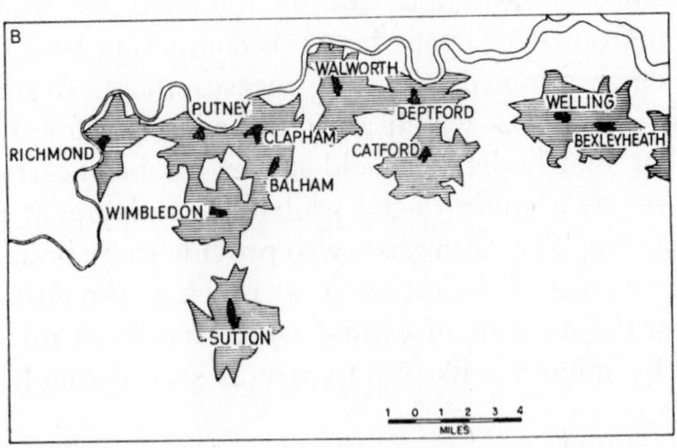

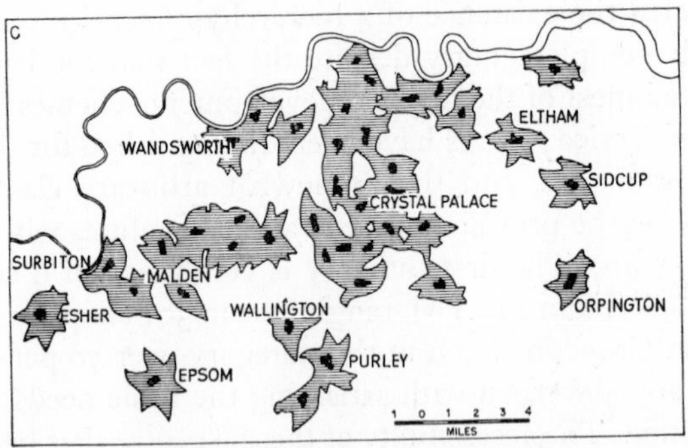

FIG. 11.1. *The location and accessibility of service centres in London south of the Thames. A: Centres with 300 or more shops. Shaded areas show zones within the twenty-minute isochrone (as determined by bus, walking, or a combination of both). B: Centres with 200 to 300 shops. Shaded areas show zones within the fifteen-minute isochrone. C: Centres with 100 to 200 shops. Shaded areas show zones within the ten-minute isochrone.*

suburbs exert a considerable regional influence (e.g. Dartford, Richmond, Sutton and Wimbledon), whereas those nearer to Central London tend to serve local rather than regional needs (e.g. Balham, Catford and Deptford). The range of shops, services and entertainment facilities mirrors that found in the larger centres, and most of South London can reach one of these centres after completing a fifteen-minute journey. The third category is comprised of 39 centres with between 100 to 200 shops (Figure 11.1C), ranging from 102 shops in Malden to 195 shops in Crystal Palace. It was noted that these centres tended to attract fewer weekend shoppers, despite the fact that their basic composition mirrors that of the centres within the second category. The number of shops engaged in the sale of foodstuffs, however, tends to dominate the character of these centres and often amounts to 50 per cent of their total composition. Most of South London can reach one of these centres after completing a ten-minute journey. The fourth category is comprised of 23 centres with less than 100 shops, ranging from 45 shops in Carshalton to 98 shops in Petts Wood. Most of these centres are located in the outer suburbs, and with one exception they serve purely local needs. The exception is the Elephant and Castle which is in course of being redeveloped as a regional centre with 200,000 square feet of new shops and nearly 700,000 square feet of new office accommodation; and it may curb the regional aspirations of neighbouring centres.

As is to be expected, the multiple and variety chain stores have paid considerable attention to the probable catchment population of neighbouring centres in South London before deciding to locate a new branch store. The following table indicates the estimated population requirements of several well-known firms:

	catchment population	
Boots the Chemists	10,000	
Mac Fisheries	25,000	
Barratts Ltd. (Shoes)	20–30,000	
Sainsbury's (Groceries)	60,000	(For a medium-sized self-service store)
Marks and Spencer's	50–100,000	
John Lewis	50,000	(For a super-market)
John Lewis	100,000	(For a department store)

These figures do not tell the whole story, however, and other important factors such as the presence of a department store, the social and economic character of the surrounding area, the levels of rentals, the availability of central sites that are ripe for redevelopment, and the problems of integrating the new branch into the firm's existing administrative and supply systems, are all taken into account before the final decision is taken. The exigencies of retail competition are so severe, on the other hand, that rival combines cannot afford not to be represented in the larger service centres and this accounts, in part, for the similarities in basic composition which have been noted already.

When examining the composition of neighbouring service centres it is important to take account of the current planning policies which govern their expansion and redevelopment, for often these explain variations in the regional distribution of shops, offices and other commercial services. An attempt should also be made to investigate the dynamics of redevelopment and establish the regional potential of rival towns as revealed, for example, by the number of applications seeking planning permission to redevelop land in the central area. Each local planning authority is required to keep a register of these applications, and these statutory registers are always available for public inspection.

The failure to define quantitatively the use of central area facilities is an outstanding weakness in the present studies of service centres, for potentially this could be a significant yardstick for measuring the regional importance of any town. The Census of Distribution gives some indication of the regional importance of the shopping centre of the town, in as much as it provides the basis for estimating the number of persons who visit the town in order to patronize the shopping facilities. These estimates are based upon the average sales per head of the resident population, which establishes the percentage of the total sales that is derived from the resident population, whilst the balance comprises the percentage which derives from the surrounding catchment area (Waide, 1963).

The catchment areas of the service centres within the metropolitan built-up area are difficult to define in precise terms, moreover, due to the increasing mobility of the resident population. Admittedly the morphology of the metropolis tends to delimit areas which look inwards for services, but the main roads now constitute the arteries of the circulation pattern. Public transportation still plays an important part in determining the probable catchment area of any one centre, and isochrones can be drawn to establish the zones of divided loyalty where personal preference rather than accessibility dictates the choice of centre. These zones require further detailed study and the residents should be questioned as to their shopping habits. This affords an opportunity to establish whether these zones possess a geographical 'sense of place'. The residents should also be questioned as to the name of the district in which they live, noting whether the discrepancies are due to the fact that the persons concerned tend to patronize different service centres. When the results of the questionnaire are plotted in map form it should be possible to define the effective catchment areas of the respective service centres. It is worth noting in this context that the local newspapers tend to confuse the issue due to the overlapping of circulation areas. Local news items and private advertisements do, however, provide some indication of the community structure, but the commercial advertisements of retail stores tend to reflect their regional *aspirations* rather than the geographical *realities* of their catchment areas. This also holds true for the retail delivery areas, due to the exigencies of modern competition.

In essence these catchment areas are zones of convenience within which the residents will probably establish their geographical sense of place in terms of the nearest service centre. However, this does not necessarily engender the urban community, as is the case in most towns, due to the close proximity of rival centres. The distribution of specialist urban institutions such as libraries, grammar and secondary schools, technical colleges and welfare centres, which all play their part in furthering the creation of communal loyalties, does not complement the distribution of service centres. This lack of unity in the functional framework of the metropolis reflects its monolithic character and has undermined the historic community structure. Exceptions do occur and usually they are to be found in areas where the 'village character' of the original settlement has been preserved. Some examples of this phenomenon can be seen in London (e.g. Dulwich, Highgate, and Hampstead), but they are noteworthy only in indicating the barren nature of the surrounding built-up area.

JOURNEY TO WORK

Having examined the basic morphology of the built-up area and the journeys which have to be made in order that the residents can shop, and utilize the social, cultural and entertainment facilities located therein, the next step is to examine the daily journey to work which dominates the life of every urban community. The census on *Usual Residence and Workplace* provides quantitative information as to where people travel in order to seek employment, but it does not indicate the means of effecting the journeys or the type of employment that is being sought. The *Industrial and Occupational Tables of the Census* will provide rudimentary data regarding the availability and demand for specified categories of employment, but it is difficult to establish a precise correlation which accounts in full for the large number of journeys that take place daily. Most of the statistical information relates to local government areas which no longer reflect the geographical realities of the urban community structure, with the result that little is certain about the journey-to-work pattern, although much can be inferred. In the case of South London, for example, certain movements are of particular interest and indicate the 'uneven' distribution of industry, commerce, offices, etc. In 1951 over 327,000 travelled daily into the three central boroughs of the City, Holborn and Westminster in order to seek employment. This movement accounted for over one-third of the total journeys which originated within South London. In six of the suburban boroughs (e.g. Banstead, Beckenham and Bromley) over 40 per cent of the persons engaged in the daily journey to work travel into these three central boroughs. Only in Crayford, Dartford and Erith does the percentage of persons thus engaged fall below 20 per cent. This is due here presumably to the predominance of heavy and more specialized industries such as paper-making, electrical engineering, chemicals and munitions which have attracted a resident labour force as a result of the manifest difficulties of travelling across South London. The outer suburban belt is especially dependent upon the employment facilities located within the County of London. In the case of eight boroughs, over 70 per cent of the persons engaged in the daily journey to work in 1951 travelled into the County of London in order to seek employment.

This type of analysis is limited to those journeys which involve crossing administrative boundaries and are hence depicted in the Census. It takes no account of the considerable number of 'hidden journeys' which take place within the confines of the respective administrative districts. In 1951 in South London some 596,150 persons lived and worked within their respective administrative districts (i.e. 39 per cent of the total number of persons seeking employment), indicating the scale and importance of these hidden journeys. Whilst the land-use survey will establish the 'desire lines' which are fundamental to the journey to work, this information will need the support of local origin and destination studies to determine how these journeys are made. In essence these studies are based upon the selection of a 'cordon line', preferably related to physical barriers which limit the number of interview or observation stations (Wood, 1963). For normal purposes simple observational studies are adequate, requiring that the registration number of each vehicle, the time, and the number of occupants should be noted at the various observation stations. When this material has been correlated it is possible to establish the number of journeys which originate and terminate within the cordon area. In order to determine the actual routes it is necessary to establish a further series of observation stations along the main thoroughfares, particularly at the major intersections. This type of study requires a considerable amount of labour and can be carried out satisfactorily by local schools, which can also conduct private questionnaire studies amongst friends and neighbours. This type of local field work can provide a valuable insight into the daily journey to work, the mobility of labour, and the employment needs of local areas.

URBAN RENEWAL

An analysis of the present morphology of the built-up area leads to the need to ascertain the morphological changes which are currently taking place. Urban renewal is a constant process which can change the existing social and economic character of an area, or simply confirm the historic pattern of land use. The business advertisements in the property journals and daily newspapers constitute a valuable source of information in this respect, for they indicate where new office, industrial and commercial developments are taking place, together with the rentals which each area can command on the open market. These developments are an important feature of the changing morphology of the built-up area and quite often they indicate the changing regional status of the local service centres. In greater London many large-scale office developments are taking place in the outer suburbs in such centres as Croydon, Harrow and Tolworth (Figure 11.2).

The question of future changes is to some extent already determined by the statutory development plans of the respective Local Planning Authorities. These plans dictate the present, as well as attempting to forecast the future, use of land. When the Town Map proposals are examined in detail it will be seen that provision is made for the future employment, educational, shopping, residential and recreational needs of the community at large. A comparison of the land-use map and the Town Map will indicate whether the Town Map has confirmed the historic pattern, whilst providing for the eradication of the more blatent absurdities, or has attempted to reshape the basic functional structure of the built-up area. This comparison may also reveal the existence of a number of factories, warehouses, builders' yards, etc., which are deemed to be badly sited from the viewpoint of residential amenity and will need to be relocated when redevelopment takes place. In most cases the economic and social consequences of these proposals have not been fully evaluated. Little is known about the regional and local significance of these dispersed centres of employment, and geographical research in this field can contribute towards a fuller understanding of the dynamics of industrial expansion and location generally. Once again the local questionnaire, which forms an integral part of the land-use survey, will be of considerable assistance in assessing the economic vitality of the local area.

THE HISTORICAL EVOLUTION OF TOWNS

Having analysed the existing land-use pattern, considered the geographical implications of the resultant morphology, delimited the zones of 'divided loyalty' and regional tribute, and studied the daily journey to work, the time has come to consider the historical evolution of these phenomena. This is in contrast to the customary geographical study of a town which strives to date the first signs of human occupation of the site in the Stone Age and follows a tortuous path through the Bronze and Iron Ages until it eventually reaches the comparative safety of the Domesday Survey. The scholarship behind such endeavour is beyond reproach but one may be pardoned for questioning whether this exercise should not be left to the archaeologist who will establish the time-scale with some degree of precision, leaving the geographer free to explain why these events took place in this particular area. All too often these geographical studies evidence an alarming preoccupation with historic succession and attempt to analyse the 'Town Plan' in terms of the changing form of the buildings and layout of the streets, without proper reference to the functions of the town (Conzen, 1960). Surely the geographical significance of these changes lies in the fact that they reflect the changing functions of the town, which in turn reflect the changing needs and character of the surrounding catchment area. This historical study should attempt to highlight the factors which sustain and change the pattern of urban land use, and not just describe the phases of its evolution. Whilst hesitating to deny the fundamental significance of the questions 'Where?' and 'When'?, I would submit that foremost in the mind of every geographer should be the question 'Why?'. This is particularly relevant when examining the evolution of towns, with special reference to their transition from an agricultural to an industrial economy (Dyos, 1961; Randle, 1962). The dynamics of urban growth

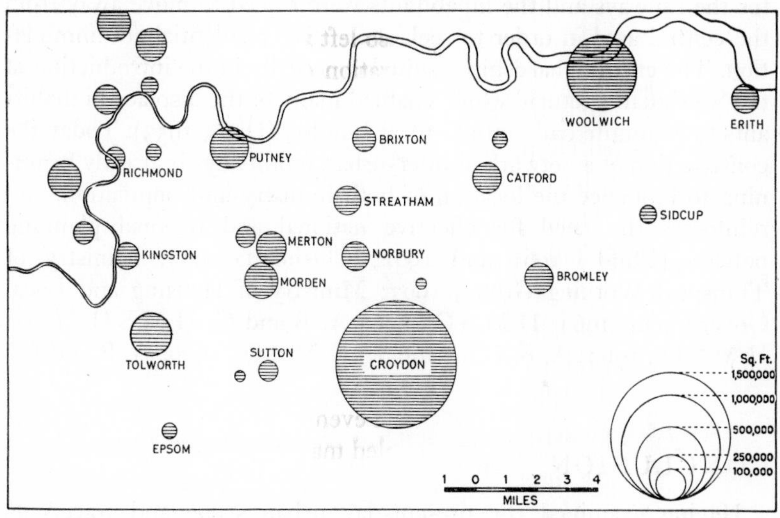

FIG. 11.2. *The location and size of office developments in London, mainly south of the Thames, 1964 (including proposals which have been approved in principle).*

are still something of a mystery, and recourse to Local Acts of Parliament, Vestry Records, Rating Returns, etc., will illuminate the processes and phases of growth without necessarily explaining them. There are manifold dangers in this type of analysis, for it is always tempting to seek geographical reasons for discernible patterns of growth and incidence. Often this amounts to hindsight, and many of the relationships which appear to be established so convincingly in terms of geographical determinism are in reality spurious. This is especially true when examining the mid-nineteenth-century growth of South London (Dyos, 1961). For often it was the *availability* of a strategically-sited country estate, due to the death or bankruptcy of its owner, which precipitated speculative housing developments.

The advent of the railway era also left its mark during this period, by facilitating the increasing separation of workplace and residence (Carter, 1959; Dyos, 1961; Sekon, 1938; Simmons, 1961; Thomas, 1928). The actual construction of the railways in South London made this separation inevitable, for they were deliberately routed through the densely-peopled poorer districts, where the land values were low, in order to reduce the cost of establishing a right-of-way (Dyos, 1955). Thousands of dwellings were destroyed to make way for the railways and the inhabitants were forced to move away from the central area in order to seek alternative residential accommodation. The ensuing hardship resulted eventually in the introduction of the 'workman's return' which enabled many of the displaced inhabitants to commute daily into Central London (Dyos, 1953). Today the construction of a network of inter-urban motorways is already beginning to influence the location of both industry and population, and reinforces the need for effective national and regional planning policies (Childs, 1961 and 1962; Chisholm, 1962; Ministry of Transport Working Group, 1963; Ministry of Housing and Local Government, 1964; H.M.S.O., 1963 A, B and C; H.M.S.O., 1964; H.M.S.O., 1965; A, B, C, D and E; H.M.S.O., 1966, A, B and C.)

CONCLUSION

Having examined the present-day urban scene and traced its historical evolution, all that remains is to identify any major defects in the present local and regional distribution of land use with a view to suggesting improvements in the basic pattern. This is, perhaps, the point where urban geography merges into town and country planning, for the planner is primarily concerned with re-shaping the existing morphology of urban settlements into a more efficient and socially acceptable form, whilst providing for their future as well as their present needs. The planner has tended to be more than somewhat conservative in his approach to this task and has attached undue significance to the historic pattern of land use. The 'Buchanan Report' (Ministry of Transport Working Group, 1963) has sounded a warning note by highlighting the hazards of attempting to determine present, let alone future, patterns of redevelopment in terms of outmoded concepts and an outdated urban inheritance. The call for a more dynamic approach cannot be allowed to remain unanswered now that the Standing Conference on London Regional Planning has forecasted the scale of future events, with an estimated increase of 550,000 jobs and 800,000 inhabitants by 1971 (London County Council, 1963). The recently published *South-East Study: 1961–1981* (Ministry of Housing and Local Government, 1964) has revealed the dearth of both national and regional planning thought in this country, and evidences the need for further research into the very nature of urban growth.

The geographer, by virtue of his training and basic philosophy, is well equipped to undertake the regional studies which are absent in current government publications dealing with the redevelopment of town centres (Ministry of Housing and Local Government, 1962, 1963 B, and 1963 C). Unfortunately, as discussed earlier, the geographer has failed to establish definitive techniques which stand the test of universal application, and further research is of paramount importance for events in the south-east of England indicate that time is short. Local field work studies will do much to reveal the composition, functions and character of the respective urban areas, and provide a valuable insight into the spatial relationships of the component land uses. Too much reliance has been placed on the decennial census reports which are years out of date by the time they are published. The government has at last recognized this defect, however, and announced its intention to hold a census every five years starting from 1966. A more serious drawback is the fact that the census statistics tend to mask the local changes which take place within the confines of the local government areas. The investigations which have been described in this chapter can all be undertaken by urban schools, and the pupils can conduct the land-use surveys, traffic counts and random sample surveys with only a limited amount of direction and supervision. Field work is an essential part of any geographical study and the pupils will learn much by having to devise survey techniques, methods of presentation, questionnaires, and analyse the resultant findings. These local field work studies are essential to a fuller understanding of the dynamics of urban regionalism, for only when the geographer has come to grips with the urban microcosm can he hope to arrive at any significant conclusions regarding the nature of towns and the areas that they serve.

References and Further Reading

ANDERSON, N., 1960, *The Urban Community* (London).

BEST, R. H. and COPPOCK, J. T., 1962, *The Changing Use of Land in Britain* (London).

CARRUTHERS, W. I., 1957, 'A Classification of Service Centres in England and Wales', *Geog. Jour.*, **123**, 371–85.

— 1962, 'Service Centres in Greater London', *Town Planning Review*, **33**, 5–27.

CARTER, E. F., 1959, *An Historical Geography of the Railways of the British Isles* (London) (see especially pp. 352–3).

CENTRAL STATISTICAL OFFICE, 1958, *Standard Industrial Classification* (revised edn) (H.M.S.O.).

CHADWICK, J. G., 1961, 'Correlation between Geographical Distributions', *Geog.*, **46**, 25–30.

CHILDS, D. R., 1961, 'Urban Balance', *Jour. Town Planning Inst.*, *47* (6), 170–1.

— 1962, 'Counterdrift', *Jour. Town Planning Inst.*, *48* (7), 215–25.

CHISHOLM, M., 1962, 'The Common Market and British Industry and Transport', *Jour. Town Planning Inst.*, *48* (1), 10–13.

COLE, H. R., 1966, 'Shopping Assessments at Haydock and Elsewhere', *Urban Studies*, **3**, No. 2, 147–57.

COLLINS, M. P., 1960 A, 'A Study of Urban Complexes within the Southern Half of the Greater London Region', in *Report of the Royal Commission on Local Government in Greater London, 1957–60*, V. 5, 731–7 (Cmnd. Paper No. 1164) (H.M.S.O.).

— 1960 B, 'South London: Its Metropolitan Evolution and Town Planning Requirements, (Unpublished Thesis, Dept. Town Planning, Univ. Coll., London).

CONZEN, G., 1960, 'Alnwick, Northumberland', *Trans. Inst. Brit. Geog.*, Pub. No. 27 (London).

DIAMOND, D. R. and GIBBS, E. B., 1962, 'Development of New Shopping Centres: Area Estimation', *Scottish Journal of Political Economy No.* 9.

DYOS, H. J., 1953, 'Workmen's Fares in South London 1860–1914', *Jour. Transport Hist.*, **1**, 3–19.

— 1954, 'The Growth of a pre-Victorian Suburb; South London 1580–1836', *Town Planning Review*, **25**, 59–78.

— 1955, 'Railways and Housing in Victorian London', *Jour. Transport Hist.*, **2**, 11–21 and 90–100.

— 1961, *The Victorian Suburb: a Study of the Growth of Camberwell* (London).

FORSHAW, H. J. and ABERCROMBIE, P., 1943, *County of London Plan* (London) (see paras. 96–111, pp. 26–29).

HERBERT, D. T., 1961, 'An Approach to the Study of the Town as a Central Place', *Sociological Rev.*, *9* (3) (NS), 273–92.

HERBERT, D. T., 1963, 'Some Aspects of Central Area Redevelopments', *Jour. Town Planning Inst.*, 49 (4), 92–99.

H.M.S.O., 1963 A, *The North-East: a Programme for Regional Development and Growth* (Cmnd. Paper No. 2206).

— 1963 B, *Central Scotland: a Programme for Development and Growth* (Cmnd. Paper No. 2188).

— 1963 C, *Growth of the U.K. Economy 1961–1966*, National Economic Development Council.

— 1964, *The Growth of the Economy*, National Economic Development Council.

— 1965 A, *The National Plan* (Cmnd. Paper No. 2764).

— 1965 B, *The West Midlands—A Regional Study*, D.E.A.

— 1965 C, *The Problems of Merseyside*, D.E.A.

— 1965 D, *Wales 1965* (Cmnd Paper No. 2918).

— 1965 E, *The North-West—A Regional Study*, D.E.A.

— 1966 A, *Investment Incentives* (Cmnd. Paper No. 2874).

— 1966 B, *The Scottish Economy* (Cmnd. Paper No. 2864).

— 1966 C, *A New City*, Ministry of Housing and Local Government.

HOOKWAY, R. J. S. *et al.*, 1963, 'Current Practice Notes', *Jour. Town Planning Inst.*, 49 (7), 234–6.

HUDSON, A. C., 1963, 'Brixton Central Area Redevelopment (Map)', *Jour. Town Planning Inst.*, 49 (8), 277.

LOMAS, G. M., 1964, 'Retail Trading Centres in the Midlands', *Jour. Town Planning Inst.*, 50 (3), 104–19.

LONDON COUNTY COUNCIL, 1951, *Administrative County of London Development Plan* (London) (see 'Analysis', Chap. 7).

— 1957, *A Plan to Combat Congestion in Central London* (London).

— 1961, *Administrative County of London Development Plan: First Review*, County Planning Dept., Vol. 1, Chap. 11.

— 1963, *Reports on Population, Employment and Transport in the London Region*, Standing Conference on London Regional Planning, dated 28 March, 31 May, 3 October and 4 December.

MADIN, JOHN H. D. AND PTNRS., 1964, *Worcester Expansion Study*, published by Ministry of Housing and Local Government.

MANCHESTER UNIVERSITY, 1964, *Regional Shopping Centres—A Planning Report on N.W. England*, Department of Town Planning.

MINISTRY OF HOUSING AND LOCAL GOVERNMENT, 1962, *Town Centres – Approach to Renewal*, Planning Bull. No. 1 (H.M.S.O.).

— 1963 A, *The Town and Country Planning (Classes) Order*, Statutory Instrument No. 708 (H.M.S.O.).

— 1963 B, *Town Centres – Cost and Control of Redevelopment*, Planning Bull. No. 3 (H.M.S.O.).

— 1963 C, *Town Centres – Current Practice*, Planning Bull. No. 4 (H.M.S.O.).

— 1964, *The South-East Study: 1961–1981* (H.M.S.O.).

MINISTRY OF TOWN AND COUNTRY PLANNING, 1949 A, *Report of the Survey*, Circular No. 63 (H.M.S.O.).

— 1949 B, *The Redevelopment of Central Areas* (H.M.S.O.).

— 1951, *Reproduction of Survey and Development Plans*, Circular No. 92 (H.M.S.O.).

MINISTRY OF TRANSPORT WORKING GROUP, 1963, *Traffic in Towns* ('Buchanan Report') (H.M.S.O.).

MOSER, C. A. and SCOTT, W., 1961, *British Towns: a Study of Their Social and Economic Differences* (Edinburgh).

NATIONAL CASH REGISTER COY. LTD. (undated), *Thoughts on Future Shopping Requirements.* Issued by Modern Merchandising Methods Dept., N.C.R. Co. Ltd., 206–216 Marylebone Road, London, N.W.1.

PARKER, H. R. 'Suburban Shopping Facilities in Liverpool', *Town Planning Review*, April 1962, **xxxiii,** 197–223.

RANDLE, P. H., 1962, 'The Uses of Historical Data', *Jour. Town Planning Inst., 48* (8), 247–50.

REYNOLDS, D. J., 1961, 'Planning, Transport and Economic Forces', *Jour. Town Planning Inst., 47* (9), 282–6.

ROYAL COMMISSION, 1940, *Report on the Distribution of the Industrial Population* ('Barlow Report') (Cmnd. Paper No. 6153) (H.M.S.O.).

SEKON, G. A. (Pseud. for NOKES, G. A.), 1938, *Locomotion in Victorian London* (London).

SIMMONS, J., 1961, *The Railways of Britain: an Historical Introduction* (London).

SMAILES, A. E., 1944, 'The Urban Hierarchy in England and Wales', *Geog.*, **29,** 41–51.

— 1953, *The Geography of Towns* (London) (see Chap. 8).

SMAILES, A. E. and HARTLEY, G., 1961, 'Shopping Centres in the Greater London Area', *Proceedings of Institute of British Geographers* No. 29, 201–213.

STAMP, L. D., 1950, *The Land of Britain, Its Use and Misuse*, 2nd Edn (London).

THOMAS, J. P., 1928, *Handling London's Underground Traffic* (London Underground Railways Publication).

TIMES, THE, 1963, 'Few Mourners as Soho Buries the Past', 18 November, p. 7.

— 1965, 'The Future Economic Shape of Britain, 15th September, p. 16.

WAIDE, W. L., 1963, 'The Changing Shopping Habits and Their Impact on Town Planning', *Jour. Town Planning Inst., 49* (8), 254–64.

WELLS, HENRY W., *Peterborough—An Expansion Study*, 116 Kensington High Street, London, W.8.

WOOD, P., 1963, 'Studying Traffic in Towns', *Jour. Town Planning Inst., 49* (8), 265–71.

CHAPTER TWELVE

Quantitative Techniques in Urban Social Geography

D. TIMMS

Lecturer in Sociology, University of Queensland

Geography has often been accused of barrenness in terms of the meaningful generalizations and models it has produced. Much of this infertility may be attributed to the crude and subjective measuring instruments which have been generally employed by geographers. In the absence of precise and objective measurement and statement, accurate comparison and abstraction become impossible, and without abstraction the construction of explanatory models can be little more than guesswork.

The last two decades have witnessed a considerable increase in studies concerned with areal variation. The traditional geographical interest in areal studies has been joined by that of several other disciplines. The incursion of human ecologists, regional scientists, and students from many of the systematic sciences, has resulted in the development of a wide range of measures for the accurate description, analysis, and generalization of the facts of areal variation. Human ecology has been the most fertile source of new techniques concerned with the measurement of the areal pattern of urban social phenomena, but significant contributions have also come from regional science and from psychology. All the techniques to be described have as their aim the accurate and objective description and analysis of the areal variation of sociological phenomena within and between urban areas.

THE DESCRIPTION OF GEOGRAPHICAL DISTRIBUTIONS

In *The Nature of Geography* Hartshorne (1949, p. 376) states that: 'The scientific ideal of certainty commands that the terms and concepts of description and relationships be made as specific as possible – we cannot develop a sound structure on a marsh foundation. . . .'

Any research concerned with the spatial structure of urban areas is confronted by the necessity of providing a systematic description of various areal distributions and of their interrelationships. Three main groups of techniques have been developed in response to this need: the various cartographic techniques, measures of spatial association, and the so-called centrographic measures.

CARTOGRAPHIC TECHNIQUES

Maps are often referred to as *the* geographical tool, although their use is by no means confined to geographers. It is assumed that readers are familiar with the elements of cartographic representation and with the considerable degree of elaboration possible therein. Much research identified as social geography has consisted of the compilation of massive inventories of the observable characteristics of community life and of the plotting of these facts on various types of dot or choropleth maps. With this operation 'analysis' has often ceased, the final report of such research generally consisting of no more than a set of maps with an accompanying descriptive text. But the plotting of distributions is only the first step in the analysis of areal variation. Analysis which is based on rule-of-thumb assessments of the similarities revealed on successive maps or on the degrees of overlap present in two distributions is open to grave errors of approximation. Subjective judgement may produce comparisons which are not real and may ignore 'awkward' relationships. As illustrative devices maps have few peers, but as analystical instruments they suffer from many fundamental weaknesses. Mapped distributions can only provide the raw material for analysis and the success of this operation depends on the use of more concise and specific measures of distribution, which are capable of quantitative statement and allow precise comparison.

MEASURES OF SPATIAL ASSOCIATION

The most widely-used and in many ways the most useful instruments yet devised for the quantitative description of geographical patterns are the various measures derived from the *Index of Dissimilarity* (I_D), a statement of the evenness of distribution of two statistical populations.[1] The index is calculated from data giving for both populations the percentage of the total living in each areal sub-unit. The index of dissimilarity is then one-half the sum of the absolute differences between the two populations, taken area by area. In the hypothetical example (Table 9) the index of dissimilarity between the 'A' and 'B' populations is 25 per cent.

Table 9. Hypothetical Data for Computation of Index of Dissimilarity

Area	*'A' Population*	*'B' Population*	*Difference*
1	15%	10%	5%
2	20	10	10
3	20	20	0
4	30	20	10
5	15	40	25
Totals:	100	100	50

The index of dissimilarity may be interpreted as a measure of net displacement, showing the percentage of the one population who would have to move into other areas in order to reproduce the percentage distribution of the other population. Thus, in the hypothetical example, 25 per cent of the 'A' population would have to move in order to reproduce the percentage distribution of the 'B' population.

The basic formula for the index of dissimilarity is given in formula 1, with x_i representing the percentage of the 'x' population in the i'th areal sub-unit, y_i representing the percentage of the 'y' population in the i'th sub-unit, and the summation being over all the k sub-units making up the given universe of territory, such as a city.

$$\text{Formula 1: } I_D = \frac{1}{2} \sum_{i=1}^{k} \left| x_i - y_i \right|$$

The basic form of the index of dissimilarity may be applied to a wide variety of phenomena, depending on the definition given to the two populations. One of its most well-known guises is the so-called *Index of Concentration*, a measure of the degree of correspondence between population units and area. Applied to urban population, x_i becomes the percentage of city population living in the i'th areal sub-unit and y_i becomes the percentage of city area contained in that unit. If the population is distributed evenly through the city

[1] Throughout this chapter the term population is to be interpreted in its statistical sense as a number of distributed objects.

area, each territorial division will contain a proportion of population equal to its proportion of total area. In this case the index of concentration will be 0. Conversely, if all the population is concentrated into one small area, the remaining parts of the city being uninhabited, the index of concentration will equal almost 100 (specifically 100 minus the percentage of city area contained in the populated sub-unit). Another application of the index of dissimilarity yields the *Index of Redistribution*, the net percentage of population which would have to change its area of residence in any given year in order to reproduce the distributional pattern of an earlier year. In this case x_i is the percentage of city population contained in the i'th sub-unit in the earlier year and y_i is the percentage in the given year. In conjunction with data on the demographic characteristics of the city population, the index of redistribution is of considerable value in studies of intra-urban migration. Florence's (1948) Coefficient of Localization, a measure of the relative regional concentration of a given industry compared to all industry, is functionally identical to the Index of Dissimilarity, but uses a different divisor.

If the index of dissimilarity is computed between a sub-group of the population and the remainder of that population (i.e., total population minus those in the specified group) the resulting measure is referred to as an *Index of Segregation* (I_S) and shows the extent to which the specified sub-group is residentially separated from the rest of the population. The most convenient means of computing the index is given in formula 2, where I_D represents the index of dissimilarity between the sub-group and the total population (including the sub-group), $\sum x_{ai}$ represents the total number of the sub-group in the city, and $\sum x_{ni}$ represents the total population of the city.

$$\textit{Formula 2: } I_S = \frac{I_D}{1 - \dfrac{\sum x_{ai}}{\sum x_{ni}}}$$

A large number of indices have been developed to deal with various aspects of segregation, but few possess the general utility and clear meaning of the present measure. For an exhaustive discussion of the various segregation indices the reader is referred to a recent methodological article by O. D. and B. Duncan (1955 A).

The index of dissimilarity provides a description of the association between two distributions over the several sub-areas comprising a city. The relative concentration of a population within any one of these sub-areas is most readily measured by the *Location Quotient* (L_Q), an index with which most geographers are doubtless familiar. The location quotient is simply defined as the ratio between the percentage of one population occurring in a given area and the percentage of another population in that area. In formula 3, x_i is the percentage of the total 'x' population occurring in the i'th area and y_i is the percentage of the 'y' population in the i'th area.

$$\textit{Formula 3:}\quad L_Q = \frac{x_i}{y_i}$$

A location quotient of one indicates that the two populations are proportionately equally represented in the area. An index of less than one indicates an under-representation of the 'x' population, while an index of more than one indicates an over-representation of the 'x' population.

All measures depending on a comparison of proportionate distributions over a series of areal sub-units are very dependent on the nature of those sub-units. In their book *Statistical Geography*, Duncan, Cuzzort and Duncan (1961, pp. 80–94) devote considerable space to a consideration of the problems raised by this dependence. There is no way of 'adjusting' index values to give a value which is independent of the territorial subdivision used. In general, the smaller the average size of areal unit, the larger is the index value. More precisely, if one system of sub-units is derived by the subdivision of the units of another system, then the index computed for the former can be no smaller than the index for the latter and will generally be larger. In Brisbane, the index of net redistribution of population between 1954 and 1961 is 24.7 when calculated on the basis of census collector's districts, which have an average population of 760, and 18·0 when calculated on the basis of statistical divisions, having an average population of nearly 10,000. No unique value can ever be attached to any measure which is derived from the comparison of percentage distributions and a statement as to the nature of the sub-units on which they are based should always accompany any description of index values.

CENTROGRAPHIC AND POTENTIAL MEASURES OF DISTRIBUTION

Centrographic and potential measures of distribution are formally more satisfying than the index of dissimilarity and its variants, but.

as yet, have proved operationally more lean. The great advantage which they possess is that they are largely independent of the territorial subdivisions on which they are computed. Their values, in general, will be more precise if based on many small units rather than a few large ones, but otherwise the significance of the values is unaltered by changes in territorial base.

The centrographic technique consists of the application of conventional statistical measures to the data of areal distributions. A distribution may differ from other distributions in two basic ways, in addition to variations in the number of units involved in its population. First, it may differ in terms of the position or positions around which it tends to cluster (i.e., its centre), and second, it may differ in the way its component units are scattered around this centre. Both features are readily amenable to measurement (Warntz and Neft, 1960).

The average position or *Mean Centre* of a distribution is the exact equivalent of the arithmetic mean of conventional statistics. It is the 'balancing point' or 'centre of gravity' of the distribution. Several alternative methods are available for the computation of the mean centre of a distribution, but the most convenient system depends on the use of grid co-ordinates. A square grid is superimposed on a map showing the base areal units for which data is available. On the assumption of even distribution within each base area, the geographical centre of each unit is determined and expressed in terms of the two co-ordinates of the grid, x and y. Each x and y measurement is weighted by the population of the base unit concerned. The sum of these terms for the whole city, divided by the total city population, gives weighted mean positions along each of the co-ordinates. The mean centre of the distribution may be found at the intersection of the mean of the x's and the mean of the y's (formula 4).

Formula 4:

$$\bar{\Delta} = \left\{ \bar{x} = \frac{\sum_i (x_i P_i)}{\sum_i P_i}, \quad \bar{y} = \frac{\sum_i (y_i P_i)}{\sum_i P_i} \right.$$

where x_i *= vertical co-ordinate of the i'th area*

P_i *= population of the i'th area*

y_i *= horizontal co-ordinate of the i'th area*

Two measures are available to describe the scatter of a distribution around its centre. Both may be computed with reference to any given point, e.g. the centre of the Central Business District or the point of highest land value. The simple measure, the *Mean Distance Deviation* (MD_Δ), is an expression of the average distance separating the units of the distribution from the given centre. The formula for computing the mean distance deviation is given in formula 5, where d_i represents the linear distance between the geographical centre of the i'th base unit and any given point, O, and P_i is the population of the i'th unit.

$$\textit{Formula 5:}\quad MD_\Delta = \frac{\sum_i (P_i d_i)}{\sum_i P_i}$$

The second measure of dispersion, the *Standard Distance Deviation* (s_Δ), is strictly analogous to the standard deviation of conventional statistics. If a system of grid co-ordinates has been used to compute the mean centre of a population it is a simple matter to extend the tabulation in order to derive the standard distance by either version of formula 6:

$$(a)\ s_\Delta = \sqrt{\frac{\sum(P_i[x_i - \bar{x}]^2)}{\sum P_i} + \frac{\sum(P_i[y_i - \bar{y}]^2)}{\sum P_i}}$$

Formula 6:

$$(b)\ s_\Delta = \sqrt{\frac{\sum P_i x_i^2}{\sum P_i} - \bar{x}^2 + \frac{\sum P_i y_i^2}{\sum P_i} - \bar{y}^2}$$

Formula 6 gives both the standard distance, s_Δ, and the two component 'latitude' and 'longitude' deviations, s_x and s_y, which comprise it. The standard distance itself is invariant under the rotation of co-ordinate axes, but this is not true of its x and y components. For this reason no truly satisfactory means has yet been devised for plotting the standard distance on a map, although a possible solution to this problem, involving respectively the maximization and minimization of the x and y deviations, has recently been proposed (Bachi, 1963).

An interesting theorem connects the standard deviation to the average root-mean-square deviation of a total frequency distribution. The proof of the theorem can be readily generalized to include areal distributions and it may then be shown that the *General Standard*

Distance Deviation (S_Δ) is equal to the standard distance deviation times the square root of two (formula 7). Thus, the amount of scatter revealed in the distances separating all the units of a distribution can be shown to be related to the amount of scatter of those units around the mean centre of the distribution.

$$\textit{Formula 7: } S_\Delta = s_\Delta \sqrt{2}$$

Closely related to centrographic measures of distribution are the concepts variously known as population potential, market potential, and workplace potential. For a full discussion of the population potential concept the reader is referred to an article by Stewart and Warntz (1958 B). The *Population Potential* (V_o) at a point may be considered as a measure of the influence of total population on that point. The initial formulation of the concept was closely modelled on analogies with mechanics, but its most general use has been as a measure of general accessibility. According to its strict definition, the potential at a point, *O*, is obtained by measuring the distance from that point of each individual in the population inhabiting the universe of territory under study, computing the reciprocals of each distance, and summing the reciprocals. In practice a simpler procedure is followed. The universe of territory is subdivided into a manageable number of sub-units, the population of each unit is ascertained and, on the assumption that this population is concentrated at a single central point within each unit, the distance is measured between this centre and point *O*. The population potential at point *O* is then given by formula 8.

$$\textit{Formula 8: } V_o = \sum_{1}^{k} \frac{P_i}{d_i}$$

Where d_i is the distance of the i'th areal unit from point O

Conversion of population potential into market potential is simply attained by the use of an income or expenditure multiplier. Workplace potential (Duncan and Duncan, 1960) is defined, analogously to population potential, as the sum over all workplaces in the city of the reciprocals of the distances separating each workplace from any given point. The measure is interpreted as an index of the accessibility of a point to workplaces, on the assumption that accessibility declines as distance increases. Exploratory work using the workplace potential in the analysis of intra-urban residential structure is part of the Chicago Urban Analysis Project. The population potential measure has primarily been used in inter-urban studies of economic activity, but as a measure of accessibility and of 'sociological intensity' offers several possibilities for intra-urban research.

THE EMPIRICAL VALUE OF MEASURES OF DISTRIBUTION

The various percentage, centrographic and potential measures used in the description of areal distributions, have as their aim the identification of general regularities underlying the details of local spatial arrangement. The utility of all the measures mentioned has been illustrated in a number of empirical studies. The use of a few of the measures can be illustrated by a brief survey of the residential distribution and social status of various birthplace and occupation groups in the area of Greater Brisbane. The study is modelled, in part, on a paper by Duncan and Duncan (1955 B), in which they examine the residential pattern of occupation groups in Chicago.

Four main aspects of distribution are dealt with. The first aspect is the degree of residential dissimilarity between each of the birthplace groups and each of the occupation categories. In this way an indication of the residential social status of each birthplace group is obtained. The second aspect to be considered is the degree of residential segregation of each birthplace and occupation group, the extent to which each given group is residentially distinct from the rest of the Brisbane population. The third aspect is the degree of residential dissimilarity between pairs of birthplace and occupation groups, the extent to which they isolate themselves from one another. The fourth aspect of distribution to be considered is the centralization of each group, the 'mean' distance separating the group from the centre of Brisbane, taken to be a point in the city's C.B.D. In each case attention is focused on the relationship between residential distribution and social status.

The data of the study are from a 6 per cent household sample carried out for the Brisbane City Council in January 1960. The territorial subdivisions used as the basis of the calculations are a series of quarter-mile zones concentric to the city centre.

Although the local newspapers devote much space to the topic, the social ranking of the various birthplace groups in Brisbane is far from obvious. An indication of the hierarchy, as it operates residentially, is provided by the data of Table 10, which show the indices of residential dissimilarity between each of the birthplace groups and those occupational categories generally accepted as providing a measure of social status. The index values represent the net percentage of the one group who would have to move from their present zones of residence in order to reproduce the percentage distribution of the other group.

Table 10. Indices of Dissimilarity between Birthplace Groups and Occupations

				Manual Workers			
	Prof.	*Manag.*	*Sales/ Cler.*	*Skilled*	*Semi- skld.*	*Un- skilled*	*Pers. Serv.*
Rest Queensland	12	12	7	7	10	20	28
Brisbane-born	13	15	11	9	12	19	30
Rest Australia	14	10	10	11	13	19	28
U.K.-born	22	20	15	12	11	18	23
Europe-born	46	44	40	34	32	26	20

The smaller the index of dissimilarity, the more similarly are the two populations distributed over Brisbane. A net total of 12 per cent of the 'rest of Queensland-born' would have to move from their present zones of residence into other zones in order to give the group the same proportionate distribution as that of the professional workers. To achieve the same balance, some 46 per cent of the European-born group would have to move. Within the limits of the sample, no significant difference can be observed in the indices of dissimilarity of the three Australian-born groups, but the United Kingdom-born and the Europeans are each clearly distinguishable. The suggested social ranking has the Australian groups at its head and the European group at its base.

The social hierarchy evident in the residential distribution of the various birthplace groups is similar, but not identical, to that revealed in their occupational structure (Table 11). The main difference revealed in the two tables is the place of the United Kingdom birthplace group. On the basis of its occupational structure the group has the same social status as the three Australian-born groups. On the basis of its residential pattern, however, the group has to be assigned to a lower status level. Although the United Kingdom-born residents have a high status occupational structure, they tend to live in areas of low status. The difference is of considerable import in the social

Table 11. Occupation of Employed Males as a Percentage of Those in Birthplace Group

	Prof.	Manag.	Sales/ Cler.	Manual Workers Skilled	Semi-skld.	Un-skilled	Pers. Serv.
Rest Queensland	9	11	23	25	20	7	5
Brisbane	8	11	21	30	20	8	2
Rest Australia	6	17	18	25	17	10	4
U.K.-born	9	13	14	35	16	10	4
Europe-born	6	9	4	39	16	21	4

ecology of Brisbane and also, presumably, in the life of the persons concerned.

In their work on Chicago, Duncan and Duncan show that a clear relationship exists between social status and residential segregation. Evidence from Brisbane largely substantiates the Duncans' conclusion. Table 12 shows the indices of segregation for the various birthplace and occupation groups in Brisbane. The higher the index the more segregated is the group concerned.

Notable features of Table 12 are the U-shaped pattern of the occupation indices, repeating a Chicago finding, and the high degrees of segregation experienced by the European-born and by those employed in the personal service occupations.

On the assumption that spatial distance parallels social distance, the greater the social disparity between any two populations the more dissimilar should be their residential distribution. The hypothesis is

Table 12. Indices of Residential Segregation for Birthplace and Occupation Groups

	Rest Queensland-born	Brisbane-born	Rest Australia	U.K.-born	Europe-born
I_S =	5	12	10	14	40

	Prof.	Manag.	Sales/ Cler.	Skilled	Semi-skld.	Un-skilled	Pers. Serv.
I_S =	16	14	7	5	9	17	26

readily tested by computing the index of dissimilarity between pairs of the birthplace and occupation categories (Tables 13 and 14).

The evidence clearly supports the hypothesis that spatial distance and social distance are closely related. A notable feature of the tables is the very distinct pattern revealed for the European-born and for the personal service category.

Table 13. Indices of Residential Dissimilarity for Birthplace Groups in Brisbane

	Rest Queensland	*Brisbane*	*Rest Australia*	*U.K.*	
Brisbane	9	—			Index show % needing to move into other zones to balance distributions
Rest Australia	9	17	—		
U.K.	14	17	15	—	
Europe	40	42	38	32	

Table 14. Indices of Residential Dissimilarity for Occupation Groups in Brisbane

	Prof.	*Manag.*	*Sales/ Cler.*	*Skilled M.*	*Semi-Skil. M.*	*Unskilled Manual.*
Managerial	11	—				
Sales/Clerical	12	11	—			
Skilled/Manual	17	16	11	—		
Semi-skilled	19	18	13	8	—	
Unskilled	27	25	20	16	14	—
Pers. Service	32	28	25	24	23	20

The final aspect of distribution to be considered is the relative centralization of the various groups. According to the well-known Burgess zonal hypothesis of city growth there is an upward gradient in socio-economic status with increasing distance from the city centre. The application of the model to Brisbane may be tested by computing the mean distance deviation of the birthplace and occupation groups from the city centre (Table 15).

The data of Table 15 are clearly at variance with the Burgess hypothesis. The arrangement of the occupational groups even suggests an inverted form of the model. More detailed knowledge of the status structure of the city in relationship to the 'ideal' models developed by human ecology must await the conclusion of further studies.

No amount of cartographic skill could elicit the information about the association between residential distribution and social status revealed by the indices of dissimilarity and segregation and by the mean distance deviation. The first two measures are implicitly dependent on the territorial subdivisions on which they are computed, but this has very little effect on their empirical value. It is unlikely that a sufficient range of centrographic measures will be developed to render the indices of association empirically redundant. In the meantime, the index of dissimilarity and its derivatives have a very considerable potential as descriptive-analytical tools for research in urban social geography.

Table 15. Mean Distance Deviations for Birthplace and Occupation Groups in Brisbane

U.K.-born	*Brisbane*	*Rest Queensland*	*Rest Australia*	*Europe-born*
4·29 miles	4·22 miles	4·08 miles	3·96 miles	3·68 miles

Semi-Skilled Man.	4·15	*Managerial*	3·87
Skilled	4·07	*Sales and Clerical*	3·83
Unskilled	4·02	*Professional*	3·76

GEOGRAPHICAL ANALYSIS AND THE PROBLEM OF REGIONALIZATION

The recognition of valid regions is the basis of geographical research. Urban social geography is dependent on the delimitation of meaningful social areas, regions characterized by a stated combination of social conditions. An aggregation of social areas forms the total social structure and provides the framework for further analyses which attempt to relate the distribution of individual phenomena to the characteristics of the local community.

The choice of a regional system appropriate to the project in hand is the greatest single problem which faces the investigator of urban social distributions. The concept of region as an abstract or as a partial model of reality has been widely explored by geographical theorists, but much less attention has been given to the definition of methods by which the requirements of a formally-valid regional system may be satisfied. The criteria by which the meaningfulness of a region may be evaluated are in dispute and largely depend on the methodological position adopted by the investigator concerned. At the same time, the requirements of a regional system delimited for illustrative purposes or for the description of 'regional character' are very different from those of a system conceived primarily as an analytical device. The present discussion is largely concerned with regions in the latter category, in particular as they are relevant to the analysis of urban social structure and as they provide a frame for the study of urban areal variation.

In order to satisfy the requirements of analytical manipulation, a system of regions should fulfil three main conditions: first, the individual regions should be strictly comparable one with another; secondly, each region should be so formulated as to provide a maximum of external variation and a minimum of internal variation; thirdly, the item or items on which the regions are delimited should be directly related to the distributional problem being studied.

The traditional geographical approach to regional delimitation has been based on the comparison of mapped distributions. The relevance of the criteria to the problem being studied or the statistical validity of the eventual system have generally been left to the subjective judgement of the individual investigator. This judgement has doubtless often been correct, but the dangers of the approach are manifold. It is in this respect that the recent interest of students from other disciplines in the methodology of regionalism has been of greatest benefit to the progress of geographical research. The problems of 'regionalization' have been brought into new prominence and the concern of human ecologists and regional scientists to provide a meaningful framework for the analysis of spatial structure has not only revived many old arguments about the reality of regions, but has also greatly increased the amount of attention accorded to the formulation of regional systems which are formally satisfactory.

The most promising new techniques of regionalization in an urban context are the various multi-variable systems associated with the names of E. Shevky, R. C. Tryon, and C. F. Schmid. All three workers have attempted to identify social area 'types'[1] which are of primary significance in the social structure of the urban population.

[1] In all cases concern is with the stratification of the variables into meaningful classes. Although the regional types have an areal base and can be mapped, the factor of areal contiguity is absent from their formulation.

The eventual regional typologies produced by the various methods are very similar, but there are important differences in their procedures.

SOCIAL AREA ANALYSIS

The theoretical basis of the social area typology has been ably outlined in a monograph by E. Shevky and W. Bell (1955). Urban aggregations are viewed not as unique phenomena, with their own organizing principles, but as component parts of the wider modern society. From a review of the major structural trends taking place within that society, the authors of the social area typology identify a number of constructs which serve as discriminatory factors in the analysis of urban differentiation and stratification. The scheme lends especial emphasis to the postulates of C. Clark, on changes in the structure of productive activity and of employment, of L. Wirth, on the increasing scale of society and of social interaction, and of W. F. Ogburn *et al.*, on changes in the importance and role of the family. The subsequent argument can best be followed with the aid of Table 16 (from Shevky and Bell, 1955). From the initial postulates, a series of interrelated trends is recognized (column 2), which are believed to be major underlying factors of modern society. At particular points in time, any given social system may be described in terms of its differential relationships to these major trends (column 3). The temporal argument can be applied with equal validity, at any one time, to sub-groups within society. Three constructs can therefore be identified (column 4) which allow a classification and stratification of population sub-groups.

> Thus, from certain broad postulates concerning modern society and from the analysis of temporal trends, we have selected three structural reflections of change which can be used as factors for the study of social differentiation and stratification at a particular time in modern society (Shevky and Bell, in *Theodorson*, 1961, p. 227).

The three factors identified in the social area typology were termed by Shevky; social rank, urbanization, and segregation. Bell prefers the more explicit terms social status, family status, and ethnic status, respectively. From a list of sample statistics believed to be related to the three constructs (column 5) a brief selection is derived which most adequately define the indices (column 6). Census tract populations

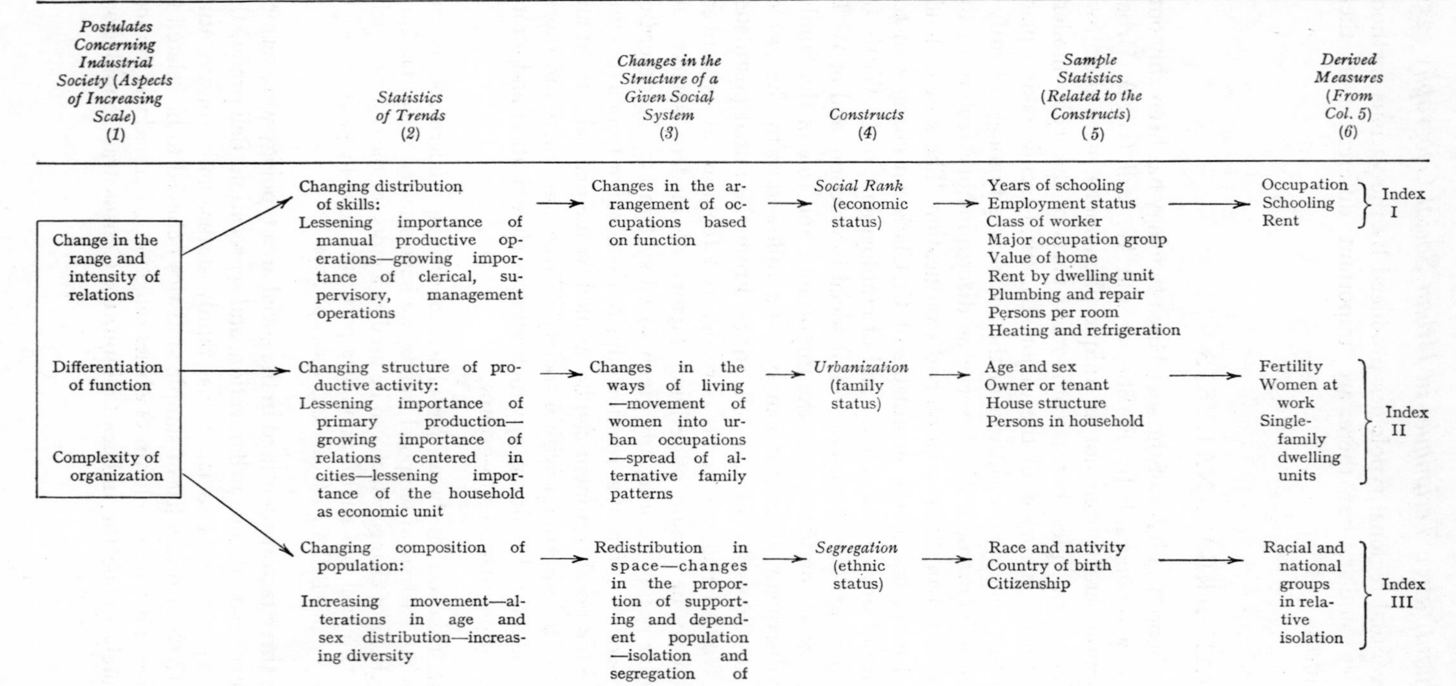

Postulates Concerning Industrial Society (Aspects of Increasing Scale) (1)	*Statistics of Trends* (2)	*Changes in the Structure of a Given Social System* (3)	*Constructs* (4)	*Sample Statistics (Related to the Constructs)* (5)	*Derived Measures (From Col. 5)* (6)	
Change in the range and intensity of relations	Changing distribution of skills: Lessening importance of manual productive operations—growing importance of clerical, supervisory, management operations	Changes in the arrangement of occupations based on function	*Social Rank* (economic status)	Years of schooling Employment status Class of worker Major occupation group Value of home Rent by dwelling unit Plumbing and repair Persons per room Heating and refrigeration	Occupation Schooling Rent	Index I
Differentiation of function	Changing structure of productive activity: Lessening importance of primary production—growing importance of relations centered in cities—lessening importance of the household as economic unit	Changes in the ways of living—movement of women into urban occupations—spread of alternative family patterns	*Urbanization* (family status)	Age and sex Owner or tenant House structure Persons in household	Fertility Women at work Single-family dwelling units	Index II
Complexity of organization	Changing composition of population: Increasing movement—alterations in age and sex distribution—increasing diversity	Redistribution in space—changes in the proportion of supporting and dependent population—isolation and segregation of groups	*Segregation* (ethnic status)	Race and nativity Country of birth Citizenship	Racial and national groups in relative isolation	Index III

Table 16. Steps in Construct Formation and Index Construction

can be described in terms of the indices and grouped into types on the basis of similar configurations of scores in the three-dimensional attribute space formed by the three status factors.

> Thus, the typological analysis . . . is a logically demonstrable reflection of those major changes which have produced modern, urban society (Shevky and Bell, in *Theodorson*, 1961, p. 230).

The social area typology is based on *a priori* reasoning from certain broad postulates concerning the nature of modern urban society. The formal validity of the system has been demonstrated by a series of American studies which suggest that not only is the typology generally applicable to urban society (Van Arsdol Jr. *et al.*, 1958), but that its key indices are both sufficient and necessary to account for the observed variation occurring in social phenomena (Bell, 1952; Tryon, 1955).

Less agreement has been forthcoming on the theoretical underpinnings of the typology or on the integration of its key indices with that theory (e.g. Hawley and Duncan, 1957). None the less, the technique has attracted wide attention as a stratification instrument and as a frame for the design and interpretation of urban sub-area field studies (Bell, 1958; Greer, 1956; Schmid, 1960 B; Shevky and Bell, 1955, pp. 20–22).

The general significance and utility of the social area typology can only be established by an extension of comparative studies, but it is readily apparent that the technique represents one of the most promising attempts yet available to provide a coherent and logically-demonstrable frame for the analysis of urban social structure.

FACTORIAL DESIGNS

In contrast to the social areas derived by the Shevky technique, regional systems developed on the basis of factor analysis are largely free of *a priori* considerations. Factor analysis allows the identification of the underlying order in a set of data and can reduce a large number of interrelated variables to a few basic independent factors which are sufficient in themselves to account for practically all the observed variation in the phenomena concerned. The technique of factor analysis is rigorous and demands a considerable degree of mathematical expertise. Readers interested in applying the technique should consult any of the standard texts listed in the bibliography.

Factor analysis was originally developed in psychology, for the purpose of identifying the principal dimensions or categories of mentality, but it has since proved of general application to a wide range of scientific problems. The application of factorial design to the delineation of geographical regions was pioneered by Hagood (1941 and 1943) who utilized data on population and agriculture. Price (1942) used the technique to evolve a classification of metropolitan centres. Application of factor analysis to the problems of intra-urban regionalization is primarily associated with the work of C. F. Schmid. The development of Schmid's technique can be followed in a series of articles dealing with various aspects of urban structure (Schmid, 1950, 1958, 1960). The most advanced form of his technique appears in a paper on the crime areas of Seattle (1960). An original correlation matrix containing twenty variables relating to crime and eighteen indices of demographic, economic, and social conditions, is reduced to eight principal factors which account for almost all the variation observed in the original data. Standard scores for each of the factors, computed tract by tract, form the basis of a classification system used to derive a sample of 'typical' tracts for further intensive study. At the same time, Schmid demonstrates that certain combinations of the principal factors 'represent the urban crime dimension par excellence'.

Closely related to factorial design is the technique of cluster analysis developed by R. C. Tryon (1939, 1955) who states that its purpose, when applied to urban social structure, is to identify the number, type and location of the sub-cultural groups which form the basis of the structure. The analysis reduces a large number of observations to three social dimensions which are both necessary and sufficient to describe all that generally differentiates the observed census tracts. In an application of the technique to San Francisco, Tryon shows that the 243 census tracts of the city fall into about eight general social areas.

> Within each social area the neighbourhoods are relatively homogeneous in terms of their values in these social dimensions, and each social area is described by its set of values in the social dimensions. The total configuration of all the social areas provides a final metric, objective description of the social structure of the population [Tryon, 1955, p. 3].

The stratification is primarily based on demographic and economic characteristics, but Tryon believes that the social areas are far more than merely regions of homogeneous demographic phenomena:

> There are . . . theoretical grounds and empirical evidence to support the belief that areas of people substantially different in demographic patterns . . . are also critically different in patterns of social ways [Loc. cit.].

The advantages of factorial and multi-variable designs are many and important. All the techniques are firmly based on empirical observations and are readily verifiable. Regional systems which are based on the factorial analysis of a set of areal data are, by formulation, basic to the understanding of the variation occurring within the data. A multitude of observations can be reduced to a manageable number of factors. In the process, previously unsuspected relationships may be revealed, while others, previously assumed, may be shown to be spurious. Particularly in the case of the social area typology, the analytical technique is intimately related to a body of more general theory and allows the formulation of readily-tested structural hypotheses. Finally, the use of certain types of factorial designs allows the objective measurement and statement of the amount of detailed variation in the data which has to be sacrificed at successive levels of regional generalization (Berry, 1961).

The disadvantages of multi-factor analysis are largely a result of its conceptual and operational complexity. The techniques demand a considerable mathematical prowess. Particularly in urban areas, they also demand a sophistication of data which is rarely attained outside the United States and Scandinavia. Perhaps the most difficult problem of all to overcome is that factorial designs depend on the use of costly computer time and presuppose, as Berry (1961) points out, considerable financial assistance. This is rarely available to the individual student of areal variation.

The adoption of factorial methods of regionalization is obviously desirable if urban social geography is ever to attain the 'scientific ideal of certainty'. In view of the demands of the techniques, however, their widespread adoption must inevitably be long delayed.

SCALOGRAM ANALYSIS

In those parts of the world which are less well endowed with statistics than the United States there is considerable scope for a method of regionalization which, while being as objective and meaningful as possible, is less complicated than factorial and social area designs and can utilize less sophisticated data. A technique which may satisfy some of these requirements is based on a modification of the Guttman scalogram method, which has been widely used in attitude surveys (see References). The technique was devised in the course of a study dealing with the geographical pattern of criminality and mental illness in two British cities (Timms, 1962).

In essence the scalogram technique is a means of ordering ranked data in such a way that a single unidimensional scale is produced along which effective measurement is possible. As in multi-factor analysis a number of observations may be reduced to a single dimension, which may then be used as a stratification and correlative instrument. The criterion of unidimensionality lies in the pattern of scale responses, the rank position of the respondents in each of the items included in the scale. A perfect scale takes the form of a parallelogram (Table 17).

Table 17. Dichotomized three item scalogram

	High Values (Score 1)			*Low Values* (Score 0)			*Scale Type.*	*Scale Score*
Items:	*A*	*B*	*C*	*A*	*B*	*C*		
Area 1	x	x	x				I	3
Area 2		x	x	x			II	2
Area 3			x	x	x		III	1
Area 4				x	x	x	IV	0

x indicates area response to item.

Knowledge of the scale type or score of an area allows a complete description of all the attributes which it possesses. Two areas with the same score have identical characteristics in terms of the scale items and categories. Random departures from the perfect form are allowable in so far as they comprise less than 10 per cent. of all responses. Certain other criteria have also to be satisfied before a scale can be accepted as valid (Ford, 1950).

The scalogram technique constitutes a valuable analytical instrument. One of its most important properties is that a correlation between scale scores and any external variable is equivalent to a multiple correlation between all the items constituting the scale dimension and whichever external variable is concerned. The scale can be constructed from data which can be ranked but not precisely measured. At the same time, a sample of items from a valid scale dimension reproduces the scale pattern of the whole.

The use of the technique in the delimitation of regions may be demonstrated by an example from Luton.

A trichotomized ranking of a number of socio-economic and social defect data, available street by street, was used to evolve two series of scale types. The initial selection of items for inclusion in the dimensions was based on a random area sample of 150 base units. From the sample two valid scalograms were constructed containing (1) the variables rateable value per adult, net population density, and percentage of jurors,[1] and (2) the variables adult and juvenile crime rate and mental illness rate. In both cases all the criteria of scalability were met.

Once the scale content was determined by the sample data, the process of regionalization was straightforward. Information was tabulated for each street in terms of the scale items and categories previously defined. An example of tabulation and of the subsequent identification of regional types is given in figure 12.1 and its associated tables.

Each of the scales yielded seven regional types and these were utilized as the cells for a conventional correlation analysis with sub-area populations entered as the frequencies. A generalized map of the resulting two-factor regional types is given in figure 12.2.

The procedure of regionalization was relatively objective. Given the prior decisions to define regional types in terms of certain dimensions relevant to the project and to use a particular division of the data, the only subjective element in the scheme concerns the allocation of sub-areas exhibiting non-conforming scale patterns. In these circumstances the allocation is rarely left ambiguous after a consideration of the scale pattern of contiguous areas and of the internal ordering of the scale items.

The scale analysis technique provides a convenient and precise means for studying the internal structure of urban areas. The technique is by no means perfect, but it is a considerable improvement over the more conventional tools of regionalization.

[1] The statistic percentage of jurors has been widely used as an idex of socio-economic status, viz. Gray, Corlett and Jones, *The Proportion of Jurors as an Index of the Economic Status of a District*, Govt. Social Survey, London, 1951.

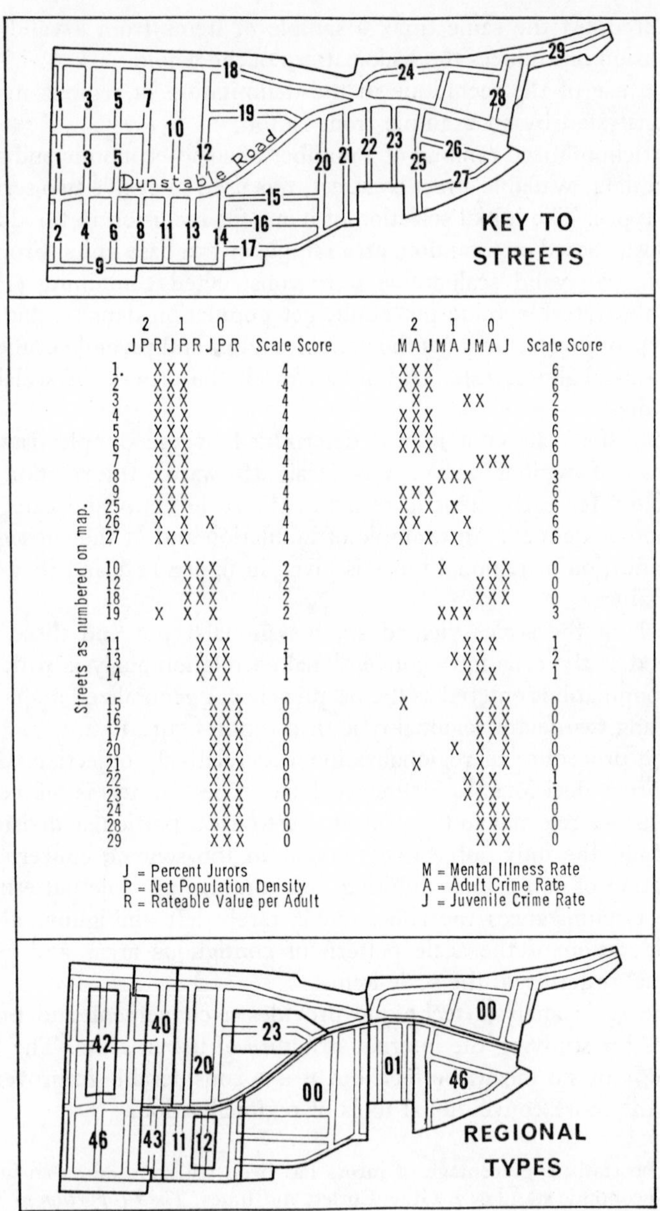

FIG. 12.1. *Luton, England: Identification of regional types based on the trichotomized ranking of a number of socio-economic and social defect data.*

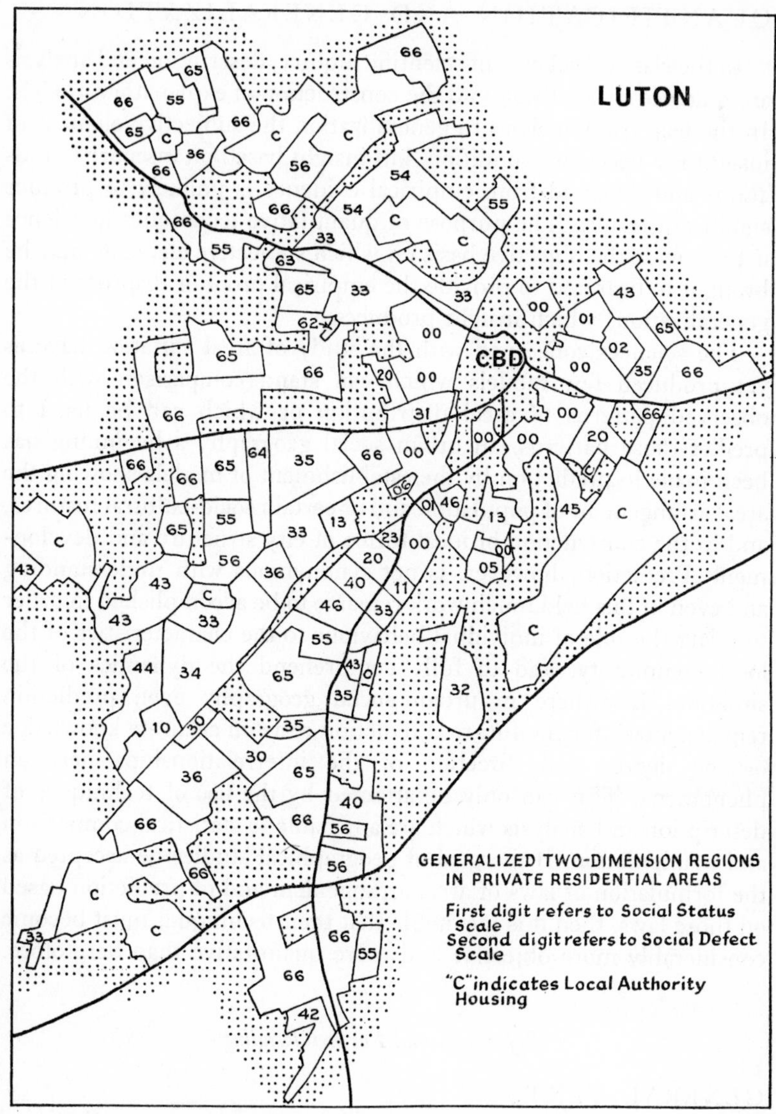

FIG. 12.2. *Luton, England: Generalized map of two-factor residential status/defect regions.*

QUANTIFICATION AND GENERALIZATION

In the classical scheme of scientific inquiry, description and analysis are followed by synthesis and the construction of explanatory models. In the leap from analysis to generalization the subjective element of insight is a necessary condition, but insight based on insecure foundations and divorced from empirical evidence is unlikely to produce significant results. The purpose of quantitative techniques in science is to provide the objective basis on which subjective elements may be brought to bear and to provide the empirical proof or disproof of the generalizations which insight produces.

The sciences concerned with the study of areal variation have as yet produced few models which can stand comparison with the observed patterns of areal distribution or which can be used to predict those patterns. In urban social geography a beginning has been made in prediction by the establishment of models showing the areal arrangement of population and of certain socio-cultural features, and in the construction of ideal forms of city structure and development. Prediction, however, is not synonymous with understanding and even in this field much work remains to be accomplished in order to relate the role of individual behaviour to the characteristics of the local community, and to fully comprehend the dynamics of the situation. Elsewhere in urban social geography even prediction remains a task for the future. Prediction rests on accurate knowledge of the degree and direction of the interrelationships between phenomena. This can only be attained by the use of techniques of description and analysis which are amenable to statistical comparison and manipulation. If the goal of geographical studies be accepted as the formulation of laws of areal arrangement and of prediction based on those laws, then it is inevitable that their techniques must become considerably more objective and more quantitative than heretofore.

References and Further Reading

GENERAL TEXTS

DUNCAN, O. D., CUZZORT, R. P. and DUNCAN, B., 1961, *Statistical Geography: Problems in Analysing Areal Data* (Glencoe, Ill.).

ISARD, W., 1960, *Methods of Regional Analysis* (Cambridge, Mass.) (see Chapters 7 and 11).

REYNOLDS, R. B., 1956, 'Statistical Methods in Geographical Research', *Geog. Rev.*, **46**, 129–31.

INDICES OF GEOGRAPHICAL ASSOCIATION

DUNCAN, O. D., 1957, 'The Measurement of Population Distribution', *Population Studies*, **11**, 27–45.

DUNCAN, O. D. and DUNCAN, B., 1955 A, 'A Methodological Analysis of Segregation Indices', *American Sociological Review*, **20**, 210–17.

— 1955 B, 'Residential Distribution and Occupational Stratification', *American Journal of Sociology*, **60**, 493–503.

DUNCAN, O. D. and LIEBERSON, S., 1959, 'Ethnic Segregation and Assimilation', *American Journal of Sociology*, **64**, 364–74.

FLORENCE, P. S., 1948, *Investment, Location and Size of Plant* (Cambridge).

PETERS, W. S., 1946, 'A Method of deriving Geographic Patterns of Associated Demographic Characteristics within Urban Areas', *Social Forces*, **35**, 62–8.

ROBINSON, A. H., 1957, 'A Method for describing Quantitatively the Correspondence of Geographical Distributions', *Ann. Assn. Amer. Geog.*, **47**, 379–91.

CENTROGRAPHIC TECHNIQUES

BACHI, R., 1963, 'Standard Distance Measures and Related Methods for Spatial Analysis', *Papers and Proceedings of the Regional Science Association*, **10**, 83–132.

HART, J. F., 1954, 'Central Tendency in Geographical Distributions', *Econ. Geog.*, **30**, 48–59.

SVIATLOVSKY, E. E. and EELLS, W. C., 1937, 'The Centrographical Method and Regional Analysis', *Geog. Rev.*, **27**, 240–54.

WARNTZ, W. and NEFT, D., 1960, 'Contributions to a Statistical Methodology for Areal Distributions', *Journal of Regional Science*, **2**, 47–66.

POPULATION POTENTIAL

ANDERSON, T., 1956, 'Potential Models and Spatial Distribution of Population', *Papers and Proceedings of the Regional Science Association*, **2**, 175–82.

CARROLL, J. D., 1955, 'Spatial Interaction and the Urban-Metropolitan Description', *Papers and Proceedings of the Regional Science Association*, **1**, 0.1–0.14.

CARROTHERS, G., 1956, 'An Historical Review of the Gravity and Potential Concepts of Human Interactions', *Journal American Institute of Planners*, **22**, 94–102.

DUNCAN, B. and DUNCAN, O. D., 1960, 'The Measurement of Intra-city Locational and Residential Patterns', *Journal of Regional Science*, **2**, 37–54.

STEWART, J. Q., 1947, 'Empirical Mathematical Rules concerning the Distribution and Equilibrium of Population', *Geog. Rev.*, **37**, 461–85.

— 1948, 'Demographic Gravitation: Evidence and Applications', *Sociometry*, **11**, 31–58.

STEWART, J. Q. and WARNTZ, W., 1958 A, 'Macrogeography and Social Science', *Geog. Rev.*, **48**, 167–84.

— 1958 B, 'Physics of Population Distribution', *Journal of Regional Science*, **1**, 99–123.

— 1959, 'Some Parameters of the Geographical Distribution of Population', *Geog. Rev.*, **49**, 270–2.

THE SOCIAL AREA TYPOLOGY

BELL, W., 1953, 'The Social Areas of the San Francisco Bay Region', *American Sociological Review*, **18**, 39–47.

— 1958, 'The Utility of the Shevky Typology for the Design of Urban Sub-area Field Studies', *Journal of Social Psychology*, **47**, 71–83.

GREER, S., 1956, 'Urbanism Reconsidered: a Comparative Study of Local Areas in a Metropolis', *American Sociological Review*, **21**, 19–25.

HAWLEY, A. H. and DUNCAN, O. D., 1957, 'Social Area Analysis: a Critical Appraisal', *Land Economics*, **33**, 337–45.

SHEVKY, E. and BELL, W., 1955, *Social Area Analysis: Theory, Illustrative Application, and Computational Procedures* (Stanford).

SHEVKY, E. and WILLIAMS, M., 1948, *The Social Areas of Los Angeles: Analysis and Typology* (Berkeley).

VAN ARSDOL, M. D., CAMILLERI, S. F. and SCHMID, C. F., 1958 A, 'The Generality of Urban Social Area Indices', *American Sociological Review*, **23**, 277–84.

— 1958 B, 'An Application of the Shevky Social Area Indexes to a Model of Urban Society', *Social Forces*, **37**, 26–32.

THEODORSON, G. A., 1961, *Studies in Human Ecology* (Evanston, Ill.).

FACTOR ANALYSIS

BELL, W., 1955, 'Economic, Family, and Ethnic Status: Empirical Test', *American Sociological Review*, **20**, 45–52.

BERRY, B. J. L., 1961, 'A Method for Deriving Multi-factor Uniform Regions', *Przeglad Geograficzny*, t. 33, z. 2.

CATTELL, R. B., 1952, *Factor Analysis: an Introduction and Manual for the Psychologist and Social Scientist* (New York).

FRUCHTER, B., 1954, *Introduction to Factor Analysis* (New York).

HAGOOD, M. J., 1943, 'Statistical Methods for Delineation of Regions applied to Data on Agriculture and Population', *Social Forces*, **21**, 287–97.

— *et al.*, 1941, 'An Examination of the Use of Factor Analysis in the Problem of Sub-regional Delineation', *Rural Sociology*, **6**, 216–33.

HAGOOD, M. J. and PRICE, D. O., 1952, *Statistics for Sociologists* (New York).

HARMAN, H. H., 1960, *Modern Factor Analysis* (Chicago).

MOSER, C. A. and SCOTT, W., 1961, *British Towns: a Statistical Study of Their Social and Economic Differences* (London).

PRICE, D. O., 1942, 'Factor Analysis in the Study of Metropolitan Centres', *Social Forces*, **20**, 449–55.

SCHMID, C. F., 1950, 'Generalizations Concerning the Ecology of the American City', *American Sociological Review*, **15**, 264–81.

— 1960 A, 'Urban Crime Areas: Part I', *American Sociological Review*, **25**, 527–42.

— 1960 B, 'Urban Crime Areas: Part II', *American Sociological Review*, **25**, 655–78.

SCHMID, C. F., MACCANNELL, E. H. and VAN ARSDOL, M. D., 1958, 'The Ecology of the American City: Further Comparison and Validation of Generalizations', *American Sociological Review*, **23**, 392–401.

SWEETSER, F. L., 1965, 'Factorial Ecology: Helsinki, 1960', *Demography*, **2**, 372–85.

THURSTONE, L. L., 1947, *Multiple Factor Analysis: a Development and Expansion of the Vectors of Mind* (Chicago).

TRYON, R. C., 1939, *Cluster Analysis* (Ann Arbor).

— 1955, *Identification of Social Areas by Cluster Analysis: a General Method with an Application to the San Francisco Bay Area* (Berkeley).

SCALOGRAM ANALYSIS

FORD, R. N., 1950, 'A Rapid Scoring Procedure for Scaling Attitude Questions', *Public Opinion Quarterly*, **14**, 507–32.

GREEN, N. E., 1956, 'Scale Analysis of Urban Structures: a Study of Birmingham, Alabama', *American Sociological Review*, **21**, 8–13.

HAGOOD, M. J. and PRICE, D. O., 1952, *Statistics for Sociologists* (New York).

MOSER, C. A., 1958, *Survey Methods in Social Investigation* (London).

STOUFFER, S. A. *et. al.*, 1950, *Measurement and Prediction* (Princeton).

TIMMS, D. W. G., 1962, *The Distribution of Social Defectiveness in Two British Cities: a Study in Human Ecology* (Ph.D. dissertation, University of Cambridge).

CHAPTER THIRTEEN

Geographical Techniques in Physical Planning

E. C. WILLATTS

Principal Planner, Ministry of Housing and Local Government

A quarter of a century ago Britain had almost no professional geographers outside the field of teaching, but since then they have established themselves in various other fields and particularly that of physical, or town and country, planning. There was an early recognition of their contribution in carrying out and presenting surveys, not merely topographical, but social and economic. More slowly came the realization that their understanding of many of the complex factors which confront physical planners entitles them not only to make surveys on the basis of which others would prepare plans, but to analyse the problems which require to be solved and thus to indicate their solutions. It was his early appreciation of this which led Lord Justice Scott, a revered Honorary Vice-President of the Town Planning Institute, to write 'Town Planning is the Art of which Geography is the Science'.

This paper is concerned with a few examples of the way in which some of those geographers in central government who are concerned with physical planning on a national level have been applying their techniques with the whole country as their field of review, and with major problems which cannot be decided at a local level. Their first duty is to array the primary facts about the land; its configuration, its geological nature, vegetation and climate, its soils and its quality, for these are vital to the physical planner. These facts require to be readily available in map form, preferably in a co-ordinated series of maps. So do the changing facts about the use man makes of the land, for work and recreation, the way population is distributed over it, the changes in that distribution, and the many factors which influence those changes, such as the patterns of industry and employment, of power, offices, shops and of communications. It is an essential part of

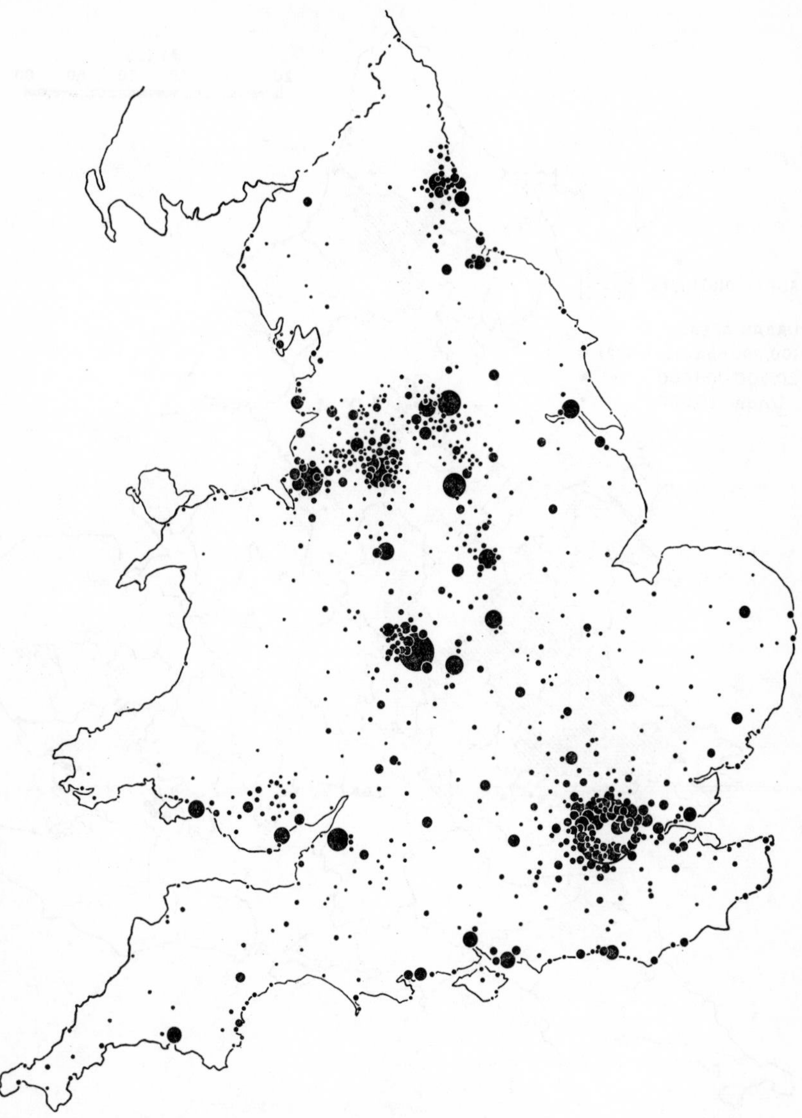

FIG. 13.1. *England and Wales: Urban population, 1961. The County of London is represented by a single open circle.*

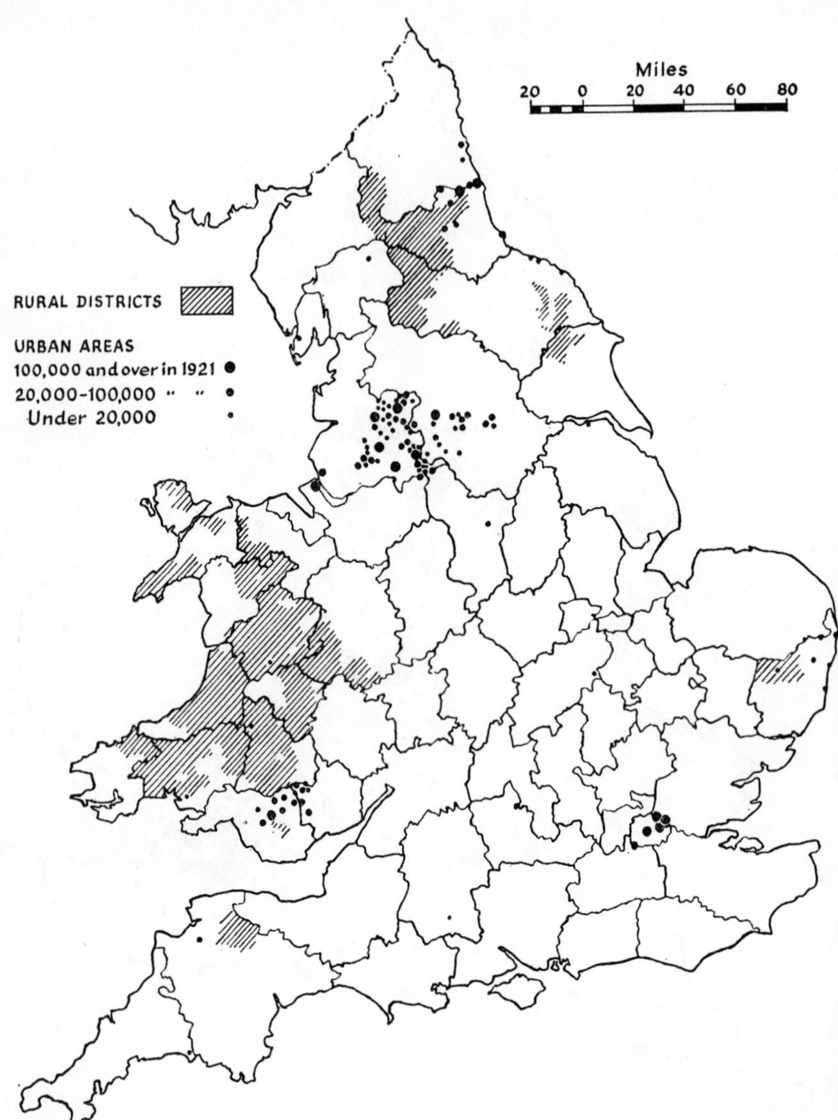

FIG. 13.2. *England and Wales: Areas of persistent population decrease, 1921–31, 1931–9, and 1939–47.*

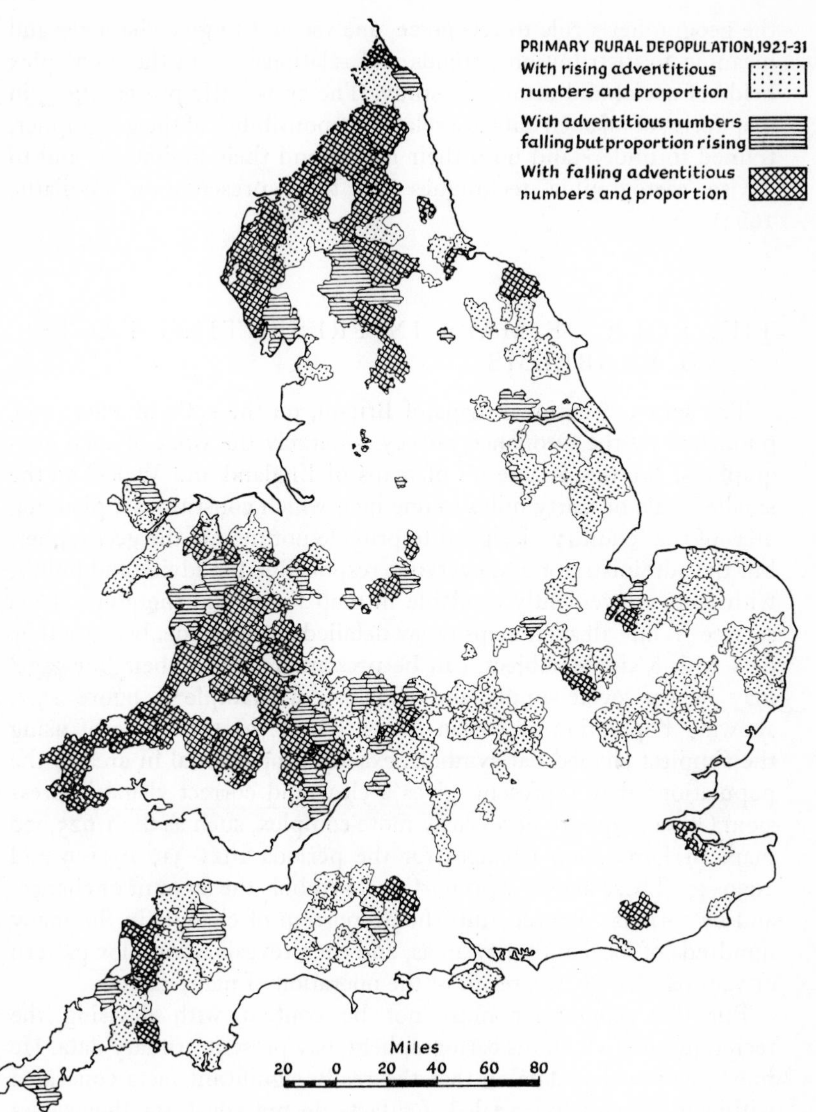

FIG. 13.3. *England and Wales: Different types of rural population decline, 1921–31.*

the geographer's role to recognize, analyse and to give character and meaning to distributions, trends and relationships in these complex fields of social and economic study. The systematic presentation, in map form, of all such data is a related responsibility of the geographer, trained to understand both their nature and their limitations and to devise cartographic techniques for their presentation (Willatts, 1963).

THE ROLE OF MAPS IN PRESENTING FACTS AND PROBLEMS

The series of national maps of Britain, on the scale of 1:625,000, published by the Ordnance Survey, is largely the work of such geographers. So, too, is a series of maps of England and Wales[1] on the smaller scale of thirty miles to one inch which constitutes a planning atlas of that country, designed to provide not merely the geographer, but the administrator and everyone responsible for advice and policy, with facts conveniently available in map form, and therefore at his service. While all such maps repay detailed study, some, because they deal with a single subject, can be presented so that their 'message' may be conveyed simply and clearly. An example is figure 13.1, showing Population of Urban Areas, where the technique of using the simplest shaped flat symbols, exactly proportional in area to the populations they represent, gives a clear and correct visual impression. Other maps are necessarily more complex, such as the 1/625,000 maps of Population Changes for the periods 1921–31, 1931–9 and 1939–47. These show by proportional symbols the amount of change, and by various coloured tints the proportion of change, in the many hundreds of local authority areas, and each reveals a complex pattern of varying change, the result of the operation of many factors.

But the geographer must not be content with devising the techniques by which his cartographers may present primary data. He must analyse that data so that the really significant facts concealed within it are clearly revealed, for facts do not speak for themselves until they are correctly marshalled. Trends in population change may be taken as an example. The detailed maps already referred to may each reveal much by close study, but further laboratory techniques must be applied to them before significant and persistent

[1] To be published by H.M.S.O. as 'Desk Atlas of Planning Maps of England and Wales'.

patterns emerge. Such techniques, adopted by workers with a trained curiosity, have resulted in the preparation of such a map as that shown in figure 13.2 (Willatts and Newson, 1953). This map is very simple, a ruling for rural areas and graded dots for the towns in England and Wales which were found to be persistently losing population in each of the three consecutive periods covered by the published series of maps. The map is a typical one prepared not primarily for geographers but for intelligent lay readers, a class of user to whom the essential truths need to be effectively and speedily conveyed. The simplicity of such a map is therefore its strength and merit. Its 'message' is clear to readers who only with the utmost difficulty might have discovered the facts from volumes of arid statistics and then would still not have seen them placed on a map where the patterns are so clear that they imprint a photographic impression on the mind. Having thus answered the question 'what?' and aroused curiosity by the presentation of significant patterns of distribution the geographer should face the obvious challenge and answer the inevitable question, 'why?'.

Not that a simple answer is usually possible, particularly with a subject so complex as population changes. Vince (1952) demonstrated this in a study of the structure and distribution of rural population in England and Wales, 1921–31, in which he divided the rural population into the *primary* population, directly dependent on the land,[1] the *secondary*, which serves the needs of the primary, and the *adventitious* which neither depends on the land nor serves those who do. He developed a detailed analysis by which he was able to show, among other things, not merely where both the total occupied population and the primary rural population had declined, but to recognize and delimit different types of decline, as shown on figure 13.3 which is reproduced from his paper. It reveals, in his words, that: 'While rural depopulation is almost everywhere a function of remoteness there is a significant difference between upland and lowland Britain. Even in lowland eastern England some primary depopulation occurred, but mostly it was accompanied by adventitious increases on a scale less than sufficient to maintain the total level. In Wales and the Highland zone generally, however, the adventitious numbers on the whole fell to a greater extent than the primary population.' His analysis of the three components of rural population has facilitated a clearer understanding of various aspects of the problem of rural population changes.

[1] In this context: agriculture, horticulture and forestry.

SYNOPTIC MAPS

Techniques which result in a complicated and laborious analysis of a subject being presented with graphic simplicity are suited to certain subjects, especially those where it is desirable that the results should be continually borne in the minds of planners and others. But there is another class of problem which calls, not for a simplified summary of the results of an analysis, but for a synoptic presentation of all the complex factors involved in a problem. The cartographic technique which has been evolved for such purposes, though commonly called a 'sieve' map, is, at its best, much more than that title implies. A good example of its use is in the selection of a suitable locality for some major project such as a new town. The method pre-supposes that the location factors are known and can be translated into terms of geographical data which can be marshalled. Its object is not to eliminate all areas where unfavourable factors operate, but to indicate the relative suitability of different parts of the 'search area' for the construction of a new town, and to do so in such a way that each of several factors operating in any locality may be clearly distinguished. In this way those concerned with taking decisions are enabled to appreciate the issues involved. These may be so many that they cannot be presented on one map without one or more overlays being constructed. A map of this type for the West Midlands of England, prepared in connexion with the selection of a site for a new town, covers 6,000 square miles and consists of a 1/250,000 topographic base map and two transparent overlays on which are shown areas subject to various disadvantages. The notations in which these are presented are carefully designed so that none is opaque and all allow two or more factors to be shown as operating in the same area.

The major factors shown on the base map are: physical; land too high, land too steep, and land which for other reasons (such as liability to flooding) is unsuitable for building; locational and accessibility factors, such as areas more than five miles from a major road or railway. Competing land uses are indicated, including areas used or allocated for mineral working; areas of the highest quality agricultural land; Green Belt land; land reserved for afforestation or for its natural beauty. Further information on competing uses is shown on a first overlay which shows land liable to subsidence arising from the underground working of minerals; other good agricultural land; areas of high landscape value; and land occupied by government uses.

The second overlay deals with other aspects of physical limitations, position and accessibility, such as the restrictions on the supply of underground water, the tracts which lie within certain arbitrary radii of the main conurbation and other large cities and towns, and areas more than two miles from a major road or railway.

The map and overlays together thus present a cumulative picture of the various difficulties to which the building of a new town at any location within the study area would be subject and obversely suggest areas which it would seem most profitable to examine in detail. The technique is applicable to geographical factors which can be expressed in relation to the land which they affect. It is not possible to use it to portray imponderable economic and social factors.

MAPS OF LOCATIONAL PROBLEMS

Similar techniques have been applied to other subjects, such as the location of nuclear power stations, projects which by their size and nature must make a significant impact on the landscape and life of the area in which they are placed. Once the necessary maps have been compiled they are often found to be of great assistance in the examination of the problems of siting other developments which must somehow be fitted into a crowded land in which there are many societies and individuals ready to object to the alleged ill-chosen siting of any new project. Such champions of natural beauty are often in as much, or more, doubt than technical experts as to the visible effect of some proposal. Artists and photographers have often combined their skills to show what some new building or other proposed erection, such as a radio or television mast, would look like in a town or country landscape. Their illustrations raise, but do not answer, the question 'from how much countryside would it be visible?' A geographical technique can be used to supply the answer.

On a close-contoured map, such as the 1:25,000, which has a contour interval of 25 feet, the position of the proposed project is marked and radial lines are drawn from this at intervals of, usually, 5 degrees with intermediate rays at greater distances. Along each of these a profile is plotted. The project is then marked on each as a vertical feature and from it sight lines are projected against the profile which is then marked to show from which parts of the surface there is unobstructed vision of the project. Due allowance must be made for trees and other features, which can often be studied, and their heights assessed, from air photos. The portions of each radial line from which the feature may be seen are then marked on the map and this is then studied to enable an interpolation to be made of the intervening areas from which sight lines would be uninterrupted and so a map is compiled to show the ground from which a view of the proposed erection would be obtained. In practice it has been found useful to divide this into two classes, the areas from which, say, more than 75 per cent and less than 25 per cent of it would be visible. In the case of a structure whose bulk and colour would render it obtrusive from a considerable distance, allowance must be made for refraction and the curvature of the earth. With minor adaptations the technique has been used to suggest locations where certain projects would be likely to give rise to much less objection on grounds of prominence than on the sites first proposed by their sponsors.

The problem of where to site the many new developments which are inevitable in a changing industrial economy does indeed provide a challenge to the geographer. He must not merely find the answer, but must demonstrate both the truth and the consequences of his conclusion. An example of such a problem is the location of oil refineries. In recent years radical changes have occurred with the building of refineries in consumer rather than producer countries, with the vast expansion in their capacity and the development of the so-called 'super-tanker' for importing the crude oil. The use of these vast vessels, of from 65,000 to 100,000 or more tons, remarkably reduces the cost of conveying oil. But they demand harbours with a draught of more than 45 ft, a sheltered anchorage and a quick turn-round. At the unloading point interference by wind, tidal streams and other vessels must be avoided, while nearness to markets is important. The ideal refinery site is one of 1,000 or more acres on flat land, affording good foundations, with ample fresh water and from which distribution can be easily and economically made, and it must be adjacent to a sheltered anchorage where super-tankers can unload quickly and safely.

A careful analysis of the chief factor, deep water in sheltered anchorages, has been a vital prerequisite to the discussion of the siting of new refineries. Once made, it can be expressed quite simply in map form on large or small scale, enabling the significance of the real scarcity of such localities to be appreciated. Figure 13.4 shows the close relationship of sheltered anchorages, suitable for large tankers, to oil refineries of which the eight largest have been built, or greatly enlarged, since the Second World War (James, Scott and Willatts, 1961).

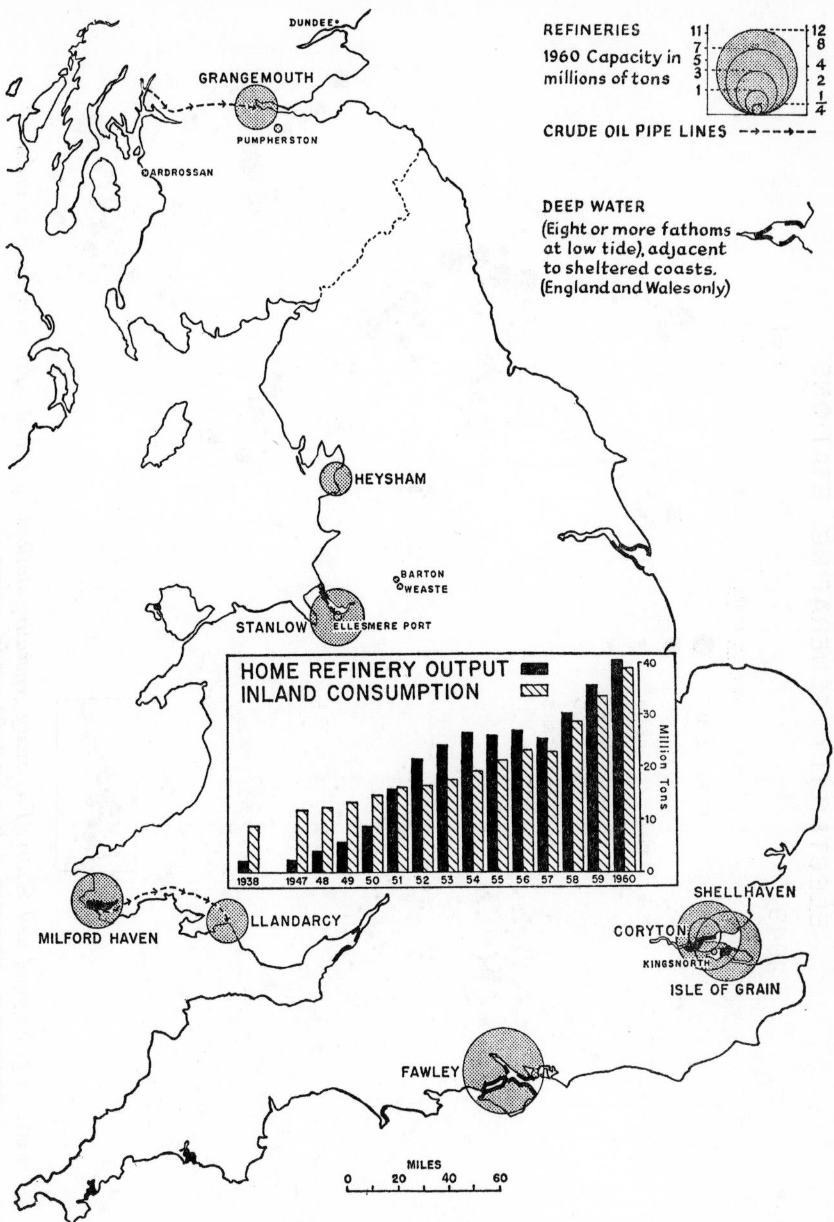

FIG. 13.4. *United Kingdom: Oil refineries, 1960, in relation to deep, sheltered anchorages. Another large refinery on Milford Haven has subsequently been completed.*

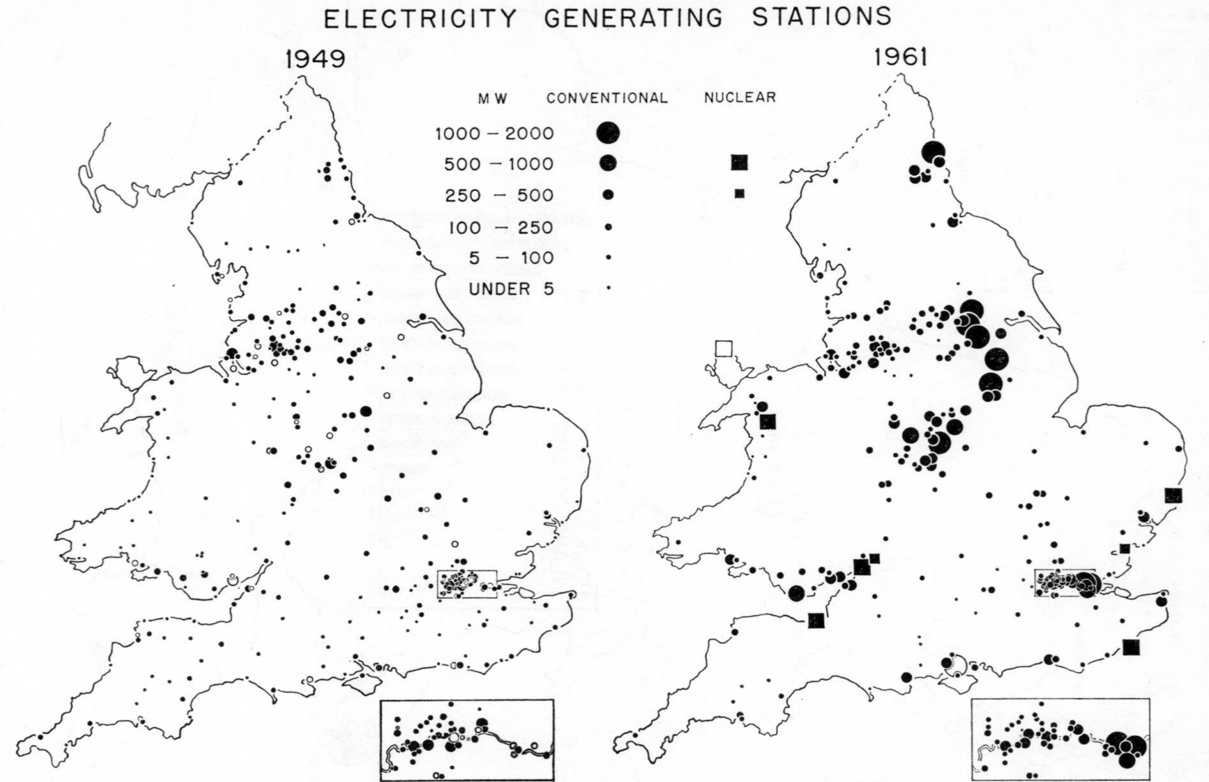

FIG. 13.5. *England and Wales: Electricity generating stations 1949 and 1961, completed or under construction. Projected stations in 1961 are in outline.*

Another power industry has provided a very different siting problem in which geographers have helped to evolve a solution. As shown in figure 13.5, the pattern of generation of electric power from coal-burning stations is changing from one of many small stations near the points of consumption to a few very large ones on or near the coalfields, whence the current is conveyed by high voltage cables. The greatest concentration of these in Britain is now in the Trent Valley where low grade small coal is readily available in proximity to the river from which cooling water is drawn.

Vast quantities of coal are burned (over 20,000 tons a day at one site) and from the group of power stations in this valley about three and a half million cubic yards of powdered fuel ash will soon be produced each year. This presents an enormous and growing problem of disposal, in spite of attempts to find economic uses for the material. However, the valley is also a very important source of gravel, worked from shallow wet pits, yielding 4½ million cubic yards per year. A careful study of the separate problems of the disposal of fuel ash, which tends to create a dust nuisance, and the phasing of gravel working, has led to arrangements whereby much of the ash is used to fill gravel pits, largely by pumping it as a slurry, while future gravel workings are planned in relation to the needs of the power stations for space in which to dispose of their ash. Thus it is proving possible to make the best and most economical use of land: the mineral is obtained and a waste product disposed of on the same sites, which can subsequently be restored to agriculture and additional land is not sterilized by being covered with heaps of waste ash. This solution was bevolved y consultation and team work, in which geographers played a significant role, an important part, but only one of which, was expressing the facts quantitatively in map form so that the nature of the problem and the means of solving it were clearly evident to the policy makers. Figure 13.6 shows the relationship between the existing and proposed power stations in the middle Trent valley, their sources of fuel and the gravel pits to which some of the powdered fuel ash is conveyed. (The large station shown as 'projected' is not being constructed because the Ministry of Power subsequently rejected the application for consent to build it on that site.)

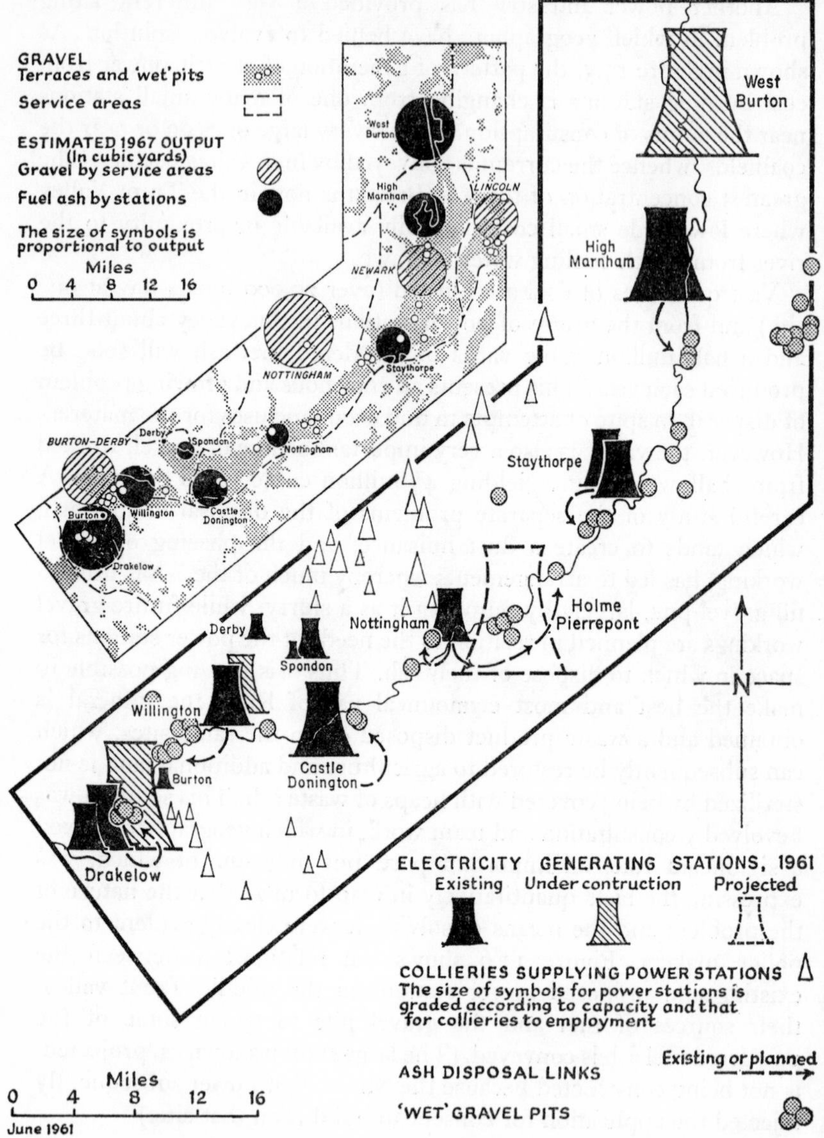

FIG. 13.6. *Middle Trent Valley: The relationship between power stations, fuel supplies and gravel pits, 1961.*

PROBLEMS IN LAND-USE COMPETITION

The extraction of minerals from underground or from the surface of our country, of which over 3,000 acres are annually made derelict by surface mineral working, gives rise to problems in the field of competing land uses which call for continuous study by both geographers and their colleagues, the geologists, and has been given close attention by S. H. Beaver and others (Beaver, 1944; Wooldridge and Beaver, 1950; Beaver, 1949). The problems raised can and do change with technical and social changes, but they still require to be studied. For example the development of modern earth-moving machinery has made it possible, and legislation has made it necessary, to restore to agricultural use land from which ironstone has been won by opencast workings. On the other hand, with the winning of gravel (now second only to coal in value and tonnage) the opposite trend has rapidly occurred. Only a few years ago it was generally thought to be very desirable to find means of filling the large lagoons left around parts of London and other large towns as a result of the extraction of river valley gravel. But changes in leisure habits have brought a rapid revolution. Lagoons or other stretches of water near large centres of population are now in great demand for recreational use particularly by sailing clubs, so that there has even been talk of a 'blue belt', as well as a green belt, around the metropolis.

The working of minerals which give rise to subsidence, particularly but not only, coal, gives rise to obvious problems in the siting of new developments in a country where 5,000,000 people still live on active coalfields. Large-scale development requires not only special structural precautions in building but in the detailed complementary phasing of mining and building. On the site of the New Town of Peterlee, Co. Durham, this was particularly important, for when its plans were first prepared there were about 30 million tons of coal beneath the site and although a cover of glacial drift tends to reduce the shock of subsidence, the fracturing of the underlying magnesian limestone causes serious surface distortion. Figure 13.7 shows, for the northern part of the town, how its development was phased in relation to the availability of land after the extraction of the underlying coal. A million tons had to be sterilized under the town centre, the early construction of which was imperative, but elsewhere detailed programmes were evolved so that building could take place, area by area, as the land became available after the coal had been mined.

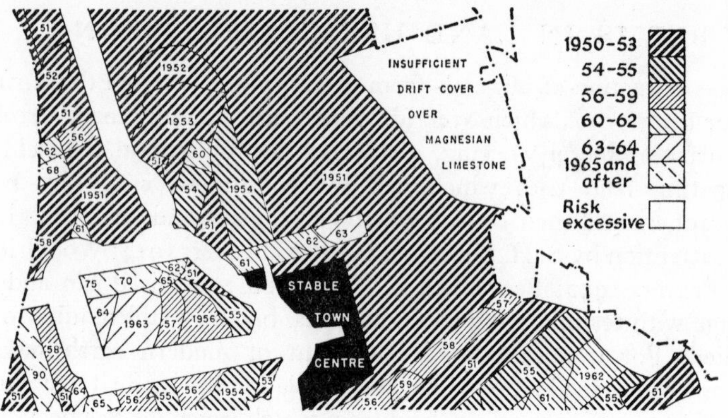

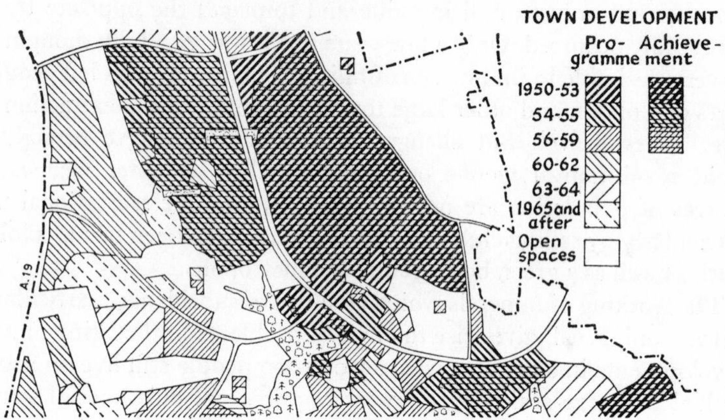

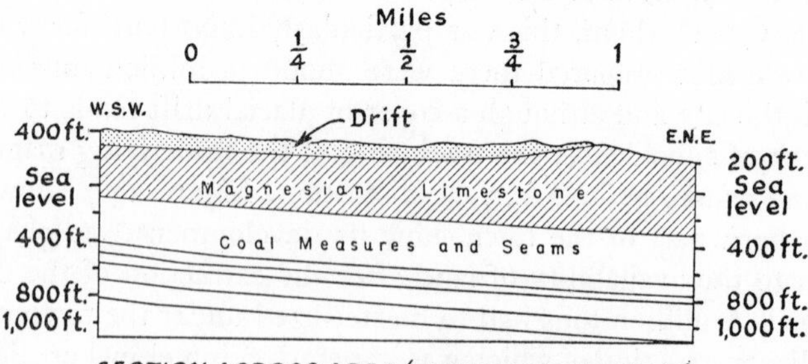

FIG. 13.7. *Peterlee, Co. Durham: Phasing of development in the northern part of the town in relation to the availability of land after the extraction of coal.*

MAPS OF URBAN RELATIONSHIPS

Many aspects of human geography fall to be investigated by geographers in connexion with planning. One which frequently occurs is the problem of 'journey to work', a study of which has many immediate practical applications. A general study of the numbers of persons who live in one place and work in another, as in the West Midlands conurbation, may reveal the very varying 'pull' of certain centres for employment. In such a complex area, a picture which simultaneously reveals the attraction exerted by a number of different towns is best achieved by the use of coloured symbols and the methods used in the Belgian *Atlas du Survey National* constitute an admirable technique. But there are many facets to the problem, and many false impressions which require correction by objective mapping. That a very large proportion of people working in the centres of large cities travel considerable distances to work is a common impression among the suburban 'commuters' to central London. But the facts revealed by the simple technique of showing the numbers who do make the journey from each locality suggests that popular impressions can be very misleading.

Figure 13.8 shows the pattern of commuting to central London as revealed by the 1951 census and emphasises that in general the largest numbers of workers travel relatively short distances. But, as Powell (1960) has pointed out, it also 'shows clearly that the daily influence of London in 1951 extended in some force as far afield as Brighton and Southend and in lesser degree to Luton and Reading'. It forms part of his study of the recent development of the London Region (bounded by the outer heavy line, the inner is the conurbation) which demonstrated that its planning, based on pre-war thinking, was outdated and that 'the expanding conurbation is the product of geographical and economic forces too powerful for man to reverse. He can only, within limits, direct them into convenient channels'.

The relationship between towns and the surrounding countryside which looks to the town for various services is a subject of great interest and importance to planners and many others, and is one in which geographical research and mapping techniques have been profitably applied to the recognition of centres, the delimitation of hinterlands and the estimation of their populations (Green, 1950). Such information is, for example, highly relevant to the development of central shopping and business districts of service centres. The subject now has a considerable literature.

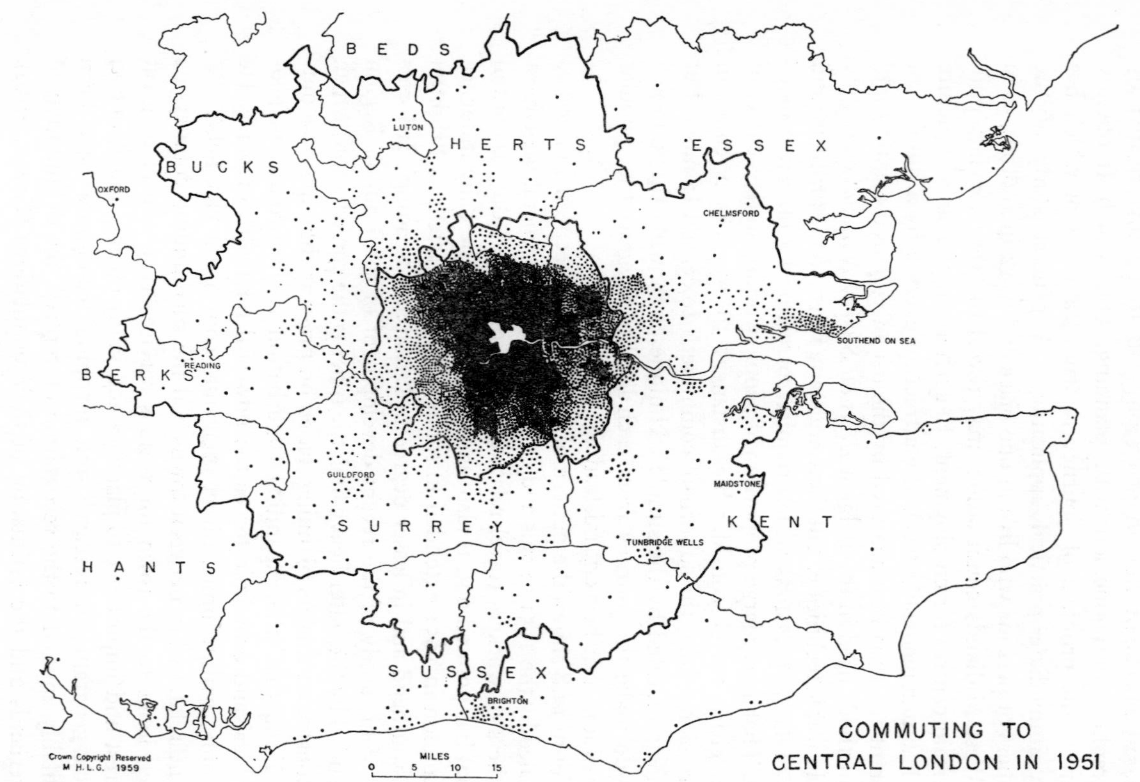

FIG. 13.8. *Commuting to central London in 1951. Numbers of daily travellers to the City and to the boroughs of Finsbury, Holborn, St Marylebone, and Westminster are indicated. One dot represents one hundred*

For the whole of Britain a map (Ordnance Survey, 1955) has been constructed and published showing the hinterlands of towns as determined by an analysis of bus services. The basic technique is to plot the complete bus route network for every centre, defined as a town having at least one regular service operating only to and from smaller places (which therefore look to it as a centre). By superimposing the diagrams on one another the hinterland boundaries are drawn in the same way that watershed boundaries may be drawn on a map showing the drainage pattern. Figure 13.9, reproduced from the

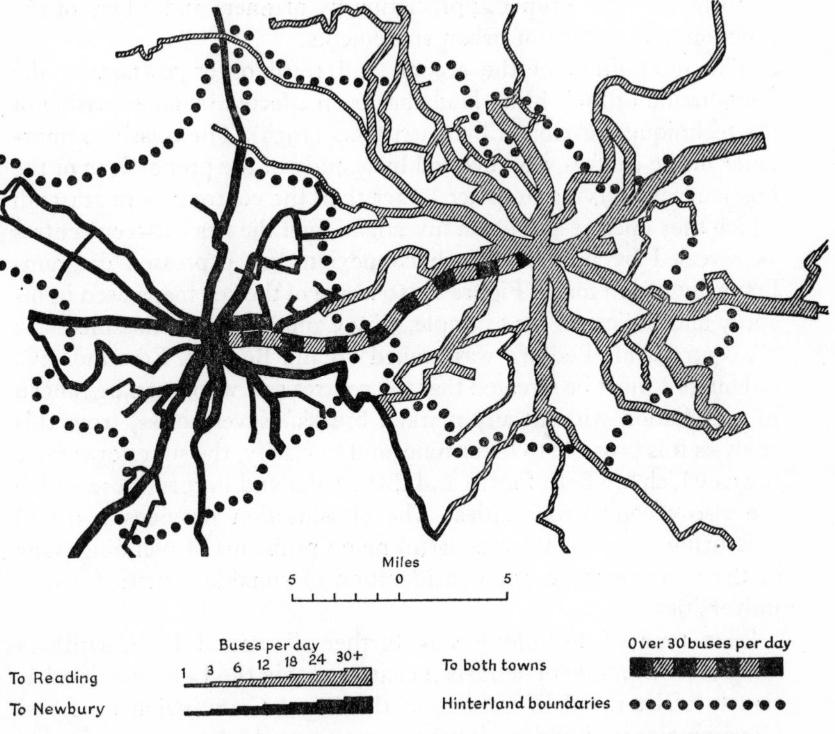

FIG. 13.9. *Reading and Newbury: A method of delimiting hinterlands.*

explanatory text published by the Ordnance Survey to accompany the ten-mile map, shows the method of delimiting the hinterland between Reading and Newbury. The population of each hinterland is calculated and on the final map, showing the boundaries of hinterlands, the population of each centre and its hinterland are shown by proportional concentric circles. The resultant map serves many purposes including the reconsideration of administrative boundaries, commercial sales organizations and the administration of medical, social and many other services.

It relates essentially to local accessibility, the relationship to their 'umland' of towns of local importance. These are towns of the 'Fourth Order' of importance, the single 'First Order' centre being the metropolis, the 'Second Order' the 'Provincial' cities such as Bristol, Birmingham and Leeds and the 'Third Order' the lesser towns of greatest regional importance such as Exeter and Lincoln. A realization, and understanding, of the hierarchy of service centres is important for the proper appreciation by planners and others of the function and pattern of urban settlements.

The recognition of the second and third order towns and the delimitation of their hinterlands has been effected by an extension of the technique, developed by Carruthers (1957). The relative importance of the centres was revealed by a study of the proportions of the bus journeys serving no place larger than the centre into or through which they operate and by an investigation of the ties between centres as revealed by bus and coach connexions, all expressed diagrammatically and in maps. Figure 13.10 is one of the key maps used in his study and indicates, for example, why Cambridge was classified as a 3A centre while Bedford was ranked 3B and Bury St. Edmunds 3C, although it must be stressed that the centres as a whole can be ranged in a gradation without any marked breaks. Nevertheless, from this analysis it is possible to determine, and to clarify, the superior service towns which are both fourth and third order and in turn those which are also second order cities. The classification provides a useful measuring rod which can assist in many problems of planning. One of the most recent is the consideration of suitable centres for new universities.

This type of technique was further developed by Carruthers (1962) for the study of suburban centres within the London conurbation to assist the deliberations of the Royal Commission on Local Government in London. Three separate analyses were made and presented on a series of maps. A study of the intensity of the equipment of banking, entertainment and selected shopping facilities in the central area of each 'centre' provides a 'status' classification. Figure 13.11 is a fragment of the map showing the status of the centres as determined by the provision of selected facilities. It shows how the varying grades of intensity of the six selected facilities graphically reveal the differences in status of these centres. The rateable value of

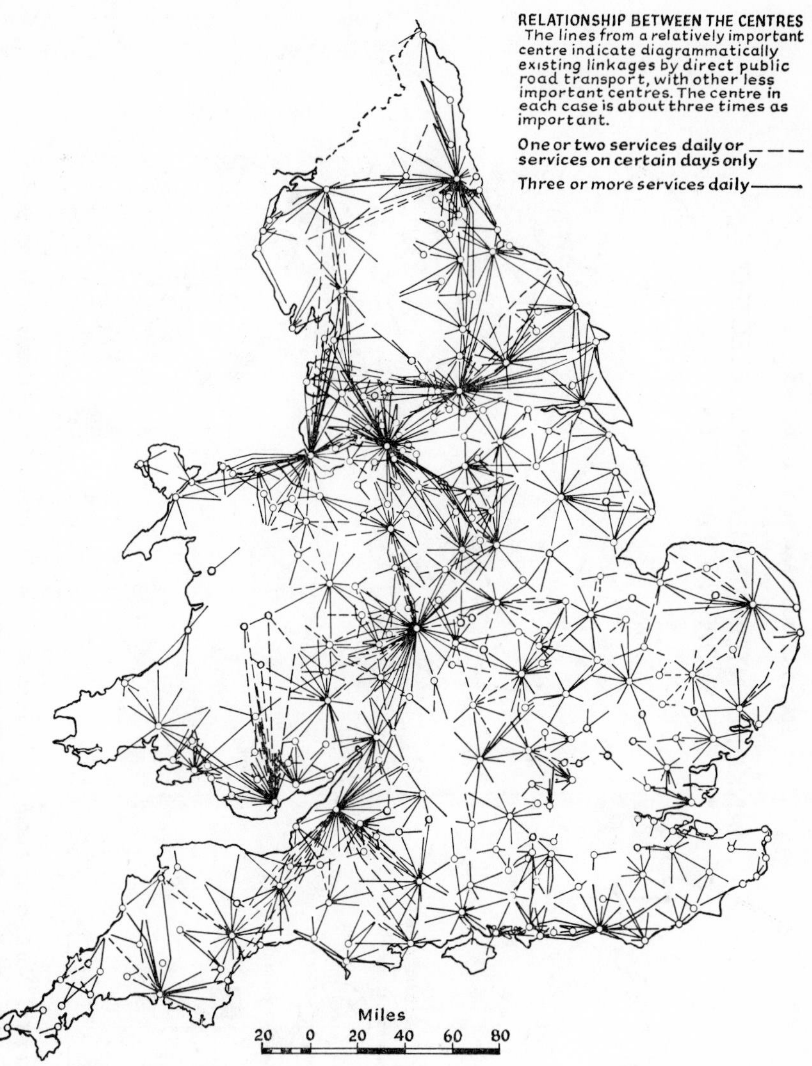
RELATIONSHIP BETWEEN THE CENTRES
The lines from a relatively important centre indicate diagrammatically existing linkages by direct public road transport, with other less important centres. The centre in each case is about three times as important.
One or two services daily or services on certain days only
Three or more services daily
Miles
20 0 20 40 60 80

FIG. 13.10. *England and Wales: The more important urban centres indicated by the public road transport links from other centres which are dependent on them.*

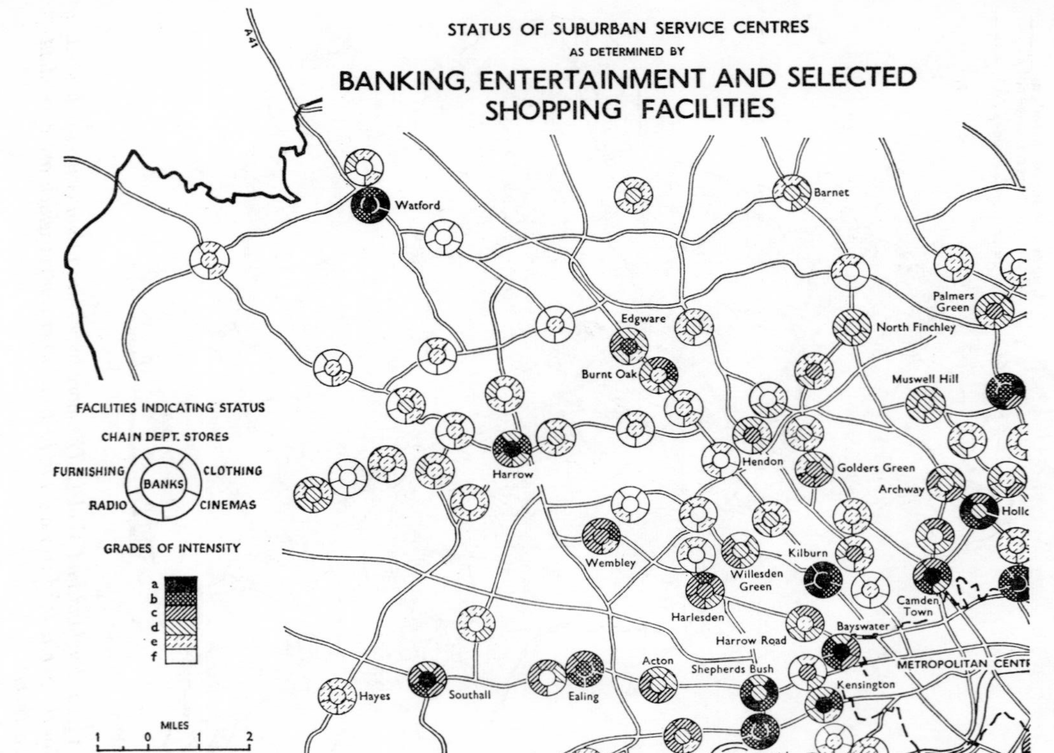

FIG. 13.11. *North-west London: An extract from Carruthers' map showing the method of determining the status of suburban service centres.*

those shops was also recorded to give a qualitative evaluation and an analysis of the off-peak hour nodal bus journeys furnished another measure of the use of the centre.

A correlation of all three analyses furnished both a graded classification of centres and an indication of whether their provision and use are broadly equated, or whether they are under- or over-equipped, which is shown by whether the provision is low or high relative to value and nodality. As their author says, '. . . it is not sufficient to investigate only the provision and availability of services; an assessment of the intensity with which existing services are used offers, if anything, a more telling and sensitive index. This makes it possible to appreciate the true position of the centres in a constantly changing environment.'

MAPS OF REGIONAL RELATIONSHIPS

The recognition and analysis of distributions and relationships by geographers must take place at all levels from local to national, and studies on the latter scale can be particularly valuable. An example of a recent study of the whole of England and Wales is shown in figure 13.12, which shows the location and relative importance of almost all major post-war developments affecting land-use planning in England and Wales except increases in population and employment, which could not be effectively represented here in black and white.

The assembly of the facts is the first task in the preparation of such a map. But to present them so that their relative importance may be seen is less easy. For the purpose of this map the common factor of cost was used and a capital cost of £5 million was in general taken as qualifying for inclusion. The cost of certain projects, such as collieries, is published, but for others, e.g. manufacturing industry, an average cost per square foot for building and plant was applied. Each subject was normally graded into three sizes according to the relative magnitude of the projects in that subject and the size of the symbols used to represent the items was made exactly proportional to the average capital cost of the projects in the relevant grade of the particular subject. Thus the area of the symbols is a reasonably valid measure of the importance of the data they represent: the largest steel works, costing about £100 million, appears as ten times the area of the largest grade of collieries, which cost about £10 million.

Such a map reveals the extent to which many large new (or

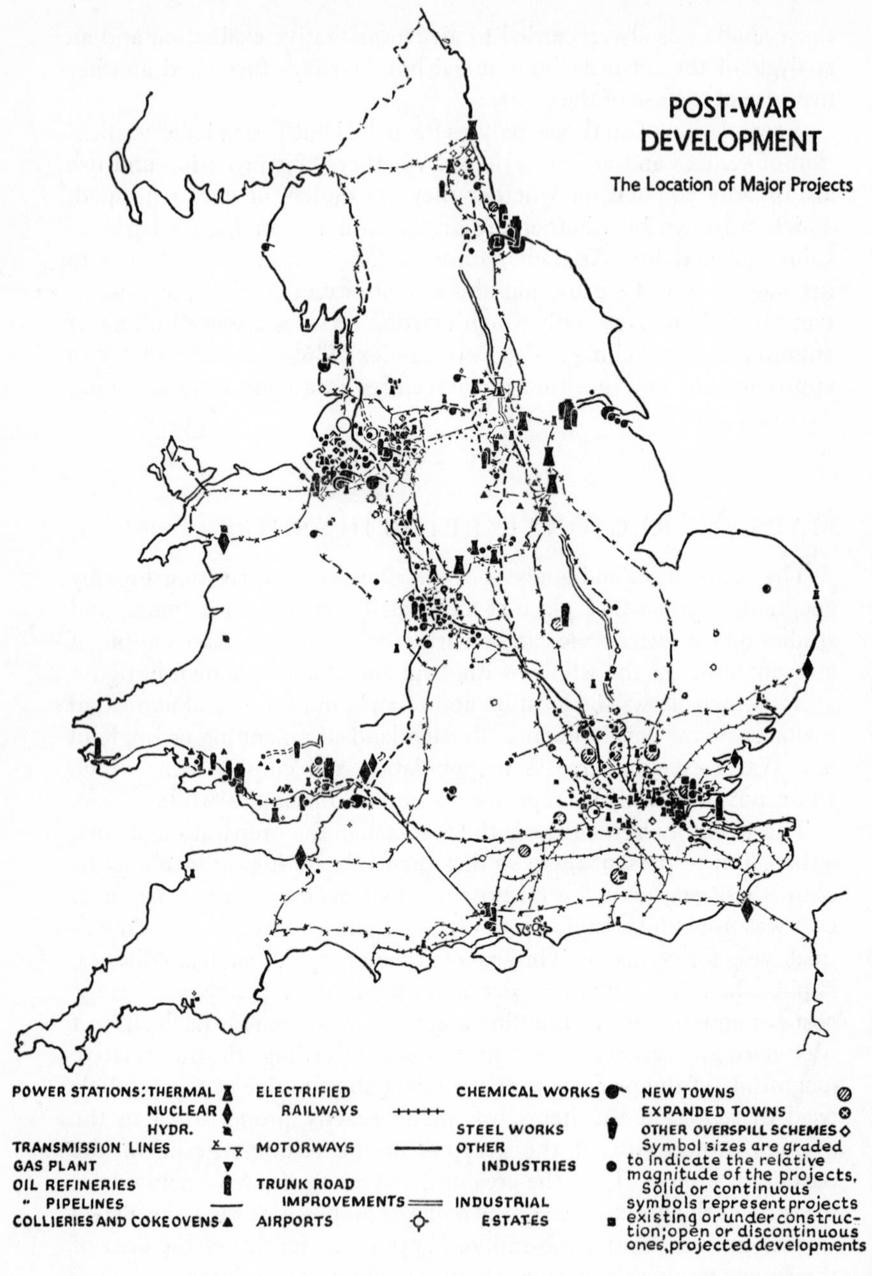

FIG. 13.12. *England and Wales: Location of major post-war projects, 1961.*

expanded) projects have been concentrated in comparatively few localities linked by improved communications and power lines, although there have been some individual projects in more isolated localities (Willatts, 1962).

The uses to which such a map may be put are too numerous to discuss here but the point must be made that although the nature of the patterns on such a map may be very obvious, their significance is usually only to be understood after further study and analysis.

Figure 13.13 uses rating statistics to show the country-wide distribution of all industry. A symbol directly proportional to the value has been drawn in each local authority where the industrial rateable value exceeded £100,000. Thus, using what is perhaps the only useful comparable data, the map gives a quantitative impression of the general distribution of industry and comparison with the industrial symbols on figure 13.12 shows that there are significant differences between the total pattern of industry and that made by major post-war developments. This is even more true if the new projects of the latter are separated from the extensions of former developments. This has been done in figure 13.14 in which the new developments, whose nature can be read from figure 13.12, are shown to be making new patterns on the industrial map. Thus it may be seen that there is a very significant concentration, particularly by chemical, oil and steel plants, along the major estuaries and deep water havens such as the Humber, Southampton Water, Severn and Milford Haven, which hitherto had not been very attractive to industry. At the same time there has been an intensification of development on the Tees, Thames and Mersey. In the Greater London area, except in the New Towns and Luton, there have been few other new developments in the belt fifteen to forty miles from the centre. Otherwise in most of the major industrial axis from London to Lancashire and the West Riding most of the major post-war developments are expansions of older industrial plants. The reasons for these changes are complex and include the forces of geographical advantages, of government policy, the need for large new sites and the developments in the production and distribution of electric power which have emancipated industry from the bondage of the coalfields.

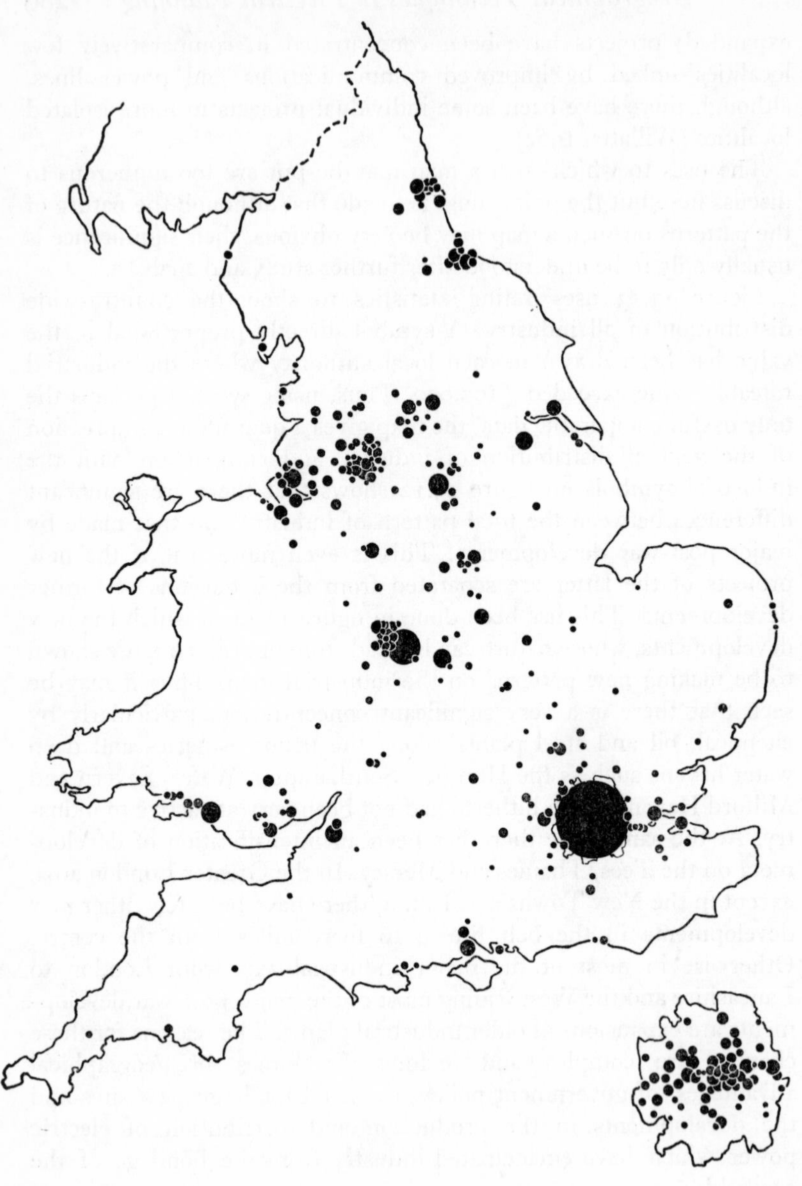

FIG. 13.13. *England and Wales: Distribution of industry as shown by rating statistics.*

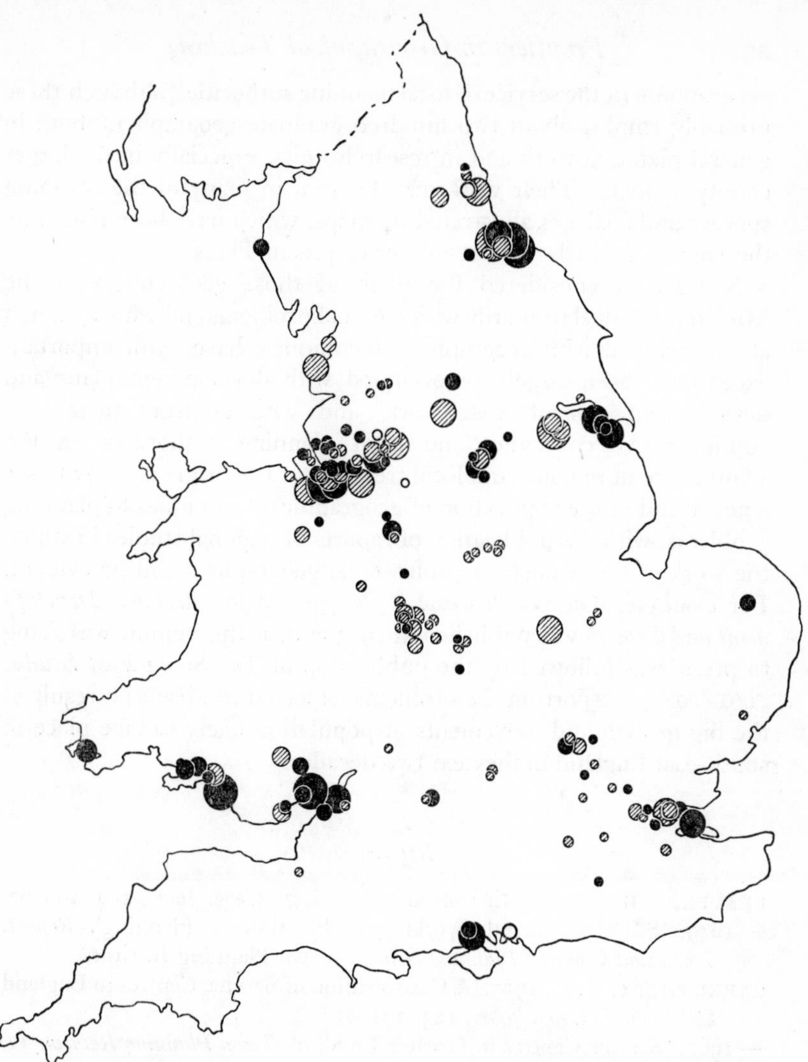

FIG. 13.14. *England and Wales: Major post-war industrial developments. New projects are shown by solid circles, extensions by shaded circles, and proposals by open circles, graded according to capital cost.*

CONCLUSION

This paper has been essentially concerned with examples of work carried out by a small, and changing, group of geographers employed in a headquarters section of the Ministry of Housing and Local Government. It has not attempted to deal with the work of geographers in the service of local planning authorities, although these probably employ about two hundred graduate geographers, both in general planning work and in research units, especially in the larger county councils. Their work may be seen in many of the planning surveys and analyses, illustrated by maps, which have been issued by the county councils with their Development Plans.

Nor has it considered the work of those geographers in the Ministry who deal primarily with regional problems and who, although also concerned with geographical techniques, have, with important exceptions, been largely preoccupied with development plans and development control 'case work' and with advising their own administrative colleagues and local planning authorities on the significance of regional and local trends. But the future is likely to see a new trend in the application of geographical techniques to planning problems, with the publication of reports of regional studies in which the work of the Ministry's professional geographers will be evident. For example, *The North-East: a Programme for Regional Development and Growth* was published in 1963 and, as this volume was going to press, was followed by the publication of *The South-East Study: 1961–1981*, a report on the problems expected to arise as a result of the big growth and movements of population likely to take place in south-east England in the next two decades.

References

BEAVER, S. H., 1944, 'Minerals and Planning', *Geog. Jour.*, **104**, 166–93.

— 1949, 'Surface Mineral Working in Relation to Planning', *Report. Town and Country Planning School.* Town Planning Institute.

CARRUTHERS, W. I., 1957, 'A Classification of Service Centres in England and Wales', *Geog. Jour.*, **123**, 371–85.

— 1962, 'Service Centres in Greater London', *Town Planning Review*, **33**, 5–31.

GREEN, F. H. W., 1950, 'Urban Hinterlands in England and Wales: an Analysis of Bus Services', *Geog. Jour.*, **116**, 64–81.

JAMES, J. R., SCOTT, S. F. and WILLATTS, E. C., 1961, 'Land Use and the Changing Power Industry of England and Wales', *Geog. Jour.*, **127**, 286–309.

ORDNANCE SURVEY, 1955, *Local Accessibility: the Hinterlands of Town and Other Centres as determined by an Analysis of Bus Services, 1/625,000* (London).

POWELL, A. G., 1960, 'The Recent Development of Greater London', *Adv. Sci.*, **17**, 76–86.

VINCE, S. W. E., 1952, 'Reflections on the Structure and Distribution of Population in England and Wales, 1921–31', *Trans. Inst. Brit. Geog.*, Pub. No. **18**, 53–76.

WILLATTS, E. C., 1962, 'Post-war Development: the Location of Major Projects in England and Wales', *Chartered Surveyor*, **94**, 356–63.

— 1963, 'Some Principles and Problems of preparing Thematic Maps', *Proceedings. Conference of Commonwealth Survey Officers*, Cambridge.

WILLATTS, E. C. and NEWSON, M. G. C., 1953, 'The Geographical Pattern of Population Changes in England and Wales, 1921–51', *Geog. Jour.*, **119**, 431–50.

WOOLDRIDGE, S. W. and BEAVER, S. H., 1950, 'The Working of Sand and Gravel in Britain: a Problem in Land Use', *Geog. Jour.*, **115**, 42–57.

PART THREE
TEACHING

CHAPTER FOURTEEN

Geography in the Universities

PART 1: GEOGRAPHY IN THE OLDER UNIVERSITIES

C. BOARD

Lecturer in Geography, London School of Economics

An examination of calendars and prospectuses for British universities in the early 1960's reveals that Honours courses in Geography make a recognized contribution to the work of all but a very few universities and colleges in Britain. The pattern of specialisms and papers taken at the various examinations held at the end of these courses differs considerably from place to place. But it is nevertheless possible and moreover instructive to perceive some general pattern; a pattern through which some hundreds of graduates are yearly introduced to geography as an academic discipline. Thus, in some measure, it is reasonable to assert that the dominant characteristics of instruction in geography in Britain may be seen in the pattern outlined below.

It is important first of all to distinguish between English and Welsh Universities on the one hand, and Scottish and Irish on the other. The latter offer courses of much broader scope and permit the taking of Honours degrees only after four years full-time instruction. English and Welsh courses leading to Honours degrees take only three years and are generally much more specialized from the start. For this reason, the two groups are not strictly comparable. Furthermore, Keele University, in common with the newer universities, has features which also set it apart from other universities, rendering comparisons less useful.

Questions of comparability do not really affect the analysis of optional or specialized subjects offered by different institutions so that a broader sample will be used here in order to establish a pattern which is beginning to show the way in which Geography is being treated in depth. It is more than likely that these specialisms may, in the not too distant future, lead to more formal post-graduate courses, which are intended to supply the elements of a professional qualification to geographers who have already been trained in a broad and general way.

A further division must be made between two systems of progression to the degree of Honours in Geography in England and Wales. First, and more commonly, the course is split into two parts with examinations of university status at the end of each. It is common for the first part to be a means of qualifying for the second, or sometimes to be taken into account when final degree classes are being discussed. Only the Cambridge system, with its two completely independent examinations (the Tripos) differs significantly from these generalizations. In that case, the results of examinations at both Part I and Part II are classed separately and without reference to previous assessments. Since students may combine a Part I in one subject with a Part II in another, results for both parts of the Tripos should be taken into account when describing the attainment of Cambridge graduates. This is all the more important when it is realized that Part II of the Geographical Tripos is much more specialized than Part I, in which all papers are compulsory.

The second and less common system is to take one university degree examination, at the end of the third year of the Honours course. Where this is the case the number of papers is frequently ten or more; whereas if courses are formally split into two parts, fewer papers are taken at the end of the third year. Specialization on certain aspects of geography is not normally so advanced when all topics have to be carried into the third year for a final examination.

All university Departments of Geography now include field work in their Honours courses, although there are a few remaining exceptions to the rule that students attend a compulsory field course lasting about a week. Sometimes evidence of having completed field work to a satisfactory standard is included in degree regulations.

Allied to field work and often, but not exclusively, resulting from it, is the dissertation which most universities require of Honours candidates. Whether they are studies of pieces of country or essays on some geographical topic, they are intended to be essentially the individual work of the student, and provide a useful guide to his research capabilities. There is a long tradition of studies of this kind in Britain. Oxford and the London School of Economics were among the pioneers of this form of test and their studies provided some of the models for the regional surveys of the 1920's. Only one university requires neither field work nor a dissertation as part of its Honours course.

Most Honours courses in Geography lead to final examinations which include both compulsory and optional papers. Those which have a Part I or second year examination of mandatory character generally ensure that it is composed largely of compulsory papers. Physical, human and regional geography and a map work or practical paper are most commonly taken. One university, even at this stage, sets a compulsory paper in the history of geographical thought. On the other hand, there are few universities which have no optional papers in the final examinations. This does not preclude specialization, but it inevitably restricts it. As the number of options made available by individual departments has increased considerably since the war, the degree of choice open to an Honours student has correspondingly expanded. There is sometimes a complementary tendency for the number of compulsory papers to be reduced, with the result that there is now little chance that the Honours student can aspire to a knowledge of all continents and aspects of Geography.

Compulsory papers in the final year of Honours courses generally include both systematic and regional geography, although the latter forms a substantial part of them. This is in tune with the supposition that regional geography is the core and culmination of the subject and that systematic geographies are subservient to it. Most universities consider it necessary for Honours students to know something of the history and nature of geographical thought. The geography of the British Isles is, somewhat surprisingly, a requirement in fewer than half of English and Welsh universities. Rather more than half of the courses include map work or practical geography as compulsory papers. The apparent reluctance of roughly half our universities to examine physical geography in the final year is partly explained by their relegating it to the second year, where it is usually compulsory.

Much of the richness and value of British Honours courses in Geography lies in the wealth of specialized subjects offered. These are normally regarded as options, students choosing one or two from a list. Three universities have liberalized their syllabus to the extent of requiring three or more options to be chosen.

From an examination of the subjects offered by twenty-one universities in the British Isles, it is easy to obtain some idea of the options that have become well established. The frequency with which options are offered is also a guide to the relative strengths of the specialisms in university departments. Physiography, physical geography or geomorphology and historical geography are each offered by nearly all of the universities. Economic geography is almost as popular, with climatology coming a close fourth. Cartography, or mathematical geography and surveying, and political geography are not far behind, being offered by more than half of the departments having options. All other options are relatively rare in that only a few universities teach them. Included here are plant geography, biogeography, 'human geography', and the history of geographical ideas and discovery (a subject which goes under a bewildering variety of names).

It may come as a surprise to readers outside Britain that urban geography is available at so few universities, that only two universities offer settlement geography and only one social geography. In a number of cases it is possible for Honours students to take a non-geographical subject, such as social anthropology, as an alternative to a geographical option. This does not exhaust the list of options because there frequently is a choice of continent in the regional papers in the so-called compulsory part of the examination.

Among the more unusual options are the geography of planning and the geography of population, each offered by the same university. Some departments also have extremely specialized courses, more, it seems, in the nature of technical training. These include photogrammetry, map design and compilation, and atlas planning. One university finds it necessary to specify as a topic for study in the final year 'some consideration of current writings in geographical literature'! It should be pointed out that the options mentioned above are those which are formally examined by papers in final examinations. It is clear from an inspection of details provided by individual university Departments that the list of options *taught* is longer. This is quite characteristic of the development of university studies in Britain, where changes in the teaching programme are much more easy to effect than are changes in the formal structure of examinations. Since such changes are almost always dependent on the policies of other departments included in the faculty (whether Arts, Science, Social Science), they generally lag behind changes in the character of geography teaching and research within departments. [See notes on courses available at Cambridge, University College, London, Aberystwyth and Swansea in the *Geographical Journal*, volume 128, p. 568; volume 129, pp. 243, 573, 572.]

An indication of compulsory papers taken and options examined in the final year of some Honours courses is given in Tables 18 and 19. (Because they refer to the early 1960's changes in detail may be expected.)

Table 18. English and Welsh Universities, c. 1962: Summary of Compulsory Papers taken in the Final Year of the Honours School in Geography

	UNIVERSITIES																
PAPERS	*Birmingham*	*Bristol*	*Cambridge*[1]	*Durham/Newcastle*	*Exeter*	*Hull*	*Leeds*	*Leicester*	*Liverpool*	*London*	*Nottingham*	*Oxford*	*Reading*	*Sheffield*	*Southampton*	*Wales (Swansea)*	*Total*
Physical[2]	0	0	0	1	0	0	2	0	1	1	0	0	1	2	0	1	9
Practical[3]	1	0	0	1	0	0	0	0	0	1	0	1	1	1	0	2	8
Human	1	0	0	1	1	0	2	0	0	0	0	2	1	2	0	2	12
Economic	0	0	0	0	0	0	0	0	0	0	0	0	2	0	0	0	2
Regional I[4]	2	2	0	2	1	2	3	1	2	3	2	4	2	2	0	2	30
Regional II[5]	1	0	0	0	1	0	0	1	0	1	0	1	1	0	1	1	8
General[6]	0	0	2	1	1	1	1	1	1	1	1	0	1	1	1	0	13
Local stud.	0	0	0	0	0	0	0	1	0	0	0	0	0	0	0	0	1
English essay	0	1	1	0	0	0	0	0	0	0	0	0	0	0	1	0	3
Dissertation	1	1	1	0	0	0	1	1	1	0	1	1	0	1	1	1	11
No. of compulsory papers	6	4	4	6	4	3	9	5	5	7	4	9	9	9	4	9	
Total No. of papers	7	7	8	6	6	4	11	6	6	9	5	11[7]	9	11	6	10	

[1] The new Part II of the Geographical Tripos (1964–) makes only the English Essay and Dissertation compulsory.

[2] Includes geomorphology, climatology, etc.

[3] Includes cartography.

[4] Regions other than the British Isles.

[5] British Isles.

[6] Includes papers on the history and nature of geographical thought.

[7] Includes two papers in a special geographical subject which may be omitted by candidates not aiming at 1st or 2nd Class Honours.

Table 19. Some British Universities, c. 1962: Summary of Optional Subjects taken in the Final Year of the Honours School in Geography

UNIVERSITIES

	Aberystwyth	*Birmingham*	*Bristol*	*Cambridge*	*Durham/Newcastle*	*Exeter*	*Hull*	*Leeds*	*Leicester*	*Liverpool*	*London*	*Nottingham*	*Oxford*	*Reading*	*Sheffield*	*Southampton*	*Swansea*	*Belfast*	*Dundee*	*Edinburgh*	*Glasgow*	*Number offering each subject*
Geomorphology	x	x	x	x	o	x	x	x	x	x	x	x	x	o	x	x	x	x	o	o	x	17
Climatology	x	o	x	o	o	x	x	o	o	x	x	x	x	o	x	x	x	x	o	x	o	13
Plant Geography	o	o	o	o	o	o	o	o	o	o	x	o	o	o	o	o	o	o	o	o	o	1
Biogeography	x	o	x	o	o	o	o	o	o	o	o	o	o	o	o	o	o	o	o	x	x	4
Cartography	o	o	o	o	o	o	x	x	o	x	o	o	o	o	o	o	x	o	o	x	o	5
Math. Geog. and Survey	x	o	x	x	o	o	o	o	o	o	x	o	x	o	o	o	o	o	o	o	x	6
Human Geography	o	x	o	x	o	o	o	o	o	o	o	o	o	o	o	o	o	x	o	o	x	4
Economic Geography	x	o	x	o	o	o	x	x	x	x	x	x	x	o	o	x	x	x	x	x	o	14
Political Geography	x	o	o	o	o	x	o	x	x	o	x	x	o	o	o	x	o	x	x	x	o	10
Historical Geography	x	x	x	x	o	x	x	x	x	x	x	x	x	o	o	x	x	x	x	x	x	18
Urban Geography	x	o	o	o	o	o	o	o	o	o	o	x	o	o	x	x	o	o	o	x	o	5
Settlement Geography	o	o	o	o	o	o	o	o	o	o	x	o	x	o	o	o	o	o	o	o	o	2
Social Geography	o	o	o	o	o	o	o	o	o	o	o	o	x	o	o	o	o	o	o	o	o	1
History and Methodology	o	o	o	o	o	x	o	o	o	o	o	o	x	o	o	o	o	o	o	o	o	2
Regional Geography	o	o	o	o	o	o	o	o	o	o	o	o	o	o	o	o	o	x	o	x	o	2
Geography of Planning	o	o	o	o	o	o	o	x	o	o	o	o	o	o	o	o	o	o	o	o	o	1
Population Geography	o	o	o	o	o	o	o	x	o	o	o	o	o	o	o	o	o	o	o	o	o	1
Photogrammetry	x	o	o	o	o	o	o	o	o	o	o	o	o	o	o	o	o	o	o	o	x	2
Hydrology	x	o	o	o	o	o	o	o	o	o	o	o	o	o	o	o	o	o	o	o	o	1
No. of subjects	10	3	6	4	0	5	5	7	4	5	8	6	8	0	3	6	5	7	3	8	6	109

PART 2: GEOGRAPHY IN A NEW UNIVERSITY

T. H. ELKINS

Professor of Geography, University of Sussex

Why should there be geography in the new universities? This question has had to be faced by the Academic Planning Committee of each of the universities that are being created in the current decade. So far, although a number have announced teaching appointments, only one, the University of Sussex, appears to have plunged unambiguously into the field of geography. Two others are introducing 'Schools of Environmental Studies' and in one of these, at the University of Lancaster, a geographer is Dean.

Why should there be geography in any university? One answer is that it is customary; according to a statement issued in 1961 by the Geographical Association, all the previously existing universities in Great Britain had fully-staffed departments with a professor in charge. Precedent was certainly followed by the post-war 'new university' of Keele, which has had a department under a professor since its foundation. There is certainly no lack of respectable precedent for establishing further university departments.

Neither can it be said that there is any obvious lack of demand for the products of these departments. Before the war, geography was rather an introverted subject. The schools taught geography to pupils, a few of whom read geography in the universities, and then returned to the schools as teachers. Now the situation is very different. The majority of graduates still pass on to teaching, mainly in the schools, but also in university departments at home and abroad. But many now go on to use their subject professionally in a variety of fields ranging from soils survey to market research. A gratifying development here has been the increasing acceptance of geographers in town and country planning. Increasingly, too, geographers are using their training as a general education prior to work in quite unrelated fields in administration, industry and commerce.

But custom and the existence of a market for the product are not in themselves compelling reasons for teaching a subject at the university. The justification must be on academic grounds. This is not the place for a long discourse on the academic claims of geography; indeed, on the evidence of the hundreds of departments of geography in universities throughout the world, we may regard the case as proved. But we are concerned with new universities, some with distinctly new ideas on university teaching. It is essential to examine the claims of geography anew, to see if the subject is as appropriate to the new systems of teaching, as it was evidently appropriate to the old.

At this point it is instructive to turn for a moment to the institution that was, until recently, Britain's only post-war university, the University of Keele. From Keele came much that was hopeful and stimulating in British university development in the dark and difficult years after the war. All the present generation of new universities are indebted to Keele, and quite as much for Keele innovations that they have decided to reject as for those that they have adopted. One trend, at least, Keele and several of the new universities have in common, and this is the reaction against what is felt to be the excessive narrowness of the normal honours degree. In this respect the Keele solution was very radical. The course was extended to four years. A foundation year of broad general studies was introduced, and in the remaining three years each undergraduate had to study several subjects, including one from the opposite side of the arts/science gap. The Keele structure may be said to have very definite advantage for geography, a subject which unites aspects of arts and science, and in which research workers need expertise in one of the bordering disciplines, geology, history, economics or the like.

A similar concern with over-specialization is found in at least some of the new universities of the present decade, including the 'first of the new', the University of Sussex. If most of the rest of this paper is taken up with this university, it is not because it is the first and most rapidly developing of the group, but because it was the first to establish a Chair of Geography. But at Sussex and elsewhere criticism of the traditional honours degree is combined with a realization that the way out does not lie in the direction of the traditional general degree. The defects of the general degree are well known. The various subjects studied are too often unrelated; even if related, they are usually taught in isolation by distinct and independent departments. At least for the more able students, breadth of study is no substitute for the challenge and excitement of being taught to honours level by men actively engaged in advancing the frontiers of knowledge. There is much truth in the view that an honours degree properly taught to able students provides breadth as well as depth, is a general education in itself. The ideal modern language student, we are told, will necessarily go beyond his linguistic and literary studies to learn something also of the politics, the social structure, and the philosophical systems of the country whose language he is studying. The historian must necessarily concern himself with philosophy and sociology. But how often is this so? It has to be admitted that many students of languages, history, or geography, are quite as limited in their outlook as the traditionally narrow-minded scientist with his nose in a test-tube; indeed perhaps more so. Is it not then possible to formalize this pattern of background and contextual reading, so as to oblige even the dullest undergraduate to do what should be done eagerly and naturally?

Something of this sort is provided by the system now used at Sussex, and apparently to be adopted at some of the other 'new universities'. The university is organized not in departments but in Schools of Studies, each with a Dean as chairman. The Faculty of Arts has five Schools, English and American Studies, Social Studies, European Studies, African and Asian Studies and Educational Studies. The Faculty of Science consists of the Schools of Mathematical and Physical Sciences, Molecular Sciences, Applied Sciences and Biological Sciences. Geography is at present available as a major subject within all the 'Arts' schools except English and American Studies, and also in the School of Biological Sciences. The latter is considered a more relevant Science 'link' for geography than the customary geology.

Undergraduates join a particular school rather than a department but they nevertheless specialize in one particular discipline, their 'major subject'. To this extent the system approaches the standard Honours course, but the time devoted to the major subject is slightly less. The ideal here is advance in depth, towards the most vigorously growing part of the subject, rather than advance in breadth. But in addition the undergraduate spends about the same amount of time studying a group of common or contextual papers. The important point is that these are not unrelated subjects selected at the whim of the undergraduate, as they are liable to be in the ordinary general degree. The common papers serve, as it were, to anchor the major subject in its academic context, showing how it links with other subjects taught within the School of Study. There is an attempt to show how the various specialist subjects within the School of Study approach common problems, and an endeavour to demonstrate the manner in which an undergraduate's own, and neighbouring, subjects may be applied in gaining an understanding of the contemporary world. That this degree of integration and common purpose should be very real is ensured by the system of university organization, which draws teachers of varying disciplines together in the common field of interest represented by the School of Study. A geographer must find this pattern attractive. If integration is the ideal, then geography should have no difficulty in fitting into the scheme, since no subject offers a greater wealth of links with adjoining subjects. This view was adopted by the academic planners of the University of Sussex, where the teaching of the subject began in April 1963, but so far, with the partial exception of Lancaster, there has been no great rush of the other new universities along the same path.

At this stage it may be appropriate to see how geography fits into the Sussex pattern, taking to begin with the School of Social Studies, within which most geography teaching at present takes place. On entering the University, the undergraduate spends his first two terms studying for his Preliminary Examination, which consists of three papers. Two of these papers, studied for a term each, are common to all the Faculty of Arts. One, called Language and Values, is concerned with philosophical problems that arise in decision-making and judgement. It is very much involved with questions of the use of language. In a second course, Introduction to History, one or two major historical works are critically examined in a manner aimed at developing insights into the problems of academic investigation, rather than the conveying of historical fact. The third course, lasting two terms, is specific to the School; in Social Studies it deals with the Economic and Social Framework. The fact that two of these subjects are shared by the whole Faculty of Arts, and the third by a whole School gives to the undergraduates a palpable sense of shared foundations, of a common approach to learning.

The remainder of the undergraduate course is devoted to the Final examination. In this, as well as the major subject, there are four papers which are common, with certain modifications, to all members of the School. Once more philosophy makes its appearance, to sharpen the undergraduate's critical faculties. Then there are two papers that attempt to draw together the approaches of various disciplines to the problems of the present-day world. In the first of these members of both the School of Social Studies and of the School of English Studies join in seminars which draw on both factual sources and imaginative literature in a study of Contemporary Britain. The second paper unites geographers, economists and international historians in a study of World Population and Resources. Finally a paper on 'Concepts, Methods and Values in the Social Sciences' is concerned with the different methods of investigation used in the various branches of the Social Sciences. It will be apparent that this is only one of a number of contexts that could be provided for work in social studies. A different group of scholars might come up with a different set of papers, more ecological perhaps, or more economic. But the pattern set out above is reasonably coherent, and is proving stimulating to teach.

The remaining five papers are devoted to the major subject. Apart from this specialist teaching, geographers as such make their major contribution in the course on 'World Population and Resources'. Because subject lines in the University are so fluid, they are also to be found collaborating with colleagues in a wide range of other courses. Individual geographers on the University Faculty contribute to courses on Contemporary Britain, Contemporary Europe, Urban and Rural Sociology, Community Studies, Statistics, and various others.

It will be recalled that certain of the Schools of Study at Sussex are of a regional nature. It is obvious that the School of English and American Studies will primarily attract undergraduates majoring in English. It is not difficult also to see that geography might be studied in a context of European Studies. But what of an exotic regional school such as African and Asian Studies? Few undergraduates will wish to follow courses in an African or Asian language, and indeed it is not the intention of the University to provide them. It is, however, intended that some undergraduates each year will major in geography or sociology or politics, but will do so not in the familiar context of European Studies or Social Studies but in the context of papers with such intriguing titles as Cultures and Society; Westernization and Modernization; Imperialism and Nationalism; or The Tropical Environment. The idea is to turn out graduates who are fully competent in their major subject, but with the advantage of having studied it in this very different context. One can see that as a training ground for the geographer, this School has very interesting possibilities.

What of the links with Science? Here both the problem and the opportunity at Sussex was the lack of a Department of Geology and the small prospects of there ever being one. With this orthodox link cut off, new and exciting prospects opened up in connexion with the School of Biological Sciences. Those responsible for planning the School had independently decided that opportunity would need to be given in their course for the study of ecology and physical geography. In this exciting collaboration, the University's field laboratory and hostel at the Isle of Thorns, in the heart of Ashdown Forest, will undoubtedly play a major part. In addition to this teaching of those who remain essentially biologists, arrangements have also been made for a major in geography within the School of Biological Sciences, in which the contextual 'frame' of geography will be a biological one.

Lastly, a word about teaching. As is fairly well known by now, the standard teaching method at Sussex is by tutorial. Naturally with a subject such as geography, which has practical aspects, some supplementation of the tutorial system is necessary. There is always some material that must be provided by lectures. There must obviously be practical classes, for there are skills to be mastered. But we do try at Sussex to avoid those unrelated and meaningless practical exercises designed solely to teach techniques. The time to acquire a technique is when it is needed for some practical investigation. So at Sussex we try to put the emphasis on projects, on real investigations. The effect on the syllabus is untidy; techniques are not learned in a logical order, but they are learned willingly. The method is matched by the new laboratories that are being created, where instead of lines of fixed benches there are substantial but mobile tables which can be grouped informally by collaborators on a particular project. And when the moment comes for work in the field, then it is only necessary to open the door and to walk up through the great beeches of Stanmer Park to arrive at the crest of the South Downs, and to see the open-air laboratory of the Weald spread out below. For work in Social Geography, at present our liveliest speciality, there is all the range of subject matter from the youthful industrial community of Crawley New Town to the areas of retirement in the coastal resorts. The Newhaven Ferry and Gatwick Airport link us to the object of our European Studies. Those who have to teach geography in the University of Sussex are indeed fortunate in the opportunities provided by its setting.

CHAPTER FIFTEEN

Geography in the Colleges

PART 1: GEOGRAPHY IN THE TRAINING COLLEGES

S. M. BRAZIER

Lecturer in Geography, Homerton College, Cambridge

THE DIVERSITY INHERENT IN THE TRAINING COLLEGE COURSE FOR GEOGRAPHERS

In order to understand clearly the position of geography teaching in the training college, it is perhaps important first to clarify the particular role of this segment of higher education. The function of the training college is twofold: to continue the student's own personal education and to prepare the student, as far as possible, for his or her future career of teaching. Students from such colleges will go into a range of schools from infant and primary schools to secondary modern, technical and grammar schools, and the individual training college will in all probability offer 'curriculum courses' relevant to all these possibilities. The students' own development continues, both within this Education course and through their study of usually one or two advanced subjects; here they may continue in increasing depth a study of subjects already begun in their school career, or they may explore new disciplines and fields of activity. This pattern of parallel courses, on the one hand in teaching techniques and Education – a 'professional course' – and on the other in specialist academic subjects studied for their own sake, has become known as a 'concurrent course'.

It will be seen that within this 'concurrent course' there will be, as far as the teaching of geography is concerned, two distinct and overlapping groups of students, those following a specialist course in geography, and those, often in considerable numbers, following a 'curriculum' course, i.e. concerned with the curriculum content and techniques of teaching geography in school. Students come to the training college with a considerable diversity of what may be called 'geographical background': some students have studied geography to Advanced and Scholarship levels, others have dropped it early on in their school careers; some have done a considerable amount of field-work, others none. Those with advanced knowledge have covered a diversity of regions within the geographical field, within a wide range of syllabuses and Examining Boards; many have considerable gaps in their regional knowledge – the U.S.S.R. and the Southern Continents frequently are relatively unknown. Sometimes the students' 'geographical foundations' are extremely shaky and, for example, concepts involving hydrology or climatology are strikingly vague.

Another problem in the teaching of an academic subject in the training college is, of course, lack of time. Unlike the university student reading in an Honours course, the training college student has frequently to study another subject at advanced level, a full course in Education and to participate in considerable periods of teaching practice. However, this apparent diversity may frequently be a strength in the training college course, in that the interlocking relationships between these fields of activity become increasingly apparent to the student and the experiences in one course contribute to the others also.

OBJECTIVES OF THE TRAINING COLLEGE COURSE

From what has been indicated so far, the aims of geography teaching in the training college may begin to be apparent. In part, the geography course[1] aims at stabilizing the basic principles of geography – the physical and climatological bases of world distributions or the economic factors in human activity, for example – as well as filling in some of the gaps in the students' knowledge of the regional geography of the world. It also aims, however, to open up new fields of interest and new approaches to geographical work; the historical geography of North America, for example, often proving a new and exciting approach to a continent already fairly intensively studied at school. In addition, one is hoping to help the student to realize and use the immense range of source material for geographical study, from the journals of the learned societies, to films, museums, and text-

[1] Reference is here made mainly to the 'specialist' course, as distinct from the 'curriculum' course to which reference will be made later.

books – both advanced texts dealing usually with a single specialist topic or continent, as well as school textbooks.

And where does such a course lead? At best, we hope that our geography students will begin to understand something of the basic philosophy of the geographer – of the interrelationships which exist within the environment and between man's activities and his environment. Students are encouraged to investigate critically such concepts, for instance, as 'environmentalism' and 'possibilism', and W. M. Davis' explanatory description of landscapes as a 'function of structure, process and stage'. Above all one hopes that the student, here in the training college as in other institutions, will begin to ask questions about the world around him and will develop what has been called the 'seeing eye' as he moves into different environments, whether for purposes of work or leisure. Most important of all, however, the lecturer in geography hopes to excite in his students, as future teachers, a 'geographical imagination', but one accurately based, in the sense in which James Fairgrieve used the phrase: 'The function of geography is to train future citizens to imagine accurately the conditions of the great world stage and so help them to think sanely about political and social problems in the world around' (Fairgrieve, 1926, p. 18). Such 'accuracy of imagination' must be linked with 'geographical enthusiasm' – for other peoples of the world, for landscape, for the geography of the past – for it is only when our students experience such enthusaism themselves that they will be able to transmit the delights of geography to their pupils. A teacher without enthusiasm for what he teaches is impotent.

Such aims may well be idealistic, but the training college lecturer is unusually fortunate in one respect, and that is that he is not, as yet (and one hopes will never be), subjected to the exigencies of a syllabus dictated by an external body, as are those teaching in schools.[1] Thus the lecturer concerned may plan a course which tries to satisfy the aims stated above, while allowing for the students' own previous experience and interest. Thus it is likely that courses may differ in detail, though not basically, from one year to the next, in an attempt to satisfy the needs of a particular group of students.

[1] The Training Colleges in England and Wales 'are all members of Institutes of Education which, with one exception, are institutes of the neighbouring university' (*Higher Education*. Report under chairmanship of Lord Robbins. H.M.S.O., October 1963, p. 109). Common examinations, or common parts of papers, may cover all colleges in the Institute, but these are set by the lecturers concerned, under the guidance of external examiners.

CONTENT OF THE GEOGRAPHY COURSE

It remains to describe the content of the geography course in the training college. This analysis is based on the experience of the writer, but would appear to be similar in general outline, though perhaps not in detail, to the geographical work in other training colleges. First and foremost, considerable emphasis must be given to field work. 'Geographical knowledge that is not born of direct contact with mother earth or direct observation and investigation, and is not refreshed constantly by the springs of research in the field, is practically worthless' (Geographical Association, 1959). Only thus can the theoretical concepts learnt in the lecture room be conceived in actuality. For example, the true character of blow-out dunes, spring-line settlements, and dry valleys are much more fully understood when they have been seen in the field as well as in the textbook or blackboard diagram. Field work allows time and opportunity for the practice of many skills – of map-reading, clinometer measurements, soil analyses, chain surveys, compass traverses, for example, as well as the making of transect diagrams, field diagrams, field sketches and land utilization surveys. Here, too, can be learnt and practised the 'seeing eye'. In this connexion, it is important that students be trained to attempt their own description and analysis of landscape and settlement patterns. All too frequently, they stand with pencils and notebooks poised, waiting to 'catch' all that the lecturer says, but fearing to open their own mouths and sometimes, metaphorically, their eyes! (Hutchings, 1962). Some of the most valuable lessons and exercises in observation and deduction can be learnt informally in the field, where the formality of the classroom is absent and where students gain increasing confidence in making their own investigations.

Ideally field work should include a detailed and close study of a relatively confined area where variety exists within that area. In the Cambridge region, for example, the Fenlands, the Brecklands, Chalklands and Claylands offer contrasts in rural landscapes and land utilization, while Cambridge itself offers a particularly interesting study in an evolving settlement site and urban landscape. Other areas, however, will prove to contain as sharp contrasts. If possible, a prolonged and concentrated study, in the nature of a field week, may be attempted of an area sharply different from the region adjacent to the college – for example of a coastal or glaciated highland area.

It is largely through such local field studies that physical geographical principles are learnt or re-established; the same is true of the regional study which is also an essential element of the course. By noting regional divisions within the local area, one hopes that the students will gain an insight into the concept of 'pays' or 'regions', and that such understanding will be transferred to their more formal regional studies. Here there is need to fill the many gaps in the students' world, so that one finds a demand for courses in, for example, Africa, Latin America, the U.S.S.R. and sometimes Australasia and S.E. Asia, rather than Europe, N. America and the British Isles, already frequently fairly intensively studied at school. Here, too, one finds that a new approach gains student interest, as for example in the historical geography of North America (for which a wealth of contemporary materials exists) and in problems of the modern world, including references to overpopulation, problems of rapid economic development, transport problems, soil erosion and so on, related to particular areas.

Along with both local and regional courses in geography in the training college goes the learning of 'geographical skills', i.e. those concerned with the presentation of geographical information. Few students (though there are some) have learnt, before coming up to college, how to make dot maps, density maps, cross-sections, simple field traverses and even landscape field sketches; many need further help in the interpretation of topographical and geological maps, air photographs and pictures. For many, too, the information to be gleaned from museums, and exhibitions such as the Geological Museum, the Commonwealth Institute, archaeological and anthropological collections, is a new experience.

Such training in skills and interpretation may be concentrated ultimately in what most geography lecturers regard as an essential part of the course – an independent piece of geographical investigation. Frequently this takes the form of a local study of the home area, or some aspect of the home area such as a parish study, a study of the geology and related topography, investigations of local traffic problems, urban sprawl, a study of a river valley or a section of a coastline. Many students seem to value this work and through it discover the use of a wealth of 'primary' sources of information contained in topographical maps, geological maps, meteorological statistics, census returns, Ministry of Agriculture parish statistics of land utilization, the Industrial and Occupational Tables of the Census as well as direct observation. Such 'original' work, as distinct from that culled from secondary or tertiary sources, is found to be exacting, but satisfying and enlightening, and generally serves to concentrate or focus most of what has been learnt in the academic or specialist course.

On the teaching, or 'curriculum' side of the course, it seems essential, as has been said earlier, both to kindle the students' own enthusiasm and also to broaden their usually fairly limited knowledge. (Many students, for example, have little understanding of such basic principles as the earth's rotation, the seasons, day and night, even latitude and longitude.) Clearly, however, it is impossible to teach students all the geography they themselves will teach in school, nor should this be our aim, but rather to give them confidence in setting about their own preparation. Such confidence can only be acquired by knowing both where to look for such material – i.e. the major texts (e.g. Strahler, 1963), both advanced and at school level – and how to use libraries and bibliographies. Students also need to be introduced to, and encouraged to read, reputable travel books, 'geographical' novels and anthologies such as M. S. Anderson's *Splendour of Earth*. We would concur wholeheartedly with this author's introduction: 'For I do firmly believe that no deadly accurate, purely technical description can bring to life a mountain, a great river, or even a climate, can make it our own to love and remember, as an imaginative description by a great writer can do' (Anderson, 1954, p. xxv). Students need help also with the finding of source materials and visual aids; much good material can be obtained, for example, from commercial firms, but the handling of such material and the manipulation of the mechanical aids to teaching, need instruction and practice – and not least the use of the blackboard, so much essential to the work of the classroom geographer (Cons and Honeybone, 1960).

CONCLUSION

The particular relevance of geography has been noted many times elsewhere. Suffice it to say, here, that for many students it serves as a new perspective, both on the immediate locality around them, as well as on a world of which they at student age are becoming increasingly aware, through their investigation of international affairs, through their travels abroad and through their contact with nationals of a wide range of countries in student groups. Students will talk, for instance, of discovering more interest in their railway journeys to and from college; 'I noticed the dry valleys for the first time'; 'I never realized before the significance of the place-names in my home area'. One student on teaching practice persuaded an Indian friend to come and talk to her class of ten-year olds; another used her experiences to, from and at an international work camp in Yugoslavia as a basis for lessons in a Secondary school.

When they go into teaching, our students quickly find that geography has an increasingly valued place in the school curriculum, since its relevance is immediate and unmistakable. We in the training college field hope through our courses in Geography to send out 'real geographers' who will pass on their newly found enthusiasm to geographers of the future.

References

ANDERSON, M. S., 1954, *Splendour of Earth* (London).
CONS, G. J. and HONEYBONE, R. C., 1960, *Handbook for Geography Teachers* (London). (5th Edn. with LONG, M., 1964.)
FAIRGRIEVE, J., 1926, *Geography in Schools* (4th edn, 1937) (London).
GEOGRAPHICAL ASSOCIATION, 1959, *Geography Departments in Training Colleges* (Sheffield).
HUTCHINGS, G. M., 1962, 'Geographical Field Teaching', *Geog.*, **47**, 1–14.
STRAHLER, A. M., 1963, *The Earth Sciences* (New York).

PART 2: GEOGRAPHY IN THE TECHNICAL COLLEGES

W. ISLIP

Lecturer in Geography, Cambridge College of Arts and Technology

Of the three stages of education envisaged in the 1944 (Butler) Act, Further Education is the branch which is showing the most rapid changes. In a complex pattern it makes provision for all types of education beyond the official school-leaving age. The particular concern here is the section of Further Education which comes under the heading of Technical Colleges. A brief explanation of their nature and organization is necessary to make it clear what place geography can take in their curricula.

PROBLEMS OF TECHNICAL COLLEGE WORK

As shown in Table 20 there are four main types of college, graded according to the level of work offered and the type of students on the courses. In a different category are the six National Colleges providing highly specialized post-graduate industrial courses which cannot be taken on a local basis for reasons of economy.

Table 20. Classification of Technical Colleges

		Number in 1963	*Types of Courses Available*
I	Colleges of Advanced Technology	10	Diplomas in Technology; London University Degrees (first degrees and higher degrees); post-graduate diplomas; professional qualifications.
II	Regional Colleges	25	Some Diplomas in Technology and London Degrees; Higher National Certificates; City and Guilds Final Certificates. Some G.C.E. A levels.
III	Area Colleges	About 155	Some Higher National Certificates; City and Guilds Intermediate and Finals; G.C.E. O and A levels.
IV	Local Colleges	About 275	Ordinary National Certificate; City and Guilds Intermediate and Finals; G.C.E. O and A levels.

N.B. There is no clear line of demarcation between II and III, and III and IV.

A good short survey of the organization, scope and opportunities offered by Technical Colleges is given in the special publications of the National Union of Teachers (1963) and Cornmarket Press (1964). Today, almost the whole range of British industry, commerce, art and agriculture benefits from the developments in technical education, and most students entering Technical Colleges will be seeking vocational training, probably in manufacturing industry. These students are classified as technologists, technicians or craftsmen. In addition there may be numbers of academic students. They are all above school-leaving age and their range of ability extends from semi-literacy to post-graduate research level. All types of students may have to be catered for in one college; alternatively there may be a concentration on only one activity. It is impossible to generalize about conditions and standards found in Technical Colleges. It is safer to say that each college is the unique expression of local conditions – both social and geographical.

Vocational courses may be full-time or part-time. The aim is always to help the student to understand the basic principles of what he will be doing in his working career. Full-time students, naturally, receive a more broadly-based programme which may include General Studies (and perhaps some geography); part-time students have little time, or even inclination, to read outside their subject. 'Sandwich' courses leading to the Diploma in Technology are usually in the form of six months in industry and six months in college for a period of four years.

Each college is organized in departments. For example, in the Cambridgeshire College of Arts and Technology appear Art, Building, Catering and Domestic Science, Commerce and Professional Studies, Science, Engineering, and English and Social Studies. The activities of these reflect the varying needs of a University city. In this college geography, rather unusually, operates within the English and General Studies Department. This kind of organization should explain the difficulty of including geography, or any academic subject apart from English, in the curriculum. Thus teaching problems are not the same as those in schools, training colleges or universities, for the demand for a particular course may be only short-lived and continuity of any academic programme cannot be guaranteed.

The place that geography is taking in this system is unusual. In some respects the picture is one which gives rise to concern because of the wide variations between conditions in colleges. Each institution draws up a prospectus or calendar of courses for the session, based on a careful survey of the likely requirements of industry, business and commerce within the region. An examination of a wide selection of calendars in any year will reveal that there are many places in which geography is not taught at all. In the colleges where it is given the classes fall into three main groups:

1. Academic, ranging from Royal Society of Arts, G.C.E. O and A and S levels to the Diploma in Geography, General, Honours and Special external degrees of London University. They are often

attended by part-time students although the numbers in full-time classes are increasing. In view of the difficulty in finding sufficient places in Universities for the students who have passed the minimum entrance requirements this is a process which is likely to continue.

2. Professional qualifications for the various Institutes such as Banking, Transport, Marketing and Export; Civil Service Clerical and Executive grades; Local Government bodies; National Certificates and Diplomas. Many of these are on a part-time basis but here too numbers in full-time classes are increasing.

3. Liberal studies and general courses. The prospectuses give no clear picture of the content of these. Many lecturers draw up their own syllabuses to fit local requirements and to suit the needs and aptitudes of their students.

This classification should make it clear that the whole range of geography, from University degrees to pre-Ordinary level G.C.E., is being taught somewhere or other in Technical Colleges as a whole, and, indeed, it may be taught within one institution. In some colleges, on the other hand, it is not taught at all. As suggested earlier the present departmental structure of Technical Colleges raises problems of teaching which do not exist to the same extent in schools where the demand for a subject is relatively inelastic. In technical education there may be a good deal of fluctuation in the demand for a geographer's services, possibly arising from purely arbitrary causes or decisions. A particular series of craft (or technical) courses may suddenly dwindle because of changing national or local conditions of industry, and if geography has been one of the subjects in the curriculum the need for it vanishes. A new Head of Department perhaps may decide that a subject other than geography shall be included in a given course and the full-time geography specialist may then find himself having to teach outside his specialism as there may not be enough specialist work for him to do.

In the past, geographers practising in Further Education have tended to work in isolation, somewhat out of touch with one another. More recently, the Geographical Association, living up to its main purpose of furthering the knowledge of geography and of the teaching of geography, has interested itself in this problem. The first practical step was the formation of a Further Education Section of the Association which has brought together scattered and solitary workers to discuss common problems and aims. This Section in 1954 and 1955 organized a survey by means of a questionnaire about the status of geography in Technical Colleges. The information required was sought from more than fifty different Colleges (not schools), and the results and findings were published as a report (Geographical Association, 1956) which showed that only about half of the institutes investigated had a full-time geographer on the staff. In some cases the teaching was being done by non-geographers, and only a few establishments had more than one geographer. The majority of the teachers were on the lowest grade of the Burnham (Technical) Scale (Grade A), and very few were of lecturer-status. Usually there was adequate space for the work, even if the rooms were not always designed as geography rooms. Only in about half the cases was the equipment sufficient for the type of work being done. There was a general complaint about the poor library facilities. Thus, the overall impressions gained from the report were depressing, and the best hope for the future appeared to lie with the possibilities of changing the attitudes of the principals and heads of departments, but more immediately, on the energies of the lecturer responsible for the teaching of geography. The report appeared in July 1956, after the publication of the White Paper on Technical Education (Ministry of Education, 1956). The survey itself related to conditions in 1954 and 1955 and is somewhat out of date now.

Subsequently, the Geographical Association sponsored a discussion on Geography in Further Education at its Annual Conference in January 1961 (Geographical Association, 1961). The opening paper given by Professor M. J. Wise briefly recalled the points made in the survey and commented on the slow rate of advance made since 1956, in spite of a further report by the Geographical Association in 1957. Their continuing interest has resulted in regular meetings of the Further Education Section and the circulation of a Newsletter in which are raised points of immediate interest and importance such as the form and content of syllabuses for professional qualifications and liberal studies classes.

The Royal Geographical Society also showed its concern about the position of geography in Further Education and published a memorandum 'Geography and Technical Education' (Royal Geographical Society, 1958). The White Paper of 1956 had stated that a place must always be found in technical studies for liberal education. The Ministry of Education Circular 323 'Liberal Education in Technical Colleges' made suggestions as to how this recommendation could be put into effect. The Royal Geographical Society's memorandum stated the case for geography as a liberal study and gave some examples of what might be done in technical courses by geographers. Its argument is still valid and might be read by all who have to formulate new courses for technologists, technicians and craftsmen.

The literature on the question of liberal studies, their composition, function and place is growing, but the matter is still highly controversial. The argument now turns not so much on whether liberal studies should be given at all but rather on what their character shall be. One view is that general studies should be an extension of the technical; another sees them as giving a general education in the humanities, stimulating people to think in new ways about the world. Both ways give scope for the geographer. Many technical students find their jobs are not all-absorbing, and that they cannot pursue them for too long at a stretch because there is no lasting satisfaction gained from their work. These students do not willingly enter into activities beyond the immediate goal of the technical qualification they are seeking. They cannot be blamed for not wanting to increase their experience, but for all that, it is essential that we offer them some new ideas, for this is likely to be the last chance we shall have of doing anything for them. The Ministry of Education recommends that in all new courses up to one-fifth of the time should be given to general studies, i.e. possibly five or six hours a week. It does not seem excessive to set aside for geographical studies one hour a week out of this time for the first two years of a four-year course.

THE POSITION OF GEOGRAPHY: A SURVEY

This picture of past conditions has not been altogether encouraging. Moreover, there have been signs that with the growth of work in Diplomas in Technology (a new qualification tending to replace university external degrees) geography was being dropped in some colleges which had previously supplied it as one of their degree subjects. An attempt was made by the writer in December 1963 to discover whether any noticeable change had taken place in the amount of teaching at degree and other levels in technical colleges since the survey of 1954–5. Inquiries were limited to the institutions in England and Wales listed by the Advisory Council for Education (A.C.E.) in October 1963 (Scottish institutions also appear in the list but no attempt was made to examine the position in Scotland). The colleges named were those offering full-time courses leading to diplomas in Technology (equivalent to degrees) and to external degrees of London University. The list was compiled from the Report of the National Council for Technological Awards and from Truman and Knightley's *Full-time Degree Courses Outside Universities*, and can be said to give a representative selection.

A short inquiry form was sent to fifty-nine colleges. There was a good and immediate response, forty-nine replies being received. The questionnaire sought to know:

1. Whether geography was offered as a degree subject, and if so, the type of degree.
2. The department in which geography was placed.
3. The numbers of full-time staff engaged in teaching it at degree level and at other levels.
4. The other types of courses classified under headings, Academic, Professional Qualifications, Liberal and General Studies, for which geography was given as a subject.

Table 21. Degree Courses Offered in Technical Colleges

Course		*Number of colleges*
B.Sc. General	Part I	6
B.Sc. General	Part II	4
B.Sc. (Economics)	Part I	13
B.Sc. (Economics)	Part II (with geography as special subject)	5
B.A. Honours B.Sc. Special		2

Only five colleges specified that a course was available in B.Sc. (Econ.) Part II as well as in Part I. In two cases a degree course (B.Sc. General) was available but it had never run because of a lack of demand.

The replies showed that 25 colleges employed 1 or more full-time teachers of geography; 23 colleges had no geography teaching at all; 1 college employed 1 part-time teacher; and 6 colleges which at the time of the inquiry had no courses in geography stated their intention of starting some in the near future (including degree courses if permission was granted by the Ministry of Education).

It is also known to the writer that of the ten colleges from which no return was received, geography was being given in at least two and to degree level in one (Table 21). Sixteen colleges were providing a degree course of one kind or another, i.e. one-third of the total. This compares favourably with the 10 per cent quoted in the 1954–5 return. The proportion of colleges with a full-time lecturer in geography seems to have remained the same – about half at each survey (Table 22).

Table 22. Size of Staffs in Technical Colleges

	Number of lecturers in geography department					
	1	*2*	*3*	*4*	*5*	*6*
No. of colleges:						
London Degree Level	8	3	1	1	—	—
Not-London Degree Level	9	11	1	—	—	1

The location of the geography department within the general structure of the technical college varied widely. As Table 23 shows, in about half of the cases it was grouped within the commercial and management departments but was less commonly grouped with the science and liberal studies groups. The Honours Degrees were being given in an Arts Department and a General Studies Department. The widest range of degree work appeared in a General Studies Department and two Departments of Chemistry and Geology, one of which also housed Biology and Geography (a useful combination of subjects for the geographer). One college dealt with degree work only – for B.Sc. General. The other twenty-five offered courses at other levels which are summarized in Table 24. A good deal of the

Table 23. Location of Geography in the Departmental Structure of Technical Colleges

*Department**	*Number of Colleges*
Commerce, Management, Business and Professional Studies	17
Liberal and/or General Studies	5
Science	4
Social Science and/or Economics	2
Arts	1
Day Continuation	1

* In some colleges geography was organized in more than one department.

work under the heading Professional Qualifications consists in preparing students for the qualifying examinations for membership of various Institutes. Administratively these classes may present some difficulty in that one course may have to supply the needs of several different examinations. Problems relating to this group are frequently on the agenda of the Further Education Section meetings of the Geographical Association.

Table 24. Non-degree Courses Offered in Technical Colleges

Course	*Number of colleges*	*Nature of the qualification*
Institute of Bankers	18	Professional
General Certificate of Education (A-Level)*	17	Academic
General Certificate of Education (O-level)*	10	Academic
National Diploma (Business Studies)	10	Professional
National Certificate	9	Professional
London Diploma in Geography	3	Academic
Royal Society of Arts	2	Academic

* Various examining boards.

POSSIBILITIES FOR EXPANSION

The Liberal and General Studies field is likely to offer the greatest opportunities for geographers. The term 'liberal study' raises many issues (including the fundamental one of defining 'liberal education'), which are outside the scope of this chapter. Yet the link nature of geography makes it well suited to be included in the curricula of all types of colleges. In the new conditions which may emerge from the implementation of the Robbins recommendations, many of these institutions will have freedom to organize their own courses as they think best. Since some College authorities think of geography as purely an Arts subject, and, as such, see no place for it in technical education there is the great initial difficulty of gaining a footing.

A technical contribution which geographers can make is the training of geographical technicians. One college is already providing a course which combines a training in cartography and cartographical methods with the basic elements of lettering, draughtsmanship and line drawing. It is an experimental course but there should be scope for similar ones in colleges in large cities in which the functions of university, national and local administration, publishing and printing, etc., are found. There is a growing need for maps of all kinds and for people trained to interpret them. The shortage of these technicians is felt not only in Britain but also in other countries of the Commonwealth. There is a lack of training facilities. Here is a chance for us to provide some strictly vocational instruction.

If a comparison be made of the conditions in 1954–5 with those in 1963 as suggested by the findings of the two surveys, the impression gained is of a steady advance, both in the status of geography and in the amount of work being done. There is still room for improvement.

The question of the department from which geography should operate is important. Which department is best depends on many factors, some physical, some personal. As suggested earlier, a good deal turns on the attitudes of Heads of Departments and others towards geography. The most suitable department is that which gives it the greatest scope. Colleges with a science department offering physics, botany, geology and biology should provide a stimulating atmosphere. In some colleges this does occur, but in others the geographer is sometimes regarded as a poor relation. Frequently, science students are working for specific technological examinations and no place for geography has been found in their curriculum. However, in colleges which provide O and A level G.C.E. courses in science subjects, geography is a very suitable alternative for students who cannot offer one or another of the pure sciences.

It is more usual to find the geographer working in the Commerce Department, or its equivalent (see Table 23). Here are great opportunities and sometimes the greatest difficulties. The Newsletters of the Further Education Section (November 1961 to June 1963) refer to numerous problems relating to geography for Business Studies or for Commercial or Economic Geography. They indicate the strength of the ties with the commercial world. Much of the work is with pre-apprentice and other preliminary courses and less frequently with higher levels. Some courses labelled geography deal with commodities only and some are even given by non-geographers. At higher levels Economic Geography can be successfully taught, especially to those students reading Economics at about A level standard. Since syllabuses for Higher National Diplomas are left to individual colleges to devise there should be ample scope for Economic Geography to appear in them. For the enterprising there is unique opportunity to experiment in a way not possible to teachers in schools.

One venture is Parker's work with technical students at Tottenham (Parker, 1962). He exploits the practical nature of geography in making an approach to human studies, linking the technical world to the humanities via practical methods and human geography. Parker would have the student begin by becoming proficient in cartographic methods. When this is accomplished the student analyses the specific sets of distributional patterns mapped and attempts to correlate the facts shown. Later, examples from other parts of the world are examined, compared and generalizations drawn. As competence grows the method is applied to particular regional studies. The virtue of this approach is that it can be made with all levels of student ability. It would seem to be a slow but thorough process.

There are other possibilities with advanced students in 'sandwich' or similar courses. One particular approach is the presentation of geography for its own sake; to explain what geographers attempt to do and to illustrate this with sample cases; to examine and make use of cartographical techniques; to appreciate critically the different types of maps and atlases, British and foreign, the *Atlas of Britain* and other national atlases; to discuss the concept of regionalism and illustrate first from a local then from a foreign region. These, and other ideas can be presented so as to offer exciting new lines of thought to the students (and they are numerous) who may never have had a chance to study geography before. Engineering students, for example, readily appreciate the factors of physical geography which influence the control of water supply, irrigation and power and are interested in new projects of this kind within and beyond Britain.

Some of these advanced students may eventually represent their firms overseas. They need to know more about the conditions in the countries of their main markets and chief competitors, and to be aware of factors which may affect the future of their industry. They also need to study conditions at home, to examine the changes which have taken place within Britain in particular industries and areas so that they may more readily meet the challenge of competition.

For this programme to be taught effectively one hour a week over a period of from three to four years is little enough to ask – yet a good deal more than is likely to be available in most institutions. If, in the short time allowed, the more advanced students can be given an understanding of the fundamentals and an introduction to the most suitable textbooks and works of reference, there is hope that they will be able to acquire the details for themselves, and will want to do so. To enable these things to be done well more facilities are needed; at least one fully equipped geography room in each college, and more books, atlases and maps in the libraries are minimum requirements. Opportunities for research are needed. Each lecturer should be allowed some sort of sabbatical leave at least once during his academic career. The nature of his work is exacting: terms are long; there is much administration and a heavy lecture programme which may include evening teaching.

CONCLUSIONS

Conditions in technical education are subject to constant changes and fluctuations because of its nature and functions. The inevitability of change must be recognized. Geographers who work in technical colleges must be ready to adapt their courses to suit the changing conditions. In theory there is no reason why a Technical College should not be as well known for the geography which is done in it as for the teaching of a particular craft or skill.

To those who work in these Institutions the prospects for the future of geography could seem discouraging. They need not be. Technical education offers a greater freedom for experiment at all levels than is possible in any other kind of establishment. The opportunities exist and need exploiting. Geographers are failing in their duties if this challenge is not taken up. The onus is on them to demonstrate the value of their discipline within technical education.

References

CORNMARKET PRESS, 1964, *Which University?* (London).

GEOGRAPHICAL ASSOCIATION, 1956, 'Geography in Institutes of Further Education', *Geog.*, **41**, 183–7.

— 1957, 'The Place of Geography in the Education of Boys and Girls of 15 to 18 Years', *Geog.*, **42**, 174–81.

— 1961, 'The Role of Geography in Technical Education', *Geog.*, **46**, 342–8.

MINISTRY OF EDUCATION, 1956, *Technical Education* (London).

NATIONAL UNION OF TEACHERS, 1963, *University and College Entrance: the Basic Facts* (London).

PARKER, G., 1962, 'Human Geography in the Technical College', *Geog.*, **47**, 278–84.

ROYAL GEOGRAPHICAL SOCIETY, 1958, 'Geography and Technical Education', *Geog. Jour.*, **124**, 232–4.

CHAPTER SIXTEEN
Geography in Schools

P. BRYAN

Senior Geography Master, Cambridgeshire High School for Boys

INTRODUCTION

The heading of this chapter will perhaps mislead the reader in two ways. First, the writer has teaching experience in grammar schools only. The comments are therefore directed primarily to grammar-school teaching, although they may not be entirely irrelevant to other types of school. Secondly, no attempt has been made to provide a comprehensive survey of teaching geography, even in a grammar school. Some topics to which I should have liked to refer, such as textbooks, have been left out for reasons of space. Others, such as the role of field work, I have omitted because I felt that I had nothing to say which had not been better expressed by others. The three topics I have chosen to discuss have little or no unity. They are subjects which have caused me to think during my teaching career and which seem to me to be important in the future of 'Geography in Schools'.

SYLLABUSES

In considering some aspects of the teaching of Geography in schools, primary consideration must be given to the question of syllabuses, since they determine what we teach, and in large measure how we teach. Many of the newer lines of thought at University level are now also beginning to reach down into the schools, at least at VIth-form level, and they should cause us to re-examine our traditional view of the geography syllabus, both from the point of view of content, and of our intellectual aims.

The traditional school syllabus is based on world regional geography, and relies on two main assumptions. First, that a 4/5-year secondary school course should attempt to cover the geography of the larger part of the world. Secondly, that the fundamental approach should be that of regional division. Both propositions are open to question. I have for some time thought that we are cramming too much regional geography into our syllabuses. On at least two grounds it can be questioned whether we need so much of it; that it leads to much bad geography teaching, and that it leads to the exclusion of much that is interesting and valuable.

To complete their world regional coverage, most teachers make a start in their first year, and follow one of the conventional arrangements of continents, culminating in the prescribed examination regions. Other topics, such as local geography, mapwork and physical geography are woven into the regional course in varying degrees, usually not by shortening the regional content, but by shortening the time allocated to each topic.

There are obvious objections to this type of syllabus. Is there any compelling reason for covering the whole of the world? Is any other subject reproached for being selective in its approach? This type of over-full syllabus also places on the teacher a compulsion to cover large areas in a short time, and this so often leads to stereotyped teaching, where the teachers pump in the facts in the oft-repeated pattern of the Regional Catechism – relief, drainage, climate, vegetation, minerals, agriculture, etc. Hence in a large number of cases, regional geography becomes a dull uninteresting routine. Obviously regional geography can be vital and interesting, but most teachers would agree that the more interesting approaches are also the most time consuming. Therefore, we so often fall back on the stereotyped approach, which is economical of time and thus enables us to complete the syllabus. One turns a blind eye on the boredom thus created. But what good has been done if the price of finishing the syllabus is the loss of your pupils' interest? What is the value of covering the whole world if it has left no time to develop topics which interest the teacher or the class? Most teachers do not have enough time to develop work which interests them, as distinct from work which the Examination syllabuses and Handbooks of Geography Teaching (Cons, 1955) suggest they ought to cover. Which is most important? The actual assembly of facts on the whole of the world; or the sorting and analysis of facts, involving mental processes other than sheer memory? Does one really need to cover every area of the world to get over the sort of approach at which most of us are basically aiming?

I suspect also that Regional Geography is often started at too early an age. Regional Geography is a fairly sophisticated idea. Whatever one's definition of it, it obviously requires the ability to discern and appreciate regional differences. These are things difficult for young children, not least because they lack the experience. The complex interweaving of different strands, often imperfectly understood by us, must largely pass them by, leaving them with little more than a catalogue of facts. Could not this type of work be left later, say until the third year of a five-years secondary school course? This would leave two years for a wider introductory course leading up to regional geography to be laid out as the individual teacher pleases. It might indeed be a challenge to geography teachers to devise such a course which was not directly dictated by external syllabuses. Interest should be a primary aim; it is so often forgotten in the rush to complete the syllabus. Too often we suppose that our adult mature approach to the subject is necessarily the one which should be imposed on the children. Should we not pay more attention to what interests them and use this as a basis for the first two years of our syllabus?

The average five-year regional course also takes little account of the intellectual development of the pupil. The continents are, it is true, arranged in order of complexity, but in the main the only concession made to intellectual maturity is to increase the amount of detail. Yet there is a lot of evidence of a definite maturing intellectually in early adolescence – say at about thirteen. Is there not something to be said for making this coincide with a new stage in our teaching; that of regional geography? That is, that after two years of introductory work, one should consciously advance a step in geographical technique and comprehension, by introducing a new and more advanced concept of the subject.

Of course the price to pay for abandoning part of the orthodox syllabus is that one's pupils will leave school with no knowledge of quite large areas of the world; whole continents in fact. We must make our individual decisions about this, but I would prefer my pupils to remember their geography as an interesting and stimulating subject, and to have absorbed something of its approach and application to world affairs, than that they should remember (or more likely forget) the facts about any one area.

It is pertinent to ask from whence comes this pressure to cover large regional areas. The answer, but not the blame, lies quite clearly with the examination syllabuses devised by the various Examining Boards. I say not the blame, since their syllabuses are drawn up by panels of geographers, and new lines of thought penetrate but slowly into examination syllabuses. This again is probably right, and again not the fault of the Board, who are in the main very ready to listen to new ideas. But the fact remains that Examination syllabuses dictate more than anything else what we teach in schools. Most Boards, explicitly or implicitly, demand knowledge of the larger part of the world, and thus to be fair to one's pupils, one has to cover these areas. Syllabuses which are intended to be a framework are, I feel, so overloaded that they become a straitjacket from which you escape at your peril. To cover them forces one into stereotyped teaching methods which are economical in time. Most syllabuses could do with some drastic pruning on the regional side; it is not the content which is in anyway wrong, but the amount of it. It would seem to me wholly reasonable to demand detailed knowledge of two major regional areas, to form the basis of two years work in the schools. How much more than this should be examined is open to debate. The more that is examined, the more the influence and dictates of the examination spread further back into the school syllabus, thereby circumscribing the teacher's freedom of choice and approach. It is perhaps possible that ordinary teachers do not have enough say in the formulation of syllabuses. It is true that they sit on subject committees, along with University teachers, but their opinions may not carry enough weight. University teachers are often appallingly ignorant of the context of O-level syllabuses. It is one thing to lay out a syllabus which makes good sense geographically; it is another thing entirely to know what this involves when translated into classroom teaching.

I would make it my thesis then, that there is a case to be made out for dropping part of the orthodox syllabus of regional teaching, and leaving this type of work until the third year. Starting then at an age when children are more likely to have developed intellectually to a point where they can appreciate the import of regional geography, a course can be planned leading up to the O-level certificate. A course in which the extended treatment of the O-level regions plays a major part.

Here again I would part company from many teachers. The attempt to cover too much regional geography earlier in the course often forces a rushed coverage of the 'examination regions' in the last year. It also frequently entails the repetition of an area covered earlier in the school. Some teachers even claim it as virtue that they pay no regard to the examination syllabus until the last year. I would deem this a virtue only if the examination were so wholly pernicious that it should be avoided until the last possible moment. But is this the case? If, as most teachers believe, regional geography is a good thing, why shouldn't the regional geography set for the examination receive careful and protracted preparation? Not in order that they may be crammed for the examination over an even longer period. Quite the reverse. I advocate this rather that these areas may be covered in an unhurried way; that time may be given to topics which are not necessarily going to be potential examination questions; that some attempt may be made to portray the geography of the country as a whole, rather than as a quick traverse through the various regions: that up-to-date information may be passed on, which isn't in the books as yet. One may note in parenthesis what a pity it is that O-level questions are generally set on the geography of a country as it was a generation ago, the excuse being that anything more up-to-date is not yet in print. This is particularly true of such rapidly changing areas as North America.

On one other ground I would urge consideration of a more protracted approach to the examination. It is that time should be made in every school for a comprehensive course on landforms, meteorology and climatology. Far too often these subjects are still taught by the 'incidental' method, i.e. that any particular topic in physical geography is dealt with when the regional geography happens to throw up a suitable context, e.g. Switzerland and glaciation. I doubt the value of this method on two grounds. First, that a regional context is rarely a good one for dealing with an entire topic in physical geography; secondly, that it lessens the chances of any overall view being obtained of landscape processes or the physical processes of the atmosphere. Yet surely it is this comprehensive view that we should be seeking to obtain. The basis of many of our troubles in teaching physical geography lies in this division of the syllabus into dispersed sections. Nowhere does this apply more than in the study of climate. I suspect that the vast majority of teachers would agree that climate is regarded with boredom by their pupils, and is less well understood than any other branch of geography. I wonder if the reason is not that we deal with it much as we do with our regional geography, i.e. regarding each climate as a separate entity, to be dealt with in turn. There are in fact several problems to be faced. The questions of levels of comprehension is an important one. We must all be conscious of the half truths taught in climatology, and yet this seems inevitable, because the full truth rests on an understanding of physical processes which junior pupils cannot have. We must accept this, and at lower levels stick to our simple patterns of world wind and pressure belts, onshore and offshore winds, and suchlike concepts. But at a later stage we must be prepared to destroy these concepts, and introduce more advanced models on which to work. Certainly by G.C.E. O-level pupils should be able to handle such concepts as depressions and anti-cyclones. In the VIth form one can advance to air masses, stability and instability and at least a nodding acquaintance with upper air phenomena such as jet streams. The idea that we should be prepared to introduce radically different models at different stages seems to be a fruitful one; that is, we should not be afraid to say to our pupils that we are going to advance to a new concept of certain facts, and that in the process we shall reject much that they have been taught as truth at an earlier stage. The same approach is certainly valid in the field of landforms. Our VIth-formers, largely raised on Davisian concepts, are now coming across much in the universities which does not sit comfortably in this framework, and we shall have done them no service if our teaching has not prepared them to expect to throw away some of their carefully nurtured earlier beliefs.

We also create problems, I think, by the type of climatology we ask our pupils to learn. We concentrate so much on descriptive climatology and climatic division, offering little in the way of analysis of underlying causes. Anyone marking A-level papers soon sees that in many cases an encyclopedic knowledge of the facts of world climate is based on an understanding of atmospheric processes of about lower school level. Yet the meteorological foundations of climate are being increasingly discovered and understood. And children find in meteorology the same fascination which they find in the evolution of landforms, and which they do not find in climatic division. Should we not therefore be approaching our climatology more through a study of processes and the distributions of processes? Many teachers are already doing this for temperate latitudes; it is more difficult to do it for tropical areas, but the time is perhaps not so far distant when we shall be able to approach our climatology entirely from meteorological concepts, and in the process make our teaching more vital and interesting.

I would hold therefore, that there is much in the orthodox school syllabus which is worth looking at critically. What are its aims? What account does it take of your own and your pupil's interests? How much challenge does it offer them as they mature intellectually? Is the time devoted to various topics being used in the most effective way? How far is it dominated by the straightjacket of the examination syllabus? Every teacher should make up his mind on these questions and formulate his syllabus accordingly.

THE RELATIONSHIP BETWEEN SCHOOL AND UNIVERSITY GEOGRAPHY

This subject presents one of the most vexed problems in the teaching of Geography. The number of VIth-formers reading Geography, and going on to the Universities, increases yearly. The status of the subject in schools has materially increased, and with it the number of really able pupils wishing to specialize in Geography. As this situation has developed, so the overlap problem has obtruded more and more. The schools are accused of overspecialization, but on our side of the fence the universities appear to be setting an ever hotter pace. A-level syllabuses have become ever more crowded in recent years, and topics which even ten years ago hardly appeared in Honours Degree courses now crop up regularly in A- and S-level papers. University teachers complain that much of this advanced work would be best left to them to teach, an argument that many schoolteachers would thankfully accept. But plainly the schools must be left with enough intellectually stimulating material to teach. The two sides have for too long glared at each other across a fence of mutual recrimination about the failings of the other. Quite plainly what is needed is more co-operation, and more frequent meetings between university and school geographers. University teachers are beginning to realize that what happens in the schools is of material interest to them and that they must be prepared for more contact with schools to explain their points of view. Teachers must be more prepared to go back to their universities and explain their problems. Our misunderstandings arise largely from a lack of comprehension of each other's aims, needs and problems. There should be far more short residential courses enabling ex-students to return to their universities and discuss with university staff the problems which arise from the school–university link. Overworked as many university staff undoubtedly are in teaching, research and administration, I venture to suggest that the maintainance of these links is perhaps of greater importance to them than they are sometimes willing to admit. When universities are expanding in numbers and size, is it an impossible dream to think that each university might 'adopt' its local grammar schools, and maintain some form of regular contact between them? Even if this cannot be put on a formal basis, one would like to think that the universities recognized some obligation to take the local educational pulse at intervals. Schemes such as the Cambridge Local Examination Syndicate's Field Work Projects are already bringing university staff into more regular contact with schools. Anything which increases these contacts is likely to be fruitful for Geography as a whole. We do not wish to blur the line between school and university geography; students should always be aware that 'going up' is a major step forward in their education. But we do need to eliminate wasteful overlapping and over-specialization and overcrowding of syllabuses. I believe that more discussion between the two sides, formal and informal, could largely eliminate these problems.

I have also become disturbed in recent years about the ever increasing demands which Geography makes on the scientific knowledge of our VIth-formers and university students. Disturbed not because this is a bad thing, but because so many of them are so completely incapable of meeting this challenge. In a subject which probably draws 90 per cent of its intake from students of 'Arts' parentage, it is disturbing to find that so many students are unable to pursue the more scientific branches of the subject beyond the fairly elementary level represented by first-year courses in landforms, climatology and meteorology. It has often been claimed that one of the virtues of Geography is that it bridges the gap between 'Science' and 'Arts'. Yet in fact the scientific knowledge of many undergraduates is rudimentary in the extreme. Small wonder that many scientists look askance at some of the claims made on behalf of the subject. Can we do anything about this in the schools? First, I would suggest that we must stress to our VIth-formers that even though the majority of them are reading Geography as an Arts subject, they must not turn their backs on Science, or consign to the limbo all their scientific knowledge. We must point out the increasing content of science and mathematics in the subject, and make it plain that these cannot be ignored or by-passed.

Secondly, as a matter of long term policy, we must strive in more schools for the situation which already exists in a minority. That is for greater flexibility of VIth-form subject combinations allowing some crossing of Arts and Science courses; or at the very least that Geography should be available to both Arts and Science sides in the VIth forms. We shall not move rapidly in this. Geography is deeply entrenched on the Arts side in the schools, and there are very great practical difficulties involved with time-tables, to say nothing of obstinate Heads and recalcitrant colleagues, who see their academic parishes threatened. But the spirit of the times is perhaps towards a more liberal approach to VIth-form studies; the Crowther Report at least supports us (Report of the Central Advisory Council for Education, England, 1959).

We shall not in fact make much progress on this issue until we convince our Science colleagues in schools and universities that Geography is respectable. In many schools the science staff would, and do, actively resist the inclusion of Geography as a Science option. A change in this situation must come partly from the universities. So long as Faculties demand rigid subject combinations as a price of entry, Geography will be at a disadvantage in respect of the Science Faculties. A student who has read any normal combination of Arts subjects in the VIth form is regarded by any Arts Faculty as being intellectually respectable. But in many (fortunately not all) universities, a pupil who reads a combination of science subjects including Geography – say Chemistry, Physics and Geography – is not regarded as being as intellectually respectable as the boy who has read Chemistry, Physics and Maths. The geographer who has read science is on very strong ground; the scientist who has read Geography is on much weaker ground. While this situation persists we shall not easily persuade school science departments that the best scientists should be allowed to take Geography in place of one of the older established sciences. It is not a matter of whether we think Geography is a Science or not, but whether other scientists think it is or not. I would venture to suggest we shall not make much progress in this while so many Geographers are so appallingly ignorant of Science.

AUDIO/VISUAL AIDS

It is curious to find that what has for long been accepted as part of the technique of teaching Geography is in fact not nearly so widely used as one thought. Surveys have shown that large numbers of schools lack these facilities, and that even where they exist, they are often under-used. Apathy and indifference are still commonly met, especially among older teachers. This plainly indicates that many teachers regard these aids purely as auxiliary teaching methods, to be used only occasionally, rather than as the basic teaching method, to be used as often as the blackboard itself. I hold that the use of these teaching aids should be as regular as that of the blackboard itself, and should be accepted in this light by teacher and class alike. It is odd how many teachers will agree on the tremendous impact of the visual image, and yet turn their backs on the best methods of using this approach. Why are visual aids not as widely used as they should be? Obviously there is in some schools still a shortage of equipment; this is, however, a fast disappearing situation. In many cases there is ignorance of the equipment available, and of how to operate it. In fact of course most of the equipment can be mastered in half an hour as far as its operation is concerned. Maintenance is more difficult, but is in any case better left to experts. It is surprising to find so many Training Colleges and Departments of Education still do not offer a comprehensive course on Visual Aids. Small wonder that so many teachers, especially women, go out to teach without any conception of what one of their basic teaching tools can do.

Teachers are also not made aware of the vast range of material available in this field, much of it free or at nominal cost. Government departments and industrial concerns pour out this material. How many schools still do not belong to the Educational Foundation for Visual Aids? – a body which can supply all the advice, equipment and information that any teacher could need. Of course not all material is ideal for any one person's needs. But here again the approach is symptomatic; the teacher expects the material to suit him. He rarely considers that it is his approach which may be in need of modification to suit the material, which is, after all, generally prepared by experts. Of course the ordering and examining of visual aid material thrusts yet another burden on to the teacher; yet without this preparation it cannot be successfully integrated into lessons. The reward is that of knowing that your lessons are likely to be more live and more lively, and that the impact made means that material is less likely to be forgotten.

Many teachers fear visual aids as time-wasters. And indeed they can be. Unless the equipment is permanently in position it is likely to take several minutes to get a classroom ready for use. If the room is properly set up it can be brought into operation in seconds, not once but several times in a lesson. Authorities are now beginning to spend large sums of money on equipment. But it is much more difficult to persuade them to give you the right rooms in which to use the equipment – rooms with blackout, plugs of uniform size, rooms designed or adapted to the use of visual aids. Schools need not one such room, but several. Efficient and widespread use of visual aids does not mean being able to command a vast array of expensive equipment. A little equipment and a lot of facilities to use it, will often go just as far, if not further, in promoting good teaching.

Like all teaching aids, visual aids are ultimately as good as the people who use them. More is the pity then that greater attention is not paid to giving new teachers a wider experience of the materials, equipment and techniques of teaching with visual aids. It is a gospel to be spread among new teachers. Old dogs do not readily learn new tricks, especially when in their own estimation the old tricks work so well. The converted must expect to meet resistance from Heads and colleagues alike; to be regarded as a crank, or as someone looking for an easy way of teaching – 'the playway'. The faith of a prophet is needed in the stony ground of many staff-rooms, but the reward will come in the interest of the pupils.

CONCLUSION

I said by way of introduction that my three topics had little in the way of unity. On reflection I think that a tenuous, but important, thread runs through them. As a subject Geography has made enormous strides in schools over the last half century. It is now taught more often to more pupils than ever before. But I sense that there is a danger in being complacent about our success. What we are teaching, and how we are teaching it have not kept pace with the tremendous surge which has occurred in the universities in the last ten years. The scope and techniques of the subject have broadened enormously, so that even a post-war graduate is likely to be out of date and out of his depth in many topics. And most of our school geography is being taught by pre-war graduates. It seems to me to be vital to look constantly at the content of our subject and our teaching methods; to stimulate thought and discussion between teachers of Geography. If this chapter does nothing more than provoke disagreement it will have served its purpose.

References

INCORPORATED ASSOCIATION OF ASSISTANT MASTERS IN SECONDARY SCHOOLS, 1954, *The Teaching of Geography in Secondary Schools* (London).

CONS, G. J., 1955, *Handbook for Geography Teachers* (London). (See more recently CONS, G. J., HONEYBONE, R. C. and LONG, M., 1964, *Handbook for Geography Teachers*, 5th Edn., London).

REPORT OF THE CENTRAL ADVISORY COUNCIL FOR EDUCATION, ENGLAND, 1959, *15–18* ('The Crowther Report'), Vol. I (London).

CHAPTER SEVENTEEN

Teaching the New Africa

R. J. HARRISON CHURCH

Professor of Geography, London School of Economics

Few if any parts of the world are changing so fast as Africa. The 'wind of change' has become both a cliché and a typical British understatement, for a veritable whirlwind is roaring through Africa, bringing formidable changes in the economy and the political scene. These demand our attention, and cannot and should not be ignored; to do so would be to turn our backs on an outstanding aspect of the twentieth century. In any case, Africa occupies between one-fifth and one-quarter of the inhabited area of the world, the African states compose one-third the membership of the United Nations, and the Afro-Asian group about one-half.

POLITICAL EVOLUTION

The colonial era has almost ended and most of Africa became independent with little or no bloodshed between 1956 and 1961, except for the long and bloody prelude to independence in Algeria, the similar aftermath to it in the Congo, and the uncertain future for Portuguese Africa. The federations of French West and Equatorial Africa and that of Rhodesia and Nyasaland have broken up, and Africa is now the second most politically-divided continent. In part this is a natural reaction to excessive centralization in the ex-French lands, and to the failure to develop true partnership in the formative years of the Rhodesian Federation, but there are other reasons. Separate independence means a voice at the United Nations with which to seek aid, and ministries and embassies with posts for those who have been prominent in the independence movement. Africa appears to be following in the footsteps of South America; indeed, Africa is even more divided and its nations the successors of more varied colonizers than those of most of South America, while racial divisions and hatreds are far stronger in Africa.

FIG. 17.1. *Political divisions of Africa.*

On the other hand the trends to co-operation and integration in Africa are not inconsiderable. Pan-Africanism is a potent force which has brought the African nations together at many conferences, and unites them in their hatred of *Apartheid* and Colonialism. Part of the former United Kingdom Trusteeship of the Cameroons has joined the ex-French Trusteeship in the Cameroon Federation, while ex-British Somaliland has joined the ex-Italian Trusteeship of Somaliland in the Somali Republic. An East African Federation is a possibility, and the Gambia may eventually join Senegal in some kind of federation or association.

Certain countries, while remaining independent, have come into new associations. In 1958 Guinea alone answered 'No' to De Gaulle's referendum and all French aid, personnel and equipment were withdrawn as a reprisal. Ghana was the only non-communist country to offer help to Guinea in her distress, granted her a loan of £10 million, and a 'union' of the countries was signed at the end of 1958. In 1960, when Senegal left the Mali Federation formed only in 1959, the Mali Republic joined the union, then renamed the Union of African States in the hope that other states might join. Meanwhile, the union unified nothing, but was merely an association of radicals. It demonstrated the difficulties of unifying former colonies of two or more powers with different official languages, currencies, constitutions and administrative methods – the more so as they are not all contiguous. The union, never effective, is defunct. Yet these countries have the potential for economic collaboration since they produce different commodities and Guinea's alumina might be used in Ghana's future aluminium industry (Figure 17.1).

The Benin–Sahel Entente formed in 1959 and now grouping the Ivory Coast, the Upper Volta, Dahomey, Niger and Togo has less ambitious aims but has achieved more because all four countries are ex-French administered and so have the same official language, currency, similar constitutions and election dates, and were already joined in a customs union. Even more significant is the fact that the first two and the next two already had close economic ties. The Ivory Coast is the inlet and outlet for the overseas trade of the Upper Volta, although this very poor and overpopulated country has little to offer except its labour. Men from the Upper Volta customarily work for several years on Ivory Coast coffee, cocoa and banana farms and plantations, on timber cutting and sawing, and in diamond and manganese mines. Dahomey is normally the land of transit for much of Niger's trade, although some of this also passes through Nigeria and more would do so were it not for economic nationalism which usually diverts trade through Dahomey. Both the Upper Volta and the Niger produce groundnuts, cattle, hides and skins, but the Niger more than the Upper Volta. Dahomey is mainly a producer of palm oil and kernels. The countries adjoin each other and are fairly well linked by roads, railways and air services. Collectively they have a population of 16 million, and with this and their resources are together a stronger counter than in isolation to the economically richer Ghana with 7 million or Nigeria with some 55 million people.

To both the Ghana–Guinea–Mali Union of African States and to the Benin–Sahel Entente the Upper Volta occupies a strategic position. Although poor and a member of the Entente, the country had quite as many interests with the Union, for her labourers go in even greater numbers to Ghana than to the Ivory Coast, and from Ghana the Upper Volta receives Sterling Area imports that are often cheaper and more varied than those from the Franc Zone of which she is a member. Furthermore, substantial transit trade between Mali and Ghana passes across the Upper Volta, especially kola nuts and Sterling Area imports from Ghana in exchange for cattle, hides and skins, and dried Niger River fish from Mali. Lastly, had the Upper Volta left the Entente and joined the Union, the members of the latter would have been contiguous and would have isolated the Ivory Coast from its partners, putting the Ivory Coast in the situation that Ghana now has relative to Mali and Guinea. Ghana and the Upper Volta once declared their boundary customs-free, but so long as these countries belong to different currency zones the full effect of free trade is lacking.

The Union and the Entente represent contrasted approaches to African integration, the political approach of the Union, and the economic one of the Entente. From 1961 until 1963 the former was the view of the 'Casablanca countries' which included the three of the Union, Morocco and Egypt; while the latter was the view of the 'Monrovia countries' which included almost all the other nations of Africa. At the Conference of African Heads of State in Addis Ababa in May 1963 the Monrovia view was accepted, and the rivalry of approach ended officially.

Whatever its other demerits or evils, colonialism restrained tribalism, which every now and then reappears as a disrupting influence in modern Africa. Most of its leaders are against tribal traditions and chieftaincy but these were elements in the Congo (Kinshasa) troubles, the estrangement of Togo and Ghana, the separation of the small new Mid-Western Region from the Western Region of Nigeria in 1963, in the difficult discussions prior to independence in Kenya, and the Nigerian disturbances of 1966.

Given the need for national unity in independence and the formidable problems of government and economic development, it is not surprising that African governments have become progressively more authoritarian. African societies have been accustomed to strong rule by chiefs balanced by very free discussion in councils, and 'One Party Rule' in states is often referred to as a natural evolution. Steps towards national unity where it is, perhaps, as yet not very strong, have been the incorporation of Eritrea into a unitary Ethiopia in late

1962 (Eritrea was previously in a federal relationship with Ethiopia), and the elimination of the three provinces of Libya (Fezzan, Tripolitania and Cyrenaica) in early 1963.

The colonial map of Africa was often a reflection of the 'Scramble for Africa' after 1885. Many boundaries were the limits of penetration by a European country, such as the more northerly extension of Nigeria compared with that of Ghana, or the north-westward thrust of Mozambique which reflects the farthest penetration of the Portuguese up the Zambezi Valley. Other boundaries are the result of exchange, such as African Zanzibar for European Heligoland, or of colonial administrative convenience when independence was never envisaged such as the excision of the Hodh from the then French Sudan in 1944 and its addition to Mauritania, so causing independent Mali to have a narrow waist-line in the centre and a shape like that of giant butterfly wings. Several other African states have extraordinary shapes, e.g. Zambia. African boundary disputes are likely to trouble the world, for the boundaries are not only exceptionally long but are the work of non-Africans, and they are usually utterly unrelated to ethnic or economic patterns. Africa is likely to give more trouble in this respect even than South America where, despite the numerous disputes, most of the peoples are of European descent.

There is tremendous variation in the size of African states as, indeed, there is in the World generally, but in Africa this may be the more serious given the poverty of the countries and the paucity of skilled and dedicated administrators. Many states include much desert, notably Mauritania, Algeria, Libya, Egypt, Sudan, Mali, Niger and Chad, although it has brought good fortune with iron to the first and much oil to the next two. In several states there are major problems of integration because of size, natural and human diversity or length of communications, especially in the Republic of South Africa, the Congo (Kinshasa), Ethiopia, the Sudan and Libya.

Some states are very small, especially Gambia (4,008 square miles), Rwanda (10,166), Burundi (10,744) and Lesotho (11,716). Many have small total populations, such as Gambia (315,000), Gabon (449,000), Botswana (560,000), Congo Brazzaville (790,000), Mauritania (1,000,000 and mostly nomadic), Liberia (1,016,000), Lesotho (1,025,000 including migrants), the Central African Republic (1,203,000), Libya (1,564,000 and many nomadic), Togo (1,630,000), the Somali Republic (2,250,000 – mostly nomadic), Dahomey (2,260,000), Sierra Leone (2,180,000) and Guinea (2,900,000), to mention only those under the very low figure of 3 million which Dr

Nkrumah once gave as the lowest desirable minimum population for a state. And while most of Africa is very thinly peopled, mainly because of the difficulties of the environment and the consequences of the slave trade, Malawi and the Upper Volta are over-peopled and large numbers of their men must seek work abroad for long periods.

Most of the land-locked states and territories of the world are in Africa (14 out of 25), and they comprise 25 per cent of the area and have 15 per cent of the population of Africa (Hamdan, 1963). In addition, Mauritania, Ethiopia and the Congo (Kinshasa) while not land-locked have had many of the problems of such states. The short coastline of Africa relative to its area is well known, and the small extent is made worse by its inhospitable character with few deep inlets or substantial peninsulas, surf and coral (east coast). As Africa is so overwhelmingly dependent upon overseas trade this is the more serious, and the construction of ports has been a major engineering feat in Africa, notably at Casablanca, Monrovia, Abidjan, Takoradi, Tema and Cape Town.

So many of the recently independent states of Africa rely on foreign aid that it might be objected that consideration of shape, area and population are academic points of no importance, yet it is difficult to be sure that external aid can or will continue indefinitely on the present scale, and while it continues aid tends to be unduly mixed up with the politics of the 'cold war'. Colonialism may be replaced by Aid-Dependence.

AGRICULTURAL DEVELOPMENTS

Change is no less rapid or profound in the economic field than in the political one. Africa is mainly, and will long continue to be a producer of vegetable and mineral raw materials. High or at least good prices for agricultural raw materials for many years after the Second World War put money into the hands of villagers, and made prosperity fairly widespread. Cocoa production has been increased by new plantings on the eastern fringes of the Nigerian cocoa belt, and on the western side of the Ghana and Ivory Coast areas. Improved pest and disease control have also helped, so that African countries (mainly the West African ones) now produce some three-quarters of the world's cocoa (Table 25); Ghana alone produces over one-third, and had record harvests in the early 'sixties. Almost all the West African crop comes from small farms.

Groundnut production is about 900,000 tons annually in Senegal and 1,400,000 tons in Northern Nigeria, almost all again from small African farms, such amounts being secured by the use of improved seed, artificial and animal fertilizer. Substantial quantities of groundnuts are used as food in Africa or crushed for oil. Turning to the oil palm, there are extensive foreign-owned plantations in both Congos, but Nigeria produces as much (nearly one-third of world exports) from small African-owned farms, and the quality of its oil has been vastly improved since the Second World War by extraction in small oil mills, by initial bonus payments for high grade oil, and by clarification and bulking in large holders at the ports. West Africa also supplies about four-fifths of the palm-kernel exports of the world.

Table 25.[1] *Production of Selected Crops in Africa in Thousands of Metric Tons*

Crop	*Average 1934–8*	*Per cent of World Production*[2]	*1963–4*	*Per cent of World Production*
Cocoa	484	66	900	74
Groundnuts	1,500	17	4,670	31
Cotton Lint	590	11	880	8
Sugar Cane	800	4	32,010	7
Coffee	140	6	1,000	25

[1] Sources: *Production Yearbook*, 1951 and 1964, F.A.O., Rome.
[2] Excluding the U.S.S.R.

Another crop which has expanded greatly in some countries is cotton, especially in the Sudan, Tanzania, Mozambique, Chad, Cameroon, Nigeria, Egypt and the Republic of South Africa. Most of the increased production is for use in African mills, of which there are some very modern examples in several countries, e.g. in most of the above-mentioned states, the Congo (Kinshasa), Angola, Rhodesia and Ethiopia. Cotton production has, however, declined in the Congo (Kinshasa), while remaining stable in the rest of Africa, so that the percentage of African cotton production in world output has declined (Table 25). Sugar is also being grown much more widely in Africa to satisfy local needs, which increase rapidly with improved standards of living. Important areas using irrigation are in and near the Lundi Valley (Rhodesia), near Moshi and Kilombero in Tanzania, and at Wonji in the Rift Valley south of Addis Ababa in Ethiopia, while the crop is developing in the Niger Valley in Nigeria, and in the Inland Niger Delta in Mali.

The most outstanding crop development has, however, been coffee. Whereas in 1934–8 the continent accounted for only 6 per cent of world production, in 1964 the percentage was 25. This rise came about after the Second World War as the result of a guaranteed market at very high prices in France for coffee from the then French colonies, of the founding of co-operatives for African growers in Tanzania, of the demand for soluble coffees which use the robusta variety much grown in Africa, of prosperity and improved living standards in coffee-consuming countries, and of increasing demand for coffee in countries like the United Kingdom which are traditionally greater consumers of tea. The Ivory Coast is now the third or fourth world producer of coffee after Brazil and Colombia, and surpassing Angola. Uganda and Ethiopia are other important producers, so that Africa now exports a quarter of the world's coffee.

Table 26.[1] *Exports of Selected Crops from African Countries in Thousands of Metric Tons*

Crop	*Average 1934–8*	*Per cent of World Exports*	*1963*	*Per cent of World Exports*
Cocoa	460	67	738	76
Groundnuts[2]	770	42	1,180	83
Palm-oil[3]	230	50	303	56
Palm Kernels	670	92	532	99
Cotton Lint	560	18	740	22
Raw Sugar[4]	720	7	1,647	5
Coffee	130	8	684	25

[1] Sources: *Trade Yearbook*, 1951 and 1964, F.A.O., Rome.
[2] Total in shelled equivalent.
[3] 1962 figures.
[4] Africa is also a considerable importer (470,000 tons average in 1934–8 and 1,043,000 tons in 1963).

The exports of Africa are almost everywhere being further processed locally before export. Thus more than one-half (470,000 tons) of Senegal's average groundnut harvest is crushed locally and the oil exported by tanker, while some 245,000 tons of the Northern

Nigeria harvest are crushed in four Kano mills. Together with oil from small Niger Republic mills it is taken south in specially-designed saddle-back railway tankers, the interior tank being used to carry mineral oil northwards and the outer 'saddle' tanks taking groundnut oil southwards to Lagos. Likewise, far more varied forest trees are being felled than in the past, and whereas the fewer species of the past were exported entirely as logs, the now more-varied species are also exported sawn, as veneer or plywood from ultra-modern factories at Abidjan (Ivory Coast), Samreboi (Ghana), Sapele (Nigeria) and Port Gentil (Gabon), the latter claiming to be the world's largest plywood works. Nevertheless, Africa produces only some 5 per cent of the world's timber, although its value is much higher.

New methods of agricultural production are being tried. Whereas the British had rarely permitted plantations in West Africa (although freely in East, Central and South Africa), African governments are developing them as means of increasing output and improving the quality of crops. Nigeria is experimenting with Farm Settlements where farmers and their families are thoroughly instructed in new techniques over a number of years. Under Nkrumah Ghana had over a hundred State Farms on the Russian model, run with the help of Russian experts. Each farm was rather under a thousand acres in area, but most were to be larger when more mechanical equipment was available, and the aim was to produce a standard crop for processing in local factories. There were also farms run by the Workers Brigade, where the aim was to provide work for urban unemployed who had recently come from rural areas, but the farms were quite considerably mechanized. The Ghana Army was to undertake farming to provide its own food, as the Ivory Coast one is already trying to do with second-year conscripts. Ghana also has some Young Farmers Settlements for school-leavers but they are fewer and less developed than in Nigeria. Most countries have for long tried to develop co-operatives for ordinary farmers but progress is slow, partly because of the scarcity of honest and efficient clerks. In Kenya Africans are acquiring land in the former European areas of the highlands, while there has long been a vigorous effort to improve methods on African highland farms, mainly by introducing livestock and modern methods.

Many African governments are anxious to extend irrigated agriculture. The most spectacular success has been in the Sudan where the government has added 800,000 acres of irrigated land in the Managil Extension, on the north-west of the one million acres of the Gezira Scheme,[1] and where cotton is the vital crop. The Roseires Dam has been built to supply future water-needs of the Managil, which were at first met from the Sennar Dam. The height of this has been raised, and now produces hydro-electric power which is gridded away as far as Khartoum. Another remarkable scheme is for the comprehensive development of the Medjerda Valley in north-eastern Tunisia. The valley has suffered severely from soil erosion, silting, flooding, gross disparity of land-holding, and insufficiently productive farming. All these matters are now being tackled and some 27,000 acres have been irrigated for the intensive production of vegetables. Over 1,350 families have been settled on sub-divided farms. A third very notable achievement is on the Limpopo in Mozambique. A dam at Guija, which also carries the railway opened in 1956 between Bannockburn (Rhodesia) and Lourenço Marques, provides water for Portuguese and African peasant farmers who each have between 10 and 25 acres of irrigated land and 62 acres for dry farming. The main crops are rice, wheat, cotton, maize, vegetables (especially tomatoes, onions, beans, and potatoes), kenaff and alfalfa, and only here and in Angola can European and African settlers be seen living and working side by side under identical conditions for similar rewards. There have also been many irrigation developments with mechanical cultivation and paid labour, as at Richard-Toll in Senegal for rice (but at high cost), by private companies as in Ethiopia at Wonji for sugar and at Tendaho on the Awash in the Danakil Plain for cotton, and for sugar in Tanzania and Rhodesia.

CHANGES IN MINERAL PRODUCTION

The most dramatic change has been the rapid development in much less than a decade of the great oilfields of Algeria, Libya and Nigeria, the laying of oil and gas pipelines to the Algerian, Tunisian and Libyan coasts (Figure 17.2), and of an oil pipeline from the Nigerian fields to the sea at Bonny. A new world region of petroleum and gas production has been created. Oil is now by far the leading export of Algeria, Libya and Nigeria. Although North Africa produces the now

[1] There is an excellent film about this entitled *White Gold* obtainable free on loan from the Sudan Government Publicity Officer, Cleveland Row London, S.W.1.

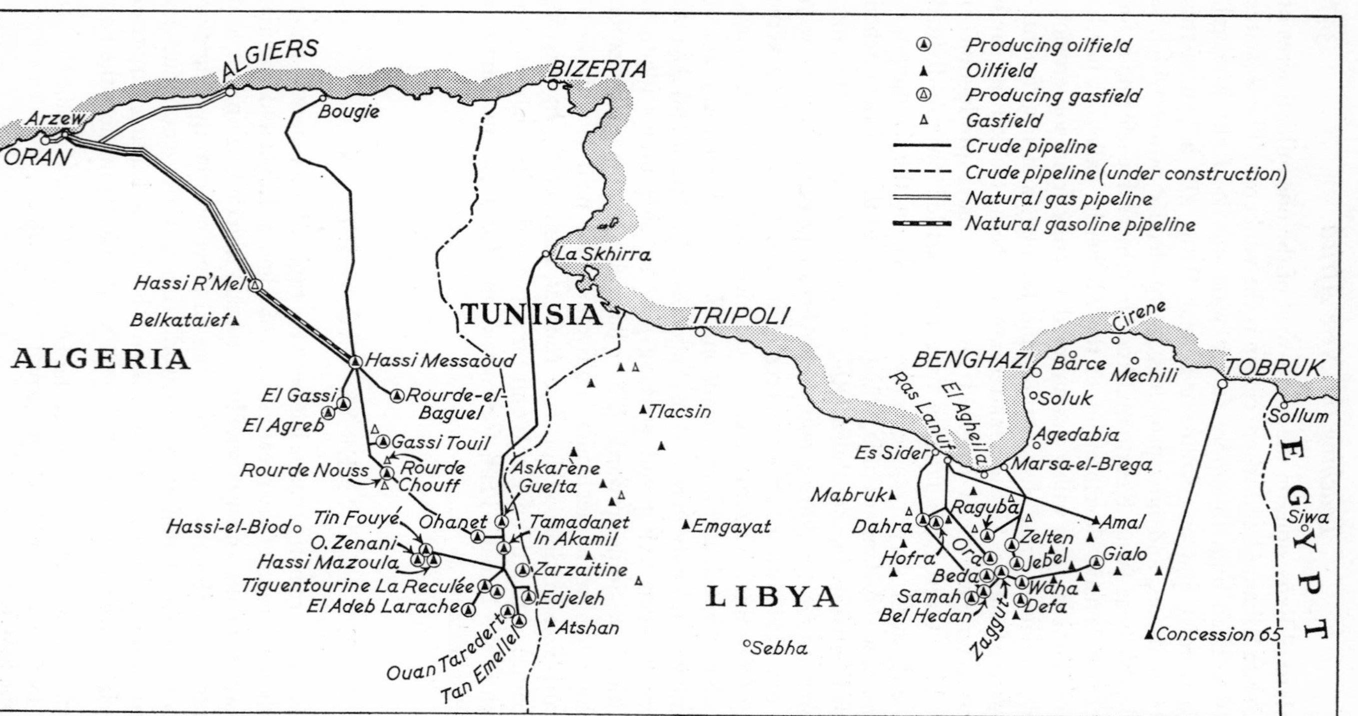

FIG. 17.2. *Saharan oilfields, showing oil and gas pipelines.*

less-needed high-octane products these fields are nearer to European markets than the Middle East or Venezuela fields. Moreover, natural gas is being brought to Europe by tanker, and this with considerable natural gas developments in Europe (e.g. in the North Sea, Netherlands, France and Italy) may also help that part of the Old World to secure some of the enormous benefits the United States has enjoyed from vast resources of natural gas.

Mineral oil has also been found in recent years on the coasts of Gabon near Port Gentil, of the Congo (Brazzaville) at Pointe Noire, and of Angola near Luanda. These are of much less significance. Very intensive prospecting is going on there and in some other producing and non-producing countries, but African oil production is still no more than some 4 per cent of World output.

Many oil refineries have been or are being built by companies in the hope of larger local markets and to stake a claim in such expanded markets (Hoyle, 1963). Refineries are sought by African governments as part of their drive for industrialization and as prestige symbols. Apart from the fairly numerous older-established ones in the more developed extremities of the continent, refineries have been built or are in progress at Luanda (Angola), Matola (opposite Lourenço Marques), Tema (Ghana), Port Harcourt (Nigeria), Mombasa (Kenya), Port Sudan, Abidjan (Ivory Coast), Dakar (Senegal), Dar es Salaam (Tanganyika), Assab (Ethiopia), Monrovia (Liberia) and Tamatave (Madagascar). Another refinery at Umtali (Rhodesia) is an African example of the world trend to establish refineries of imported crude oil nearer to inland markets. The pipe-line normally carries oil from Beira (Mozambique) across mountainous country along the international boundary.

Among the solid minerals, Africa is becoming an important supplier of iron ore. Before the Second World War only Algeria, Morocco, Tunisia, Sierra Leone and South Africa were significant producers of rich iron ore and the latter alone had an iron and steel industry. Since 1951 numerous other deposits, mostly of rich haematite ore, have been developed. In that year an American firm began operations in the Bomi Hills 50 miles north of Monrovia, and in 1962 it opened another reserve on the Mano River on the Liberian–Sierra Leone boundary (Figure 17.3). Both are linked by a mineral railway to special loading piers in the deep-water port of Monrovia, and iron-ore has displaced rubber as the leading Liberian export. A richer deposit on Mt. Nimba, at the meeting point of the Liberian, Guinea and Ivory Coast boundaries, has been developed by another

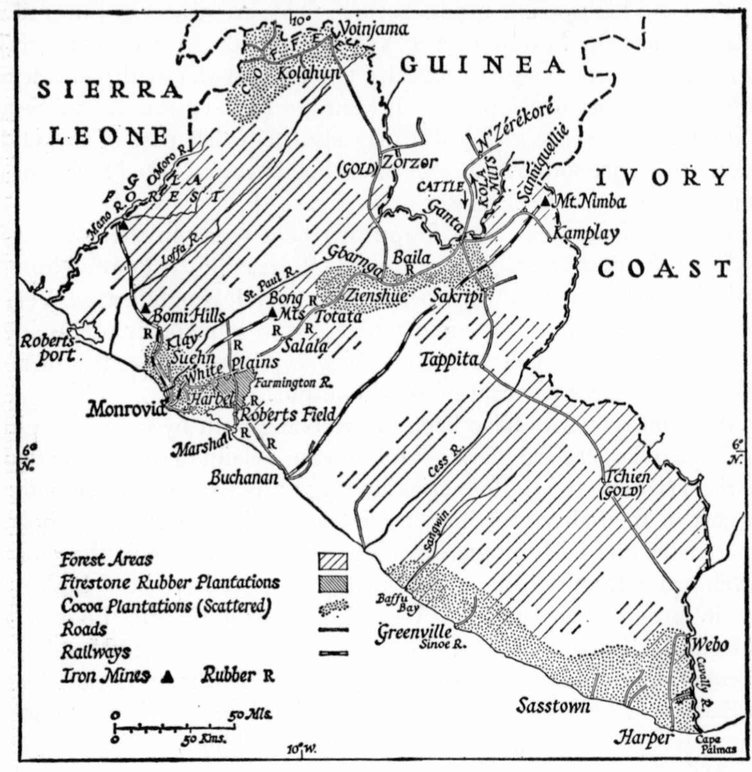

FIG. 17.3. *Liberia.*

company. This ore is taken out by a railway to a new port at Buchanan. A fourth Liberian reserve is being exploited by a third company in the Bong Hills north-east of Monrovia, and is linked with the latter by Liberia's third mineral railway.

After the development in Liberia in 1951 there followed one near Conakry of laterized ore of ferruginous magnetite. It has chrome, nickel, alumina and much moisture (which delayed its use), but it is easily quarried and near the port. In the later 'fifties rich ferro-manganese and titaniferous iron ores were developed on the Luanda Railway in northern Angola, and on or near the Benguela and Moçâmedes lines. In late 1962 commercial production began of the rich haematite deposits near Fort Gouraud in Mauritania (Figure 17.4). These are joined to the deep-water port of Port Etienne by a 400-mile mineral line through desert and keeping just within Mauritania. Ore carriers of 45,000 tons are taking the ore to participating

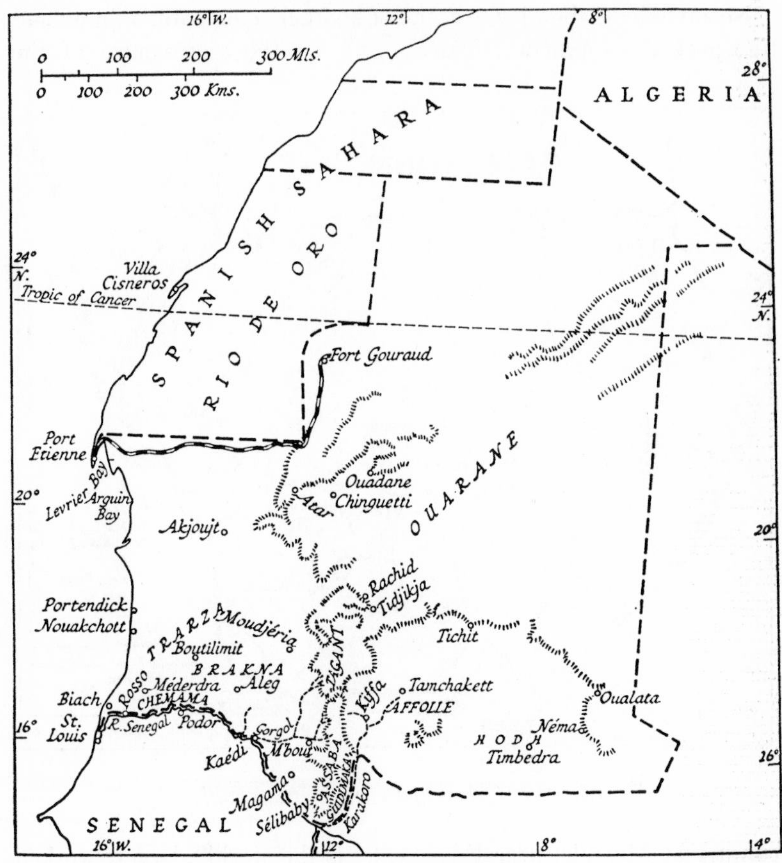

FIG. 17.4. *Mauritania.*

firms in France, Britain, Germany and Italy; in the first case to the new coastal integrated steel works at Dunkirk. Royalties on this ore are making Mauritania into a viable state. The same is happening to Swaziland, as iron ore reserves are opened up in the Bomva Ridge close to the boundary with the Republic of South Africa. To make this possible Swaziland's first railway has been built across the country to link the deposits with Lourenço Marques, in Mozambique, from where the ore goes to Japanese works. The distance from Japan well illustrates the way that modern steel industries in countries with costly fuel will go far to secure the richer ores that countries in Africa are increasingly producing. More such reserves are likely to be developed, e.g. in Gabon, given good markets for steel.

Meanwhile, Gabon is the scene of another remarkable mining and transport development (Figure 17.5). In the west-centre of the

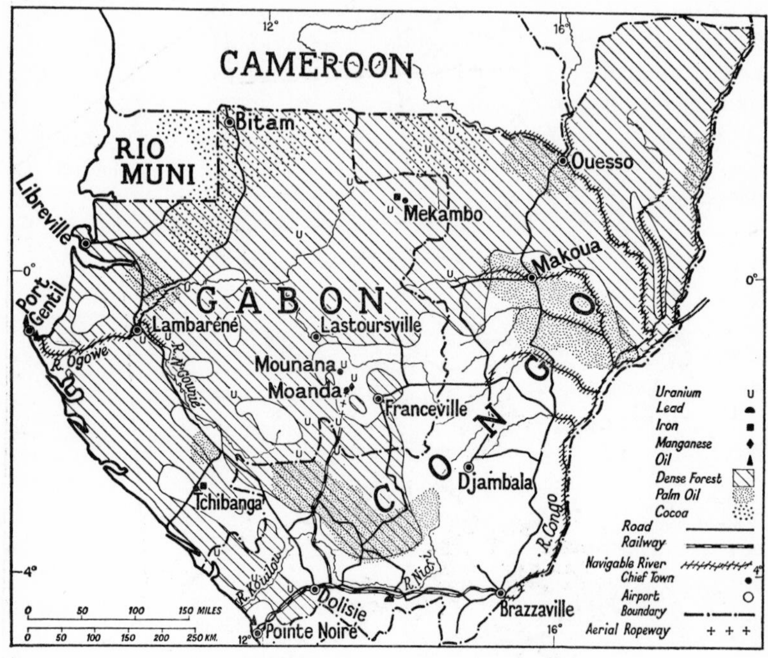

FIG. 17.5. *Gabon and Congo (Brazzaville).*

country at Moanda, near Franceville, is the world's largest worked deposit of manganese, with a content of 48 per cent. To export it from this remote area of difficult terrain the ore is carried first by a fifty-three-mile aerial ropeway, then by 2,000-ton trains of forty wagons on a 178-mile long mineral railway to the Congo-Ocean Railway, along which the ore travels a further 124 miles to Pointe Noire in the Congo (Brazzaville). All this came into full operation in 1962 and is another engineering marvel of modern Africa, made worth while by the richness of the manganese ore. Gabon quickly became the fourth world producer of manganese.

THE DEVELOPMENT OF HYDRO-ELECTRIC POWER

Africa's hydro-electric power potential has been described as equal to the amount of developed power in the world; that on the lower Congo is alone equal to that already developed in the United States. The difficulties of developing Africa's power potential are enormous – especially the greater costs of any such development in the tropics because of the seasonal rainfall, high evaporation, the need for greater insulation, and higher costs of transmission. All this is made more difficult by the smallness of the general market and the usual need to have one or more large industrial users of the power. So far Africa has only just over 2 per cent. of the world's hydro-electric power, and that has been developed mainly for mining needs – particularly in the Congo (Kinshasa). The Kariba development is in line with this, since it was conceived principally to serve the power needs of the Zambian Copperbelt; nevertheless, it has also served the industries and other needs of Rhodesia. In this latter respect it is aiding industrialization in Africa, and this was the purpose of the Owen Falls Scheme which has supplied power to new cement, textile and copper-refining industries, as well as for domestic needs in Uganda and western Kenya. The Mabubas, Biopio and Matala plants in Angola supply power for numerous industries and domestic needs in Luanda, Lobito, Benguela and Matala, while the Edea dam in Cameroon provides power for Africa's first aluminium smelter close by, as well as for industrial and domestic needs in Douala and Yaoundé. The great Volta River Dam at Akosombo on the lower Volta in Ghana (Figure 17.6) is likewise designed mainly for the provision of power to an aluminium smelter at the new port and planned town of Tema, east of Accra, although power is also available for Ghana's existing mines and southern towns. At first imported alumina will be used but Ghana bauxite may be used after the first eight or ten years. The Kainji Dam on the Niger above Jebba in Nigeria will produce power for use generally in Nigeria.

QUICKENING INDUSTRIALIZATION

Although the Republic of South Africa is the only African country with a considerable range and diversity of industry, and with substantial industrial exports, Rhodesia and, perhaps, Egypt, are within sight of the same goal, while the Maghreb lands and the Congo (Kinshasa) have a fair range. Industrialization is sought as a

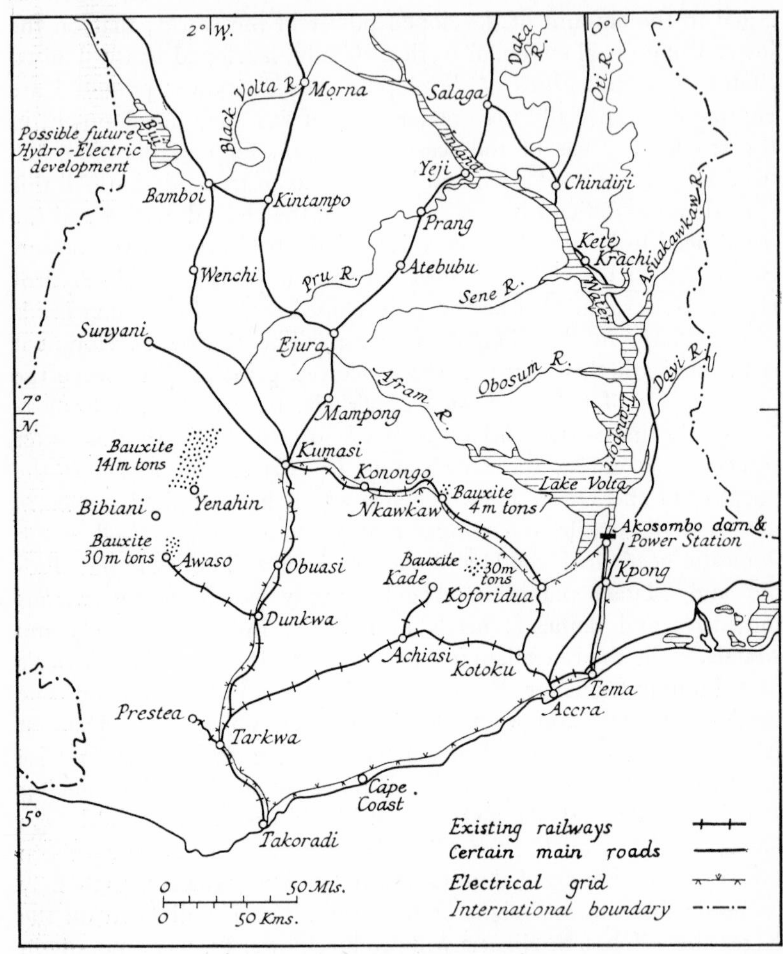

FIG. 17.6. *Volta River project.*

means of avoiding the price fluctuations of the raw materials now mainly exported, of securing further income from such materials by processing them locally (e.g. the extraction of groundnut oil described above), of broadening the wage structure and enhancing labour skills, and of providing employment in towns to which there is such a formidable exodus from the African countryside. To all these is added the prestige factor of industrialization, the belief that without an iron and steel works, oil refinery and aluminium or other smelter a country is primitive. It is not surprising, therefore, that the same industries have been set up in adjacent countries, even in the several regions of federal Nigeria, so that markets are being severely restricted to each national area or region. This problem is made more acute by the small and poor populations of many countries.

Apart from industries processing exported vegetable or mineral produce, a common type are those manufacturing simple yet common consumer goods such as textiles, soap and drinks, especially when these are costly and bulky imports. Some industries in these categories will start by using mainly imported goods, replacing them gradually by local material, e.g. tobacco and leather. There are cycle, scooter and car assembly works at ports putting together articles more costly to import in their finished form, as well as flour mills at ports to which a necessary import must come. Ports are thus significant industrial areas for this reason, and because they are often capitals they have local markets and are the termini of transport systems. Inland towns with industries are usually large centres of population, such as Ibadan with 627,379 (1963) or Kinshasa with 402,492 (1959). Industries are commonly grouped on industrial estates, such as the Trans-Amadi Estate at Port Harcourt, which is also supplied with natural gas from the Nigerian oilfields.

TRANSPORT

Outstanding characteristics of modern Africa are the importance of roads and road transport. The latter is usually far more important than rail traffic, so that atlases which emphasize railways give a false impression. Michelin, Shell and other tyre or petroleum companies issue road maps of African countries, and these are helpful in showing the number and importance of roads. They are constantly being added to or improved, and road transport is a common African enterprise. Air services are still for the relatively few, but most African countries have their own airline and there are numerous internal services as well as international ones in almost all countries. Airlines have done more than anything else to enable Africans to know and meet each other; the aeroplane is often regarded as the pacemaker of Pan-Africanism.

Railways are still vital for long-distance hauls of bulky produce, and new mineral lines in Mauritania, Liberia, Gabon and Swaziland have been mentioned. A general purpose line has been built in Nigeria from Jos to Maiduguri, the Luanda and Moçamedes lines are being extended in Angola, where branches are also being made to the Benguela line, and the line from Nacala in Mozambique is nearing Lake Malawi. Rhodesia has since 1956 had a second outlet through Mozambique to Lourenço Marques. Among important new or re-developed ports are Port Etienne (Mauritania), Monrovia (Liberia), Tema (Ghana), Cotonou (Dahomey), Nacala (Mozambique), Assab (Ethiopia) and Bougie (Algeria), the latter for export of Saharan oil.

CONCLUSION

Africa is entering another great epoch during which it will not only be transformed but will impinge increasingly upon the rest of the world. Africa cannot be ignored, and the fascination of the continent and the needs of its peoples demand that it be studied adequately. There are already a number of detailed studies of large parts of the continent (Cole, 1966; Wellington, 1955; Harrison Church, 1966 and 1963) which are available for teachers and for reference by senior pupils. Three modern A-level studies of the continent (Jarrett, 1966; Harrison Church, Clarke, Clarke and Henderson, 1967; Mountjoy and Embleton, 1965) have appeared. Articles on Africa are listed regularly in *Geography*. *The Geographical Digest*, Philip, annual, details the main developments. The Central Office of Information, Horseferry Road, London, S.W.1. publishes *The Changing Map of Africa* and leaflets on Commonwealth countries, as does the Commonwealth Institute, W.8. The Petroleum Information Bureau, 4 Brook Street, London, W.1. has useful handouts on *Oil in Africa*, *Oil in the Commonwealth*, *Saharan Oil in the French Economy*, etc. The British Iron and Steel Federation, Steel House, Tothill Street, S.W.1. provides similar material on iron ore in Africa. Both organizations have excellent films, as have Unilever, Blackfriars, E.C.4., the United Africa Company, United Africa House, S.E.1., and Shell International, Shell Centre, Waterloo, S.E.1. Several of the African Embassies or High Commissions publish gratis monthly or quarterly bulletins, notably Liberia, Ghana, Nigeria, Rhodesia, and the Republic of South Africa.

References

COLE, M. M., 1966 (2nd edit.), *South Africa* (London).

HAMDAN, G., 1963, 'The Political Map of the New Africa', *Geog. Rev.*, **129,** 418–39.

HARRISON CHURCH, R. J., 1966 (5th edit.), *West Africa* (London).

— 1963, *Environment and Policies in West Africa* (London).

HARRISON CHURCH, R. J., CLARKE, J. I., CLARKE, P. J. H. and HENDERSON, H. J. R., 1967 (2nd edit.), *Africa and the Islands* (London).

HOYLE, B. S., 1963, 'New Oil Refinery Construction in Africa', *Geog.*, **48,** 190–4.

MOUNTJOY, A. B., and EMBLETON, C., 1965, *Africa, A Geographical Study* (London).

JARRETT, H. J., 1966 (2nd edit.), *Africa* (London).

WELLINGTON, J. H., 1955, *Southern Africa*, 2 Vols. (Cambridge).

CHAPTER EIGHTEEN

Frontier Movements and the Geographical Tradition

P. HAGGETT and R. J. CHORLEY[1]

Lecturers in Geography, University of Cambridge

RETROSPECT

In preparing this volume for press we have been forced to take a retrospective view back over the months since the inception of the Madingley Symposia and are surprised by even the small measure of order which seems to have emerged from chaotic beginnings. The make-up of the contributors to the First Symposium, from which this volume was developed, reflects a combination of least effort and of randomness that would have delighted Zipf (1949). Contributors were gathered both from personal contacts and colleagues who could be persuaded to leave the safety of their prepared geographical positions and 'go over the top' into the exposed conflict of pre-university education. However, there was no guarantee that, once over the top, everyone would identify the same enemy and charge in the same direction and, indeed, this volume is in no way a concerted and radical attack by angry young men. Neither does this volume represent any special 'Cambridge approach' (whatever this means) to the problems of geographical teaching for, although the myth of distinctive 'schools' dies hard, we are certain of receiving as critical a reception in Cambridge as in any other centre of scholarship.

Perhaps the only common features linking the fifteen contributors are that they share, in Quaker terminology, a 'concern' for geography; that they are actively engaged in some aspect of geographical work in the 1960's; and that none of them believe that the best geography, like the best wine, must be necessarily both French and long-matured.

[1] The views expressed in this chapter are those of the two authors and do not necessarily reflect the opinions of the other contributors to this volume.

Indeed, a recognition of the need for a complete and radical re-evaluation of the traditional approaches both to geography and to geographical teaching in Britain characterizes many of the contributions (e.g. Wrigley, Chapter 1; Chorley, Chapter 2; Haggett, Chapter 6; Board, Chapter 10). If the average view of the contributors is radical, however, it is an average with a recognizably large standard deviation.

Indeed, we should not expect any well-defined common philosophy to emerge from these essays – nor does it. There has been no conscious attempt to dictate viewpoints, neither to integrate nor reconcile opposing ones. In some contributions sharp contrasts emerge, for example, regarding the significance of regional method (see Wrigley, Chapter 1, and Timms, Chapter 12). There are, none the less, themes which appear and reappear with variations throughout the volume and, in so far as this work presents any unified attitude towards certain key aspects of geographical teaching, it is in these themes that the coherence rests. So individually modulated are these themes, however, that to present any one of them in a simple generalized form seems to destroy much of its singularity. It is possible, however, to isolate some of them which appeared both in the written contributions and in the productive discussions with the teachers attending the Madingley Symposia. The present clash between the historical and the functional approaches to geographical matters (Wrigley, Chapter 1; Chorley, Chapters 2 and 8; Smith, Chapter 7) was commented on, and certain of the contributors pointed up the shortcomings of the largely historical treatments of physical (Chorley, Chapters 2 and 8) and human geography (Wrigley, Chapter 4; Collins, Chapter 11). The need for geographers to be at least aware of attitudes and techniques in the associated social (Pahl, Chapter 5; Haggett, Chapter 6) and physical sciences (Beckinsale, Chapter 3; Chorley, Chapter 8) in general, was allied with a concern for an increase of quantification in geography (Haggett, Chapters 6 and 9; Chorley, Chapter 8; Board, Chapter 10; Timms, Chapter 12), and an opposition to the idiographic character traditionally assumed by British geography. The recognition of 'man in society' as a focus of geographical interest emerged strongly (Wrigley, Chapter 4; Pahl, Chapter 5), with especial reference to the character of urban centres (Collins, Chapter 11; Timms, Chapter 12), in contrast with the strong current emphasis on primitive and agricultural societies fostered by the traditional man/land approaches to the subject. Attitudes to mapping also showed interesting features, not the least significant of which was the increasing regard for the map as a framework within which data can be organized as a springboard to higher and more sophisticated analysis (Haggett, Chapter 9; Board, Chapter 10; Willatts, Chapter 13), rather than as an end-product of geographical labour. This last attitude was also part of a wider concern regarding the place of geography in national planning (Pahl, Chapter 5; Collins, Chapter 11; Willatts, Chapter 13), where the contribution of geographers should be more than merely to map the information collected by others and then to surrender these maps for interpretation and analysis. At such a symposium on geographical teaching it was natural that the problems and opportunities of teaching on many different levels were also discussed (Chapters 14, 15 and 16), together with the need to keep both material and especially, teaching attitudes up-to-date (Harrison Church, Chapter 17), and this last need gave rise to many criticisms of the existing British geographical publications as the means of providing the vital 'academic retooling' necessary for practising teachers of a rapidly developing subject in a rapidly changing world.

Three other matters of interest seemed to us of such especial significance that we have attempted to develop them in the following sections. They are the use of model teaching in geography, the integration of such models with conventional regional treatments, and the need to clarify the model of geography itself which we hold.

MODEL THEORY IN GEOGRAPHY

Although it is unnecessary here to recapitulate the significance of the construction of theoretical model structures in geographical teaching and research (see Chorley, Chapter 2, and Chorley, 1964), it became readily apparent from the contributions to these essays, and, particularly, from discussions with teachers that such structures have an especial value in the understanding and presentation of information of geographical interest. The regional model (Wrigley, Chapter 1), the Davisian physiographic model (Chorley, Chapter 2), climatological models (Beckinsale, Chapter 3), the spatial locational models of economic geography (Haggett, Chapter 6), historical models (Smith, Chapter 7), urban models (Timms, Chapter 12) and those others less sharply outlined in other chapters seem to provide specially appropriate frameworks for both research and teaching. We cannot but recognize the importance of the construction of theoretical models, wherein aspects of 'geographical reality' are presented together in some organic structural relationship, the juxtaposition of which leads one to comprehend, at least, more than might appear from the information presented piecemeal and, at most, to apprehend general principles which may have much wider application than merely to the information from which they were derived. Geographical teaching has been markedly barren of such models, partly as a result of the interest which has centred largely on the unique and special qualities of geographical phenomena, and geographers have been loath to make use of these powerful frameworks, despite the teaching successes of, for example, the Davisian cyclical model. This reticence stems largely, one suspects, from a misconception of the nature of model thinking, wherein such frameworks are expected to be 'true', or 'real', or to possess other equally equivocal qualities. Models are subjective frameworks, constructed for specific purposes, relating to a limited range of reality and only possessing relevance within well-defined levels of information content, sophistication and time. On certain levels the Davisian system has a large measure of 'reality' and 'truth', on other levels it has a smaller measure. Model frameworks are like discardable cartons, very important and productive receptacles for advantageously presenting selected aspects of reality, but no one model should be expected to accommodate many aspects of reality, at different levels of information content, sophistication and time.

Figure 18.1 indicates the manner in which model thinking has a bearing on the structure of teaching. In figure 18.1A a loose framework is adopted at an early stage (perhaps some regional or simple spatial one), and through time attempts are made to incorporate progressively larger amounts of information and, to a much more restricted extent, to increase the level of sophistication of the framework. The lack of sophistication of the initial basic structure of the framework, together with the restrictive properties which any one structure most possess, however, imply that attempts to adapt it to encompass more and more information will result in further losses of internal form and cohesion, and in its degeneration into an amorphous mass of loosely-related information. Attempts to increase the level of sophistication with the passage of time by *ad hoc* tinkering with the framework in the hope of making it serviceable throughout a wide range of educational experience usually have a similar effect. It is such 'one-gear' teaching which is largely responsible for obtaining geography its poor reputation as a scholastic academic discipline

Sometimes some measure of structural coherence of knowledge is obtained by the teacher adopting a *classification* type of approach (Figure 18.1B). Classifications have the advantages of well-defined structure (i.e. the relationships between different elements of classified information are usually clear), of the ability to subsume all the apparently relevant information in a more or less satisfactory manner at any one time, and, most significantly, of interchangeability in that an embarrassing increase in the amount of information can be met by discarding some outmoded classification and leaping to a new and more sophisticated one. This last quality is attractive to teachers in that geographical material can thus be presented to different scholastic levels in contrasting and intellectually-appropriate frameworks. Classifications, however, have obvious shortcomings as teaching vehicles in that their construction involves the dissection and categorization of information such that, instead of associating related information intimately together in a suggestive and productive manner, it is disassociated and usually incapable of promoting the student towards novel speculations as to *how reality operates*. Despite these shortcomings, classifications are susceptible of much sophistication and, on higher levels, the so-called genetic classifications merge into models. Teaching by means of *model frameworks* is illustrated diagrammatically in figure 18.1C. Each model framework is tightly knit such that different pieces of information are set in provoking juxtaposition, and the whole is capable of considerable exploitation through time as experience with the model leads to more and more sophisticated handling of the information within it. One has only to recognize the difference of levels on which the Davisian cyclic scheme can operate educationally to recognize this property of exploitability as representing one of the most striking contrasts between models and classifications.

However, as has been recognized, every theoretical model framework becomes strained and less appropriate as the amount of information required to be built into it and the level of sophistication on which the model is required to operate increase, and it is therefore necessary within any vital teaching programme to make imaginative leaps from time to time from less to more sophisticated models. These leaps (which might, for example, involve transitions from simple form-space observations for young students of physiography, to the cyclic model of Davis, and to the process-form-equilibrium Gilbert model for the older students) are most difficult to accomplish, largely because of the common desire to retain a previously-successful

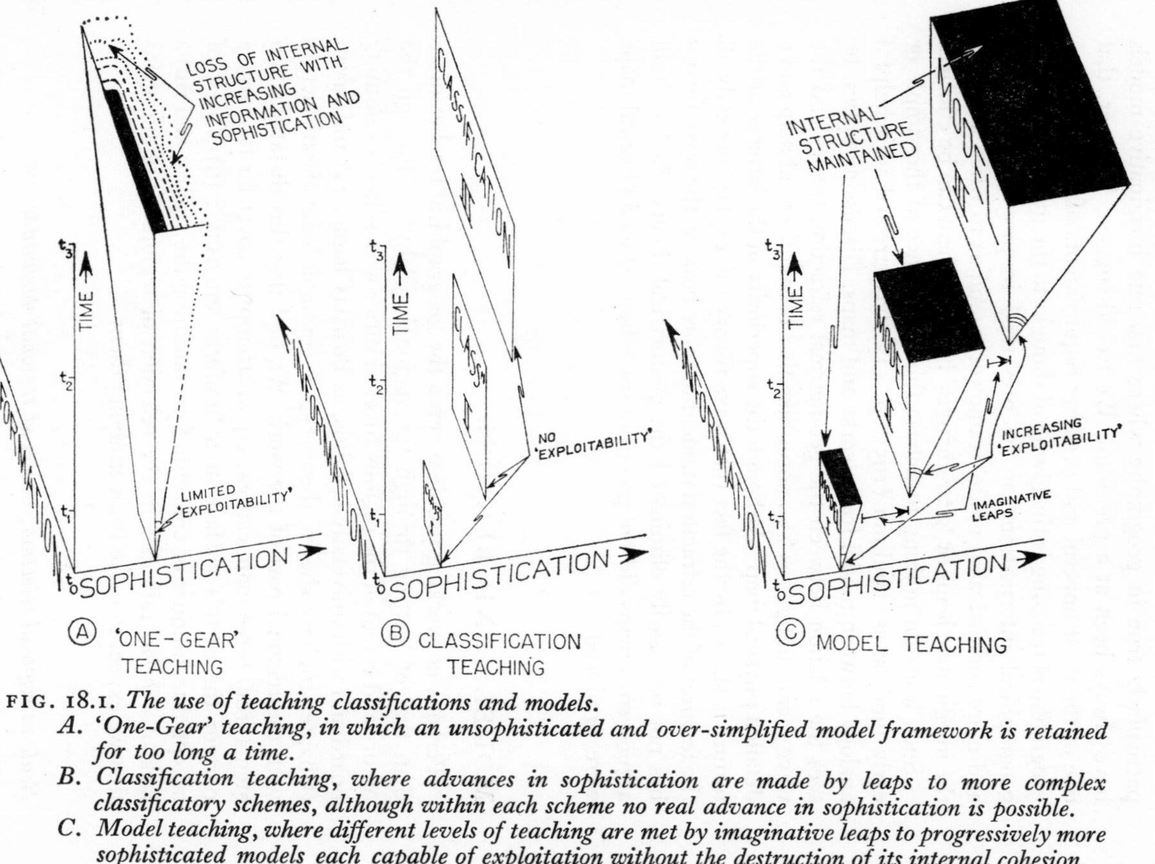

FIG. 18.1. *The use of teaching classifications and models.*

A. 'One-Gear' teaching, in which an unsophisticated and over-simplified model framework is retained for too long a time.

B. Classification teaching, where advances in sophistication are made by leaps to more complex classificatory schemes, although within each scheme no real advance in sophistication is possible.

C. Model teaching, where different levels of teaching are met by imaginative leaps to progressively more sophisticated models each capable of exploitation without the destruction of its internal cohesion.

model long after the time of its optimum utility. This has been particularly true in geography where first-rate imaginative models have always been at a premium. We would suggest, however, that one of the most hopeful and pressing requirements of contemporary geographical teaching is the giving of thought to the manner in which geographical information can be presented in progressively more and more sophisticated model frameworks, and as to when and how the imaginative leaps or 'gear changes' between them can be effected.

Such leaps do nothing to debase the importance of the simple or early models, as Toulmin (1953, p. 115) has stressed in a helpful analogy between model frameworks and maps. Discussing optics, he sees the relation between the geometrical refraction theory and the wave theory of light as '... not unlike that between a road map and a detailed physical map'. Although the superiority of the latter over the former is shown by the fact that wave theory can explain not only all the features of the refraction model but more besides, the wave theory has not necessarily eliminated the geometrical theory. As Toulmin argues, road maps did not go out of use when detailed physical maps were produced.

MODELS AND REGIONS

Perhaps no working problem vexes the geographical teacher more than that of 'getting through the regional syllabus'. Although the nature and extent of the regions covered varies with school examining board and with university syllabus, as Board (Chapter 14) and Bryan (Chapter 16) have shown, there is some common basic dissatisfaction with the regional part of the work. We find that this dissatisfaction stems from three main causes: (i) an attempt to cover far too large a part of the earth's surface at a 'uniform intensity'; (ii) a lack of accepted techniques or concepts for examining the region; and (iii) a lack of *a priori* rationale in the selection of regions or of the gains that we expect to make from studying them.

Scale in regional selection: orders of regional magnitude

There are clearly no absolute limits to the size of a region other than the limits of the earth itself, with its total land and sea area of 196,836,000 square miles. Below this the region contracts continuously towards the very small. Attempts to demarcate regions on the basis of size have been reviewed elsewhere in this book (Haggett, Chapter 9) and it is worth while here to try to draw together some of these schemes into a single regional-functional model in which the magnitude of the region is related to the functional use to which the region is to be put.

Table 27 presents a tentative order of magnitude for areal studies based on successive logarithmic subdivision of the earth's surface. It uses an index, the *G*-scale (Haggett, Chorley and Stoddart, 1965), in which *G* is a dimensionless ratio given by the formula:

$$G = log\ (G_a/R_a)$$

where G_a is the area of the earth's surface (i.e. 1.986×10^8 square miles) and R_a is the area of a regional subdivision. Large areas have small *G* values (e.g. U.S.A. = 1.82) while small areas have large values (e.g. Rutland = 6.11), and the scale may be continued down to the smallest areas of geographical investigation. The general advantages and disadvantages of the scale and rapid methods of computation have been given at length elsewhere (Haggett *et al.*, 1965), but it seems appropriate here to examine some of the implications of a logarithmic view of regions for geographical teaching. Eighteen regional works of very different characteristics have been plotted on Table 27 in relation to both their regional magnitude (*y*-axis) and their subject matter (*x*-axis) to show something of the range of interest on both axes. At best these studies are a 'grab-sample' of the relevant literature and show no distinctive pattern. Work in hand, however, suggests that the great bulk of 'classic' regional geography (represented on Table 27 by Vidal de la Blache's *La France de L'Est*) may cluster significantly around the central part of the diagram; the volume of work by geographers falling off steeply both upwards towards the world level, and downwards towards the site level. Whether this represents a fundamental characteristic of geographic writing or a convenient 'ecological niche' within the academic climax is not yet clear. Certainly, for most disciplines concerned with field study (e.g. geology, botany, agriculture, etc.) the 'centre of gravity' of their work falls well below that of geography on any table of areal magnitude.

Individual geographical monographs may show, in their internal structure, distinctive approaches to the problem of generalizing statements over a large area: e.g Platt in his *Latin America: Countrysides and United Regions* (1942) uses small sample studies ranging down to the *G* = 10–11 level, whereas James' *Latin America* (1959)

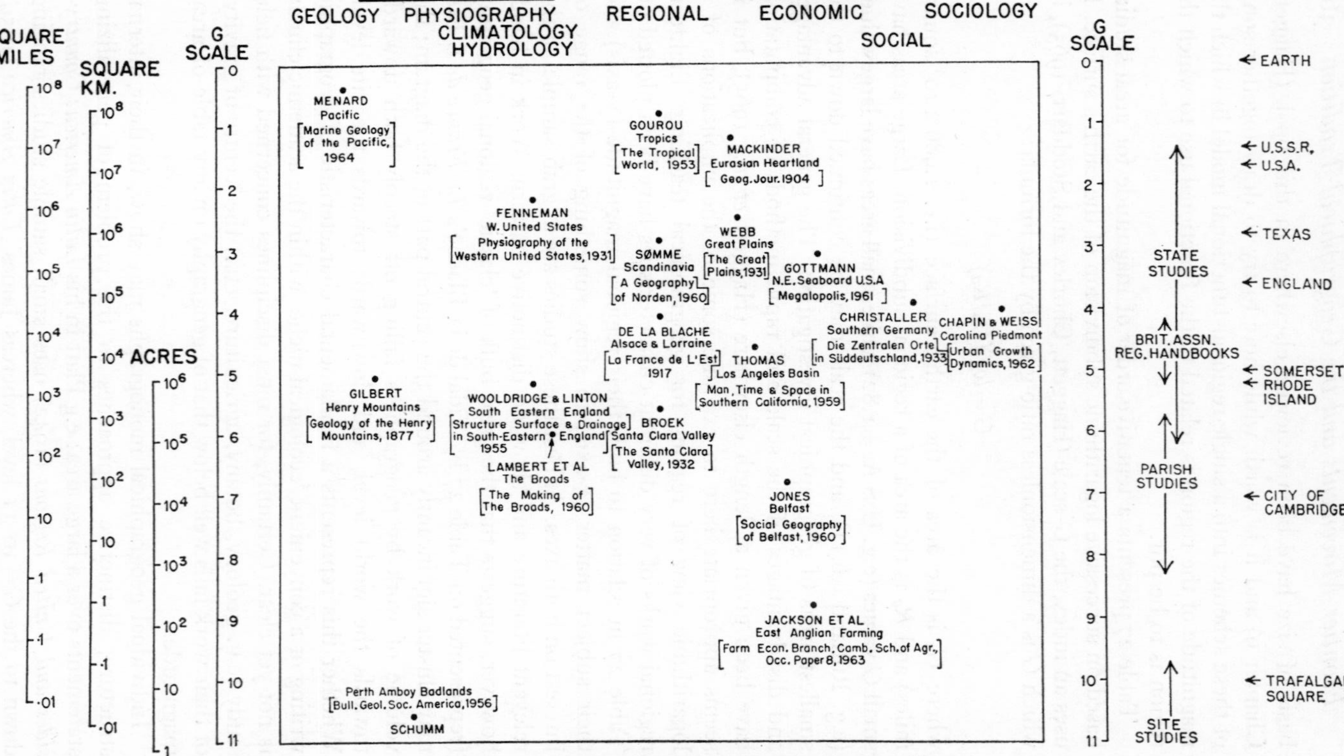

TABLE 27. *Areal studies referred to the G-scale.*

adopts a two-stage hierarchy of political and physical units ranging down to about the $G = 6$–7 level. To be accurate, therefore, geographical works should be plotted not as points but as either graph networks or zones. Some tentative work along these lines with an added time-dimension (z-axis) has been begun, and we hope that Table 27 may be useful in arousing or annoying its readers into making their own experiments with the areal structure of works with which they are themselves familiar.

Although the system may seem over-elaborate it has two major virtues. First, it supplies a simple reference scale for regional magnitude. Secondly, it allows ready comparison between regions (the area

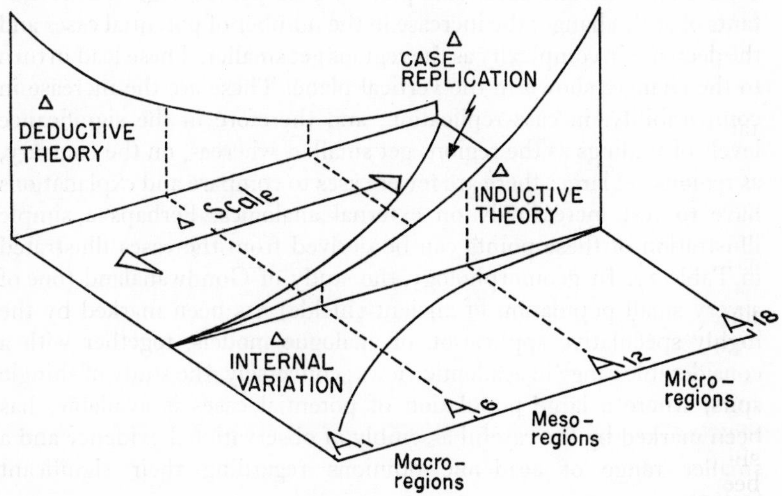

FIG. 18.2. *Implications of regional scale changes for teaching models.*

of regions in adjacent classes will vary on average by 1:10 and never by more than 1:100) on a ratio scale with the reference 'benchmark' being the whole planetary surface. Whether differences in regional magnitude have any deeper significance is a matter for debate, but it is perhaps noteworthy that the importance of dimensional differences is of key importance in classical physics where changes in one dimension (e.g. length) may be associated with disproportionate changes in area, mass, viscosity and so on. These problems in 'similitude' also have crucial importance in biology where D'Arcy Thompson (1917) devotes a considerable part of his *On Growth and Form* to a consideration of magnitude in zoological and botanical design. As geographers draw increasingly on physical models and their biological derivatives, we shall need to be increasingly alive to the dangers of spatial or dimensional 'anachronisms', if we may call them such. Measurements of distance inputs (length), boundaries (perimeters), populations (masses) in the 'gravitational' models of economic geography (Isard, 1960, pp. 493–568) may need to be successively re-cast at different areal levels if we are to retain their principles of similitude. Likewise, the models of airflow, or erosion, or migration, or regional development may not necessarily hold equally well at all magnitude levels of regional application.

Some of these implications of changes in magnitude are shown in Figure 18.2. On the horizontal plane are shown two logical concomitants of scale change: the increase in the number of potential cases and the decrease in complexity as the regions get smaller. These lead in turn to the changes shown in the vertical plane. These are the increase in comparability, in case-replication, and therefore in the significance levels of findings as the regions get smaller, whereas, on the contrary, as regions get larger there are fewer cases to compare and explanations have to rest increasingly on external analogies. Perhaps a simple illustration of these points can be derived from the cases illustrated in Table 27. In geomorphology, the study of Gondwanaland (one of a very small population of ancient shields) has been marked by the highly speculative application of analogue models, together with a considerable range in academic views; conversely, the study of shingle spits, where a large population of potential cases is available, has been marked by the careful assembly of observational evidence and a smaller range of academic opinions regarding their significant features.

Whether our reliance on external theory for explanation of the few macro-regional features of the earth is a built-in characteristic of geography is an interesting but unresolved point. Certainly in the economic units in Table 27 we rely heavily on international trade theory to explain major differences (e.g. developed versus underdeveloped areas), while at the lower levels of the meso- and microregion both field observation and field-based models (like that of Christaller (1933)) play a more important part. Geography at the moment certainly appears to be more self-supporting in theory at the lower levels of regional magnitude.

Fusion of regions and models: the search for the modular unit

The separation of regional and systematic geography may not be as great in teaching practice as is often suggested in methodological reviews (e.g. Hartshorne, 1959, Chapter 9), but it might be largely resolved if the ideas of 'models' and 'regions' could be fused. If, for example, we select the Great Plains of North America (the *region*) to examine Turner's frontier concept (*the model*), or the North Atlantic region to examine the model of sub-tropical high-pressure systems, the lower Mekong region to examine models of river-basin development, the London region to examine alternative models of urban growth, or South Germany to examine settlement-spacing models, we are intuitively fusing our regions and our models. The advantage of such an approach is twofold. Regional delimitation is based on the 'modular unit' (the pressure system, river basin, migration field, hinterland, etc.) and the resulting regional treatment, far from being a mystical amalgam of 'landforms and life', is geared to the specific demands of the systematic model.

One disadvantage we see in such a fusion is some loss of the 'regional integration' of material irrelevant to the chosen systematic model. However, the feature of focal interest (e.g. settlement spacing in South Germany) must obviously be studied in relation to local variations in history, terrain and the conventional range of regional variables. Through such correlation the 'vertical integration' which characterized the classic French regional geographies will be preserved. Equally important in our view is the possibility for inter-regional or 'horizontal integration' through comparison of the region under study with other regions in the same 'set'. (See later in this chapter.) Thus we should expect a study of the Turnerian concept in the Great Plains to raise important questions of comparison in relation to the other major mid-latitude grasslands (i.e. the Pampas, Steppes, Veld, Murray-Darling plains, etc.)

To object to a fused region-model system on the grounds that it failed to cover the whole world at a uniform intensity we regard as a trivial criticism. There are few more discouraging teaching experiences than attempting to teach the 'flat' regional geography of an area which has no outstanding systematic problem or which has played no part in the development of any systematic geographical models, and we doubt the wisdom of selecting such regions for syllabus and examination purposes. Certainly Wooldridge's castigation on 'the eyes of the fool . . .' (Board, Chapter 10) is well justified if we direct our prime efforts towards remote and little-studied areas rather than towards equally appropriate nearby examples having also strong systematic significance. To ask for facts and nothing but the facts is, as Wittgenstein clearly saw, to demand the impossible; regions without theories are like maps without scale or projection.

If we compare geography with such sister disciplines as botany, geology, chemistry and history, we see with Bunge (1962, pp. 14–26) that regional geography may be regarded as a type of classification, a 'taxonomy of the earth's surface'. If this analogy is correct then an examination of these parallel taxonomies suggests the following implications:

1. No unique taxonomy is likely. The history of classification in each of the subjects we cite is a history of continuous reappraisal (i.e. the 'leaps' previously referred to), and indeed of conflicting reappraisal. Classifications merely represent useful working frameworks at a given state of information.

2. No attempt is made in teaching such taxonomies to 'go right through the card'. From the zoologist to the student of English literature the basis of taxonomic teaching is rational sampling. Whether the object is *Chlamydomonas* or *Othello* it is studied in part for itself and in part as a sample of a larger population.

3. The most efficient classifications are often those with model tendencies or affiliations.

Finally, one assumption which is worth challenging in such a reappraisal is that conventional regions for study should be 'nodular areas'. 'Linear' regions such as that along the 100° W meridian in the United States, the fringe of sub-Arctic settlement, the moorland edge in Britain, the London green belt, also come to mind as topics appropriate to regional study.

A SYNOPTIC MODEL OF GEOGRAPHY

In each of the preceding chapters the views argued have been moulded by the general view of geography held by the author. In some chapters, notably those by Wrigley (Chapter 1), Smith (Chapter 7), Board (Chapter 10), and Timms (Chapter 12), and in all the contributions to Part III, these views have been explicit and in others implicit. What kind of picture do these views give of geography? What light do they throw on such thorny problems as 'VI Arts' or 'VI Science', in which our educational administrators are continually faced with the perplexing problem of 'locating' geography within some broader academic framework? In short, what sort of subject is geography?

The Problem of Classification

One initial problem is quickly resolved. To ask whether geography is or is not a science is like asking whether sports are games. Geography can, like any other subject, be studied scientifically or aesthetically, and occasionally both. The study of birds can yield either sonnets or genetics, human suffering may provoke a Goethe or a Pasteur, rocks may inspire an Epstein or a Lyell. Our subject matter is passive and it is we who make the decisions as to how we shall study it or indeed if we shall study it at all. A more appropriate question is then: Can geography be studied as a science? It is in response to this question that we hope this book will show that in part it can, and, more important, that to an increasing extent it is being studied scientifically.

To say that geography can be studied scientifically is to say that it shares a trend common to many academic disciplines, a trend from which even history is not entirely exempt (Postan, 1962), but what kind of a science is it? Here we have two alternative approaches to an answer. One is to follow Hartshorne who, in his classic study on *The Nature of Geography* (1939), urged us to comprehend the nature of geography by studying what it had shown itself to be from the Greek geographers on, such that '. . . we must first look back of us to see in what direction that track has led' (Hartshorne, 1939, p. 31). A second source is to deduce the nature of geography from as few basic assumptions as possible, constructing a deductive-logical skeleton for what geography should be – even if it isn't. Bunge (1962) in his provoking *Theoretical Geography* argues this view and suggests we should be wary of our great forbears '. . . because the great men of the past might now, in view of more recent events, hold opinions different from those they then held' (Bunge, 1962, p. 1). In fact, our views of geography stem from both sources. These views must be, to a greater extent than we perhaps realize, influenced by the manner in which geography has evolved – partly because we ourselves have been involved in its evolution. At the same time outside influences enable us to see this solution in some perspective and to envisage the kind of changes which we would like to achieve.

We suggest that confused evaluations of geography stem not from any evasiveness or lack of thought on the part of geographers but from its very complexity. In the past it has been variously classified as an '*earth science*' (at Cambridge it is part of the Faculty of Geograpny and Geology, which include geophysics, mineralogy and petrology), a '*social science*' (as in most universities in the United States), and less commonly as a '*geometrical science*', a position it held in Greek times and which a few workers, notably members of the Michigan Inter-University Community of Mathematical Geographers, would like it to resume. These alternative placings arise largely from the different growth of the subject in Germany (the *landschäft* school), in France (the *human ecology* school), and in the United States (the *locational* school), although of course its evolution has been more complex than these facile associations of nation and school would suggest.

An Attempt at Fusion

An attempt to fuse these alternative views using the basic approach of set theory and Boolean algebra has been made by Haggett (In Press). Briefly, each of the three sciences into which geography has been placed can be viewed as a *set*, and each separate subject is an *element* within that set. Three sets can be defined: an earth sciences set (α), a social sciences set (β), and a geometrical sciences set (γ). Set α contains geography (1), geology (2) and other earth sciences and can be written as

$$\alpha = \{1, 2\}$$

Similarly we can define the other two sets:

$$\beta = \{1, 3\}$$

$$\gamma = \{1, 4\}$$

where 3 is demography together with other social sciences, and where 4 is topology together with other geometric sciences. This situation can also be shown diagrammatically by the use of Venn diagrams as in Figure 18.3.

We can show the relations between any two sets by overlapping the diagrams. Thus geography is by definition part of both α and β sets, and its position is shown in the overlapping area of Figure 18.3B. Overlap of the three sets in pairs also suggests the position of the human ecology view of geography ('man in relation to his environment') in 5, of geomorphology (6) and of surveying (7) at the overlap of the α and γ sets, and of locational analysis (8) at the overlap of the β and γ sets. We can write these intersections as

$$\alpha \cap \beta = \{1, 5\}$$

$$\alpha \cap \gamma = \{1, 6, 7\}$$

$$\beta \cap \gamma = \{1, 8\}$$

More complicated relationships between the three sets are shown in Figure 18.3C where geography (1) is seen to occupy the central position at the intersection of all three sets, that is,

$$\alpha \cap \beta \cap \gamma = \{1\}$$

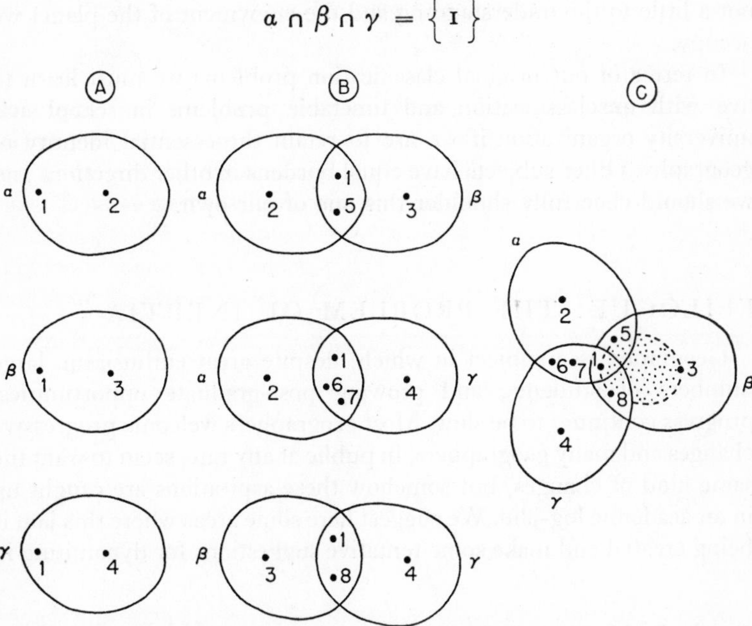

FIG. 18.3. *Set-theory approach to the location of geography within alternative definitions.*

with the cognate subjects, geomorphology, human ecology, surveying, and locational analysis occupying two-set intersections about it. The position of the newly emerging regional science (Haggett, Chapter 6) with its strong connections with geography locational studies, human ecology and the systematic social studies like economics is shown by the shaded area.

It is not suggested here that this type of analysis solves our problems of definition but it does suggest, if our analysis is correct, just why it is so difficult to 'locate' geography or to define it simply. To describe it as 'the study of the earth's surface', or 'man in relation to his environment', or 'the science of distribution', or 'areal differentiation' is to grasp only part of its real complexity. As in geomorphology (Chorley, Chapter 2) these are just some of the alternative views of the elephant. Geography can be defined not solely in terms of *what* it studies or of *how* it studies but by the intersection of the two. It is what Sauer (1952, p. 1) has called a 'focused curiosity' which has created techniques, traditions, and a literature of its own and which in our more hopeful moments we believe has contributed not a little to the understanding and the enjoyment of the planet we occupy.

In terms of our original classification problems we must learn to live with misclassification and timetable problems in school and university organization if we are to retain the essential identity of geography. Other subjects have equal burdens in other directions and we should cheerfully shoulder this one of our own.

EPILOGUE: THE PROBLEM OF INERTIA

Geography is a subject in which, despite great enthusiasm, large numbers of students, and growing post-graduate opportunities, progress continues to be slow. Most geographers welcome progressive changes and many geographers, in public at any rate, seem to want the same kind of changes, but somehow these aspirations are caught up in an academic log-jam. We suggest here some areas where this jam is being created and make some tentative suggestions for dynamiting it.

On the Problem

Perhaps the most immediate inertial problem faced by geographers is the constriction imposed by geography's past growth. We inhabit a Victorian academic structure every bit as solid and constraining as its architectural counterpart. Despite the vigorous growth of the subject in universities and schools over the last fifty years, the everyday popular image of geography is an antique one dogged by 'exploration, description and capes and bays' and this in turn influences the character of our intake of young minds, together with the amount, sources and destination of research funds. The influences of earlier models (of cyclic erosion, of environmental determinism, or of regional 'character') have been taken up in earlier chapters, but perhaps the saddest aspect of this apparent reverence for the past is that this has too infrequently been accompanied by genuine historical research. Old models, like old myths, have been handed down to us, while the original intentions and aims of their constructors lay, reverenced but unread, in some corner of the geographical library (see Wrigley on Vidal de la Blache, Chapter 1).

A second major problem is that of conflicting objectives. Paradoxically, this may stem in reality from the popularity of geography as a school and university subject in an era of educational expansion. Disregarding those students attracted to geography through a belief in its elementary character, much of its popularity derives not from its possessing any satisfying basic academic discipline as from the valuable 'side-products' which are believed to spring from its study. We would not deny that a deeper understanding of international affairs, of planning problems, or of the 'bridge' between the 'two cultures' may stem from a study of geography. We would argue, however, that if geography trims its sails to the vicissitudes of every profitable wind of social and educational demand that blows it is likely to lose any sense of distinctive intellectual purpose, will fail to attract its most necessary growth ingredient (the research student) and is likely to be eventually replaced by or amalgamated with other subjects which serve the purposes of society as well as possessing some intellectual identity. Geography's most important contribution to society will in the long-run result from its producing good geographical research, not by over-extending itself in fields of immediate educational profit.

A third inertial problem unquestionably results from the kind of academic isolationism which has been fostered by the idiographic and artistic preoccupation of so much past geographical work (see Ackerman, 1963, and Chorley, Chapter 8). There is no doubt that the most sterile aspects of present geography are the result of such academic in-breeding, whereas the most virile work is proceeding in fields which have been most willing to draw upon general intellectual advances (which, of course, are today mainly the scientific ones). It is not sufficient to dismiss this reality by the belief that 'the best research always goes on along inter-disciplinary boundaries'. Worthwhile disciplines do not develop like a coral atoll by the outward growth of active margins around a dead or atrophied centre. What should characterize the living heart of geography, and thereby justify its academic identity, is neither any single simple methodology nor any immutable body of subject matter, but the kind of physical/social/geometrical fusion which we have attempted to explain in terms of set theory.

On the Solution

Easy solutions are rarely the optimum ones. There has been no lack of thought or effort in British geography in the last half-century and if basic solutions had been at hand they would long since have been applied. Perhaps, with Chesterton's Father Brown, we find our greatest difficulty in recognizing the exact nature of the problem rather than in solving it. Here, however, we must be willing to commit ourselves on both the problem and the solution. On the problem, we suggest that the features of inertia recognized above are symptoms of a deeper malaise – the failure to recognize the multivariate nature of geography (as shown in Fig. 18.3C). In particular there has been the neglect of the strong geometrical tradition in geography.

The geometrical tradition was basic to the original Greek conception of the subject, and many of the more successful attempts at geographical models have stemmed from this type of analysis. The geometry of Christaller hexagons, of Lewis' shoreline curves, of Wooldridge's erosion surfaces, of Hägerstrand's diffusion waves, of Breisemeister's projections come vividly to mind. Indeed from one point of view, much of the new statistical work relating to regression analysis (Chapter 8) and generalized surfaces (Chapter 9) represent merely more abstract geometries. Much of the most exciting geographical work in the 1960's is emerging from applications of higher-order geometries; for example, the multidimensional geometry of Dacey's settlement models and the graph-theory and topology of Kansky's network analysis. It is an interesting reflection that the increasing separation of geomorphology and human geography may have come just at the time when each has most to offer to the other. Sauer (1925) in his *Morphology of Landscape* drew basic parallels between the two, but it was unfortunate, as Board (Chapter 10) so clearly shows, that 'landscape' was seized upon and 'morphology' neglected by those who drew inspiration from this important paper. The topographic surface is only one of the many three-dimensional surfaces that geographers analyse and there is no fundamental reason why, for example, the analysis of landform and population-density surfaces should not proceed along very similar lines. Geometry not only offers a chance of welding aspects of human and physical geography in a new working partnership, but revives the central role of cartography in relation to the two.

Our immediate solution then is to press for a re-establishment of the tripartite balance in geography by building up the neglected geometrical side of the discipline. Research is already swinging strongly into this field and the problem of implementation may be more acute in the schools than in the universities. Here we are continually impressed by the vigour and reforming zeal of 'ginger groups' like the School Mathematics Association which have shared in a fundamental review of mathematics teaching in schools. There the inertia problems – established textbooks, syllabuses, examinations – are being successfully overcome and a new wave of interest is sweeping through the schools. The need in geography is just as great and we see no good reason why changes here should not yield results equally rewarding. Better that geography should explode in an excess of reform than bask in the watery sunset of its former glories; for, in an age of rising standards in school and university, to maintain the present standards is not enough – to stand still is to retreat, to move forward hesitantly is to fall back from the frontier. If we move with that frontier new horizons emerge into our view, and we find new territories to be explored as exciting and demanding as the dark continents that beckoned an earlier generation of geographers. This is the teaching frontier of geography.

References

ACKERMAN, E. A., 1963, 'Where is a Research Frontier?', *Ann. Assn. Amer. Geog.*, **53**, 429–40.

BUNGE, W., 1962, *Theoretical Geography* (Lund).

CHORLEY, R. J., 1964, 'Geography and Analogue Theory', *Ann. Assn. Amer. Geog.*, **54**, 127–37.

CHRISTALLER, W., 1933, *Die zentralen Orte in Süddeutschland* (Jena).

HAGGETT, P., CHORLEY, R. J. and STODDART, D. R., 1965, 'Scale standards in geographical research: A new measure of a real magnitude', *Nature*, **205**, 844–47.

HAGGETT, P., In Press, *Locational Analysis in Human Geography* (London).

HARTSHORNE, R., 1939, *The Nature of Geography* (Lancaster, Pa.).

— 1959, *Perspective on the Nature of Geography* (London).

ISARD, W., 1960, *Methods of Regional Analysis* (New York).
POSTAN, M., 1962, 'Function and Dialectic in History', *Econ. Hist. Rev., 2nd Series*, **14**, 397–407.
SAUER, C. O., 1925, 'The Morphology of Landscape', *Univ. of Calif. Pubs. in Geog.*, **2**, 19–53.
— 1952, *Agricultural Origins and Dispersals* (New York).
THOMPSON, D'ARCY W., 1917, *On Growth and Form* (Cambridge).
TOULMIN, S., 1953, *The Philosophy of Science* (London).
ZIPF, G., 1949, *Human Behaviour and the Principle of Least Effort* (New York).

Selective Index

It has seemed to the editors that the detailed table of contents (pp. *v–x*) and chapter references have obviated the necessity for extensive general and author indexes. There follows a short index in which some major methodological themes running through the volume are indicated. Where the whole of a chapter is concerned with such a theme it is shown by Roman numerals.

SPECIAL MESSAGE TO READERS

THE ULVERSCROFT FOUNDATION
(registered UK charity number 264873)
was established in 1972 to provide funds for research, diagnosis and treatment of eye diseases. Examples of major projects funded by the Ulverscroft Foundation are:-

- The Children's Eye Unit at Moorfields Eye Hospital, London
- The Ulverscroft Children's Eye Unit at Great Ormond Street Hospital for Sick Children
- Funding research into eye diseases and treatment at the Department of Ophthalmology, University of Leicester
- The Ulverscroft Vision Research Group, Institute of Child Health
- Twin operating theatres at the Western Ophthalmic Hospital, London
- The Chair of Ophthalmology at the Royal Australian College of Ophthalmologists

You can help further the work of the Foundation by making a donation or leaving a legacy. Every contribution is gratefully received. If you would like to help support the Foundation or require further information, please contact:

THE ULVERSCROFT FOUNDATION
The Green, Bradgate Road, Anstey
Leicester LE7 7FU, England
Tel: (0116) 236 4325

website: www.foundation.ulverscroft.com

Sarah J. Naughton grew up in Dorset, on a diet of tales of imperiled heroines and wolves in disguise. As an adult, her reading matter changed, but those dark fairytales had deep roots. Her debut children's thriller, *The Hanged Man Rises*, featured a fiend from beyond the grave menacing the streets of Victorian London, and was shortlisted for the 2013 Costa award. *Tattletale* is her first adult novel, and has a monster of a different kind. She lives in central London with her husband and two sons.

TATTLETALE

One day changes Jody's life forever. She has shut herself down, haunted by her memories and unable to trust anyone. And then she meets Abe, the perfect stranger next door, and suddenly life seems full of possibilities and hope . . . One day changes Mags's life forever, too. After years of estrangement from her family, she receives a shocking phone call. Her brother Abe is in hospital, and no one knows what happened to him. She meets his fiancée Jody, and gradually pieces together the ruins of the life she left behind. But the pieces don't quite seem to fit. . .

SARAH J. NAUGHTON

TATTLETALE

Complete and Unabridged

CHARNWOOD
Leicester

First published in Great Britain in 2017 by
Orion Books
an imprint of The Orion Publishing Group Ltd
London

First Charnwood Edition
published 2017
by arrangement with
The Orion Publishing Group Ltd
London

All characters in this book are fictitious, and any resemblance to actual persons, living or dead, is purely coincidental.

A catalogue record for this book is available from the British Library.

ISBN 978–1–4448–3466–6

Published by
F. A. Thorpe (Publishing)
Anstey, Leicestershire

Set by Words & Graphics Ltd.
Anstey, Leicestershire
Printed and bound in Great Britain by
T. J. International Ltd., Padstow, Cornwall

This book is printed on acid-free paper

For my husband, Vince.

'You know that place between sleep and awake, that place where you still remember dreaming? That's where I'll always love you.'

J. M. Barrie

Before

On a clear morning the sun shines so strongly through the stained glass it looks as if the concrete floor is awash with blood.

But it's past eight in the evening now and the only light comes from the wall lamps on each floor. Their dim illumination reveals a slowly spreading pool of pitch or tar.

Blood doesn't look like blood in the dark.

Now the adrenaline that powered her scramble down the stairs has drained away, she feels as if all her bones have been pulled out. She can barely stand, has to grasp the metal newel post for support as she stares and stares.

The fourth-floor landing light goes out.

It takes a long time for the brain to process a sudden accident — the nought-to-sixty acceleration from normality to calamity — to ratchet itself up to an appropriate response. She can feel it slowly building in her belly as she takes in the black spatters on the doors and walls of the ground-floor flats, the widening creep of the black pool.

At first she thought he would be OK. A few bruises. A bumped head. There is too much blood for that.

The third-floor landing light goes out.

In the few frozen moments after it happened she was dimly aware of a latch snicking shut, heavy footsteps rattling down the stairs, the creak and slam of the front door, but now everything is silent. The church is holding its breath, waiting to see what she will do.

She takes a wobbling step towards him.

There's a smell, like her purse when it's full of coppers.

He looks so uncomfortable. Why doesn't he move his leg so that his hips aren't so twisted? Why doesn't he turn his head as her shadow falls across him? Why doesn't he call out to her?

She kneels beside him and takes his hand. It's pure white against the blackness that is slowly seeping into his hair and clothes. She tries to say his name but there's a fist around her throat. Her thoughts sputter. There's something she should do. Yes. She should call 999.

The second-floor landing light goes out.

His lips are moving and his eyes are open. As she leans close to him to try and make out what he is saying her hair falls into the pool. Jerking back, the tips of her hair flick against her wrist, drawing scarlet lines on her white skin. Now she can see where the blood is coming from. A small noise escapes her lips. Horror and shock are hurtling towards her like an articulated lorry.

The first-floor landing light goes out.

She must do something for him. Now, here, in this moment, she is all he has. She must take her phone from her pocket, unlock it, and tap in the numbers. But she cannot let go of his hand; she cannot leave him adrift in all this darkness.

Her heart is racing, like the wheeling legs of a cartoon character just before it realises it's run off the cliff edge. Before it falls.

The ground-floor light goes out.

It is the sudden darkness, as much as anything else, that makes her scream. And once she's started she cannot stop.

After

The lino's slippery with spilled drinks. As he crosses the dance floor a fat girl blunders into his path and he grabs her by the flesh of her waist, making her squirm and shriek. Someone slaps him on the back and he grins, though he didn't hear what was said. The music is so loud the floor vibrates and the disco lights have turned carefully made-up faces lurid colours. All the girls are off their tits, some of the weedier blokes too. Gary and Kieran are draped over one another, bellowing 'Auld Lang Syne', though it's still two hours until midnight. But it takes more than a few double vodkas to affect him. He glances at himself in the dark window that looks over the pitch.

Not bad considering he'll be thirty this year.

In the reflection he sees a woman he doesn't recognise walking across the room behind him. Catching his eye, she pauses and smiles.

He smirks. Still got it.

The toilets stink, as usual.

He pisses for England, then shakes himself off and does up his flies, checking his reflection in the square of buckled stainless steel that passes for a mirror. The shirt is a size too small and pulls tight across his pecs. He washes his hands and runs damp fingers through his hair. He's noticed it thinning at the temples over the past few months and has been considering trying a spray from the chemist.

The new winger comes in and stands at the urinal. He's considerably shorter and weedier than Rob.

'Having a good time, mate?' Rob says.

'Brilliant,' the lad says.

'Just you wait,' Rob says. 'The ladies'll be so pissed you'll be fighting them off with a stick.' He puts ironic emphasis on *ladies*.

The boy laughs.

'See you later.' Rob thumps him so hard on the back he almost overbalances into the urinal. He's laughing as he emerges to a line of grumbling females.

'Sorry I kept you waiting!' he cries, spreading his arms.

'In your dreams,' says Elaine, Marcus's ugly wife. 'The toilet's blocked. Clive's in there trying to fix it.'

'Use the men's, then.'

'The state you lot leave it in? No thanks.'

'Well, don't be surprised if I'm booked up for the rest of the evening by the time you come out.'

'We'll take that risk.'

He bows and pushes open the door to the bar.

The air's heavy with aftershave and cigarette smoke. It's illegal to smoke in here but the lads pay no attention, though Clive keeps threatening to hand the CCTV footage to the police if they don't stop. Through the haze he can make out Sophie muttering to her little coven. Probably about him. He stares at them until she glances up, then gives her a cheery wave. She looks guilty. Bitch can get her own drink.

There's a girl at the bar but he's not in the mood to wait so he raises his twenty and Derek waddles straight up, a craven grin on his puffy face. Either he's scared of Rob or he fancies him. Rob pretends to find the latter idea funny when the boys rib him about it, but if Derek ever so much as touches him, apart from to hand him his change, he'll knock him out.

'What can I get you, mate?'

'Vodka, lime and soda. And you'd better not sweat in it, you fat bastard.'

Derek laughs.

Rob feels the gaze of the girl he queue-barged and his head snaps around, ready for a row. His scowl vanishes. It's the girl from the reflection. She's seriously hot.

'You scored the hat trick, didn't you?' she says, and her voice is smooth like chocolate.

'Guilty,' he says, putting up his hand and lowering his head modestly. Then he wonders if he's used the wrong word. The pre-party friendly had been too much like hard work on last night's hangover and the bloke he'd tackled to get the last try was still in A & E. But when he looks up she's smiling.

'Haven't seen you here before,' he says. 'You with the other team?'

She nods. 'My sister's dating one of the props.'

Good. She wasn't attached. Not that it mattered — *he* was and it wouldn't make any difference.

'You know what, I'm so pissed I can't remember his name!' she giggles.

'They all look the same anyway. Mr Potato Head!'

She laughs uproariously.

He glances over at Sophie, but now she's too busy making a twat of herself on the dance floor to notice.

Thankfully this year Clive and the rest of the old duffers aren't in charge of the music, so there's a lot less Abba and Bee Gees and a lot more hip-hop. Not that he minds a bit of 'Dancing Queen'. Him and the lads like to dress up for that one, demanding an item of clothing from all the women there. This year he'd make Sophie give him her revolting support girdle, embarrass the bitch. With a bit of luck she'll piss off home.

But when he looks back the girl is gone. He swears under his breath, knocks back his vodka, then goes for a dance.

⋆ ⋆ ⋆

It's coming up to midnight and Derek's so overwhelmed that the lads are just going behind the bar and helping themselves, occasionally pausing to flip the bird at the CCTV camera trained on the till. Boys will be boys.

Rob's dancing, his shirt soaked in sweat, his thinning hair plastered to his forehead. Occasionally he'll go up behind a girl and grind his groin into her. Some of them press back and he gets a semi. Most of them aren't attractive enough for the full nine yards. Soph's the best looking of the lot of them, and she's blubbing in the corner, surrounded by clucking mates. *He's such a b-bastard, boo hoo.* Well, she's not going to ruin his night. He grabs the nearest girl to him and gives her a proper snog, thrusting his tongue into her mouth. Her saliva is bitter with alcohol and cigarettes. She pushes him away with a playful slap and he wipes his mouth on his sleeve, swaying slightly in the glare of the lights. His eardrums throb in time to the music. His heart is racing. His muscles hum with tension.

Slim fingers caress his side as someone slips past behind him and he turns to see it's the girl from the bar.

She's even better looking than Sophie. She's — he fumbles for the word — *elegant*. None of the other girls here are elegant. They've all got identical long blonde hair, skirts up to their arses, fake tan, glitter across their tits. This one looks classy. He doesn't try to grind his pelvis into her.

'Hi,' he says. 'How are you doing?'

'Good,' she says. 'It's been fun.'

'You're not going?'

'I'm not sure I'm going to get what I came here for.'

He frowns. 'What's that?'

She speaks so softly he has to lip-read over the music. He blinks rapidly, his lips part. He might have misunderstood. He leans over.

'What did you say?'

As she tilts her head to murmur into his ear her hair brushes his cheek, sleek and cool as satin. He didn't misunderstand.

He doesn't know what to say. He's not used to

girls coming on so strong and isn't sure he likes it.

She pulls away. Her eyes hold his. His insides turn to liquid.

'M-me,' he stammers. 'I will. I can.' He sounds like a twat. He rolls his shoulders and runs his tongue across his front teeth. 'You won't be disappointed.' He still sounds like a twat. He regrets the last round of sambucas. 'There's a storage cupboard around by the toilets.' It stinks of bleach but Sophie didn't seem to mind.

'How about something more . . . al fresco?'

This one does, then.

He nods vigorously and glances over at Sophie. She's stopped crying and is doing shots.

'I'll see you outside.'

As she walks away he glances around to see if someone's setting him up and considers for a brief moment whether Sophie's arranged one of those honey-trap things. What does it matter? They're probably finished after tonight anyway.

He crosses the dance floor and passes out into the foyer. The air is cold and clean and he stands in the darkness as the inner door swings shut and the music and screeching laughter becomes muted. The evil red eye of the ancient CCTV camera watches him from the corner.

Is he too pissed to get it up? He's never failed yet, but he's never had a woman like this before.

Only one way to find out. Pushing open the main doors he strides outside into the night.

He spots her by her white top, gleaming in the shadows of the stands.

The pitch is churned and muddy so he walks

around the spectator part, breathing slowly and deeply to calm himself down. Stupid, but he feels like he's on the way to an exam. She's something special, this one, and he doesn't even know her name. That makes it more special. That's how he'll phrase it when he tells his mates later. *The mysterious beauty*.

The effect is spoiled when he reaches her and sees that she's covered in mud. It's caked all over her boots, her knees and even in her hair.

'Jeez,' he says. 'What happened to you?'

'Fell over.' She giggles.

It annoys him. She's spoiled the effect. 'You should have walked around the edge.'

'Who cares?' she says. Then she pulls off her top. She must be pissed, because she lets it drop into the muddy puddles on the concrete, then yanks down the vest so roughly the strap snaps.

She isn't wearing a bra. Her breasts are smooth and tanned, glimmering in the lights from the clubhouse. The music is just a throbbing beat now, like a heart. She leans against the bench behind, arching her back.

She's one of those who likes it rough. He puts his hand over her mouth to shut her up and she bites his fingers. She tears a couple of shirt buttons off trying to get to his pecs, kisses him so hard his lips are crushed against his teeth. She even takes a chunk out of his hair, which is not on, considering, and he punishes her for it, thrusting into her so hard she cries out in pain. Normally he's more careful — some girls tear when he does that — but she deserves it. She obviously thinks she's a bit special. The thought of her hobbling about tomorrow, bruised and torn and unable to sit down because of him, gives him a head rush of arousal. He won't last much longer.

The countdown to midnight drifts across the pitch as he's coming, and by the time the fire-cracks of the party poppers have subsided he's done up his trousers and is making his way back to the clubhouse.

The whole thing was over so quickly Soph won't know he's been away. Not that he'll be able to explain the lost buttons or the scratch marks. There's even one down the side of his face. Still, at least he'll have a laugh about it with the boys before World War Three breaks out.

At the clubhouse door he turns back. She's sitting up now, and just for a moment she raises her hand, in greeting or farewell. He doesn't wave back.

As he yanks open the door he's laughing to himself. To think he'd thought she was a notch above the others. Elegant. *Ha*. Not so elegant staggering home covered in mud with her tits hanging out of her top.

Then she starts screaming.

The sound of the TV is a lullaby, making her drowsy, despite the cold. One of the springs is poking through the musty-smelling mattress and she has to curl up at the very edge so that it doesn't scratch her. They've hung a blanket up at the window to stop the morning sun waking her too early and an orange bar of light from the street lamp outside falls through the gaps, cutting her in half.

Her stomach gives a squealing twist and she draws her knees to her chest to ease it. She wishes she had eaten more at school. The after-school club gives you biscuits and she managed to get two before the others grabbed the rest, but she's still hungry.

If she can go to sleep she'll forget about being hungry. She will forget about what Stuart Talley will say about her in front of everyone at break time tomorrow. She'll forget about the way the teachers whisper about her during assembly and how everyone knows she steals school uniform from the lost property box. Sometimes she wishes she could stay asleep forever.

There are slow footsteps on the stairs and she squeezes her eyes shut and goes very still.

The footsteps come into the room and a weight lands on the bed, making the wire mesh under the mattress twang.

'I know you're awake.'

She opens her eyes.

'Want a bedtime story?'

For a moment she just stares at him. Then she whispers, 'Yes, please.'

She had a bedtime story once before, when one of Nanny's boyfriends came up to her room and started telling her about a brother and sister whose parents left them in the forest. They were trying to find their way home when they came upon a house made of gingerbread and sweeties owned by a kindly old lady. She wanted to hear all about what each part of the house tasted like — especially the windows — but Nanny's boyfriend fell asleep, so she had to make the rest of the story up. The people that left them in the forest, she decided, weren't the children's real parents at all. The old lady was actually their grandma and had built the sweetie house all ready to welcome them, while their real mummy and daddy searched the world for them, their hearts breaking with sadness. When they got back they were so happy to see their children they thought their hearts would burst.

'Once upon a time there was a little bunny rabbit,' says the man sitting on her bed. 'She lived with her family in a burrow on a hill.'

The little girl sits up. She likes the sound of this story. There is a bunny on the pyjama top that her nan gave her.

'The mummy and daddy bunny worked very hard all the time, but the little bunny never thought about anyone but herself. She wasn't very clever and she was always disobeying her parents.'

Her eyes widen. Is something bad going to happen to the bunny?

'Whenever they were busy working she would run out of the burrow, laughing, and wander about the countryside, talking to whoever she met, telling horrible stories about her parents that weren't true, to get attention.'

The little girl frowns. This is a bad bunny.

'One day she met a man having a picnic in a field, and because she was greedy and wanted some of his food, she told a lie that she was starving because her parents didn't give her enough to eat.'

The girl pulls the blanket up to her chin and bites her bottom lip.

'The farmer gave her a little bit of bread and while she was chewing he asked her where she lived so that he could bring her round a nice big chocolate cake for her tea. She told him and thought she was very clever for tricking him.'

The man's face is in shadow but the bar of orange light falls across his hand. His skin is rough and purple, and a tattoo of a dragon's claw pokes out from his sleeve.

'But really,' he carries on, more softly, 'she had been very stupid because that night the farmer came with his gun and his dogs, and he shot the little bunny rabbit's mummy and daddy and all her brothers and sisters to make into a pie for his supper.'

The little girl starts to cry.

'As the mummy bunny died she said she wished the nasty lying bunny had never been born.'

A car goes past outside the window, its headlights sweeping across the room, casting long curled shadows from the peeling strips of wallpaper. On the other side of the room is another bed, with a motionless shape curled up under its own thin blanket. The headlights pass and the room returns to darkness.

'Do you know what happened to the little bunny who had told the tale?'

The little girl shakes her head. She doesn't want to hear but if she puts her hands over her ears she will be punished.

'The farmer cut all her skin off, while she was still alive, and then dropped her in a pan of boiling water and chopped her into bits to feed to his dogs.'

Her gasp sounds like the page of a book tearing.

The man leans in so close to her that she can smell the sweetness of cider on his breath and the cigarette smoke in his hair.

'If I hear that you've been blabbing your fucking mouth off to anyone at school again about what we do in the privacy of our own home, then that's what'll happen to you, you little bitch. Do you understand me?'

She nods.

He gets up and walks out of the room and down the stairs. The TV gets louder for a moment as the door downstairs opens, and then goes quiet as it shuts behind him.

The little girl lies perfectly still as a blood-warm wetness spreads out underneath her.

Tuesday 8 November

1

Jody

Do you remember the first night we slept together? No, not *that* bit. That's easy. The part afterwards, when the sky had darkened to that greyish orange that is as dark as it ever gets in the city, and we'd gone inside, into the warmth of your flat. Everything was quiet except for the odd distant siren, hurried footsteps down Gordon Terrace as people tried to get home without being mugged, the wind rustling the rubbish blowing around the playground.

I didn't sleep much. How could I? I watched you sleep, watched your eyes moving beneath the lids. Were you dreaming about me? I never asked. Didn't want to seem too keen.

I watched your nostrils flare gently on every inward breath, your chest rise and fall, disturbing the hair that ran in a fine line to your belly button.

Your body was so boyish, the muscles as soft as mine. I liked the way our bodies mirrored each other. You dark and slim, with wide brown eyes and long, black lashes, me fair and skinny, with the lightest of eyes and lashes that are almost invisible. You were a masculine me, and I was a feminine you. Sometimes we would press our palms together and marvel at how similar they were in size and shape.

At least your hands are still the same, resting on the starched white sheet.

You're not in pain. The doctors promised me. In an induced coma you don't even dream. Beneath the lids your eyes are perfectly still. Your lashes rest on your cheek, almost the same colour as the dark flesh. They said the bruises would fade, that the swelling would go down, that your face would become yours again. I can't help thinking (hoping): what if it isn't really you under there? That they made a mistake; that you're sleeping peacefully in another ward somewhere, wondering why I'm not there.

No. It *is* you. I saw you fall.

I twist your ring about my own finger. Press my fingertip onto the engraving so that its mirror image is etched into my flesh.

True love.

I know that they're just clichéd words, like the hokey stuff they write in greetings cards, but whoever thought of them could never have known how right they were.

There has never been a truer love. And whatever happens, Abe, whatever you're like when you wake up, my love for you will stay true forever.

I take your hand and whisper the promise into your fingertips.

2

Mags

Everyone else is asleep. Wound in their white sheets like mummies, wedged into the tiny open caskets advertised as *fully flat beds*.

God knows what time it is.

I should have changed my watch before the first glass of champagne. It was *personally selected* by some wine guru who must be famous in Britain. They handed it to me when I boarded, presumably by way of apology for the ten hours of cramped, muzzy-headed tedium I was about to endure.

My phone will tell me when we arrive; until then I'm in a timeless limbo.

The remains of the *Cromer crab cake and lime foam* sit, dissected but untouched, on the pull-out table in front of me. Considering how many hosties per pampered fat cat there are in first class, you'd think they'd have figured out that I'm not going to eat it. Even the wine tastes shit, coating my tongue with sourness. I can feel my breath going bad, and though I showered in the club lounge, I feel sticky and smelly.

I tip the vanity bag onto the table, looking for breath freshener. Toothpaste, toothbrush, moisturiser, eye mask, something called 'soothing pillow mist', earplugs and a crappy pair of velour slippers. No breath spray.

I think about putting the eye mask on and *misting* the pillow, but I'm not sure there's any point. My brain is far too wired to sleep and every time I close my eyes the same film runs through my head. I'm falling through darkness, the wind blowing my hair, the circle of light above me getting smaller by the moment.

May as well keep drinking.

The next time a hostie comes past I ask her for a large whisky.

I make another attempt to get into the novel I bought at the airport, a pulp thriller about some woman who thinks her husband has killed their son, but it turns out it was her and she's just forgotten all about it, because he's been spiking her food to protect her. I'm three quarters of the way through and I still don't give a toss about any of them. But it's probably just my state of mind.

The hostie returns and puts the drink down on a little doily.

'This is wine,' I say.

She smiles so hard the foundation at the corners of her mouth crackles. 'Yes, ma'am.'

'I asked for whisky.'

'Whisky?'

'Same first letter, but a sneaky extra syllable.'

Her eyelashes tremble, unsure whether I'm joking. I smile so she knows I'm not. Her gaze becomes glassy. *Another bitch*.

'I'll get your whisky right away.'

'You know what?' I hate it but still can't stop that American uplift at the end of my sentences. 'I'll just go to the bar.'

‘As you wish.’

She stands back to let me struggle out of my seat-bed and the smell of perfume is overpowering. Beneath it is something medicinal. Hand soap, perhaps, or those lemon wipes in the economy cutlery pack. It makes her seem entirely synthetic — but what do I expect on a Vegas flight?

I can feel her eyes on my back as I make my way up the aisle to the bar. Stepping through first class into business, the plane gives a little hiccup and I stumble, turning my ankle.

‘Careful, now,’ she calls after me, and I resist the urge to give her the finger. They can divert a plane these days for that sort of thing.

Jackson paid for the ticket. I said it was kind of him. He said, *No such thing, just another bribe to keep you at the firm*. I resisted the urge to reassure him that I wasn’t going anywhere. If you don’t keep your boss on his toes, you don’t get first-class flights and six-figure bonuses. Not that they do me much good. Now that the apartment’s paid for, I find myself throwing money away on expensive crap like the Louboutins I now slip off to massage my ankle.

There’s only one other drinker at the self-service bar, a man around my age, whose face has that flaky redness that always gets you on long haul if you don’t keep hydrated. Normally I’d have been downing Evian since the wheels left the tarmac, but tonight I don’t give a shit. It’s not as if Abe’s going to notice. I pour myself a large whisky and toss in some ice from the bucket. I think about taking it back to my seat — if I stay there's a definite risk the guy will try to talk to me — but it feels good to stretch my legs, so I lean on the bar stool and flick through the in-flight magazine. There's an article about an actress, the retouched pictures make her appear two-dimensional, and her upper lip is so stretched by collagen it looks simian.

'Going home?'

I sigh inwardly.

'Actually, I live in Vegas. Just going back to . . . see my brother.' I kick myself at the hesitation. It wouldn't have happened in court. I need to get myself together, work out the smooth lie that will stop people trying to talk to me or, worse, sympathise. There hasn't been time yet. I only heard this morning. It's taken me all day to sort out the flights and hotels and hand my cases over to Jackson. Though I've spoken to them all in person and promised I'll be back within a fortnight, my clients aren't happy. No one else in the firm has my track record for helping guilty people get away with it. Jackson is taking over IRS vs Graziano. If the case goes badly, Antonio will spend the rest of his days in a federal correctional institution, trading his ass for cigarettes. Ass. I sound like a true yank. British people say *arse*. Nice arse. It sounds oddly polite with an English accent.

'London?' the man across the bar says.

Beneath the ravages of the flight he's good-looking. Square jaw, broad shoulders, blond hair cropped tight to minimise a receding hairline. A man's man. Banker, I think. Or another lawyer. Probably the former if he's travelling in first.

‘Yes.’

‘Me too. Looking forward to seeing him?’

That hesitation again. The whisky is fugging my reactions. I nod, then spin on the stool until I’m at a forty-five degree angle from him.

‘That’s not an English accent, is it?’

I spin back, with a polite smile that, if he’s smart enough, he’ll translate as *get lost*.

‘Scottish.’

He isn’t smart enough. ‘Not strong, though, so I’m guessing you were gone by . . . hm . . . eighteen?’

I raise an eyebrow and, despite myself, say, ‘Not bad. Sixteen.’

‘Straight to Vegas from Bonnie Scotland? That takes balls.’

‘They took a while to drop. I went to London first.’

‘College?’

‘Yes.’

‘You know, you should carry one of those twenty questions gadgets around with you. It could do the talking. Save you the hassle.’

‘Yes,’ I say. Then a moment later, ‘So, what am I?’

I kick myself again. I’ve let myself be drawn in. I must be drunk.

‘Hmm . . . ’ He pretends to think. ‘Are you . . . a hedgehog?’

I laugh loudly enough to draw a disapproving grunt from the fat guy wedged into the casket nearest the bar. ‘Yeah. Spiky. Flea-infested.’

‘Not a hedgehog. You’re travelling in first. Are you an oligarch’s wife?’

He waits for me to bite. I shake my head calmly. 'That's nine questions. Twelve left.'

'Jesus, you're counting?'

'Don't take the Lord's name in vain.' I drain my glass and pour myself another.

It takes him a while but eventually he gets there.

'So, how do you get to be a first-class-travelling American lawyer when you left home at sixteen?'

'A levels at night school. Law degree at King's, my juris doctor at Columbia, then straight to Nevada because it looked like fun. Cheers.'

He clinks my glass and we drink. 'You make it sound so easy.'

It wasn't. One term I had five different jobs.

'So, what kind of law?'

'Corporate.'

'Seriously? I had you down for something more exciting.' He gives me an appreciative up and down look, but I don't think he means to be sleazy. I think he's just drunk. Actually, I'm beginning to like him. Maybe I won't rush off just yet.

'I work for gangsters.'

'Defence or prosecution?'

'Defence. I would have got Al Capone off.'

He has a nice smile. My drowsiness is wearing off. I add a Coke to my whisky. A bit of flirtation will be a good distraction from the horror film in my head.

We talk some more. The Coke kicks in and I revive. He asks me how I would have got Capone off and I tell him some of the tricks of the trade:

undermining the accused, exploiting technical loopholes, coaching your witnesses. The film is still playing but I'm not watching.

Until he says, 'So, tell me about your family.'

I almost close up on the spot, but perhaps the topic can be deflected.

'What do you want to know?'

'The truth, I guess.'

'I'm a lawyer. I don't do truth.'

'Well, I'm a banker, so I should know there's no such thing as truth. Only what you can make people believe. If I can make you believe shares in that whisky are about to go up five hundred per cent, you go and buy them — and the shares go up. Belief becomes truth.' He waggles his eyebrows devilishly. 'OK, I'll start. My kids live in Islington.'

'You don't have to tell me.'

'I want to. I want you to know. They live there, I live in Vegas.'

'So, you're a bad father. I don't give a shit.'

'Ah, but you should if we're going to date.' He sips his drink, peering over the glass at me archly.

I laugh again. 'I don't date guys with kids.'

'Why not?'

'Too complicated.'

He drinks before he answers, and when he puts his glass down the flippancy has gone. 'Life's complicated. If you think it's simple, you're not really living.'

'Goodnight.' I get up.

'Wait.' He puts his hand on my arm as I pass him. 'I'm sorry. My head always goes when I'm about to see them. I just keep thinking about how bad it'll be when I have to say goodbye.'

I sit down on the stool next to him. He's put on weight since he bought that shirt. It strains across his stomach. I imagine what his skin would feel like beneath the cotton. Warm and slightly tacky, downy blond hair running from his navel to his groin. 'What are their names?'

'Josh and Alfie. And I'm Daniel.'

'I'm Mags.' I shake his hand. 'And my brother's in a coma.'

3

Jody

They've contacted your next of kin. Your sister, Mary. I wonder why it's not your parents. We never spoke about them. We never spoke about mine either. Didn't want anything to cast a shadow over our happiness. I try to imagine what she will look like. Dark, like you. Slim. Black eyelashes even longer than yours. She'll speak softly like you do. She'll hold my hand and look into my eyes and she'll just *know*. That I'm The One for you, that you're The One for me. That whatever happens I'll stay by your side. I'll be with you while you learn to walk and talk again. Through the tears and the despair, and then the first stirrings of hope. I don't care if you're very changed, or even if you've forgotten me. I'll learn to love the you you become.

My heart clenches when you make a little gurgling noise. As if you've read my mind.

I lean in to kiss your earlobe and my tears fall into the clump of hair they didn't shave off for the operation. They nestle there, like the pearls on the dress I was wearing the day we first met. Do you remember? Is that part of your mind still whole? Maybe you've forgotten. We can remember it together.

* * *

I moved in at the end of the summer. The café job they lined up for me had gone badly. The manager was a bully. I used to spend my lunchtimes crying in the toilet, and then I just stopped going in. I lay in the bedsit for hours, unable to eat or sleep.

Then Tabby told me about St Jerome's. She sorted it all out for me, came and picked me up on a Sunday afternoon.

She wouldn't tell me much, just that the place was a deconsecrated church, owned by a Christian charity that let out the flats at piecemeal rent to vulnerable people — asylum seekers, people with mental health issues or family problems, former care home kids like me.

As the car pulled up in the little patch of tarmac by the grass I saw you. You must have been on your way out to the high road. You'd paused to watch the kids playing in the playground. It wasn't love at first sight, but it was close.

We were on the same floor. At the time it seemed like a happy coincidence; now I know it was fate. You smiled when we passed on the stairwell.

When you go into a church you don't realise how high it is. All that dusty air, just drifting in the huge empty space above the pews. They fitted four floors in there; we were at the top, looking out across the shops and takeaways to the green parks beyond. Each flat was unique: a mishmash of funny angles and sloping roofs, a gargoyle on the balcony, a column rising through the living room like a huge tree trunk. Some floors cut a stained-glass window in half, so you might have the angel Gabriel's face and the flat downstairs would have his open hands.

I've always had an imagination, and a night in a deconsecrated church should have left me paralysed with terror, especially as it was so quiet compared with the bedsit, where there was always shouting or doors banging. But as midnight came around I could hear music. A smooth woman's voice singing the blues. It was coming from your flat. It lulled me to sleep.

Tabby was good, coming in every day to make sure I was settling in, that my prescriptions were all up to date, that I'd filled in all the benefits forms, that I had enough food.

In the day I pottered around the flat, laying out all my special things, drawing, occasionally popping out to the high road where there were three charity shops, one with just books. I bought a whole set of romance novels and read one every evening. Your music was my lullaby at night.

Then one day you spoke to me.

It was a Monday afternoon. It had been raining heavily and my new book (*The Firefighter's Secret Heartbreak*) had turned pulpy in the carrier bag on the way home. I was wearing a dress from the charity shop, grey silk with little pearl beads around the neckline, and the hem was sopping wet where it hung down below my raincoat. It slapped against my legs as I ran towards St Jerome's. You were going in ahead of me and you stopped and held the security door.

‘So much for our Indian Summer,’ you said, with a smile that made one of your cheeks dimple.

I told you that my book had been ruined and you showed me how the blue dye of the carrier bag had stained your loaf of bread. You told me your name and I told you mine. Abe and Jody. Jody and Abe.

As we walked up the stairs together I said that I had just moved into Flat Twelve and you said it was nice to have a new neighbour, as the flat had been empty since the last occupant died. That frightened me, and you must have noticed because you laughed and said, ‘Oh, don’t worry, he didn’t die *in* the flat! He was staggering around in the road, drunk, and got hit by a car.’

‘Poor man.’

‘He was seventy-eight. Not a bad run for a raging alcoholic. Hope I make it that far.’

‘You will,’ I said, then blushed furiously, because I meant that you looked so young and fit and full of life, with your bright brown eyes and quick smile.

‘Lovely dress,’ you said as I unbuttoned my coat. ‘It looks like the rain.’

And then you said goodbye and went into your flat. I stood outside mine for ages afterwards, thinking *what a beautiful thing to say*.

Wednesday 9 November

4

Mags

I wake at four and can't get back to sleep so I get up and turn on my laptop, sitting in the faux leather club chair by the window that looks out over Hyde Park. Even at this time the traffic on Park Lane is nose to tail, though the double-glazing ensures the room is blanketed in an unnatural hush. The night sky reflects the glow of the city's lights, making it seem neither night nor day.

In Vegas the sun will have gone down over the desert. All the heat and dust will be vanishing straight up into the clear night sky. I'll be opening my first bottle of beer, licking the dribbles of icy perspiration off my fingers.

There are a couple of emails from angry clients. I knock off the usual pat reassurances, ending with a line about my brother to make them feel guilty. As if they're capable of an emotion other than greed.

Then I log in to my social media: an invitation to a gallery opening, angry posts about the latest gun rampage, my timeline clogged with endless *Happy Birthday Stu!*s for an ex-boyfriend's thirtieth. I don't know why we're still 'friends': we slept with each other for three months max, and then I finished it. He cried.

I sigh and switch off.

The police are coming here at ten. Six hours to kill. I can't even turn on the news in case it wakes Daniel, who, like me, didn't sleep a wink for the whole ten-hour flight. As I sit, staring down at the brake lights of the cars, I begin to feel irrationally annoyed that he is still here, spreadeagled on my bed, my sheets in a tight twist beneath him.

In the end I run myself a bath.

Catching sight of myself in the steamed mirror, I wonder why he was even interested. My hair's lank and dull, my lips are pinched, my tanned skin has become sallow. The loss of appetite has sucked the flesh from my stomach and my hipbones protrude, making me look rickety and frail, ninety instead of thirty.

The noise of the gushing water must have woken him because a moment after I get in, he enters without knocking.

'Hey. How's your head?'

'Do you mind?' I say.

He blinks at me. 'I've, er, seen it all before. Last night. If you remember.'

'I'm washing,' I say coldly.

'Sorry.' He backs out of the door and closes it softly.

When I come out he is dressed. We gather our things in silence.

'Why are you being like this?' he says finally.

'Like what?'

'I thought we had a good time last night.'

'We did. And now it's not last night any more and I've got to speak to the police about my dying brother.'

'Of course, I'm sorry.'

I stand stiffly as he tries to embrace me.

'This is a bad time for you,' he says, stepping away. 'We probably shouldn't have — but I'm still glad we did.'

'Me too,' I say, more gently. I've been a bitch. Mostly down to dread of what I'm going to have to face today, and the start of a raging hangover.

'Take my number and let me know how it goes with Abe.'

'Sure.' I pocket the scrap of cardboard he gives me. It's the corner of a condom packet he ordered from reception along with the bottle of Jack Daniels. 'Good luck with Jake and . . . '

'Josh and Alfie.'

'Yeah. Hope your wife's not too much of a cow.'

He looks at me and raises his eyebrows, and I laugh despite myself. 'Yeah, well, *I've* got a good reason to be.'

He comes over and kisses me. 'You were lovely. *It* was lovely. I'd like to do it again sometime.'

His breath is sour with booze and his skin still looks patchy from the flight.

'When you've sorted things out with your brother.'

'You mean when I've turned him off?'

He has the balls not to look away. Raising his hand to my face he passes his thumb across my cheek as if to brush away a tear that isn't there.

He seems like a decent enough person, for a banker. Although that isn't hard. For a corporate lawyer I'm an angel. Then he slings his jacket over his shoulder and picks up his case.

'Goodbye, Mags.' He turns at the door. 'Is it short for Margaret?'

I shake my head, hesitate, then say, 'Mary Magdalene.'

He looks at me quizzically, waiting for me to explain. When I don't, he opens the door.

'What did your wife leave you for?' I say suddenly.

He turns and smiles. 'Alfie's fencing coach.'

⋆ ⋆ ⋆

We sit in the breakfast room of the hotel, looking out over a rubbish-strewn side street. The squad car is tucked discreetly behind a four-by-four. There are two of them, a solidly built middle-aged blonde woman, and a thin, lantern-jawed youth, young enough to be her son. Apparently it was her who called me to tell me what had happened. I was at home catching up with work emails before I headed to the office. It felt strange, sitting on my sun-drenched balcony in a playsuit and shades, with a mouth full of blueberry pancake, listening to her talk about *induced comas* and *cranial haemorrhages*.

Now she tells me more, speaking in a soft London accent as the boy takes notes. At first she gives me the logistics, timings, the distance fallen, the hours on the operating table, then slowly spirals back to the night itself, as if it's the only way I will be able to bear it.

'Your brother's fiancée, Jody Currie, was the only witness to the accident.'

'So, you're sure it was an accident?'

At what I considered to be a throwaway comment, I'm surprised to see the boy raise his head and fix his gaze on his boss.

She pauses before answering. 'We've got no reason to suspect otherwise.'

I wait.

She sips her coffee. It's a standoff.

'But?' I say finally.

'There's no evidence to suspect foul play: no CCTV footage and no other witnesses.'

'So, why the question mark?'

I wait for her to fob me off — *What question mark?* — but to her credit she doesn't. 'Relationships are private things. Jody and your brother had both lost contact with their parents and were living quite isolated existences. We have to believe her that the relationship was a happy one.'

'As opposed to a murderous, push-you-down-a-stairwell sort of one?'

She shrugs. *Whatever*.

'So, you won't be investigating further?'

'Like I said, there's no evidence of foul play, so there's no reason not to take her word for what happened.'

'Which was?'

'On the night in question, Miss Currie had booked a meal out. She felt that your brother seemed down and wanted to cheer him up.'

'*That* must be on CCTV, right?'

'It's not police policy to waste resources going through general CCTV footage when we don't think a crime has been committed. Can I go on?'

I give a curt nod.

'They returned to St Jerome's, the church where both of them live, at about eight p.m.'

'Isn't that a bit early?'

'Miss Currie thought your brother was tired as he had been quiet all evening and had suggested they leave early. Both her and your brother's flats are on the fourth floor and she told us that they had almost reached this floor when your brother stated that he wanted to go down to check the door was securely closed. There's criminal gang activity in the area and he was concerned that if it wasn't closed properly, someone might get in. Miss Currie went into your brother's flat and, after hearing a noise, came out to find him lying at the bottom of the stairwell. It's her belief that he jumped, due to depression brought about by work pressures.'

She folds her hands in her lap, her face tactfully averted as she waits for me to process the images that have been flowing through my mind.

'Her belief? So there was no note?'

'No.'

'Couldn't these criminal gangs you mentioned have got in and attacked him?'

'If that was the case then either Miss Currie or their neighbours on the top floor would have heard something, and, aside from the injuries sustained in the fall, there were no other wounds. Also, he had no valuables on his person as Miss Currie had taken his jacket inside.'

'Why?' I said.

'Why what?'

'Why did she take his jacket?'

The policewoman smiles. 'You're a lawyer, right?'

'Yes.'

'I can see you must be good at your job. Since they were coming into a warm environment from a cold one he might have taken it off and handed it to her for convenience as he went down to check the door.'

'But he wasn't going to check the door, was he? He was going to jump. So why bother taking off the jacket in the first place?'

'In a police investigation,' she says after a pause, 'there are some questions that are vital to help us judge whether or not a crime has been committed, and some that aren't. I suggest you speak to Miss Currie yourself so that she can give you a clearer picture of what happened that night.'

They get up, leaving two unfinished cups of bland hotel coffee on the glass table.

'If you have any concerns please do get in touch.' As she hands me her card my fingertips brush hers. They feel unpleasantly soft: the nails are bitten halfway down to the cuticle and flesh bulges over the top of the remaining sliver of nail. I glance at the card. Her name is Amanda Derbyshire. A PC. Lowest of the low.

'Thanks,' I say coldly, shaking her hand and the clammy paw of her underling.

'I know you deal with criminals a lot yourself, Miss Mackenzie,' she says, turning to leave, 'but not every tragedy is a crime. Will you be seeing your brother today?'

’m going straightaway.’

…hope the doctors can give you some good …ews.’

I give her a dry smile — we both know these are empty words — and she turns away.

Sitting by the window sipping my coffee, I watch them get back into their squad car. They are too stupid to realise that the twists and turns of the hotel lobby have led them out directly beside the window they were, until a moment ago, looking straight out of.

The woman says something and the boy gives an open-mouthed guffaw, displaying rows of silver fillings. In his hand he has one of the Danish pastries from the buffet bar and as he climbs in, eating it, I hear her warn him not to get crumbs in her car.

To clear my head I’d swum for an hour in the hotel pool before our meeting and, thanks to that and my burgeoning hangover, I am finally hungry. I’m glad, as I load up my plate with hash browns, that the policewoman isn’t here to see this inappropriate show of gluttony. I should be too grief-stricken to eat, but instead I pour ketchup over my breakfast, head for a table near a TV and scroll through the channels for CNN.

* * *

I haven’t been in a British hospital for twenty-two years. In Vegas someone would be escorting me through the labyrinth of corridors to the ICU, telling me about my brother’s condition as we go, preparing me for what to expect, but here I must find my own way and will have to wait until the doctor does his rounds to hear my brother's prognosis.

He fell twelve metres. It can't be good.

I try to imagine what he must have looked like before the accident. He was always slight. Slim-boned, with narrow shoulders. A child's body even after puberty. I wonder what he does for a living. Did. I wonder how he found me. The picture on the company website would be utterly unrecognisable to anyone who knew me as a child.

I was shocked to get the Christmas card. Sent to the office, to *Mary*, so it took ages to arrive at my desk. *From Abe*, and an address in London. I sent one back — embarrassingly late. *From Mags*. A line of communication, as fine and tight as a wire. I don't know if I thought we would become closer as we got older, that we would forgive one another for the things they made us do. I suppose I did. But now it's too late. There's nothing to miss.

The hospital walls are crowded with bad art. Tasteless collages and insipid watercolours, metal twisted into the shapes of fish and birds. I pass a door marked Room for Reflection and through the half-open blinds make out empty plastic chairs facing a table with a wooden cross.

A bed clatters by. On it an old lady is curled like a chrysalis. Beneath her translucent paper skin dark veins pulse, as if there's something beautiful and new ready to squirm out. She is yellow with liver failure. Perhaps our mother looks like this now. Perhaps she is already dead.

I pass through the door marked ICU. It opens on a small reception area where a nurse frowns at a computer screen. Behind her is a set of double doors, that must lead to the beds. A wave of guilt washes through me. I could easily have afforded to put Abe on my medical insurance policy. Then he would have had his own room.

'I'm Mary Mackenzie. My brother Abraham is here.'

The fat nurse doesn't reply, just holds up her hand: *wait*.

Bristling, I step away from the desk and stare blindly at the huge painting of a peony on the wall. Surely all those blood reds and flesh pinks are inappropriate here. The whole place stinks of piss and disinfectant, that British-hospital smell that screams of underpaid cleaners, harassed nurses, and patients left to stew in their own filth. And then, abruptly, absurdly, tears spring to my eyes. The peony blurs, becoming an open wound.

As unobtrusively as possible I blink them away and breathe deeply. I'm not crying for Abe. I'm crying for myself. Stuck here in this shitty hospital, in this shitty country, away from my friends, my job, the warmth of a Vegas autumn. I will have to wait for him to die. Damn it, I almost feel like ringing Daniel, but a good lay doesn't buy you the right to snivel on someone's shoulder.

'Miss Mackenzie?'

I blink to clear my eyes and turn.

'Your brother's fiancée is with him at the

moment. I can ask her to give you some time alone with him?'

For some reason I don't want this nurse knowing that we are such a dysfunctional family I've never even met my brother's fiancée.

'It's OK,' I say. 'Just take me to him, please.'

Despite the bleeps and wheezes of the machinery, the sensation I feel when the doors swing shut behind me is of a heavy, suffocating hush. For a moment I can't take a step. Every nerve in my body is tensed, to stop me bolting, and I stand rigid as the nurse waddles up the room to disappear behind a blue curtain on the left.

There are six beds in all, each separated by a curtain, though most of them are open. The occupants lie on their backs, motionless, pale as wax, everything that makes them human concealed or distorted by pipes and masks and coloured stickers. Most of them are old; sparse white hair slicked across crêpe paper foreheads, gnarled fingers resting on the sheets like the shed husks of spiders.

A wave of nausea reminds me how much I drank last night. I can't be sick here: it would be the ultimate insult.

There are low voices, and a moment later the nurse emerges, gives me a tight smile, and passes back out through the doors.

The fiancée is waiting for me.

My heels click across the lino and the rings clatter loudly when I pull back the curtain.

The girl — and she is just a girl — sits on a plastic chair pulled very close to the bed. She raises her head and attempts a smile. Older than I first thought, in her late twenties perhaps, but her manner is that of a child: shrunken shoulders, nervous eyes that cannot hold my gaze. In appearance she is like my brother in negative: the same birdlike build, an elfin face with a high forehead, large eyes, a small rosebud mouth. But where Abe is dark she is shockingly fair, almost albino, with eyes the colour of dishwater.

For a moment I'm disappointed. I suppose I had hoped for someone like me. Someone I could talk to. I can tell immediately that all conversation with this girl will be punctuated by weeping. I will have to reassure her endlessly that it wasn't her fault, and ply her with cups of tea and tissues.

She gets to her feet unsteadily. 'I'm Jody,' she says, then adds, 'I'm so sorry,' and her face crumples.

Swallowing a sigh, I wait patiently while she composes herself, then extend my hand. 'Mags.' Her handshake is predictably limp and she inhales when I squeeze her knuckles.

Finally I look down at my brother.

At least I assume the swollen, blackened lump of flesh and bone on the pillow is my brother.

The top of his head is swathed in bandages that various lines pass into. Another bandage covers his nose and cheeks and a neck brace compresses the lower part of his face. Only his eyes and mouth are visible, the lips purple with swelling. He is naked to the waist and his body bristles with tubes leading to bags and bottles of clear liquid.

I breathe slowly and steadily, feeling Jody's eyes on me.

Finally I'm ready to speak. 'So, can you tell me what happened?'

But before she can answer, a nurse comes over and begins checking the monitors. I take Jody gently but firmly by the elbow. 'Let's talk about it over a cup of tea.'

I buy us drinks from the vending machine and lead her out into a small garden that looks out over the main road. A brass plaque on the wall of the empty fountain announces that this is the Queen Mother Memorial Garden.

Jody takes the lid off her tea and the steam curls up into the damp air. The garden is slightly below ground level and the air is leaden with cold. The sun is too weak to melt the night frost and the blades of grass are stiff and white as icicles. I sip the scalding black water that advertised itself as Americano. It is so far from American I want to weep.

As she stares at the dead fountain I wonder how far her thought process has progressed. Has she yet faced up to the prospect that Abe will die? If not immediately, then at some point in the future when the time comes to turn him off. No one gets up from a fall like that.

'It's my fault,' she says.

I wait for her to continue. Her irises are so pale that, seen from profile, they are no more than water surrounding the pupil, large in the gloom of the garden.

'I should have seen the signs. He was working

too hard. Sometimes he wouldn't get in until nine or ten. And it's such a stressful job, being a carer.'

I try not to look disappointed at the revelation that my brother cleared up piss and shit for a living; microwaved ready meals, changed incontinence pants, baby-talked sponge-brained geriatrics. I don't know what I was expecting — advertising? Graphic design? — something like me I suppose. God, what a narcissist.

'There was never enough time to get anything done, to do a good enough job, and you know how much of a perfectionist Abe is.'

I nod, *knowing*.

'And how kind he is. He couldn't bear leaving people when he knew he was the only company they would have for days. He would stay and make sure they were all right, which would make him late for the next appointment. Sometimes he had to miss one entirely. They wouldn't pay him for travelling time, and we were hoping to get married next year, so of course money worries just added to the pressure. It was really getting to him. I could see it. We barely saw each other.' She twists the ring on her engagement finger.

'That must have been difficult.'

'I understood, of course I did. But I hated to see him so stressed.'

'Tell me what happened the night he fell,' I say as gently as possible, laying my hand on hers in a gesture I hope will be reassuring and encouraging. Her skin is rough, chapped from the cold, the nails bitten ragged. I hold it there as long as I can bear, then release it into her lap.

‘I wanted to try and cheer him up,’ she says, looking away, across the garden to the city beyond. ‘One of his patients had been taken into a nursing home, and she was very upset about it. So I booked a table in Cosmo — that’s our favourite restaurant. He was quiet during the meal, but I thought he was just tired, so I suggested skipping dessert and having an early night. I should have known. I should have guessed there was something wrong.’

‘It’s not your fault.’

‘We should never have had that second glass of wine. He always gets sad when he’s had a drink or two. On the way home he didn’t say a word, just held my hand really tightly. We came in and started going up to his flat.’

‘You don’t live together, then?’

‘We’ve asked the housing association for a bigger place, but we thought we’d keep both flats on in the meantime. When we got to the third floor Abe said he couldn’t remember if he had closed the security door properly. Sometimes it sticks and there have been break-ins. He told me to go ahead, so I did. I wanted to get the heating on and light a few candles to try and help him relax.’

The ghost-grey irises swim with tears.

‘If I had known how bad he was . . . I’m s-sorry.’ Her voice rises tremulously. If she starts to sob I’ll never get any more out of her.

‘Then what happened?’ I say, firmly, as if facing an overwrought witness.

‘I opened the door of the flat and went into the hall and then I heard this . . . ’

This time I can't bring myself to make her go on.

'It was such a horrible noise.' Her voice goes up again, on the way to a wail. 'It was so loud. Like there wasn't anything soft about him. Like he was a piece of wood or something.'

I close my eyes.

'I ran out of the flat and . . . '

A lorry trundles by, its tarpaulin billowing in the wind. She waits for the roar of its engine to subside and in those few moments all the life seems to have been sucked out of her.

'I'm so sorry,' she says as the normal traffic noise resumes.

'It wasn't your fault,' I say.

* * *

Eventually the doctor graces us with his presence. It's twenty past five and I'm so on edge that every thick, shuddering breath Jody takes is making me want to grab her by the hair and smash her head into the wall. At least the weeping wives who show up in court to plead ignorance of their husband's misdemeanours are faking it. Beneath the act they're hard-nosed businesswomen, doing their utmost to prevent the IRS discovering the little offshore hoards in their names. Jody is something else. She holds my brother's hand the entire time, gazing into his pulped face, occasionally brushing the tube that protrudes from his mouth with her lips. The sight, along with the alcohol tang of disinfection, intensifies my nausea.

I pace to the window and back, trying not to look at the other cadavers, wondering what on earth the point is of spending all this money and effort to keep them in this parody of life. Presumably my brother, if he wakes at all, will be a drooling, infantilised wreck. Jody will lovingly feed him with purees and porridge, wiping the gloop from his slimy chin. At least Alzheimer's or dementia patients have the decency to be old. Abe could go on like that for decades.

Dr Bonville is very young, shorter than me, with the floppy-haired arrogance born out of the British public school system. He takes us to a shabby little room with a blue sofa so small that Jody and I must sit hip to hip.

'Well,' he says, and gives that pressed-lipped smile people use to express empathy. 'The swelling has gone down.'

Jody turns to me and I can almost feel her itching to squeeze my hand. I keep looking at the doctor. I know what's coming.

'So we've been able to assess the damage to Abe's brain.'

He pauses then, rustling the papers on his lap. He doesn't sit behind the table but pulls out the chair to sit opposite us, a more informal, human position that can only mean the worst.

'Abe's cerebral cortex has suffered major trauma. The cortex is responsible for thinking and action. For this reason, when we take him out of the coma, I'm afraid Abe will be in a vegetative state.'

As he waits to let the news sink in I can hear ambulances pulling in and out, their sirens gradually diminishing to be absorbed in the traffic.

The material of the sofa is loosely woven, like slack skin, and the arm is blotched with watermarks. How many tears, I wonder, have been shed here? My fingers are hypersensitive, as if I can feel the microscopic granules of salt beneath their tips.

'Will it heal by itself?' Jody says, her voice clumsy in the silence, making me wince.

'Of course not,' I say.

'No,' says Dr Bonville. 'I'm afraid that can't happen. I'm afraid you have to face the possibility that, in the very unlikely event that Abe ever regains consciousness, he will be very different from the man you knew.'

'He blinked,' Jody says. 'I saw his eyelids move.'

'Abe's lower brain stem is intact, so reflexes like breathing, swallowing, reacting to pain, even blinking, can still be present. Abe might even — '

'Stop saying his name.'

He turns his surprised gaze on me.

'Stop saying his name because you think it gives you *the human touch*.'

He looks at me steadily. 'I understand that you must be very upset,' he says quietly. 'Let me give you a minute.'

He goes to get up, but I get up faster. 'Talk to *her*,' I say. 'I've heard enough. Just tell me when it's time to turn him off.'

I walk out of the room and, without a glance at the doors of the ICU, stride down the corridor that leads to the exit. It seems to take years, and when I finally emerge into the grimy London air I gulp it down like water from an Alpine stream.

The traffic roars past and I'm buffeted by blank-faced office workers rushing to get home. It's at times like this that the anonymity of a city is a blessing. Nobody knows I have just walked out on a doctor trying to tell me whether my brother is going to live or die. Nobody cares.

With a glance back at the hospital to confirm Jody isn't coming after me, I join the flow of people heading for the Tube.

* * *

It feels like a year has passed when I finally arrive back at the hotel. I go for another swim, try — and fail — to read my book, pick at my room service order: a very poor imitation of a club sandwich. At six I hit the mini bar.

The room darkens.

On the pretext of a work chat, I call Jackson and when he picks up I can hear the hubbub of a restaurant behind him. I want to ask where he is but it might sound like an accusation. I imagine them at Ginelli's down in Paradise, drinking cold beers on the veranda with the smell of the desert on the wind. My heart aches.

'Let me go somewhere quieter,' he says.

'No, it's fine,' I say, desperate for the sounds of home. 'It's just to check in, really. How's Antonio?'

'We made the plea bargain and they're

thinking it through. I told him they'll probably go for it.'

'Great. Send him my love.'

Jackson laughs. 'He'll be wanking all night over that one.' He kills the laugh and says, 'How's your brother?'

I exhale. 'Not great. Brain-dead, it looks like. A botched suicide attempt.'

The muted TV at the end of the bed strobes images of a war zone — old women and children crying, grey corpses rotting in the road, an abandoned teddy.

Jackson tells me he's sorry. Then, after a seemly pause, asks me, 'Do you think . . . dying will happen . . . naturally?'

I know what he's really asking. *When will you be back at work?*

'Potentially. But it might come down to turning the machines off and it's a bit early to think about that.'

'Of course, of course.'

'I'd be happy to leave any decision to his girlfriend, but as next of kin I'm supposed to have the final say.'

There's silence on the other end of the line and I can almost hear Jackson trying to frame the words.

'How long will you . . . er . . . wait?'

Suppressing the flash of irritation I keep my tone light. 'It'll depend on the doctors.'

'Take as long as you need, Mags.'

'Thanks. Listen, go and enjoy your lunch. What are you having?'

He clears his throat. 'Lobster thermidor.'

I groan with envy.

'There's one with your name on it, when you get back.'

'Send me a photo. I'll choose him myself. You can put a deposit down.'

'Will do. Take care, Mags. Lots of love.'

I hang up, then open the mini bar. Three gin and tonics later I'm sitting on the bed with the TV blaring to try and numb my head. PC Derbyshire was right, not all tragedies are crimes, but I'm a lawyer, so all I can think of now is questions.

Why haven't they checked the CCTV?

Why did Abe take his coat off?

Did someone get in through the door that he was going down to check?

With such a plainly devoted girlfriend, why on earth would he decide to kill himself?

And if so, didn't she deserve a note?

She gazes out at the gnarled faces of the trees. They are like people, stretching their arms towards the car to pluck her from the back seat and spirit her away into the darkness. But the car is moving too fast: twigs rattle vainly against the windows as the headlights sweep relentlessly onwards. She twists her head to watch them recede. For a moment each trunk is washed in red, before falling away to blackness.

They have made this journey before. She knows that the forest bordering the road will end abruptly, the land flattening out to fields. An occasional house will dot the landscape, its windows butter yellow. But the house they are heading for cannot be seen from the road, and its lights are the cold glare of fluorescent strips. All the better to see you with, my dear.

Are there wolves in this forest? The thought does not scare her. An animal has simple desires: to eat and sleep and protect its territory. It mates only to produce children. It loves its children with such fierce passion it would tear your throat out if you harmed one.

A wolf would eat her, perhaps, despite her boniness. It would be a quick death. She can imagine her corpse being squabbled over by tumbling cubs, play-snarling at one another, little claws tangling in her hair, needle teeth chewing her finger bones.

She slides her eyes across to the child beside her. Like her he stares blankly from the window, the trees throwing moving bars across his face. The air in the car is thick with cigarette smoke and the woman in front of her winds down the window to toss out a butt. A chill wind lifts the hem of her dress and creeps up her thighs. She shivers. She needs a wee but knows better than to ask to stop.

In the front they are discussing who will be there tonight. She hears a name she knows and a little bit of wee comes out and wets her knickers. She glances at the boy again. His hands sit limply in his lap, fingers upwards, like dead crabs.

Now there is a red light up ahead, a vertical sliver of sunset. The trees are coming to an end. They are nearly there.

She wishes she could have a drink. The children at her last school said it's illegal to drink wine when you're only seven. They told the teacher and she had to pretend she was joking. A few sips of cider now would smooth the jagged edge of panic that makes sweat prickle her armpits despite the cold.

The clear patch of sky widens as they reach the edge of the forest. The man in front exclaims. There is a branch in the road. He slows down.

She unclicks her belt and opens the door.

The tarmac slams into her, making her bones crunch as she rolls over and over, coming to rest on the edge of the camber, before the road falls away to forest. The car screeches to a halt, then starts reversing. On hands and knees she scrambles down the incline, cutting her shins on needles and pinecones, then she is up and running.

The wolves watch her from the shadows as she flies through the darkness, her hair streaming behind her.

She is a good runner. Thin and long-legged, like the antelopes on the nature programmes at school. There is just enough light to see her path ahead. She ducks left and right, following the natural instinct of prey animals. The light dims as the trees grown denser, enfolding her. Her footfalls are muffled by the spongy forest floor. She will find a bush to creep into, or else she will shimmy up a tree and conceal herself among its leaves.

A huge white shape makes her steps falter. At first she thinks it might be an angel, its wings spread to enfold her, but it is only an owl. A magnificent barn owl, its black eyes glinting. As it sweeps by she feels a whoosh of cold air on her cheek. She keeps running. The darkness deepens and there is a pain in her side. Her shoelace has come undone. She should retie it but she cannot risk the hesitation. She keeps running.

There's rustling all around her, like the trees whispering to one another. She is not afraid. The stitch subsides and the air is cold and clean, scouring her out from the inside, taking away all the filth. A snatch of sky. The blood red has been replaced by velvet blue. She can see a star. Moonlight silvers the uppermost leaves.

Then her foot comes down on the loose lace,

and she is thrown forward, her breath escaping in a grunt as she lands heavily on her stomach.

She lies there, catching her breath, the pine needles tickling her thighs. If she lies here long enough a blanket of leaves will cover her. Her fingers will become roots, delving into the dark soil. Insects will make nests in her hair. No one will ever find her.

A crack.

Her consciousness sharpens, her hearing becoming hypersensitive.

There is no more rustling in the undergrowth, no whisper of wings. The creatures of the forest are afraid. Clouds scud across the moon, and the silver light winks out.

Another crack. Louder.

She would pray, but her lips won't move.

The wolf lands on her back.

She is wrenched up and spun around. His body is silhouetted against the glare of the car headlights behind him. She had come such a short distance. How foolish to think she had a chance.

He hauls her onto his shoulder like a dead stag.

'Please,' she whimpers, but her voice is drowned by the crash and crack of the undergrowth as he plunges through the trees back to the road.

5

Jody

My family would have loved you. My dad may have been a forces man, but he was never a bully. He respected gentleness; he knew that strength isn't about muscles and fists, that it comes from inside. He would have seen the strength inside you.

Mum loved him so much she couldn't go on without him. I'm not angry with her for that. I can understand. I feel that way about you — if you die I won't want to go on.

Your sister is so hard. The way she talked about . . . well, about what the doctor said. It was horrible to listen to. Like she doesn't care about you at all and just wants to get it all over and done with. I won't let her, though, don't worry. I won't let them hurt you, Abe. They'd have to get a special court order before they can do anything like that anyway. I read about it once, a case in America where a woman had a stroke and was in a coma. The husband wanted to turn her machines off but her family didn't want to. They went with the husband in the end, which makes me scared because we're not married yet, but also hopeful that they take into account the wishes of the people closest to you. Your sister hasn't seen you in years but she barely looks at you. She doesn't love you. I can't imagine her loving anyone. I'm not surprised she's on her own, even though she's really attractive.

She looks so much like you. The same slim face and wide, dark eyes. The same straight dark hair. You could be twins. How did your hearts turn out so different?

I came straight back to your flat after the doctor went away, and just being near the things you've touched is making me feel better.

I've lain on your bed for hours, gazing at the photograph of the two of us in that bar in the West End, but now I get up and open the wardrobe. As I run my fingers through your clothes the scent of you drifts out, and I close my eyes and breathe deeply. Then I take out one of your cardigans to put on after my shower, a cashmere one, soft as rabbit fur.

I use your shampoo, to keep my hair smelling like yours, and then I clean my teeth with your toothbrush and dry myself with the towel from the heated rail. A single black pubic hair curls from the weave. Yours. Mine are fair.

I put on your T-shirt and cardigan, and when I close my eyes it's almost like the ghost of you is all around me, embracing me. I wonder if your spirit can move from your body, because of the state you're in, or whether someone has to be dead for that to happen. Even if you die, Abe, it won't be the end — I promise. When two spirits like ours meet and forge such a strong and powerful love it can't just blink out like a light. Something has to remain.

Your flat is so much nicer than mine, and not

just because it's filled with you. It's so bright and modern, all greys and whites and the type of wood they call 'blond'. Your window looks down on the grass at the front and the bright colours of the children's playground. Even the kitchen, which is the same in all the flats, looks nicer, somehow. I think it's because of how you've 'accessorised' it. The glass jars of pasta, the silver coffee maker, and the corkscrew that looks like a lady in a dress. It's Alessi, which I know is expensive, because in the charity shop they keep that sort of stuff in a locked cabinet.

It's silly but at dinnertime I lay two places and dish out two bowlfuls of pasta, and then I talk to you as if you're still there.

'How was work?'

Oh, you know. Tiring.

'You work too hard.'

They need me. Mrs Evans was so relieved to see me. I don't think she'd spoken to anyone since my last visit. How was your day?

'Better now.' I close my eyes and reach across the table and imagine your hand in mine. I can almost feel it, the light touch of your warm fingers against my palm, and then the table starts to vibrate. I jump so hard my fork clatters off my plate and a blob of tomato sauce spatters the sleeve of your cardigan.

It's only my phone vibrating before the ringtone kicks in.

For a moment I think it's going to be you on the other end. But it's not. It's your sister.

'Hello?' I say, warily, wondering if she's going to be nasty.

‘Listen, I’m sorry about earlier. I just hate the way these people patronise you.’

‘Yes,’ I murmur, but I don’t really agree. Doctors have always made me feel safe.

‘I’ve been thinking. It looks like I might be hanging around for a bit longer and it’s silly to live out of a hotel room, especially when I’m so far from the hospital. I’d much rather have a bit of space and be able to cook for myself, so I’m going to move into Abe’s flat. The police haven’t returned his belongings yet so I wondered if you had a key I could have.’

My breath catches. She wants to come here?

‘I’m not stepping on your toes, am I? I mean, feel free to come around and collect any stuff you’ve left there.’

‘It’s . . . not that,’ I stammer. ‘It’s just that . . . ’ My mind goes blank, but eventually I come up with something. ‘I’m not sure the housing association would allow it.’

‘Oh, right. Well, can you give me the number and I’ll talk to them?’

‘Umm . . . wait a minute.’

I put the phone down on the table and stare at it for a moment, my skin creeping. I could say I’ve lost the number, but she’d be able to find it easily enough. I could give her the wrong one and then stop answering my phone, but she would just come and look for me at the hospital.

In the end I get up and head back to my flat, running in my socks so she can’t hear my footsteps. As I run past Flat Eleven I can feel the spyhole watching me, black as a shark’s eye. Sometimes I think I can sense someone hiding behind the door. Pushing the thought from my mind I go into my flat, find the number on an old letter, and run back.

But the spyhole has given me an idea and, after I read it out to her, I say, 'I don't know if you know, Mags, but this place is run by a charity. So as well as care home kids like me, there are other people, with worse conditions. You know, *mental* issues. I'm used to it, so I know to be careful, but you . . . ' I tail off meaningfully.

She hesitates a moment, and I think that she might change her mind.

But then she says she'll call the association and if they say it's OK, she'll come by sometime tomorrow morning to pick up the keys. She adds, conversationally, that the police will be popping round to return Abe's stuff sometime over the next few days, so if I remember anything I haven't mentioned to them already, that would be my chance to tell them.

I put the phone down and stare at your untouched plate of food, my heart thudding.

What does she mean?

Thursday 10 November

6

Mags

I dial the number Jody gave me and a young man with a heavy Arabic accent answers. After I've explained the situation he says he'll put me through to the charity's director, Peter Selby. It rings for a long time before it's finally picked up, by what sounds like a very old man, very posh, and slightly camp.

When I explain what's happened he gasps and his voice trembles when he says how sorry he is. For the first time, the clichéd words sound genuine.

'Did you know Abe?'

'Of course,' he says. 'We interview all our prospective tenants, to make sure they're eligible for our help.'

My interest pricks. 'And Abe was? Eligible?'

He hesitates. 'Well, clearly.' I can hear the surprise in his tone. I'm Abe's sister. How can I not know this about him?

'Abe and I haven't spoken for many years. We had a difficult upbringing. It created a . . . distance.' I hate talking about my family.

There's a pause and then the old man says, 'The St Jerome's Foundation offers assistance in the form of subsidised accommodation for minority or vulnerable groups. People who have been let down by society and need a helping hand to raise themselves up again.' I get the feeling he has parroted this line many times before. If they have charitable status, he must have to reapply each year.

I assume by 'vulnerable groups' he's talking about people with mental health issues.

For the first time it occurs to me that perhaps Abe had some kind of a breakdown when he left home. I managed to avoid one by self-medication with alcohol and narcotics, but only just. Perhaps that's when his depression began.

'In that case, you must have seen his medical notes, in order to assess his eligibility, right? Was he clinically depressed back then?'

There's a long pause, during which I hear a creak, as if he's sitting in a leather armchair. Unless it's his bones. Finally he speaks. 'Abe moved into St Jerome's ten years ago, when he was very young. We put him in touch with a support group, and organised vocational training that enabled him to embark on his career — a career which he seems to have been eminently suited for. A charming young man. He will be much missed.'

That's not an answer, but it's clear it's all I'm going to get.

With Jody's words in mind I ask, 'Are they dangerous?'

'To whom are you referring?'

'The people in St Jerome's. What sort of mental health problems are we talking about?'

He hesitates again before replying and I hear the wheeze of his breath through ancient lungs.

'Miss Mackenzie, as I'm sure you can understand, I am unable to share confidential information about our residents; suffice to say that in the twenty-seven years this foundation has been in operation, no resident has ever attacked or otherwise harmed another.'

'There's always a first time.'

He sighs in irritation. 'Whatever you may have read in the tabloid press, those suffering with mental health difficulties are far more likely to be a danger to themselves than others. Now, the foundation would be perfectly amenable to your staying at St Jerome's while your brother recuperates, but it is of course your choice.'

Ignoring his implication that I'm a gullible idiot who believes the mentally ill are all knife-wielding maniacs, I tell him I'd like to move in straightaway. He says the building manager will call to let me know all the various rules and regulations but as I put down the phone I wonder what I'm letting myself in for.

An hour later I check out of the hotel, bumping my wheelie case down the steps, and the doorman hails me a cab. I've dressed down — jeans and Converses and a black rain jacket that looks pretty uninspiring but cost six hundred dollars — but as we travel north, moving closer and closer to the little blue pin on my phone map, I'm glad I did. Edgware Road and Regent's Park are bright and bustling but as we pass through Camden and Chalk Farm the buildings and people become shabbier. Kentish Town is about the last bastion of civilisation before we enter a no man's land of boarded-up shops and run-down council estates.

Even the sky seems dirtier out here. The high-rises stretch away into brown clouds, their walls leprous with rot, plastic bags whirling around their bases.

My phone rings, giving me the chance to excuse myself from the cabbie's monologue about his daughter who has just moved to New Zealand.

It's the building manager, José Ribeiro. He offers to get a spare set of keys cut for me but I tell him I can use Abe's, so he moves on to the building regs. The first lot are simple enough: no pets, no smoking, no subletting the flat, but what with the traffic noise and his heavy South American accent, it takes several painful minutes for me to understand when the bins should be taken out, how to programme the hot water and the account to pay the rent into. He's about to say more but I've had enough: I tell him I'm losing signal and drop the call.

We're close now. According to my blue pin this high street we're crawling down is just around the corner from St Jerome's. There are a few independent shops and cafés, the obligatory charity shop, an Internet café and a place that promises to unlock any phone. Handwritten signs in grubby windows announce *Best Kebab in London!* or *No Groups of Children*. The fruit and vegetables in crates outside are dirtied by traffic but a Greek bakery looks promising, and there's a Food and Wine for basics.

We're stuck behind a bus emblazoned with an

advert for the local Baptist church. Shiny faces beam out of the grime, their glow of health and happiness out of place here.

All the passers-by seem bent with age or sickness; they shuffle along, dragging wheelie trolleys overflowing with the blue plastic bags favoured by all down-at-heel shops. There are few white faces, and more full-face veils than I have seen outside news footage.

The cabbie has stopped talking about his daughter and, as we wait for an elderly woman to shuffle across a zebra crossing, he taps the wheel impatiently. He seems as tense as I am. Perhaps I should have stayed at the hotel. I will stand out here like a sore thumb. Or perhaps there's a trendy part — where media types have started to move in and gentrify the place.

We turn into Gordon Terrace, a street of low-rise concrete bunkers with weed-choked front gardens. A teenager lumbers by with a dog so muscular it looks like a screwed fist.

At the end of the terrace is a patch of bumpy waste ground and then I see it. St Jerome's church, its spire silhouetted against the darkening sky.

Goosebumps trickle down my arms. This will be the first time I've been in a church for almost twenty years and, though the original wooden doors have been replaced by a faceless security door, an irrational panic rises in my throat at the thought of passing through it.

The cab stops by the pavement, under a flickering street lamp.

'Twenty-three fifty, love.'

I hand over the unfamiliar notes and, without waiting for change, I get out. A concrete path crosses an expanse of patchy grass, which seems mainly to be used as a dog toilet. It is hemmed in on all sides by a chain-link fence with buildings pressing close on the other side. On the left-hand side of the path, shadowed by a nearby high-rise, is a playground. The sole occupant, a boy of eight or nine, looks up from his swing.

It's colder here, much colder than in the city centre, and a gritty wind snatches at my jacket as I trundle the case down the path. My progress is impeded by annoying ridges in the path, like speed bumps or buried tree roots. Along the top of each ridge the tarmac is cracked like a loaf cake, exposing the black glittering crystals beneath. Soil seeps from the tear.

A moment later I am swallowed by the jagged shadow of the church.

It's constructed of grey brick, in that austere Victorian style designed to intimidate, to make you feel small.

I straighten my back and stare at the two empty arches at the base of the spire, but they gaze impassively out across the shops and tower blocks.

The central section of the building is flanked by a wing either side, with its own small arched windows, the lower of which are covered by security grilles.

From this side, the stained-glass panel above the door is just a sliver of grey, but I feel the eyes of the ghostly figures watch my approach, the case rumbling along behind me, announcing my presence.

And then this sense of being watched grows suddenly more powerful and I stop on the path, my heart thudding.

My head snaps around but I'm too late to catch any more than the flutter of a net curtain.

Someone was watching me from the ground-floor window of the left wing.

I stay where I am for a moment, to see if they will return to the window, but the curtain is still, the unlit room beyond giving nothing away.

The trill of my phone makes me jump. It's PC Derbyshire, asking when she can come and drop off Abe's belongings. I suggest it might be more appropriate for them to go to Jody but evidently, as his next of kin, I must sign for them.

As I hang up it's just starting to rain, that ice-cold drizzle Britain specialises in, creeping down the collar of my raincoat, chilling my hands and soaking through my canvas shoes. I pick up the case and run the rest of the way to the door.

The security panel glows green and next to button ten is a label with Abe's and my surname, written in biro. I buzz Flat Twelve: Currie, and a moment later the door clicks open. It's heavy and gives a loud creak that reverberates across the open ground as I pass through to a dingy foyer with a table piled with post. It all seems to be takeaway menus and flyers.

I go through the inner door, into darkness.

A glowing button at eye level must be the light

switch. I press it and a wall lamp stammers into life.

I am standing at the bottom of the stairwell.

Though the light is too weak to reach past the first landing I can feel the weight of the air above my head. It presses on my ears, setting off a high-pitched whine of tinnitus.

I breathe deeply, half expecting to see a spiderweb crack in the polished concrete, some sign of the calamity that occurred here. But there is nothing. Not a single speck of blood on the banisters. The air smells of dust and the ghost of incense.

Leaving the case by the door I walk forward, the rubber soles of my shoes sending out whispering echoes. Now I'm in the centre of the well, staircases rising dimly up on either side of me. In the shadows beneath them there are other doors: presumably Flats One, Two and Three. I was being watched, I am sure, by someone in Flat One and I'm tempted to knock on the door. For what reason, I'm not really sure myself; the occupant is probably just nosy, or lonely, or looking out for a visitor.

My nerves are on edge. I'm standing where my brother fell — to his death. I can face that fact even if Jody can't.

The room turns scarlet.

If it weren't for my black shadow looming before me, I would think something had gone wrong with my vision. Then I remember the stained-glass window.

I turn.

Long time no see, Jesus.

The sun must have come out from behind a cloud because his crimson cloak is casting a strong red wash over everything.

And then the sun goes in again and I'm plunged into gloom. The light has gone off too. It must be on a timer.

I flinch as a door clicks open somewhere above, half expecting a body to come hurtling down through the darkness. Then a wall lamp on an upper floor comes on and the full height of the building is revealed.

I inhale sharply. It is so far. So far to fall. It must have taken long, long seconds.

Soft footsteps on the stairs. At first I think it's Jody come to meet me, but the figure that descends wears a headscarf. A young woman, dressed in a loose, black abaya. A Muslim, but her face is pale, so perhaps she is Eastern European.

I have been too quiet. When she steps out onto the ground floor she starts and I apologise. There is a flicker of a smile, then she dips her head and moves past me towards the foyer.

Jody is waiting for me on the fourth floor. She's still in her nightdress, covered with a pale pink hoodie. As I step out onto the landing, breathless from the climb, I have to grip the banister, dizzied by proximity to the long drop into darkness. Now that the only light is on this floor, it has become bottomless. The rail is about waist height. Could Abe have leaned on it to catch his breath then overbalanced?

'It was here that he fell?' I say.

'Yes. That's where he jumped.'

I walk across to the rail and close my fingers around the slim metal bar. As I lean into it I feel it give a little. Such a fragile barrier between life and death. But I'm leaning too far — my heels are coming off the floor — and I pull myself back.

'Nobody saw him?'

'No.'

'Where were you?'

'In his flat.' She points behind me to a door marked with a metal ten. It's as faceless as all the others I've passed on the way up, painted a dreary grey-blue, presumably designed not to excite the spirits of the fragile inhabitants.

'You came out and . . . ' I pause to take a breath, as if the air is thinner up here. 'And you saw him. Lying down there, on the concrete?'

'Yes,' she says, her voice echoing down the column of darkness. 'I saw him and I ran down to him. I held his hand until the ambulance came. He wasn't alone.'

'You said that you screamed. Didn't anyone hear and come out?'

'I don't know. Sorry. I was paying attention to Abe.'

I want her to tell me something different — that someone saw him climb over the banisters, or saw him overbalance — but of course she can't. She can only tell me what she knows.

'Do you want me to show you around the flat?' she says.

'I'm sure I can manage.'

The small sound of the key sliding into the lock echoes through the stairwell: the acoustics of the original church haven't been entirely dampened by the renovations. Unless they've bothered with the expense of soundproofing the flats, you must be able to hear every TV theme tune, sharp word, or intimate murmur of your neighbours. I'm glad I brought the earplugs from the flight bag.

The door swings open and I step into a darkened hallway. Pictures gleam at me from the shadows. My hand tightens on the handle of the case — someone is standing at the end of the corridor.

But flipping on the light I see it's just a navy blue parka dangling from a coat hook. Leaving the case propped against the wall, I go up to it and run my hand across the shiny fabric. With its fur hood and bright orange lining it reminds me of the coats the boys used to wear in our school playground. We were never allowed one — far too common. It looks warm, though, and I've underestimated how cold a British November can be.

At a noise behind me I spin around. Jody is silhouetted against the doorway. I hide my flash of anger. What now?

'S-sorry, shall I bring you a cup of tea?'

'I'm sure Abe has a kettle.'

'Yes, but he only drinks herbal.'

'I'll be fine,' I say, forcing a smile. 'Thanks. If I need one I'll know where to come.'

'OK, well . . . I'll leave you to it. I hope I've left it tidy enough for you. I've been sleeping here a few times to try and . . . you know . . . get closer to . . . '

‘OK,’ I say. ‘Thank you.’

When Jody softly shuts the door I feel guilty. This is her home far more than it can ever be mine. It was good of her even to relinquish the keys. Even so, I walk back and check it’s properly closed. I don’t want her sneaking up on me again, especially as I’m not sure how I’ll feel walking into Abe’s domain.

I go up to the inner door.

His whole flat is the size of my master bedroom. At one end is a poky galley kitchen with a worktop separating it from the living area. A shabby leather sofa is positioned in front of a small TV, a grey throw folded on its back. Next to the TV is an electric fire: one of those where pretend flames made of fabric ripple in the glass window when you turn it on. There’s a sticker on the side saying when it was last serviced, so it must come with the flat.

At the other end of the room a pine dining table sits by the window. It’s part of the stained-glass window I saw in the foyer, bisected by the inner wall of the flat and truncated by the floor. It looks out over the bumpy path and the street where the taxi dropped me.

Even though all the walls are white, the coloured glass gives the room an air of oppression. Through the blue-cloaked upper torso of some bearded apostle I can see the children’s playground below. A Staffordshire bull terrier is shitting on the rubber matting surrounding the roundabout, no master in sight. I rap on the window but it pays no attention.

On the back of one of the dining chairs hangs

a dark grey pea coat. I check the pockets for a spare set of keys so I can give Jody's back, but there aren't any. However, I do find a wallet, Abe's phone — dead — and some loose change. In the wallet is a measly collection of cards, two five-pound notes, and a receipt for a new pair of Gap jeans. His Oyster is in there too, and slipped into the other pocket of the card wallet is a photo ID card for his job.

I stare at the photograph for several minutes, my brain adjusting to this new, decade-older little brother. His eyes are still puppy-dog large, and dark-brown like mine. They peer out almost seductively from his floppy fringe. I see for the first time that there's something attractive about his slenderness. I used to think it was wimpy, but now I recognise something Bowie-esque about his elegant neck, the sharp shoulders under his leather jacket, the high cheekbones. His lips are shapely and full. I wonder if the stubble is an attempt to add a little masculinity.

I slip the card wallet into my pocket. The taxi driver said it would be hard to get cabs around here and I'd be better off with buses.

A door leads off to a bathroom, musty-smelling, with mildew-spotted grouting and a mirror eaten away at the edges by damp. Male toiletries are lined up neatly on the glass shelf, alongside a single tub of woman's moisturiser. There's no sign of any depression medications.

The only other door leads off from the kitchen area.

I stand on the threshold surveying a tiny bedroom, double bed crammed up against the window on one side, and just enough room to open the cupboards on the other. A patch of damp blooms like an overblown rose in the corner of the room and a few specks of plaster lie like dandruff on the dark blue carpet.

I experience a twist of guilt. The place is a dump. I could have given him the deposit for a flat of his own, or money to rent something better than this. The bed is made, but I'll have to change the sheets — if only to rinse out Jody's tears from the pillow.

There's a full-length mirror in the corner and more toiletries on the shelf beside it. I smile. So, my brother was vain.

Picking my way down the side of the bed I open the cupboard. Slim, tailored trousers and dark cotton shirts, all facing the same way, all perfectly suited to his long, lean frame. A hanger full of tasteful ties in navies and purples, a couple of cashmere-mix cardigans and, in a rack below, brogues, Chelsea boots and a pair of grubby white Converses that are the twin of the pair I'm wearing.

There's just room for a bedside table. On it is a photograph of him and Jody, and a drinking tumbler of fresh flowers: a single stalk with a dense head of white blossom, so I guess handpicked rather than shop-bought. It must be Jody's doing. How far would she have had to travel from this grotty neighbourhood to find wildflowers? I'm touched again by her devotion.

There's also a phone charger and a pulp thriller I'd been intending to read myself. The corner of a page near the end is bent all the way across to the gutter. I do that too.

I'm struck by this and other parallels between us: the taste in books, in clothes, in the colours and textures of our homes. The only mystery is Jody. I just can't see what he saw in such a wet blanket. Was she always like this or has she been broken by grief? I suppose it's my problem to find out. I must try harder with her if I want to get to know Abe. I find that I want to feel . . . yes, I want to feel closer to him. Before the end.

I open the drawer of the bedside table, and step back. Then I laugh.

Way to go, Jody, you dark horse.

Handcuffs. And now I can see the rub marks on the rungs of the bedhead.

I really *must* wash the sheets. But first, a herbal tea. Three green teabags might give me a half-hearted kick of caffeine.

At a glance the gleaming white kitchenette had looked promising, but walking into it I see that the counters are laminate and the doors MDF, their edges swollen with damp. But he has done his best with cheap materials and everything is scrupulously clean. The cupboards contain jars of olives and sundried tomatoes; different kinds of oil in slim, elegant bottles; pesto and artichokes and a pack of bake-at-home French bread, still in date. Food I would have chosen myself.

There are the herbal teas Jody warned me about, but there's also a shelf full of hard liquor and — joy of joys — the same brand of knock-your-head-off coffee I drink at home.

In the corner of the worktop is a basic coffee

machine and ten minutes later I'm leaning against the countertop, eyes closed, inhaling the scent of home. It makes my eyes prick in a way that seeing his body lying in the hospital couldn't.

I sip my coffee and survey the tiny flat.

Three rooms.

Was he ever happy here? I couldn't be, but whatever the surface similarities, we are very different people. He was a carer. And he loved Jody. At least he *found* love. If it wasn't for her, and the obvious happiness they shared, I couldn't forgive myself for abandoning my baby brother. She's saved me from that guilt, and I should be on my knees in gratitude. So why aren't I? Because I'm a self-centred bitch, probably.

'I promise to try harder, Abe,' I murmur, knocking back the last bitter grinds.

Then I hear a loud buzz. There's an entry panel by the front door, and on the little screen I make out the distorted face of PC Derbyshire. I try the button marked with a speaker and tell her to come up, then press the one marked with a key. A moment later there's the distant creak of the main door opening, followed by the slam as it closes, and then footsteps on the stairs. I hear the scrape of shoes against concrete, the ting of metal against the banister rail — a wedding ring perhaps — the chime of a text message coming through.

How is it possible that *no one* heard Abe fall?

* * *

I sit down at the table by the window, with Derbyshire and her pasty underling.

'Did she set your mind at rest?' Derbyshire says, laying out various forms I have to sign before they can release the stuff. 'Miss Currie?'

I repeat what Jody said about them coming home half pissed. 'I suppose it might have affected his balance, and the rail is quite low. It could have been accidental rather than suicide.'

'There was no alcohol in his bloodstream,' the policewoman says, gathering the signed forms. 'Though he might have had one glass that was already metabolised before we tested him.'

I frown. Was one glass enough to throw his balance off so fatally? He's pretty slim, so I guess it's possible that he really couldn't handle his drink.

She hands the forms to the underling, who tucks them into his manbag. 'Did Miss Currie mention how they had been getting along? Any relationship problems?'

'No. They sounded very in love. Why do you ask?'

'There was bruising to Miss Currie's mouth on the night of the fall. She told us she slipped. On the blood.'

I stare at her. Then I say, 'If you're implying that Abe might have hit her, I don't think my brother was like that.' As I say it I realise I have no idea what he's like, but it's too late to backtrack because they're getting up to go.

They leave me with a clear carrier bag containing Abe's personal effects. I tip the clothes out onto the carpet, unfolding the parts that are stiff with blood, until I have something vaguely man-shaped.

So, this was my brother.

A pair of brogues, slim black jeans, a dark purple shirt and a cardigan — neat and elegant. Beneath the metallic scent of blood, I can smell aftershave. I dimly recognise the complex and spicy undertones, which means I must have slept with someone who wore it, which means it's probably expensive. I like it that Abe chose to spend the little spare money he had on luxuries.

Then I notice that one of the shirtsleeves is torn at the shoulder seam.

Could the fall have done that? Or did it happen when they stripped him in the hospital?

I pack the clothes back into the bag and stow them away in the wardrobe until I can figure out what to do with them. Then I wash my hands to get rid of the smell of stale blood and set about stripping the sheets, bundling them up with the towel from the bathroom. I stand there at a loss. There is no washing machine. José would probably have told me where the nearest launderette was if I'd given him the chance, but I can't face another conversation peppered with *I'm sorry could you repeat that*s, so I go out of the flat, cross the landing and knock on Jody's door.

It rattles loosely. There's something wrong with the lock. It looks like the wood has splintered behind it and it's just hanging on by one screw.

She opens the door a crack, leaving the chain

on, and I glimpse a gloomy hallway the twin of Abe's.

'Just wondering if there's a launderette around here?'

'Oh, sorry. I should have said. It's down in the basement. Wait here and I'll get you some powder.'

She closes the door, and though I am glad she didn't invite me in so I won't have to make small talk with her, it seems uncharacteristically rude.

After a moment my back starts to prickle with the consciousness of that black void behind me. I stood on the edge of the Grand Canyon once and after a minute or two had to step away. The man I was with teased me, but it wasn't vertigo. I just felt this powerful urge to jump. Did Abe feel like that as he trudged up this grim stairwell? Or was it just a moment of existential clarity. *Is this really it?*

Jody returns. She has taken off the pink hoodie and as she hands me the washing powder I notice, in the indentations beneath each of her clavicles, round bruises, the size of pound coins.

'Thanks,' I say.

'Have you got change for the machines?'

'I think so.'

She smiles wanly and closes the door.

I go back to the flat, plug in Abe's phone and collect the washing pile.

* * *

The laundry room is low-ceilinged and damp, and the walls are covered with the same red lino as the floor, so that I feel as if I am in a horror film as I sit uneasily on the bench, trying to read Abe's thriller while I wait for the wash cycle to end.

A loud beep makes me jump: someone else's load in the drier has finished. A scarlet shirt is pressed against the glass door. I think of Abe's shirt, torn at the shoulder.

I close the book and place it down on the bench beside me.

I think of Jody's cut lip and the finger bruises on her shoulder.

The policewoman asked me if their relationship was stormy. What if she and Abe *did* fight that night? They'd had a bit to drink, after all. I imagine she's the clingy and demanding type. Maybe she was angry about him spending so much more time with his clients than her. What if he told her he wanted to be alone that night? What if he tried to finish the relationship entirely?

What if she pushed him?

7

Mira

I stand in the shadows under the stairs, listening as the mad girl tells the sister that nobody saw him fall. That he was alone. That he jumped.

When they have gone away I hurry up the stairs and let myself into the flat, then lean against the door until the pounding of my heart has slowed down. I shouldn't run, the doctor has told me, because of my high blood pressure. I need to be calm, to take lots of rest. But how can I be calm? How can I rest when I know what I know?

Why is she lying for you, Loran?

I go into the living room and sit down on the sofa, slipping off my shoes and lying down with my feet above my heart as the doctor has shown me. I must calm down and I must rest. Or something might happen to the baby. And then I don't know what you would do.

8

Jody

The card is still lying open on the floor where I dropped it, like a crocodile's mouth waiting to snap at my ankles. I want to put it in the bin but I can't bring myself to touch it.

It's small, half the size of a normal card, as if it doesn't want to be seen, as if it can hide behind the others. If there were any others.

On the front is a pastel picture of a rose: more a sympathy card than a birthday card. Another year alive. Poor you.

To dearest Jody
Always thinking of you and wishing you the very best on your birthday. I'll drop your present around to the flat.
Helen x

It's *Helen* now, not *Mum*.

I can't see her. I can't. I can't bear to see the pity and revulsion in her eyes. It's guilt that brings her here, twice a year, on my birthday and at Christmas. The cards are only ever signed by her.

I know we made a commitment to you but that was on condition that . . .

If we'd known . . .

We're not capable of providing for your needs . . .

Jeanie and Tom were capable. I stayed with them for two years. Jeanie taught me to knit, Tom took me fishing. They said they would have adopted me if they hadn't been too old. I said I didn't care how old they were. But Tabby cared. She said they might not be there for me during the most important periods of life, like leaving home, getting my first job, having my first relationship. Turns out she was right. Tom had a heart attack a few months after I left and Jeanie's in a home with Alzheimer's now. You would have liked them, Abe. I can still smell the scent of Tom's fingers as he tucked me in: soil and cigars. I used to beg him to stop smoking but he said he was too old to change.

Nobody's too old to change. I changed. I was better, because of you. I didn't need to rely on pills with you in my life. I was so happy I threw them away.

But now I need them. Now . . . this card . . . I can't think straight. My blood is racing. My vision is all blotchy.

If I think about you, about us, it will make me feel better.

* * *

For days and days I kept wearing the rain dress, in case I saw you, but eventually it got too smelly under the arms. I realised then that all my clothes were dull and ugly. This is a good sign. In depression questionnaires they always ask if you have lost interest in your appearance, and I had. But now, because of you, I wanted to look pretty.

That Saturday morning I pulled on my jeans and anorak and went down to the charity shop. I had to wait for it to open, stamping my feet in the cold because the sun was still down behind the buildings.

Eventually the fat lady who runs it came waddling up and we went in together and she let me look around while she set everything up. She trusted me, even when she was in the back room, which was nice. I was tempted by a cocktail dress in iridescent navy taffeta. Perhaps if I wore it with a jumper and boots it wouldn't look too over the top.

While I was deliberating the old lady from the ground floor came in and she started oohing and aahing over it, so I put it back and went to the blouses section. I was lucky: there was a lacy white top in my size, from a shop I would never have been able to afford normally. I tried it on in the tiny cubicle. The curtain didn't fit properly and the old lady peered in and announced that it really suited me. I tried to smile but she was making me nervous. I wanted to bolt from the shop, but I didn't want to leave without the blouse, so I forced myself to take some deep breaths and calm down.

I could feel her eyes on my body as I slipped the blouse off, and I yanked the curtain across and held it while I got dressed one-handed. I thought about saying something about people deserving some privacy, and had wound myself up so much that my heart was hammering when I came out, the blouse balled in my fist. But she was at the back of the shop now, picking through a basket of cheap-looking beads. Under the heavy panstick make-up she was ancient and her gnarled hand gripped a walking stick, so I changed my mind.

'Ooh, lovely,' said the lady at the till. 'Wish I could get into something like that. You're so lovely and slim!'

I forced myself to say thank you even though my face was burning. Then I pointed to the label and added, 'I like that shop but it's so expensive.'

'We do get stuff from them sometimes — the people in the big houses by the station bring them in,' she said. Her name label read *Marion*. 'If I see anything in your size — an eight, is it? — I'll keep them for you.'

'Thank you,' I said. 'That would be really nice.'

After I left the shop, swinging the bag by my side, I felt sort of electrified. Avoiding contact with people is another sign of depression, and there I was making conversation with a complete stranger. See what you had done for me already?

The sun was shining and the flat was as warm as toast. I put the blouse on and some lipstick and tied up my hair, and then I got my new book, which I'd dried out on the windowsill, and went outside. I sat on the bench by the playground and waited.

I didn't know what you did for a living then, but I knew you didn't work on Saturdays. I had seen you coming back from the high road, with a newspaper and a bag from the local bakery. Once I had seen you eating a croissant from the bag, so I knew you weren't having breakfast with anyone, which meant you might be single.

The children came out and wanted me to time them as they tried to do the mini assault course, and we were all laughing our heads off because one of the big boys got stuck in the baby swings when a voice said, 'Aren't you cold?'

I was glad you had found me like this, laughing with the children, my cheeks pink from the wind. Out of the sun it was freezing and I wasn't sure how much longer I could have lasted.

You looked terrible, tired and hollow-eyed, with a rash around your mouth. I wanted to put my arms around you and look after you.

'Oh, I don't feel the cold,' I said, trying not to shiver.

'Lucky you! I was going to get a paper, but I don't think I can face it.'

'I'll get it for you,' I said. 'I was heading that way anyway. I need to get milk. The *Guardian*, is it?'

'Is it that obvious?'

'I can push it under the door.'

'No, no, no.' You were shivering under your thin jacket. 'Don't. Borrow some of my milk.'

It was so tempting to share the same carton you had used, but I wanted to do this thing for you.

I smiled. 'It would be a pleasure.'

'You're an angel,' you said, closing your eyes and letting your long dark eyelashes rest on your cheek. 'If you really don't mind, then I think I'll go back to bed.'

I watched you jog back to the door, hunched

against the cold, but you didn't turn back.

'Time *me* now!' one of the little boys cried, but I was already halfway to Gordon Terrace.

After I'd pushed the paper under your door, which I had to do in sections because there was so much of it, I waited for a while outside my flat, with my key poised, so that if you came out to thank me it would look as if I was just on my way back in. But you didn't come out. I imagined you curled up in your bed, all warm and musty-smelling from sleep. I imagined myself curled up behind you, my arm around your waist, my face in your hair.

Later on I found a folded piece of paper pushed under my door. It was a biro drawing of an angel with a little heart over its head.

For the rest of the weekend I couldn't stop smiling.

After that day I never took another pill. I never needed to. I woke up with a song in my heart, literally. All the songs I'd learned when I was little came flooding back to me. Songs about love and trust, perfect days and endless nights. Just cheesy pop songs really, but suddenly every word meant something.

I still feel the same Abe. Even now. Even in the hospital, watching you struggle to breathe, watching the machine pump air into your lungs.

It's a perfect day because I'm spending it with you.

9

Mags

Flat Eleven is silent. As silent as Flat Twelve, though I know that Jody's in there. What's she doing? Listening, like me?

I breathe as quietly as possible, though my lungs are burning from the long climb carrying the load of folded laundry, still hot from the drier. I thought about dropping it back at the flat, but changed my mind. If I speak to them with it tucked under my arm my questions will seem more casual. *I was just passing and thought I'd introduce myself* . . .

I haven't hit any of the light switches on the way up, and the floor and walls are crazy-paved with colour from the stained-glass window. The brass latches of the doors gleam red from Jesus's cloak. The stairwell is a yawning chasm behind me. Again I feel its pull: a dark pool I can dive into and lose myself forever.

Did Abe feel the same? Was he incapable of resisting? Or did something else happen? I try to imagine his slim fingers digging into Jody's collarbone. It seems so out of character from the boy I knew, who was so self-contained, so restrained. I never knew him express any emotion, barely ever saw him smile. But I suppose he was as fucked up as I was, pushing everything down to stop himself getting hurt.

Who knows what his personality was really like under there? Once that lid, so tightly wedged on, was allowed to come off, did he become a monster?

I think of the handwriting on the Christmas card and the security panel downstairs. Small and neat, but pleasantly looping; surely not the handwriting of a bully. But what do I know? I know nothing about him. Only that we have the same taste in shoes and coffee.

The light on the ground floor goes on and I hear one of the flat doors close. Then a rhythmic tapping begins, like bones clacking together. My brain throws up an image of a skeleton shuffling across the concrete and I can't stop myself looking over the banister. A woman with a stick is making her painful way across the foyer to the door.

And then it's as if she feels me looking. She stops, and her head tips back.

When our eyes meet she starts, and her stick clatters to the floor.

Her reaction so unnerves me that I shrink back from the banisters, breathing heavily.

Agonisingly long minutes later the tapping resumes, then the outer door closes and silence falls.

The light ticks off and I stand in the dim puddles of colour, my heart pounding.

The place is playing tricks with my mind, bringing back all those fight or flight impulses from my childhood.

I force myself to calm down, employing the techniques I had to use when I was first called to the bar. Jackson would laugh if he could see me now, sweating and trembling in the dark like a child after a nightmare. My clients would be panic-stricken, my opponents in court would rub their hands with glee: the iron bitch finally brought low.

Eventually my heartbeat is back to its normal rhythm and I tap on the door of Flat Eleven.

Minutes pass, and then I hear a soft rustling behind the door. Someone is there. I tap again and the rustling stops sharply.

'Hello?' I say, as quietly as possible. 'I'm your new neighbour. Just come to say hello.'

I glance at Jody's door with its black spyhole. Is she standing behind it also? A church full of whispering and listening, everyone watching everyone else for signs of sin.

The door opens.

It's the Muslim woman I saw earlier.

'Hi,' I say and stretch out my free hand. 'I'm Mags, Abe's sister. From number ten.'

We shake, stiffly. Her eyes are alert, darting around nervously, seeing if anyone's behind me. It makes the hair at the back of my neck bristle.

'I am sorry for what happened to your brother,' she says. 'He was a good man.'

I don't pick her up on the past tense.

'Could I come in?'

Her face blanks. 'I am sorry but . . . '

'Just for a moment,' and before she can stop me, I step over the threshold.

She holds up her hands. 'But . . . '

I press on and, as if fearful of my touch, she backs up against the wall.

The inner door is open and I walk briskly up the corridor and into the flat, where Jody is less likely to be able to hear us.

A vivid blue line slashes across the carpet. It's caused by the afternoon sun shining through the robe of the woman in the stained-glass window that rises up from the floor. Around her head is a yellow disc and it takes me a moment to realise that the smaller disc, at her shoulder, must belong to the Baby Jesus, separated from his mother by the ceiling of the flat below.

This flat is at the back of the church and directly below the window is a small, empty car park. The place has the damp, heavy smell of boiled vegetables.

A naked light bulb dangling from the centre of the ceiling goes on. The woman stands in the doorway, her eyes wide with alarm. I should not have done this to her. She's clearly new to this country and though she knows what I've done isn't right, she doesn't know how to go about making me leave.

I smile, partly to reassure her that I mean no harm, partly to make her think this is all perfectly normal.

The flat is spotlessly clean and tidy, but all that does is emphasise its bleakness. They don't seem to have added anything to the original cheap furniture provided by the housing association, aside from a paisley throw on the back of the burgundy velour sofa, and a couple of mountain scenes on the wall.

A pair of furry slippers and some cement-encrusted workboots are lined up neatly by the door. I wonder why she doesn't put the slippers on since the flat is so cold, but her feet are bare.

'Sorry, what was your name?' I say.

'Mira.'

'Mira. Hi.' I try and think of a compliment, but there is nothing positive to say about her home. Even the view is dreadful. Above the car park the sky is a flat November grey. The wind whines through a cracked windowpane.

She watches me warily, as if afraid I will make some sudden movement. I may as well get to the point.

'I was just wondering,' I begin slowly, 'if you'd heard anything the night my brother fell. Any raised voices? My brother shouting, perhaps?'

Her face closes up at once. I kick myself. If there was anything to reveal she won't tell me now.

'Sorry, but English not good. Do not understand.'

Like hell you don't.

I try turning on the waterworks, screwing up my face and looking away. 'It was such a shock. I just want to know what happened. For our parents' sakes.'

This usually works with Europeans, who worship at the altar of *The Family*.

'I very sorry but I not hear anything.'

So, your English isn't all that bad then.

'Didn't you go out of the flat to see what was going on?'

She hesitates.

'It's just that someone saw you,' I hazard. 'From downstairs. They thought you might have seen what happened.'

Her eyes flash with something. Could it be fear? Has she come from a country where you are always being watched, and if someone reports you, you might be taken away and never seen again?

'Maybe they were mistaken,' I add.

'There was no one else. Just your brother and the girl.'

'Jody.'

She nods.

'What made you go outside? Did you hear something?'

'The girl screaming.'

'Jody?'

Another nod.

'You didn't hear any shouting before that? Any sounds of an argument?'

As she shakes her head her eyes slide away from mine, but I cannot think of a reason why this woman would lie for Jody. Evidently she doesn't even know her name.

'What about your husband? Could he have seen something?'

'He is at the gym then pub. All night. He sees nothing.'

'What gym?'

'Stone's Boxing Club.'

She seems tired, and as she leans back against the wall her robe settles against her stomach. She's pregnant. No wonder she was scared when I pushed past her. I shouldn't have come.

'I'm sorry,' I say, 'to barge in like this.'

'It's OK. You are upset about your brother.'

'I think he was depressed.'

'Yes. It must be that. He seemed always very sad.'

'I'll leave you to rest.'

I pass back along the darkened hallway and pause at the front door. 'What are you hoping for?' I gesture to the bump.

'A boy, of course,' she says. 'Like everybody.'

'Not me,' I say. 'I'd want a girl. I'm a feminist.'

As she laughs her face alters, like the sun coming out. 'Not many of these in Albania.'

As Mira closes the door I walk to the banister and try to imagine what she would have seen as she emerged from her flat that night. Jody running down the stairs, screaming for help, my brother splayed out on the concrete far below, in a halo of blood.

An overweight woman is puffing up the stairs, carrying a glittery gift bag that sparkles in the light from the second-floor lamp. I head for Abe's flat and am just turning the latch when the woman puffs out onto the landing. Before I can go inside Jody's door opens.

'Hello, darling,' the woman says warmly. 'I've come to give you your birthday present!' As she waggles the bag the light goes off and we are plunged into the habitual red-stained gloom.

I pause at the door to see if Jody will introduce us before she invites the woman in. She does neither. In fact, she says nothing, and doesn't even reach to take the proffered bag.

'There's a cake that Tyra made. Chocolate and orange. She reckons it's her signature dish!' The woman gives a bubbly laugh, seemingly unfazed by Jody's taciturnity. I retreat into the flat and close the door, though I can still hear the conversation that follows perfectly clearly.

'How is everything?'

'Fine,' Jody says.

I wait to hear if she will elaborate, tell her friend about Abe's prognosis, but either she has already told the woman, or doesn't want to.

'You're eating OK?'

'Yes.' There's a spike of irritation in Jody's voice that I haven't heard before.

'Everything in the flat working?'

Presumably Jody nods.

'Can I come in?'

'It's a bit of a mess,' Jody says. 'Another time.'

There's a silence, then the woman says, 'You've lost weight. Are you taking your pills?'

'I don't need them any more.'

Another silence. I imagine the woman sighing unhappily. She seems very maternal, an aunt by marriage perhaps. What pills was Jody taking? I wonder.

'Is there anyone you can spend the evening with, so you're not on your own on your birthday?'

'I'm fine,' Jody says.

'Why don't you come round to ours for supper? Kieran's got a friend over, but the more the merrier. We'll all squeeze in somehow.'

'I'm fine. Honestly. I'll just have an early night.'

'Well, OK. You pamper yourself tonight then, sweetie. You deserve it. And call me if you need anything.'

'OK.'

'Promise?'

'Yes.' She sounds so sad. 'Thanks for the present.'

'It was a pleasure.' There's a whispering rustle, as if the woman has leaned to kiss Jody, and then the door closes and the woman's footsteps descend the stairs, faster on the way down than they were on the way up.

I head for the kitchen and unplug Abe's phone.

Even though I'd expected it, I'm disappointed to find a lock code. I try a few numbers — his birthday, Jody and his flat numbers — but eventually it locks me out completely. Because it's an iPhone it'll probably be encrypted, so I won't even be able to get the dodgy shop on the high road to have a look at it. Texts between him and Jody might have revealed if there were any problems in their relationship.

Tossing the phone in a drawer I make myself a coffee, then sit by the window and watch the children playing outside. The sun is starting to go down and I can feel the cold seeping through the glass. Gradually each child peels away, until one little boy is left alone on the roundabout. He looks Somalian and can be no more than four or five. The roundabout revolves slowly, casting his long shadow onto the scrubby grass. Then some youths climb over the low railing that keeps out the dogs, and go right up to him. Their hoods are pulled up and smoke coils around their heads from the joint they're passing back and forth. I'm too far away to see his expression. Is he scared? Should I go down there? Call the police? I knock back the last dregs of coffee. Perhaps if I just come out of the building it will make them leave him alone.

But I'm too late. One of the youths stretches his arm out. The boy takes it and is pulled to his feet. They move towards the railing, with him trapped between them. But he has left his jacket on the roundabout. Then suddenly he breaks away, running back the way he had come. The youth that pulled him up turns to go after him. It is a Somali girl. The boy retrieves his jacket from the roundabout, then runs back to the group. The girl pulls him up on her hip and he lets his head settle on her shoulder, then they pass out of my sight.

It was just his sister come to collect him. I'm letting my imagination run away with me. It must be stress. I go into the kitchen and look through Abe's booze collection for something to calm me down. As well as a four-pack of lager and some hard spirits — I'm not at that stage yet — there are a few bottles of red wine. I open an own-brand Beaujolais and pour myself a glass.

As I drink, the last rays of the sunset throw an amber wash over the grass and the terrace beyond.

It's Jody's birthday, and her fiancé lies in the hospital at the edge of death.

How could I have imagined she had pushed him over the banister? She's so slight even Abe could have fought her off. Abe, who never in our whole childhood laid a finger on — or even raised his voice — to anyone. Mira heard no sounds of an argument, just the screams of my brother's lover as she ran down to try and put the pieces of his head back together.

Damn.

I promised myself I'd make more of an effort.

* * *

Jody takes so long to come to the door I think she must have gone out or fallen asleep but finally, when I have started down the stairs, she opens it.

'Oh, hey,' I say. 'I was just popping out for groceries and I wondered if you were doing anything tonight.'

Her face is in shadow. I thought I could manage without hitting the light switch, but now I wish I had.

'It's just that I couldn't help hearing that it's your birthday.'

I wait for a response. Eventually she takes a great inhalation of breath, as if she hasn't breathed for hours. 'Yes,' she says. 'Twenty-five. What a granny, eh?'

'Would you like to come round to mine — well, to Abe's — for dinner? I'd love some company.'

'Are you sure?' It's too dark to see her expression.

'Of course. I'm a pretty shit cook but I might be able to rustle up pasta and pesto.'

'That would be lovely.'

'OK, great. What colour do you drink?'

'Sorry?'

'Wine. Red or white. Or pink?'

'Oh, I don't really drink much.'

So, I was wrong about the drunken row.

'Champagne, then. Tonight, we're going to forget our troubles and celebrate, OK?'

She gives a breathy laugh. 'OK, but it'll go right to my head.'

'Excellent. I love a cheap date.'

We arrange for her to come round at eight and I head down the stairs and out of the building. The wind has dropped, but the door still slams shut when I let it go. Experimentally I open it again and this time I hold it as it closes, slowing the momentum right down, until the latch is touching the edge of the door. But when I let go it shunts into place at once. Unless someone's fixed it since the accident there would never be any need to check this door was closed properly. And yet Abe managed to convince Jody that this was exactly what he was going down to do.

I sigh and zip my coat up. Not every tragedy is a crime. And maybe a sudden death always leaves unanswered questions. Abe told her he was going to check the door to get her out of the way. So she didn't think to question him about that. She's not a lawyer, and what the hell does it matter now anyway?

As soon as I step out of the shelter of the building I wish I had worn more layers. It very rarely hits freezing in Vegas, but here it must be a few degrees below. I walk quickly to warm myself up, stumbling over the bumps on the path. Beyond the path of light formed by St Jerome's security light meeting the flickering street lamps of Gordon Terrace, all is darkness.

The wind has dropped, so I hear the rustle quite clearly.

It could be a cat. Or a fox, or a rat.

I think of the gangs Derbyshire mentioned and quicken my steps. I've just reached the relative safety of the terrace that leads on to the high road when my phone rings. An English number, so it can't be the office. Could it be the police? I frown, let it ring, then on the last ring before it goes to voicemail I pick up.

'Mags?'

'Yes.'

'It's Daniel.'

I stop. A man is coming out of one of the nearby houses, so I feel safe enough to pause here. The traffic on the high road will be too noisy to hear him properly. 'How did you get my number?'

'The hotel. But don't blame them; I was pretty sly wheedling it out of them. I just . . . I wanted to say hi, and find out how your brother is.' He continues quickly, 'I figured you might not have a chance to call me, and I thought I'd probably get your voicemail actually and was just going to leave a message. Sorry if it's a bad time.'

This is brave of him, considering how I treated him at the hotel. It deserves some courtesy.

'It's fine. It's good to hear from you. I'm just on my way to the shops. I've moved into my brother's flat.'

'How's he doing?'

'Not great.' I pause as a souped-up turquoise Ford Escort with a spoiler starts up opposite me.

The pasty-faced guy at the wheel gives me a leering smirk, then pulls away at ridiculous speed only to come to a jarring halt at the junction.

'There's been a lot of damage to the brain stem. He's basically brain-dead.'

He inhales. 'Mags, I'm so sorry.'

'It's fine. Really. I mean, not for his girlfriend, she's really cut up about it, but I hardly knew him — as an adult — so I can't pretend to be, you know . . . '

'You'd be surprised. It might hit you later. I only started grieving properly for my dad in my twenties and he died when I was twelve.'

'I *would* be surprised. I'm sorry about your dad, but this is different. We were never close. I can't pretend to have feelings just because I'm supposed to.'

There is an awkward silence, which I'm about to break by saying goodbye, when he says, 'If you want to talk about it over a drink sometime, this is my number.'

The guy has serious balls, or else he's too stupid to know when he's on to a bad thing.

'I might do that.'

He laughs, knowing full well that I'm just feeding him a line, which makes me laugh too. 'No, really. I might. Before I head back home.'

'I'll be waiting by the phone.'

'You do that. See you real soon.'

He laughs at my comedy Vegas drawl, then hangs up. I'm still smiling as I go into the Food and Wine.

It's even darker by the time I make my way back and I find myself wishing I'd just brought my cash card, rather than the entire wallet bulging out of my jeans pocket. Several of the street lights on Gordon Terrace are out of order and the weeds in the front gardens are dense enough to conceal an adult male. Most of the windows are dark, and those that aren't radiate a cold glare from eco light bulbs hanging from the ceiling. No luxuries like mood-lighting here. It's almost a relief to reach the end of the street, but then the waste ground opens up before me, the path a bridge of light across a sea of darkness.

I take out my phone and shine the torch app into the gloom. Its pathetically weak beam just manages to pick up the weave of the metal fence, then a pair of yellow eyes flash from the darkness.

I start back, ready to bolt.

But it's just a cat. With impossible agility it claws its way up and over the fence and disappears.

Can it really have been sitting there all that time?

Without pausing to ponder this, I hurry down the path towards St Jerome's, now a jagged black shape cut out of the light-polluted sky.

There are lights in a couple of the windows, including Flat One, and as I approach the main door I'm sure I can make out a face behind the veil of netting. Despite the bitter wind racing around the building, I stop and stare.

A hand reaches forward and pulls the curtain aside.

She must be in her eighties or nineties. Her face is a ruin, folds of sagging, wrinkled flesh clogged with thick make-up, like a horror film clown. I wonder if this is a joke: kids trying to scare the newbie, but then the red slash of a mouth curves into a smile and the clawed fingers bend. She's waving at me.

I can't help it — I run the last few feet to the door.

Thoroughly on edge, I can't even wait for the door to shut by itself and pull it closed with a bang that echoes around the building. The stretch of grass outside the little wire-mesh window is so black I wouldn't see anyone's approach until their face loomed up against the glass. I hammer the light switch and my own scared face jumps into the window.

Forcing myself to turn away (*calm the fuck down, Mags*), I blindly go through the post on the table. The banality of this activity eventually tricks my brain into thinking all is well. My heart settles back to its usual rhythm. I stuff the few bits of direct mail for Abe into the carrier bag and pass through the inner door to the stairwell.

10

Jody

Why did I say yes?

I crouch by the door after she's gone downstairs, trying to pluck up the courage to run out and call after her that I've changed my mind, that I'm sick, or tired, but then the front door slams and it's too late.

Maybe it won't be so bad. The card upset me, but thinking about you made me feel better, and it might be nice to have some company on my birthday. I get so lonely without you, Abe.

I remember my last birthday, in the bedsit. The girl next door tried to get me to take drugs with her. It was kind of her, really, because she had to do all sorts of horrible things to get the money for her own habit. That's why I've never taken drugs, even though sometimes I want that oblivion so much. I know what it costs.

I don't drink much either. The smell of alcohol brings back horrible memories. Plus it makes you say things you wouldn't otherwise. It makes you give things away that you shouldn't.

It will be so good to be back in your flat again. I can stand your sister's sharp eyes and loaded words for a while, just to be close to your things again. Perhaps I can bring home another memento.

She might be nicer, now that the shock is

subsiding. I want her to like me, Abe, of course I do. She's part of you. I'll bring her something. A present. That's what people do when they go to someone's house for dinner, isn't it? Chocolates or wine. But I don't want to go out again now that it's getting dark.

When the front door gives its characteristic squeak and slam, I get to my feet and creep to the spyhole.

A minute later she comes trudging up the stairs with a clanking carrier bag. She looks pale and drawn. At the top of the stairs she pauses, gazing down the stairwell as the lights on each floor click off one by one. Then suddenly she looks right at me and I am pinned to the spot by your dark eyes.

I know she can't see me through the spyhole. I made Tabby test it out with me when we first got here, but even so I can't move or breathe until the fourth-floor light clicks off and the landing is plunged into darkness.

I hear her letting herself into your flat and the door closing, then she turns on the hall light.

It still lifts my heart to see that line of yellow beneath your door.

I go into my bedroom and put on the rain dress, tying my hair up the way you like it. I even put make-up on, and when I look in the mirror another face looks back at me. A face I can hide behind.

11

Mags

The pasta is bland and gritty. I have to add far too much of Abe's rock salt to make it palatable, and then I grate half a packet of parmesan over it and leave it covered with a plate to keep warm while I heat up the garlic bread. I haven't eaten like this since university. At home I mostly get takeouts, or else I'm dining out. Sushi, usually, or Thai. My tongue pricks at the thought of wasabi and Szechwan pepper. I wonder if Jackson's out with a client, or one of the other partners. His favourite restaurant is an Aussie fusion place on the Strip. Last time we went I had tuna tartare with yuzu dashi, and a bottle of saki. We laughed so much that the next day I felt like I'd done fifty sit-ups. The sudden longing I feel to be beside him takes me by surprise. Could I be a bit in love with him? Or is it just homesickness? I should be careful. He's made no secret of his attraction to me, but he's married, with two adult kids from his last marriage, and twin seven-year-old boys from this one. Plus I'm not in the least attracted to him. He's wiry and muscular from his daily sessions with the personal trainer, he works on his tan, and I'm pretty sure he's had a brow lift. I like my men more natural, a little less prissy about their clothes and weight and 'skin regime'. Like

Daniel, I suppose. Poor old Daniel. He just caught me at a bad time. I knock back my first glass of wine. Poor old me.

There's a Bose speaker on the kitchen window so I go in search of an MP3 player and eventually find one in a drawer. Some of the bands I don't even recognise — British ones, I assume, who haven't made it over the Atlantic. But there's one female singer who's as big back home as she is here. Her voice is low and smoky, and if you listen too closely to the lyrics when you've had too much to drink they'll make you cry. Abe's got all her albums. I programme them to run on a loop and plug the player into the speaker. The voice washes over me, warm and rich as melted chocolate, and I'm just tipsy enough to sway my hips.

According to the microwave it's 20:00 on the dot when Jody rings the buzzer. The echoes reverberate through the flat, and I imagine all my neighbours' flats.

She is wearing a grey tea dress sprinkled with clear plastic beads. Its ruffled sleeves are starting to fray where the overlocking has unravelled. Her hair is tied up, with curling fronds left to dangle by her ears, and her frosted pink lipstick suits her pallid colouring. For the first time I can see that she is pretty. My sour brain adds, *if you like that 'feminine' look*.

'You look lovely,' I say, feeling, absurdly, as if I'm on a date.

'I brought you this.' She hands me a tiny velvet pouch and stands in the hall, watching me expectantly. I tip it out onto my palm. It is a tiny silver fairy for a charm bracelet.

'Thanks,' I say.

'It's a guardian angel,' she says.

I smile. 'Sweet. Come on through.'

I put the charm in my pocket and forget about it immediately.

As well as a couple of wine boxes, I've splashed out on some Veuve Cliquot — for myself more than Jody. The diminutive Indian shop assistant had to get a set of stepladders to reach it down from on top of the soft drinks fridge, where it must have been gathering dust and grease for years. Jody ducks as I pop the cork and it pings off the metal lampshade over the table.

'Happy birthday.' I clink her glass. 'To absent friends.'

Surprisingly enough, it's drinkable. I close my eyes and think of the warm crush of gallery openings and awards ceremonies. Abe would have fitted in perfectly back home; he could have been PA to a producer or, if caring really was his vocation, some ancient celebrity. He might have been left a house down in Malibu. We could have stood by the ocean sharing a bottle of bubbly, toasting our astonishing survival, our success despite the odds.

But only one of us survived.

I moved the flowers from the bedside table to the dining table and now I thank her for them.

'Where on earth did you get them?'

'I . . . Abe did. He picked them from a client's garden.'

'Oh, right. They've survived well, haven't they,

since the accident? They look so fresh. Thanks for topping up the water.'

Another beat, then a smile. 'That's OK.'

I pour the remains of the champagne into my glass then go to the kitchen for a box of white. I like to have alcohol close to hand, like a security blanket.

'So, you were talking to Mira earlier,' she says when I come back to the table. Her face is open and guileless; does she know I was checking out her story?

'I just wanted to introduce myself.'

'When that baby arrives it's going to keep the whole place awake.' Jody smiles wistfully. Her baby dreams have been dashed along with the marriage ones.

'They're Albanian, aren't they? How come they got to live here?'

'I think they're Roma. Roma people get persecuted in some of those countries, don't they?'

'I thought Roma didn't like to stay in one place. Aren't there supposed to be special sites set aside for them?'

Jody shrugs. 'I don't really know. They're pretty quiet.'

'What's he like?'

A look of distaste twists her mouth. 'I stay out of his way and he's never around much anyway. He's a builder I think. Never speaks. I don't even know if either of them speak English.'

'She does,' I say. 'Quite well, actually. She understood words like *depressed* and *feminist*.' I don't add that I still can't work out what possible reason she would have to pretend not to speak English.

Jody's staring at the flowers. She doesn't seem interested in her next-door neighbours, or perhaps she's jealous that I'm meeting new people.

Feeling my gaze she gives an almost imperceptible start.

'Sorry. It's the flowers. It's funny, but I remembered them being yellow.'

I shrug. 'Maybe they change the longer they bloom or something.'

I have no idea what I'm talking about. The only living thing to grace my terrace in Vegas is a small and boring-looking cactus that was there when I moved in and has *never* been watered.

Hoping the conversation will become less stilted by a change of subject, I ask whether there's a park nearby for jogging. Any weight I put on here will be trebled when I get back to the land of size-zero gym bunnies and I must be sinking several thousand extra calories a day on booze. But I daren't go to bed sober, for fear of what my brain will throw up for me.

'There's a square at the end of the high road, but it can be dodgy after dark. I think men use it, gay men, for uh . . . '

'I'll be careful not to slip.' I smile.

In the time it takes Jody to finish her champagne, I've sunk two more glasses of white. When I try to refill her glass she puts her hand over the top, but I talk her into a wine and orange juice. It'll be good for her to get drunk. Already I can see her loosening up. Her cheeks are pink and there's a sheen of sweat on her top lip. I've turned the heating up to thirty degrees so that I can walk around in my vest and bare feet like I do at home.

'So, tell me how you met,' I say. 'You and Abe.'

She twists the ring on her finger and smiles coyly. 'Oh, you know, we were neighbours so we used to run into one another.'

'Come *on*,' I say. 'Give me the whole juice. Who made the first move?'

'Well, he actually saved my life.'

It's not as dramatic as it sounds. Apparently Abe came and put out a fire she'd managed to start while cooking sausages.

'Wow,' I say, when she's done, 'I never took Abe for an action hero.'

She doesn't reply to this and I fear I might have offended her by sounding dismissive of her big love scene.

'I left home very young,' I say quickly. 'We didn't really get the chance to get to know one another as adults. He must be different now.'

'That's so sad,' she murmurs. Then she says, 'I grew up in a care home, actually.'

I raise my eyebrows.

'My dad was in the forces. He was killed when I was seven, and then my mum killed herself a year later.'

I am riven with shame at my own self-pity. 'Jody . . . I'm so, so sorry.'

'It's OK. They loved me, and they loved each other.'

'Too much, maybe.'

She frowns again. 'You can't love someone too much.'

'Your mother owed it to you to carry on,' I say, 'after his death. It wasn't fair on you to do what she did.'

Jody shakes her head. 'She was sick.'

I wonder how much of that sickness Jody has inherited. She must be here because of the upbringing in care, but I wouldn't be surprised if there's some mental illness at play too. She seems so fragile, though given what has happened to the man she loves, I suppose it must take formidable strength of character even to get up in the morning. Unbidden, Daniel pops into my head. The way he looked when he said goodbye at the hotel, the expression of bewildered hurt. I wish I had been kinder. I find that I'm glad his number is now on my phone.

'What was it like?' I say. 'Growing up in care?' Looking back, Abe and I would have been better off, but no one would ever have believed us if we'd asked for help. We were on our own and we knew it.

'I was happy enough.'

She sips her wine with the delicacy of a bird dipping its beak into a flower. I reassess my assumptions about her again. To have survived the death of both parents and an upbringing in care, she must be pretty resilient. I wonder if it was his depression that drew her to my brother, the bird with a broken wing. Classic white-knight syndrome. She wanted to save him because she couldn't save her mother.

She lays her glass down. 'My aunt, Helen, was supposed to be my legal guardian. It was in my parents' will. After they died I moved in with her and her husband for a while, but then they changed their minds. Their son was a drug addict and she said it would be too difficult.'

'Shit.' The wine has slowed my thoughts and I can't think of anything else to say.

She smiles. 'But let's not talk about sad things. You wanted to know about Abe and me. Shall I tell you about our first night together?' Her face is flushed. Christ, she really is pissed.

'Well, only if you're . . . umm . . . '

'It was here. On the roof.'

Automatically I glance up at the ceiling. 'You can get up there?'

'There's a door at the end of the landing. Abe took us up to watch the sunset. It was . . . it was beautiful.'

'Enough,' I say, holding up my hand. 'I think you might be about to give me too much information.'

She giggles. 'No, I meant *the sunset* was beautiful. If he . . . ' She inhales. 'If he dies I'll go back up there and wait for him.'

I swallow my mouthful too fast and cough. 'You mean his *ghost?*'

She nods guilelessly. 'If you were going to come back, wouldn't you come back to the place you were happiest?'

'I have no idea. I don't know how that sort of thing works. I thought it was all about unfinished business or . . . whatever.'

There's an awkward silence as she gazes down into her wine glass, a half smile on her lips, presumably remembering things I don't want to imagine.

''Scuse me, need a pee.' Getting up from the table I stagger a little and rebound off the sofa on my way to the bathroom.

It's cooler in here and, resting my head against the back wall, I close my eyes to test whether the room's spinning. Not yet. I'm good for another bottle or so. There's an unpleasant crawling sensation at the back of my head as the condensation dribbling down the wall creeps into my hair, so I get up and flush. My neighbours must be getting to know my toilet habits by now, although I'm careful not to flush in the middle of the night, having been woken myself by the nocturnal squeal of ancient plumbing a few times.

As I'm washing my hands my eyes are drawn to the shelf of toiletries — there was something important I needed to ask Jody.

She's staring at the flowers when I emerge, as the smoky-voiced singer croons about lost love.

'Do you know the name of the antidepressants Abe was on? I haven't been able to find any.'

If he ran out that could have been the problem. If so, Jody must take some of the blame herself. She could have spotted the signs.

She shakes her head, blinking.

'Had he been on them long, because some SSRIs have been known to cause suicidal thoughts in young men when they first start taking them.'

Blink. Blink.

'And have you spoken to his doctor since it all happened? I mean, the guy should have been on top of this. Suicide is the biggest killer of young men. We might be able to make a case for medical negligence, and get some compensation. I know it doesn't make up for what happened but these bastards shouldn't be allowed to — oh!'

Jody has spilled her wine all over her dress.

I jump to my feet. 'Quick, take it off and I'll rinse it.'

She gets up too, gathering the sopping hem into a ball around her thighs. 'It's OK, I'll go home and do it.'

'Just get changed and then head back over.'

But she's already halfway across the room. 'No, no. It's getting late. I should go to bed.' She stumbles then, like she's walking across the deck of a listing ship, and falls against the bookcase by the inner door. I smile — it's a feeling I know well — but then my smile slips. As her hands jerked up to save herself, the dress rode up and I saw, just for a moment, the unmistakable white scars of a self-harmer.

I look away quickly as she rights herself and weaves off down the hall to the door. 'Thanks for dinner,' she calls back over her shoulder. 'It was lovely.'

'That's OK. Happy birthday!'

She closes the door behind her and her footsteps recede across the landing. Then her door closes, and there's just silence but for the lilting voice of the bereft lover.

Perhaps it's not such a big deal. Self-harming is pretty common among teenagers, especially, I imagine, those who have lost both parents and grown up in a care home. Ah, well, like she says, she's over it all now. Abe saved her.

I sigh and rub my face. See, this is why I don't

have relationships. Even if you're lucky enough to meet someone you genuinely care about, someone who feels the same and isn't a complete asshole, as soon as you let your guard down and start to rely on them, then bang! Out of the blue comes some shitty tumour or terrorist attack, or a fall down a stairwell. It's not worth it. You can't miss it if it was never there.

Then I remember I'd meant to ask her for his doctor's name. I've got friends in medico-legal. If someone's fucked up, they will pay. I'll give the money to a charity or something.

I write a note to remind myself in the morning, then set about getting comatose drunk. I know I'll have a shitty hangover, but it's not as if Abe's going to notice, and I stocked up with painkillers at the Food and Wine for just that contingency.

At midnight I decide to call it a day and stagger to the bathroom to clean the black stains off my teeth before bed. There are two faces in the mirror, the second fainter, revolving around the first. I try to hold the gaze of the ghostly pair of brown eyes looking back at me, but it keeps sliding away from me.

'I'm sorry,' I whisper, my voice close in the cramped room. 'I'm sorry that I left you there. I'm sorry that you ended up with another broken person. I'm sorry you're going to die.'

I stumble to bed, relieved I had the foresight to put the sheets back on before Jody arrived, and sink heavily onto the mattress. My limbs are leaden, my brain punch drunk, and sleep hurtles towards me like a freight train.

Friday 11 November

12

Mags

I wake up, completely alert.

Isn't it great the way alcohol kicks you unconscious at midnight only to kick you awake a couple of hours later? A glance at my watch reveals it's barely 3 a.m. Urgh. Turning onto my back I prepare for the tedious slog to dawn. The radio will help numb my brain, but as I reach for my phone to hit the Radio 4 app, my hand stops mid-air.

A noise, close by.

Next-door going to the toilet? Kids messing around outside?

I hear it again. A soft rustle.

Someone's in the flat.

Jody must have a second key. Jesus Christ, she can't keep away. Has she come to dry her tears with his towel or curl up on the sofa they fucked on *just to be close to him*? I'm seriously not in the mood.

I get out of bed and stride to the door. Fortunately it doesn't creak as I yank it open to give her a piece of my mind.

Because standing by the table, his broad shape silhouetted against the window, is a man.

My sharp intake of breath is like paper tearing in the silence.

But he doesn't turn.

I am suddenly aware how short my T-shirt is, how skimpy my knickers. My eyes flick to the hall door. Could I make it before he had the chance to cross the room? But then I would somehow have to get to safety inside one of the other flats, or else run downstairs and out into the night in just my underwear, which doesn't seem any less dangerous.

He turns and my heart jumps into my throat. But he hasn't seen me. He's looking down at something in his hands.

Abe's pea coat.

He must be going through the pockets.

I can hear his stertorous breathing from here — presumably from the stair climb, which means he can't be that fit. I am, or at least was, before I came here and hit the bottle. I might be in with a chance. Without taking my eyes off him, I feel my way along the edge of the kitchen countertop, racking my brain trying to remember which is the knife drawer.

He takes something out of Abe's pocket, his wallet. The guy is big. Not just big — muscular. His torso is a black triangle against the fluorescent sky: broad shoulders, narrow waist, thick neck.

It's not worth the risk. I've seen the youths that hang around this area. They've got nothing to lose. He has the wallet — maybe he'll be happy with that. I should just creep back to bed, call the police once he's gone.

But as I back away my hand brushes the champagne bottle I left on the counter, making it rock.

Clunk, clunk.

His head turns.

I drop to the floor, just in time — I hope — and crouch there on my haunches. The open hall door is blocking my view of the left side of the room, while the kitchen counter blocks the right-hand view. Is he creeping behind the sofa, about to leap out at me from behind the worktop? But now I hear his quiet footsteps moving around the table. If I try and make it to the hall he'll certainly see me. My only option is to hide in the bedroom.

The snick of my bare feet on the kitchen lino makes me wince as I retreat back the way I have come.

I spend fruitless seconds trying to squeeze myself under the bed as all the while the soft footsteps come closer and closer. Then they pause. Has he gone into the kitchen? I have a moment. Should I hide in the wardrobe? No, it's too full. There's nothing for it but to slither under the covers, lie as flat as possible and hope he's already got what he came for. Unless he actually wants to rape or kill someone, there's no reason for him to come in here.

I lie in the darkness, panting.

Where is he?

Standing in the doorway, looking down at my curled body?

My head is at the bottom of the bed and if I lift the duvet slightly I will be able to see the doorway.

I raise it. Through this tiny sliver I can make out the dark oblong of the doorway.

All my senses are on high alert. I can hear the hum of the fridge, a pigeon cooing up on the roof: smell the musty damp blooming on the wall above me.

Little lights are exploding in my strained vision. My heart's pumping so hard my whole body trembles.

Maybe he'll think the noise was just pigeons. Maybe he's already gone.

It's too stifling to breathe and my left leg is buzzing with pins and needles. Can I come out yet? I pull the cover back a little to let in some air.

And then the darkness solidifies and he's there, in the doorway.

Dropping the duvet I press deeper into the mattress, trying to make myself completely flat. But it's no good: he knows I'm here. I can tell by his ragged breaths — not exertion as I had first thought, but arousal.

He's going to rape me.

Should have gone for the knife. *Should have gone for the fucking knife*.

Will he kill me afterwards?

As my mind contracts to a single point of terror, I realise that I will do *anything* to stay alive. I will let him do whatever he wants to me. I will beg. I want more life. I'm only half done.

He's coming into the room.

I hold my breath, wide-eyed in the blackness, waiting for the duvet to be flung back.

I hear him walk up the side of the bed then stop.

My heart jumps with hope. Maybe he'll just

check the drawers and wardrobe and go. I risk lifting the side of the duvet and see a sliver of pale hand. He's not wearing gloves. Not a professional then. An amateur. In law school they teach you that amateurs are the most dangerous. They're scared and fear makes them irrational, prone to violence.

Should I scream?

But before I can suck in enough air he sits down on the bed.

I freeze, my heart throwing itself against my ribcage.

Is he going to lie down? Is he high and just looking for a place to sleep? Did he think this place was unoccupied?

His back touches my leg.

We both cry out and he springs to his feet.

Throwing back the duvet I scramble off the other side of the bed, snatching up a glass bottle of aftershave from the shelf. But before I can hurl it he shouts something incomprehensible and stumbles from the room. A moment later I hear footsteps retreating down the hallway, followed by the surprisingly quiet snick of the front door, as if, even after all that, he's trying not to be heard.

He's gone.

As I stand there, panting, the glass bottle still raised above my head, I think of everything I could have lost in that moment: my successful life in Vegas, a life of professional and material satisfaction, of clever friends, expensive clothes, fine dining. And suddenly it doesn't seem so much.

Could it be that Jody and Abe, in this shabby little apartment in this grubby city, had something more than me?

When I've recovered enough to move, I climb over the bed and wobble to the door. The room beyond is silent, the shadows still. I make my way past the kitchen, turning on the lights as I go, and check the bathroom — empty — then the hallway. Abe's pea coat hangs on the back of the chair. I check the pockets and frown. The wallet has been replaced. I check the front door: closed firmly. No signs of forced entry. Did he have a skeleton key?

Then it occurs to me.

The building manager. José Ribeira.

He offered to get keys cut for me so he must have a master set. I couldn't understand what the guy shouted as he jumped up from the bed — could it have been Portuguese?

I call the police but, even though I'm alone in the house and whoever it was clearly had a key, they won't come over until the morning. When I protest that I may well be in imminent danger they say that it's unlikely that the person will return, but if I'm worried perhaps I could spend the rest of the night with a friend or neighbour.

I'm tempted to tell them where to stick their advice, but it's not a good idea to get on the wrong side of the British police, so I thank them and hang up.

The sofa legs make a horrible screeching noise that must echo through the whole church as I drag it across the floor and force it through the hall door, grunting with effort. It's almost the same width as the corridor and once I've wedged it up against the wall at the end, the door won't open more than a centimetre. Afterwards I sit at the table, huddled under Abe's blanket drinking strong coffee, until the flat light of dawn filters through the stained glass.

But as the tumbler's translucent shadow stretches across the table, I notice something strange.

The white flowers are gone.

13

Jody

I can't sleep. Why is she asking all these questions? Is she trying to catch me out? Are the police in on it too? Will she tell them everything I said? They always wanted to punish me for what happened before.

No no no no no no no no.

Nothing happened before.

Don't think about it.

Think about us.

Our first kiss.

Do you remember?

* * *

I've never been much of a cook. At Abbott's Manor all the meals were prepared for you, and in the bedsit I didn't even have a microwave. Besides ready meals, the only things I can really do are Spanish omelettes, which I learned from Jeanie, and lemon drizzle cake, which Helen taught me. But that evening, because my appetite had been so good, I decided to cook myself a treat — sausages and mash. I bought the mash ready-made in a plastic dish to microwave, but when it came to the sausages I decided to grill them. The pack said to prick them to let the fat out but I didn't want it going all over the grill pan, which is hard to clean as the bottom is all black and lumpy, so I put some foil down on the metal rungs, then closed the door and went to read my book for ten minutes, which is how long the packet said to leave them before turning them over.

I'd finished the new book and gone back to one I'd read two or three times before. It was one of those perfect ones you can't put down. The heroine filled with fears and insecurities. Not loads of sex. The path to love strewn with difficulty so that at the end, when they finally get together, you feel like your heart will burst out of your chest and fly up into the sky.

I was so engrossed it took me ages to realise something was wrong. The battery in my smoke alarm's dead, so when the landing one started going off I didn't realise it was anything to do with me. Then I saw the black smoke pouring out of the oven.

I know now what happened. Because the fat couldn't drain away into the bottom of the grill pan, it was too close to the heat and in the end it just caught fire.

The oven door was so hot I burned my hand trying to open it.

You must have heard my scream because a minute later you were banging on my door, asking if I was all right. I was in so much pain I could barely speak. When I didn't reply you started throwing yourself at the door, trying to break it down. That's why the lock's still broken now. I managed to tell you to stop and hobbled over to let you in, still clutching my hand, which had started to blister.

You burst in, wild-eyed. 'What's happened?'

You must have seen the flames licking from the oven, but you seemed more concerned with my hand. Pulling me to the sink, you turned on the cold tap and held my palm under the flow.

'Hold it there,' you said sternly.

Now you went to the oven. The flames were licking almost to the ceiling and in a minute the polystyrene tiles would start to melt. Using the oven glove you managed to pull the grill pan out onto the open door, but the fire didn't go out.

'Tea towel?' you shouted.

'I'll get one.'

'No! Keep your hand under the tap.' Then you started taking off your shirt. 'Wet this for me!' You tossed it to me and I caught the manly scent of fresh perspiration and your flowery deodorant that I liked so much.

I did as I was told.

Grabbing the wet shirt back you threw it over the grill pan and the flames went out at once. You went to the window and opened it, and the air started to clear. Then the smoke alarm stopped, its last ring echoing down through the stairwell.

We looked at each other and then we both laughed.

'Well, that was exciting,' you said.

'I'm so sorry. You've ruined your shirt.'

'Shirts are two a penny. Hands aren't. Let me see.' You came over to the sink, where the water still gushed over my palm. It was the right thing to do. The redness was already going down.

‘I’ve got a burn spray in my flat.’

But you didn’t move.

I tried not to look down at your bare chest. There was a tiny speck of soot in one of the narrow shadows made by your ribcage. I couldn’t help myself. I licked my finger and wiped it away. I could feel your eyes on mine, your breathing slow and deep. My own heart was beating like a humming bird’s wings.

I tried to keep my voice steady, to talk about something mundane.

‘Do you think I should tell José?’

‘There’s been no damage, so only if you want to.’

‘I don’t want to,’ I said.

And then I kissed you.

14

Mags

First thing in the morning I call Peter Selby's office and arrange to have the locks changed.

It's gone half past ten when a bored and very young-looking PC arrives to take my statement. She doesn't dust for fingerprints or footprints and doesn't even bother to write down the fact that the flowers are missing.

'So, that's it, then?' I say as she gets up to leave. 'You're not actually going to do anything, are you?'

'As nothing was actually taken . . . '

'Except the flowers.'

' . . . you weren't assaulted, and you've arranged to have the locks changed, we can hope this was an isolated incident. Make sure your door is secure at night, and call us if you experience any more trouble. We'll let you know if we get any leads.'

'Course you will.'

She's too thick or uninterested to pick up on the sarcasm.

Half an hour later there is a knock at my door. A Brazilian guy with neck tattoos introduces himself as José Ribeira. He's wiry and muscular around the shoulders but, I think, shorter than the intruder, plus he smells very pungently of aftershave, which the intruder didn't.

I talk him through what happened and he purses his shapely lips and says I must have been very frightened. When I say not particularly, he gives a gold-incisored grin and says he likes a girl with cahoonas. Despite the weather he is wearing a baggy vest, which shows off his sleek black armpit hair and ripped lats. He reminds me of the pimps that patrol the Strip back home, which makes me like him.

He waits with me until the locksmiths arrive, drinking black coffee and telling me about his cousin who lives in Sacramento, then he gives me his number and makes me promise to call if I ever need anything.

The locksmiths leave me two sets of keys. Apparently this is standard for every flat in the block. It occurs to me that, since I've got Jody's set, and I haven't found any others in the flat, someone else has a set of my brother's keys. I suppose it doesn't matter now that the locks have been changed.

I wonder if Jody's heard the men working. If so, she should have got them to fix her door too. Maybe I should have asked her. I feel guilty enough about it that I don't fancy popping over to ask for Abe's doctor's contact details and so set about calling round the local surgeries.

Abe's not registered with any of them.

A search of the flat unearths a cardboard box full of papers: bank statements and gas bills and salary slips. From these I get the name of the company he worked for — Sunnydale — and a number.

The call handler laughs in my face when I ask

if the firm runs a healthcare scheme. I ask to speak to Abe's boss and am put through to a very guarded woman who expresses dry condolences as if she's asking the time of the next bus.

'Was Abe happy in his work?' I say.

'I believe so.'

'Because his fiancée has said that he felt overwhelmed by the workload, to the extent that it brought on depression. Did he mention to you he was struggling?'

There is a long pause, and when she speaks again her tone is even more careful. 'Sunnydale treats staff and patient well-being as its highest priority. Abe made no complaints to us, either verbally or in writing, that he felt stressed or overworked. The caring profession is a challenging one, of course, but whenever he had cause to speak to us, it was with a specific, unexpected problem — turning up to find a client had had a fall, and having to cancel his next client to wait for an ambulance. That sort of thing.'

I stop myself from saying that I thought UK caring company policy for this sort of contingency was to leave them exactly where they landed.

'I must tell you,' I say, 'that if we decide to pursue a court case for, let's say, corporate manslaughter . . .'

She inhales.

' . . . you will be required by law to reveal any correspondence you had with Abe. And there are, of course, ways of retrieving information that has been deleted.'

‘I . . . ’ she stammers. ‘We . . . Sunnydale . . . ’

‘Of course, if you can tell me categorically that Abe never gave you any cause to think he was struggling with his workload, then I’ll take you at your word.’

She knows full well that I won’t, of course, and I wait for her brain to whirr. Sure enough, she decides to tell the truth — at least, it sounds like the truth and I’m usually a pretty good judge.

‘I promise you,’ she says, more human now. ‘I’m his line manager and he never ever said to me that he was struggling. We would laugh about it sometimes, how crazy it was — I did his job until a couple of years ago — but I really think he liked it. Yes, I really do. Some of his clients were friends.’

I thank her and hang up, then go back to the box in the cupboard and sift through the papers. They don’t seem to be in any particular order. His mind was as chaotic as mine when it comes to non-work stuff.

Eventually I find a letter from a ‘John Hatfield Clinic’ in Camden.

Dear Mr Mackenzie
Your appointment is now booked as per the details below.

Please arrive ten minutes before the scheduled time. The test will take fifteen to twenty minutes. You do not need to fast beforehand.

The appointment date was three weeks ago. I sift through the remaining papers looking for the results, but there's nothing.

The letter itself gives little away, just an address, a date, and a cc'd signature from a Dr Indoe.

So, Abe was having tests for something. Was it something serious?

Suddenly I think of the graveyard in our hometown. Full of Mackenzies who'd died young, in their fifties and sixties, some younger. The usual Scottish maladies of heart disease and cancer. At least two aunts died from breast cancer when I was in my teens, and an uncle from bowel cancer. Consequently I've always been paranoid, paying through the nose every year for the west coast's best oncologist to check me over, then ignoring all his advice about cutting down on the booze and upping my fibre. I've been lucky so far, but what if Abe wasn't? What if that was what the Christmas card was all about? Say he got cancer and has been keeping it from Jody to protect her.

Another wave of guilt crashes over me. He had no one to talk to. Could trying to deal with it alone have been enough to push him over the edge?

But this is just guesswork.

I think about calling Jody, but surely she would have told me if she knew. And if not me, then surely the doctors at the hospital.

At a loss, I call the number on the letterhead. There are several options but none of them allows me to speak to a real human being. I won't be able to make an appointment without filling in wads of paperwork, so I end the call, put on my jacket and head out to the bus stop.

* * *

The clinic is in a characterless brick building a few minutes from the Tube station. Myriad signs on the way up the stairs tell me to sanitise my hands, to turn off my mobile, to call the Samaritans. I do the first two, then pass through the double doors into the fullest clinic I've ever seen.

The reception desk is manned by a harassed-looking West Indian woman whose neat bun is beginning to unravel, tight black curls pinging out at every angle. When it's my turn I say I'd like to see Dr Indoe. She tells me Dr Indoe's diary is full for the day and I should make an appointment online. I thank her, then while she's distracted with the next person in the queue, slip around the corner to the seats at the end of a corridor of treatment rooms.

When the doctor comes out to call someone through I'll nab her. It's only a quick question after all: what was Abe being tested for and what were the results?

A well-preserved man in his fifties shuffles up to let me sit down. I thank him but he doesn't give me a second glance.

Every twenty minutes or so a doctor emerges to summon someone. They're dressed in normal clothes, with just an ID badge to distinguish them from the patients. After about forty minutes a short, curvy woman of around thirty emerges from a consulting room. She's subtly made-up, her suit trousers and dark green blouse unobtrusively stylish. I wait until I can read her ID badge, then stand to block her path before she can call her patient.

For a moment her eyes are fixed on her paperwork and I wait for her to look up. When she does she gives a little start and tries to get past me. 'Excuse me.'

I block her path. 'My name is Mags Mackenzie. My brother was a patient of yours. Abraham.'

She frowns. Then I remember the photocard in the Oyster wallet. I get it out and show her, and have time to see recognition flash across her face before it's replaced by a guarded blankness.

'He had some tests recently,' I say, slipping the card back into my pocket. 'I'd like to know what they were for.'

'That's confidential, I'm afraid. I would be in breach of our code of ethics. I advise you to speak to your brother — '

'My brother's in a coma.'

The hubbub of the reception area dies away as people realise what's happening. Dr Indoe's eyes flick to the desk. The West Indian woman looks back at her.

'He tried to kill himself, Dr Indoe.' I decide to go on the attack, to try and intimidate her into telling me. 'You had a duty of care towards him. Where was the counselling and the support? Where I come from that's called medical negligence.'

Her eyes never leave mine but she makes no attempt to answer me, and now I can hear footsteps echoing up the concrete stairs behind me. *Shit*. She's called security. I'm running out of time.

'Please,' I say urgently as the doors bang open. 'Please tell me what was wrong with him. Was it cancer?'

Her brow furrows for a moment, then she makes up her mind. 'Look around you, Miss Mackenzie. Does this look like a cancer ward?'

I turn. I was so busy watching out for her arrival I didn't pay any attention to my surroundings. Now I see the brightly painted walls are lined with posters. One says *Keep Calm and Carry CONdoms*. Another features a close-up of a woman's knickers printed with the slogan, *I've Got Gonorrhoea*. A third depicts two men kissing and the line *Time You Tested*. The well-preserved man glares up at me from a copy of *Heat*.

I am in a sexual health clinic.

Two burly men stride across the room and position themselves, one on each side of me. 'Time to leave, miss.'

I hold my ground, not taking my eyes from Dr Indoe's face. 'Please tell me. I'm his sister.'

They're about to drag me away when she places a hand on one of their thick forearms. Then she comes very close to me, until I can smell the medicinal freshness of her breath. 'Your brother asked for an HIV test,' she says softly. 'It was negative.'

I am escorted down the stairs and out of the building.

* * *

Back in the flat I sit at the table drinking coffee and working my way through a bar of supermarket chocolate I found in a cupboard.

Oh Abe, what have you been up to? Is it as simple and grubby as an affair? Or worse. Have you been visiting prostitutes? Mainlining drugs?

Fuck.

I thought he'd be OK. That he'd get out relatively unscathed, like me. If I'd known . . . If I'd known, then what? I'd have given up my shiny new life to come over here and drag him out of whatever shitty mire he'd got stuck into? No. No, I wouldn't. And I wouldn't have expected him to do that for me.

But as I sit there, something niggles me.

When I was as low as I ever got — in the second year of university — my digs were as squalid as a crack house. I had no motivation or energy to keep the place clean. Rubbish stacked up, food rotted in the fridge. The floor was littered with empty booze bottles.

Abe's flat is spotlessly clean and tidy (although that could be Jody's influence) and well stocked with healthy food and the odd luxury. Alcohol, too, but not the strong liquor of a depressive. Reasonable wines and the odd bottle of artisan beer.

And then there are his clothes. The ones in the wardrobe and those the police returned to me.

At my worst I didn't bother with my appearance at all. The clothes I had went unwashed and I certainly didn't buy any new ones. I'd never have thought to apply perfume. Abe was wearing aftershave when he died and, apart from the blood, his clothes looked clean, the combinations of colours and textures put together carefully.

And what about the medical evidence? Where are the sleeping pills, the SSRIs, the doctor's appointments?

I'm sure Jody means well: she's just trying to make sense of what seems like a senseless tragedy, but she never actually *saw* what happened with her own eyes.

I don't know, but it feels to me as if he wasn't depressed. Which leaves me with two options: either he fell by accident or someone pushed him. And if someone did come in after him and attack him when Jody had gone into the flat, then *surely* someone here must have heard something.

There's nothing for it. It's time to introduce myself to the occupants of St Jerome's.

* * *

The place is completely silent as I head out onto the landing. I glance at my watch. Midday. I suppose some of them must be at work. But then again, many of them won't be capable of holding down a steady job.

My footsteps are wincingly loud on the steps as I pass down the four flights to the ground floor. I have to steel myself to knock at the door of Flat One — whose occupant has been watching me from the moment I arrived. Standing in the shadows, waiting for an answer, I make out quiet sounds all around me, strands of pop music interwoven with the burble of radios, an odd rhythmic tapping, a flushing toilet, the hum of the washing machines beneath my feet.

When there's no reply after a few minutes I move on to Flat Two, which is opened eventually by a harassed-looking woman with an inch of white at the roots of her hair.

I introduce myself and ask if she saw anything the night of Abe's fall. She says that she was away that evening because her son was on a residential course. I see what has caused the black lines that run all the way down the walls when a disabled boy bumps his wheelchair through the door at the end, and scrapes his way down the too-narrow corridor towards his mother.

'Wait a minute, Dale!' she snaps. 'I'm talking.'

'It's all right. I'm sorry for disturbing you.'

She is closing the door when she sees I am making for Flat Three and calls after me. 'He was sectioned two months ago. I think they're going to relet the flat.'

I thank her and start up the stairs to the first floor.

The occupant of Flat Four takes a long time to answer and when he does it is with an explosion of indignation.

'I work nights! I cannot bear these constant disturbances!'

I can't imagine what night-job vacancies there are for powdered middle-aged queens in silk pyjamas, but when I ask whether he saw anything he practically spits at me.

'Of course I didn't! I was *trying* to get some *sleep*.'

'May I suggest earplugs?' I say, but he slams the door in my face.

There's no answer at Flat Five, though I think that's where the tapping noise is coming from. I move on to Flat Six and, at my knock, am surprised to hear the yap of a dog. A moment later the door is opened by an old man in a shirt and tie. A Yorkshire terrier the size of a kitten scampers around our feet and I lean over to pat its bouncing head.

'Well, hello! And who might you be?'

When I introduce myself as Abe's sister he takes my hand and squeezes it in his warm, dry palms.

'He seemed like a lovely boy. Always had a smile for people. Kept himself nice and smart, not like most of the slobs you get around here.'

I ask if he saw or heard anything.

'I'm afraid not. I was at the care home and by the time I got back the paramedics were here, and that poor girl, crying her heart out.'

'Jody?'

'Sweet thing, isn't she? Brings my post up for me and takes Tessy out when I'm visiting Brenda. That's my wife. Fit as a flea physically, but her poor mind . . . ' His voice trembles.

'Ah, well, thanks anyway.'

'You won't get much out of him next door.' He nods towards the door of Flat Five. 'He just plays his organ all day with his headphones on. The queer chap complained when he played out loud, which is a shame really. I liked it. Can you hear his foot tapping?'

He clearly wants to talk, but I thank him and pass on up to the third floor.

The door of Flat Seven is ajar. This is where the pop music was coming from.

'Hello?' I call down the corridor.

'Yeah?' a voice screeches from the depths of the flat.

'Oh hi,' I call. 'I live upstairs. Can I speak to you?'

'Come in then.' The voice is shrill and irritable.

I step across the threshold and recoil from the smell of rubbish and unflushed toilets.

A woman is sitting on the sofa, smoking. She is skeletally thin and her skin is almost as yellow as her hair. She could be anywhere from twenty-five to sixty, and she makes no attempt to hide the track marks on her arms.

I am immediately on my guard. Are we alone in here? If not, will someone attack or rob me? The man who broke in last night? Could he be this woman's pimp? As she bends to stub the cigarette out onto a pockmarked coffee table I dart a glance around the flat. It's filthy but seemingly deserted. On the wall is a single picture: of a smiling baby in a blue Babygro. Embossed into the mount is the name Tyler-James and two dates, separated by an achingly short number of months. No wonder she's a junkie.

A terrible thought occurs to me. Beneath the hollow eyes and sunken cheeks is a ghost of the attractive young woman she must once have been.

Could my brother have been screwing her?

Was that the reason for his HIV test?

'What did you wanna talk about? Not the radio again? If I turn it any lower I won't hear it, will I?'

'I've just moved in upstairs. Did you know my brother, Abe? The man who fell down the stairwell.'

She lights another cigarette and sucks it hard, making the end flare. Her fingertips are brown and cracked. 'Yeah, I knew him.'

As the smoke streams out through her nostrils I try and read some meaning into her words. I know not to push too hard. People like her will clam up unless they think they'll get something by co-operating.

The silence stretches.

'I don't suppose you saw anything,' I say as lightly as possible, 'the night he fell?'

She picks at a scab on her bony knee. Her greasy blonde hair has an inch of black-and-grey roots. 'Weren't even here. Couldn't even get into the building. They thought I was gonna rob stuff. Almost got arrested until that old bag from downstairs come out and spoke up for me.'

'So you didn't hear anything either?'

'What did I just say?' Her head snaps up, then her eyes narrow slyly. 'Is there a reward or something?'

A pause.

'If there was?'

She looks at me for a moment, then she gives a cackle of laughter. 'I still weren't here!'

I get up, resisting the urge to brush the filth

from her sofa off my trousers. ‘OK. Well, thanks anyway.’

My feet drag on the trudge up to the third floor. I’d prepared myself for disappointment but it’s getting me down anyway.

Flat Eight is answered by a swarthy man with a missing arm. Iraqi, perhaps, or Syrian. His eyes are bright and he seems desperate to understand me during our attempts at conversation but in the end I resort to mime.

I roll my arms for the tumble from the floor above, clap my hand to indicate the impact. The sudden splaying of my fingers at the back of my head, to indicate the head injury, makes him flinch.

I place a palm on my heart. *My brother*.

This he understands. His eyes well with tears and he pulls me into an embrace that reeks of stale sweat and nicotine.

Just as quickly he pulls away again.

‘Sorry, I sorry . . . ’ Evidently he has learned that emotional expression is forbidden in the UK.

I point to him, to my eyes, to the stairwell. *Did you see?*

‘Blood,’ he says. ‘I see blood.’ That, at least, is a word he knows.

I thank him and move on.

The other two flats are empty so I go back upstairs.

Gazing out of the window at the bleak view I think about the woman in Flat Seven. Like all junkies she is bound to be a liar. *Was* she having a relationship with Abe? Did her pimp have a problem with it and decide he needed to give Abe a warning? A warning that went wrong.

I rub my face and let out a growl of frustration. I'm not getting anywhere. Perhaps there's nowhere to get. Perhaps it's all just as Jody said it was.

But there's one thing I'm sure she's wrong about. Abe wasn't depressed. Not properly, not like I was at uni. He might have been tired, or pissed off, or hating his job, but from what I know of my brother, Abe wasn't the type to do something stupid on impulse just because he was feeling down. Something else happened that night. Something that maybe Jody didn't know anything about. Or some*one*.

Either way it's something Derbyshire should be on to.

15

Mira

She is speaking to the police.

I have to put my ear to the bathroom wall to hear properly.

'I don't think Abe had depression,' she says. 'I don't think he jumped.'

There is a silence as she listens. We both listen.

'But he had no symptoms. You're only going by what Jody told you.'

She moves deeper into the flat and I hurry out to the living room and press my ear against the wall behind the television.

'I spoke to his line manager. She said he hadn't complained.'

She listens. I hope the police officer is reassuring her that it was an accident, nothing more.

Then she says, 'What if someone else is involved?'

My bladder loosens.

There is another pause and she says, 'What makes you think they'd say if they *had* seen anything? You can't trust any of them. They're all crazy.' She gives a mirthless laugh. 'You know what? I wouldn't be surprised if it was one of them.'

My legs go so weak I have to lean against the wall to stop me falling. The child starts to kick and I stroke his foot to calm him. *It's all right. It will be all right. I will not let her find out what Daddy did.*

16

Mags

When Derbyshire asks me how Abe is, I say, 'Much the same'. The truth is I have no idea. I haven't been back to see him, even though the hospital is barely a mile from St Jerome's. I haven't even bothered to phone to find out if there's been any change in his condition, like a normal sister would do.

Jody, I assume, is there all day, every day. She's never asked me to accompany her. I guess she still considers this intimate time with her fiancé. Or maybe she just doesn't like me. I can't say I'm surprised.

The wind is howling around the building so I put on Abe's parka before I head out.

At the door I pass a woman coming in. I assume she's one of the other tenants, but when I introduce myself she replies, in a strong Polish accent, that she's just a carer, for a Mr Griffin on the third floor.

'Ah,' I say. 'I wanted to speak to him, actually. I wondered if he had seen anything the night my brother fell down the stairwell. Did you hear about that?'

She nods.

'Is Mr Griffin with it enough, you know, in his head, to be able to tell me if he heard or saw something?'

'Is not his head is problem,' she sighs. 'Mr Griffin morbidly obese. Is in bed all time. Cannot get out. If he hears, and this not likely with television — very loud all day, all day — he cannot get up to see what is happen.'

I thank her with a little grimace, and she passes through the foyer into the darkness beyond, a pair of latex gloves waggling at me from her handbag.

There are no cabs out here so I wait at the bus stop, the strip of leg between my trouser cuffs and my Converses getting colder by the second because I haven't worn socks.

By the time the bus arrives there's a crowd of us trying to get on and with no attempt to follow the famous English queue etiquette I only just manage to squeeze in. I'm wedged by an enormous Asian woman in a sari. Her fleshy armpit as she clings to the strap above us is just centimetres from my cheek. It smells of stewing meat.

Eventually I get a seat and stare out of the window, at the filthy shop fronts and broken windows.

On impulse I take out my phone.

'Daniel,' I say, when the voicemail kicks in. 'It's Mags. I don't know if you're still in the UK, but if so I was wondering if you fancied a drink. Or lots of them. I could seriously do with getting shit-faced — and not on my own for once.'

I hang up, then spend the rest of the journey bitterly regretting the message. I sounded completely pathetic — and alcoholic. Hopefully he'll have more sense than to call me back.

* * *

Jody is at Abe's bedside, in the middle of a hushed conversation with the fat nurse who doesn't like me. On my approach they immediately stop talking and the nurse waddles off.

I pretend I haven't noticed as I join Jody by the bedside.

Abe's facial swelling has gone down and my brother's features are starting to emerge from the puffed, bruised flesh. At the moment he looks a little like our father, and I want to lift his eyelids to check the irises are brown, not that searing ice blue. I can't of course: I couldn't bear to touch that waxy flesh. Jody has no such qualms. She is stroking his flaccid cheek and murmuring a song into his ear.

I watch her from the corner of my eye. She plays the role of devoted martyr perfectly. Did she know Abe had been busy getting himself an STD behind her back? Or perhaps they enjoyed shooting up together. No, Jody doesn't look like a junkie and there are none of the telltale bruises on her arms. More likely he was having an affair. But surely he could do better than the woman in Flat Seven.

Though if not her, then who?

My phone buzzes. Daniel has replied. I've booked The Ivy for 8. Hope that's OK. Stodgy old British food but fun sleb spotting.

My heart jumps and like a love-struck fourteen-year-old I can't stop myself replying straightaway.

Gr8. See you there, I type, a serious frown on my face, as if it's a professional conversation.

For a seemingly interminable stretch of time we sit there, in a silence broken only by the machines, our breathing, and the rustling of my clothing as I cross and recross my legs.

Jody is staring at my brother's face with such intensity I wonder if she's attempting some kind of telepathic communication.

His eyelashes flutter.

Jody gasps and I admit it's an uncanny sight. I can accept that the lower brainstem being intact means that he can still make reflexive movements, but without any in-depth medical knowledge I can't help wondering why a message would even be sent when all conscious thought is gone.

A sudden and surprising bubble of hope swells in my chest. Perhaps they're wrong, perhaps there will be a miracle. Abe and I will get the chance to know each other again — this is the gift we always needed.

'Abe?' Jody leans across my brother's body, a look of rapture on her face. 'Can you hear me? Give me a sign, my darling. Squeeze my hand if you can.'

He gives no response. Of course he doesn't. He's brain-dead.

But that little futile burst of hope has shifted something in my mind. For the first time, seeing her in the full grip of this self-delusion doesn't inspire contempt in me, but pity.

She has been through so much, been let down by so many people, and now, it seems, even by my saintly brother. I can never tell her my suspicions; she must go on believing this fantasy of the perfect romance. Without thinking, I lay my hand on her arm.

She looks back and smiles at me. 'Did you see?'

I manage to smile back. 'Yes. I saw. I'll go get us a coffee.'

The fat nurse is sitting at her workstation outside the doors. I remember that whispered conversation. Has something happened that they're not telling me? Again that bubble of hope.

'What were you two talking about when I came in?' I say to her.

She opens her mouth and shuts it again.

'I'm his sister. I have a right to know.'

It's clearly distasteful for her to speak to me, and her lips purse so much she can barely squeeze the words through the sphincter of her mouth. 'Jody is very anxious that Abe's life support should carry on for the foreseeable future.'

The bubble deflates. Disappointment brings a rush of anger.

'You can override that, though, even if *I* wanted it too, right? You can go through the courts to get permission if the doctors thought treating him was pointless.'

She looks at me with eyes threaded with burst blood vessels. 'He's your brother, Miss Mackenzie.'

'What's that supposed to mean?'

'Jody loves Abe very much.'

'What, and I don't?'

'Yours is not the typical way of expressing love.'

My hands close into fists at my side. The nurse audibly exhales as I take the side exit out into the memorial garden and stamp up the steps to the pavement that runs alongside the main road. I stand there panting.

What the hell were you doing making me your next of kin, Abe? Didn't the fact that I left you to Dad's tender mercies tell you that I was a callous bitch?

I need to go home. I'll book the tickets as soon as I get back to my computer, and Jody can take over with the doctors. Poor, sweet, kind Jody, the saint to my villain, who will at least give Abe a fighting chance at recovery.

I hail a cab and am back at St Jerome's by midday. Letting myself into the main door I glance at the piles of post. There's another sheaf of leaflets for Abe, fastened with an elastic band. But peeping from the garish yellows and reds of the takeaway menus is a corner of white that might be a personal letter. I slide it out.

A blank white envelope. Probably another flyer.

I open it and unfold the single piece of paper.

For a moment I just stare at it. Then somewhere above me a door opens. I screw the paper into my pocket and pass through the inner door, ducking my head as I hurry up the stairs, not pausing until I am inside the flat with the door securely closed behind me.

Written on the paper are just three words.

She is lying

17

Jody

Our English teacher used to say that cavemen had invented stories to make sense of a brutal and chaotic world. She said that was where God came from. I don't believe that — I have faith — but I do believe stories are important to make sense of things we can't understand. Like dreams.

Dreams can come true.

That was another song from my childhood. I would lie on my bed in Abbot's Manor, eyes closed, dreaming of the future I would make for myself. The past wasn't the truth; it wasn't even a dream. It was nothing, gone, forgotten; only real if I let it be. Like the fairies in *Peter Pan* who die when you stop believing in them.

Today at the hospital I believed so hard that you were going to wake up — and then your eyes opened! Just for a moment, but it showed me that all I have to do is believe harder and I can make it happen. Like us. I believed we would be together and now we are.

You'll open your eyes again, and this time your beautiful brown irises will drift over to my face. It will take a moment for you to recognise me. At first you'll think I'm just part of the strange dream you have been dreaming for weeks. Then you'll remember. Or perhaps you won't. Perhaps you'll just understand in your heart that I love you, that you're safe, that I'll never leave you.

When I get back from the hospital I draw a picture of you, in the hospital bed, but I draw your eyes open, and when I get to your mouth I make your lips parted, as if they are just about to speak my name.

Jody? Is that you?

It's not very good, even though I'm copying it from a photograph, but it doesn't have to be. I know your face so well that I can picture it happening. I can hear the crack in your voice because it's been so long since you've spoken, the whisper of your hair against the pillow as you turn your head to look at me.

But just then, as if she's doing it deliberately to ruin my perfect moment, my phone rings. Though I deleted her number long ago, I still recognise it at once. I let it ring and ring until I can't stand it, and press my hands over my ears and scream silently. Then abruptly it stops, leaving echoes like the waves of pain after you stub your toe.

I stare at the phone for what feels like ages, but still jump when it lights up with a message. I can't delete it until I've listened to three seconds of it.

'Hello, Jody. I've got a birthday present for you, but it won't fit through the letterbox so I thought I'd drop — '

Message deleted.

Why won't she leave me alone? Why must she force herself back into my life every year, like tearing open the scar of a freshly healed wound?

I don't want to think about her, about any of them. I try to push them away by thinking about you, but I can feel the dark waters rising and rising, and then I can't stop them and they break over me.

18

Mags

What does it mean?

Is the *she* Jody? Or the junkie? Or could it be Mira?

And what are they supposed to be lying about? Abe's fall?

But then what possible reason would this writer have for sending a note to me rather than going to the police?

I pick up my phone to call PC Derbyshire, then change my mind.

I'm living in a church full of the damaged and mentally ill. Perhaps this is just malice, or delusion, or plain old racism. Someone with a grudge against one of these women. A jealous lover? Before I go to the police and stir up any trouble for them I need to at least make an attempt to get to the bottom of who sent it.

The handwriting is neat — possibly female? — but there's no full stop, which could suggest the writer is poorly educated.

There was no name on the front of the envelope, which suggests that the person who wrote it slipped it inside the rest of his post. This means either they managed to get into St Jerome's to do so, or they were already in. A resident. Letting myself quietly out of the flat I go back downstairs and out of the main door.

Huddled against the wind that races around the building I study the entry panel of names. None of the handwriting seems to match the note, but some of the labels are printed, and others might have been written by partners. It occurs to me that I've never seen Mira's husband, only heard their door shutting quietly late at night. She must be very lonely.

Lonely enough to seek comfort with my brother? Surely not right on Jody's doorstep — literally. Plus she's married and, judging by her dress, a devout Muslim. Get a grip, Mags.

As I trudge back upstairs it occurs to me that perhaps the note wasn't intended for me at all, but for Abe. Could there be others?

Back in the flat I forage through the chaotic box of papers but find nothing aside from bank statements, bills and receipts. Then I notice the laptop, tucked on top of a stack of shoe-boxes, its charger still plugged into the socket at the back of the cupboard. Green. Fully charged. And when I open the lid the screen springs into life.

A stroke of luck, but also more evidence that he didn't intend to commit suicide? Unless there's a note, or email.

His email account is open and I find an inbox stuffed with junk and mailings from theatres and listings magazines. There are a few messages from Sunnydale reminding him about time sheets and a drug recall, but not much else, certainly no suicide note and, weirdly, nothing at all from Jody. When I'm seeing someone I seem to spend half my time composing clever texts and emails. But it wouldn't surprise me if Jody didn't even have a computer. Abe's is a pretty battered old Dell model. Maybe she can't afford one. The phone would have been more revealing — I know she has one of those — but it's locked and I can't exactly ask to look over her sweet nothings.

Minimising the email I notice that his wallpaper is a picture of a Disney-style castle against a background of misty lilac hills. It seems oddly impersonal. Mine features me and Jackson knocking back tequila shots with a movie star friend of a client.

Then it hits me: the castle is called Eilean Donan. Abe and I went there on a school trip. The whole of the lower school went, seventy-five of us, in two coaches. My best friend was sick and so I agreed to let Abe sit next to me on the bus, mainly because my other girlfriends thought he was *adorable*. This meant that though he was in the year below they still fancied him, but had to pretend they only wanted to mother him.

We travelled north, skirting endless stretches of dark loch, studded by tiny islands that held a single tree or ruined bothy. Abe gazed out of the window. When we went under an avenue of trees his reflection swam up into the glass, dark eyes meeting mine, then looking away. He answered their questions but never offered any of his own thoughts.

We riffled through his packed lunch and took the bits we wanted, leaving him our cheese triangles and pieces of fruit, which he took without complaint.

It must have been autumn or spring because

the sun glared through the bus windows, reddening our faces, but when we finally got out in the car park we stamped our feet and swore and wished we hadn't been too vain to bring our coats. My girlfriends carried on complaining as we traipsed out of the car park and the weary teachers allowed them to go straight to the visitor centre.

But I didn't go with them.

I had never seen anything so beautiful in all my life as that castle shimmering into view out of the loch mist.

I followed the younger kids on the tour, through tartan rooms filled with antlers and thirteenth-century books, and then out to the battlements.

When everyone else had gone off to eat their packed lunches I was still standing there, looking out to the western sky, shot with red as the early sunset crept across the hills.

I became aware of Abe standing beside me. Just the two of us beneath that vast, uncaring sky.

We stood in silence until the sun started going down behind the hills and the cold seeped into my bones.

When I turned to leave I saw he was crying.

The tears had been caught by the setting sun and turned to little droplets of blood on his cheek. Something about the place, its beauty and stillness and silence, loosened my heart just a little bit and I put my arm around him. I had never done it before, and I never did it again, but for a few minutes before the sun disappeared behind the hills and the air grew numbingly cold, we stood together as brother and sister, realising, perhaps for the first time, that the world was beautiful.

How could I have forgotten?

Then we got back on the coach and went home. Thinking about it now, I wonder if that trip was the moment things changed, when we stopped trying to get the other one into trouble, started lying for one another, warning each other when our father was on the warpath.

Perhaps. It all seems so long ago.

When I try to blow it up the image pixelates into meaningless blocks of colour.

Then I spot a folder in the corner of the screen, called *People*. I click on it.

It's divided up into five subfolders and each filename is a surname: Bridger, Khan, Okeke, Perkins, Lyons. I click on the first and a word document comes up entitled *Freddie Bridger*. He's eighty-three, with diabetes and early-stage dementia. There's a photograph of a slack-jawed, bald man in a tank top, alongside an address and a list of dates Abe visited and the medicines he administered. I close it and glance through the other five folders. Aroon Khan: sixty-five, prostate cancer and Parkinson's. Kone Okeke: seventy-nine, diabetes and deafness. Molly Perkins: sixty-eight, arthritis and incontinence. Lula Lyons: ninety, rheumatism, heart disease and high blood pressure. Lula's photo must have been taken in the fifties. She has flame-red hair and scarlet lips and wide green eyes framed by impossibly long lashes. God knows what she must look like now.

Then I see her address.

Flat One, St Jerome's Church, N19.

The creepy old lady from downstairs. And according to the file he visited her twice a week, sometimes spending more than two hours with her. She must have known him very well. Well enough to be able to tell me more about his state of mind?

Then I notice, in the corner of the screen, another folder. This one with a more unusual surname: Redhorse.

I double click.

Instead of word documents, this folder contains a series of jpegs. I click one.

It is an image of two men. They are naked. The man standing is big and muscular, his skin whitish pink and glistening with perspiration, his pubic hair dark blond. On his hip is a tattoo of a red horse.

The man kneeling in front of him, elegant fingers splayed around the other's buttocks, mouth stretched wide to receive his cock, is my brother.

19

Jody

The panic attack goes on for an hour. I try to be as quiet as possible, but I can hear the church listening.

Afterwards my stomach aches with the pain of crying so hard, but also with hunger. I haven't eaten all day. I should go out and get something. Tabby always says it's important for me to eat properly because blood sugar affects your mood, but if I went out now I'd have to cross the grass in the dark.

If I still had your keys I could make myself a sandwich with some of the seeded bread you keep in the freezer. It's half finished so I know you've touched it. I think of what I would order if we were going to Cosmo tonight, of what you would have, the wine you would choose, the way we would clink glasses, looking over the rims at one another. How we would come home together, helping each other up the stairs, a little bit tired and giggly from the wine.

Looking out of the spyhole I can almost see us, your arm around me as you let us into your flat.

Then your door opens. For a moment — for the split second until I remember what has happened — I'm paralysed, my breath frozen in my throat as I expect you to walk out onto the landing.

And then you do.

My heart stutters to a halt.

But of course it's not you. It's your sister, in a manly black suit, her hair pulled back into a tight bun. It's a hard look, the sort that would put most men off. Her eyes are so smoky that from the shadows of my doorway they look like deep black holes in her face. Her lips are red like a vampire's.

Where is she going?

She glances over then and I'm pinned down by her dark eyes, the pupils catching a splinter of light from the main entrance far below. Then she starts descending the stairs, her heels clicking.

Where is she going? When will she be back?

She asks so many questions. They're like fingers picking at the edges of my life, trying to peel away the layers to get to the tender part beneath. I daren't go out until she comes back, in case I bump into her.

I wait by the door and eventually fall asleep on the hall chair. I'm woken by laughter and stumbling footsteps on the stairs, and then a white glare slashes under my door as the landing light goes on.

'Shhh,' Mags whispers, giggling, 'they can hear everything here.'

'Not this, they can't,' says a man's voice.

I creep to the spyhole. The man is tall and broad, his blond hair cropped tight. He's grinning as he grabs her around the waist and jerks her into him. My body goes rigid as I wait to see if she will be able to get away. But she doesn't try to. Instead, her arm slides down and disappears between their bodies. He gasps, then gives a breathless laugh.

A trill of repulsion passes through me as she leans into him and their mouths press together.

Wet snuffles and rustles echo through the stairwell. I wonder what time it is, that they can be so brazen, so unafraid that someone will come out of their flat or hear the noise and look through their spyhole. It's long after pub closing time so the man next door will be back soon.

He's kissing her so roughly that she stumbles back, coming up sharply against the banister: a hand slams down to steady herself, making the metal ring. But it doesn't make them stop. She raises her knee high up his thigh and slides her hand under his shirt, exposing a ridge of fat above his waistband.

One hand supporting herself, the other pulling his head down, she arches her back as he buries his face in her chest, like an animal at a feeding trough.

And then, in a flash, it is you bent over the banister like that, gripping the arms that held you, and I open my mouth to scream at him to stop, *STOP!*

But then it's her again.

Just when I think she's going to overbalance and fall backwards, she lunges forward to bite his neck and they lurch away from the banister. Her legs are around his waist now and he's supporting her whole weight. I wait for him to slam her against the wall and force himself into her, but he doesn't. They stay where they are in the middle of the landing, just kissing. I move closer to the door to listen for the words: words that have always sounded more like hate than love — *you want it, you whore, you dirty bitch, you slut* — but there aren't any. They kiss in silence.

My breath steams up the spyhole and I move away to clear it, squatting in the darkness, trying to keep my breath shallow and quiet. Cold air trickles under my door. The scent of perfume mingled with that smell all men have at the end of an evening — sweat and alcohol and harsh medicinal deodorant. It makes the muscles between my legs contract. When I return to the spyhole he has her up against your door and his trousers are around his thighs. His naked buttocks clench as he pushes himself into her. Her black ankle boots, so small around his broad hips, twitch at every thrust and her hands grip his back as if she is in pain.

And then the light flicks off and I can just hear rustling, animal grunts, and a soft knocking, as if someone is trying to get into your flat.

It goes on and on and then suddenly there is a sharp intake of breath, as if he has hurt her properly. The rustling rises in pitch, the knocking gets louder, the man pants like a dog. I should go out and help her, Abe. Like I should have helped you, but I'm scared. Not as scared as I was that night, but so scared that it takes me several seconds to pluck up the courage to reach for the door latch.

But as my fingers close around the cold brass,

the rustling and grunting abruptly ceases and there is a beat of absolute silence.

Then Mags speaks.

'Jesus fucking Christ,' she says, in a loud clear voice that echoes through the darkness.

The classroom is warm. She's trying to concentrate on what Miss Jarvis is saying, but she keeps falling asleep, her chin slipping off her hand to jerk her awake, producing titters from her classmates.

'So, somebody tell me one way we can use an apostrophe.'

Zoe Hill puts up her hand. 'When we shorten something, like 'that's' instead of 'that is'. The apostrophe goes before the 's'.'

She frowns, trying to understand, but her head is muzzy and the pain in her back is distracting. She shifts in her seat, making the wood creak, and Miss Jarvis's eyes flick towards the sound. When she realises where it's coming from the teacher gives her a ghost of a smile, then looks away. Sometimes, when the teachers do this — offer her some sign of friendship or sympathy — she feels like a zoo animal, a chimp behind a glass wall. The visitors who walk by feel so sorry for her being trapped in there; they make sad faces at her and shake their heads, then they move on.

She shifts again and the pain in her back makes her gasp.

Behind her, Emily Bright mimics the sound. She and Emily used to be friends, but then Emily's mum told her they weren't allowed to play together. Emily told everyone that she was dirty and her parents were criminals and that she and her brother did things with each other after bedtime. Emily got into serious trouble for that one, so she keeps quiet now, only tripping the little girl up when no teachers are looking, bumping her tray at lunch so her food falls on the floor, or scribbling over her best drawings, the ones she keeps in her desk to look at when she's feeling sad. Her favourite one was of an angel sitting beside a little girl in a field of pink flowers. She'd done the sky really well, shading in the whole area above the flowers, instead of just putting a blue line at the top like some of the others did. But Emily drew a big brown blob coming out of the angel's mouth that was obviously supposed to be poo, and she drew cuts and purple bruises all over the little girl. It makes her feel sad to look at it now, sad and scared, and when the teacher isn't watching she will throw it away.

For a while Emily had boasted that her family were moving to a bigger house in a nicer town and the little girl's heart had swelled with hope, but Emily had stopped talking about it now.

Even if Emily does leave, there is still Zoe and Melissa and Stevie Daniels. Stevie Daniels hit her because he said she was laughing when he was talking about his mum's operation. She wasn't. She was just smiling because she wanted to be friends. He got into a little bit of trouble, but not too much, because of the operation. The hit didn't hurt, but now he jumps out at her when she's walking past, slamming his foot down to make her jump and scowling as if he wants to murder her.

Sometimes she wishes he just would.

With a sense of panic she realises she needs the toilet. For the past few days it has been hurting so much to wee and she has noticed the yellow in the bowl has streaks of pink in it. Perhaps she is properly sick and will die.

'Now,' says Miss Jarvis. 'Adjectives. Who can tell me what an adjective is?'

'A describing word,' says Jamie.

'Correct. Now, I'm going to go around the class and we're going to come up with some words to describe a person. Jamie, you begin.'

'Strong,' says Jamie.

'Good. Imran?'

'Clever.'

The list went on: tall, hairy, blond, nice (disallowed), friendly, naughty, noisy, brave.

It came to the little girl's turn. 'Beautiful,' she said. 'Kind.'

'You're supposed to say one, der,' hisses Stevie beside her.

'Good,' says Miss Jarvis. 'Stevie?'

'Stupid.'

'Ugly,' says Emily, and she can feel her former friend's eyes boring into her back.

'Smelly,' Melissa says, and the whole class laughs.

Then Jason Hicks cries out, 'The police are here!' and everyone rushes to the window.

'I didn't know we were having a visit today,' Miss Jarvis says, joining them. 'Perhaps it's an Internet safety thing for year six.'

Both police officers are women. One is young and pretty, the other older and grey-haired;

neither is smiling at the faces pressed to the windows. They cross the playground with silent purposefulnes, and disappear through the door that leads to the headmistress's office.

'Has Mrs Harrison committed a crime, Miss Jarvis?' says Zoe.

'Is it because her car's too dirty to read the number plate?'

'Has she been murdering children?'

Laughter.

'Quiet!' Miss Jarvis snaps. Her face has gone white. She, like the children, is watching the police officers come back out of the door, accompanied by a grim-faced Mrs Harrison. The headteacher glances up in the direction of Miss Jarvis's classroom and for a moment a look passes between them.

'Back to your desks,' Miss Jarvis says quietly, and the children, subdued, do as they are told.

The little girl can barely put one foot in front of the other, she is so scared.

They know what she did last night.

And so many nights before that.

She knows it's illegal. Her parents told her. They said if anyone ever found out the things she had done she would go to prison, forever. She would be in prison with the same men she did the illegal things with, but her parents wouldn't be there to stop them if they tried to really hurt her. They could do anything they wanted to her.

And now the police are here. They will take her to prison and the men will be waiting for her.

Footsteps thud down the corridor. There are no voices. The children around her look at one another. Excitement has turned to trepidation. Is one of them in trouble? Has something bad happened to someone they love?

The footsteps draw closer.

She leaps up, making the desk legs screech, and flies for the door.

'Where are you going?' exclaims Miss Jarvis, but she is already running down the corridor, in the opposite direction from the three adults who have stopped in surprise.

She bursts through the emergency exit and the alarm starts up, an ear-splitting ringing that makes her teeth vibrate. There are shouts behind her, running footsteps.

She runs through the playground, dimly aware of faces pressed to windows and hands banging on glass. As she skids around the corner of the building the main gates come into view.

She is just tall enough, now, to hit the green button that releases them.

Someone is calling her name but she doesn't turn, just squeezes through the widening gap with a moan of pain as the metal bars cause pressure on her bladder and scrape the welts on her back.

The belt buckle was square and brass. When she saw him taking it off her heart had sunk, but it lifted again when he didn't undo his trousers. She hadn't understood because she had never met him before. She didn't know it was her pain that gave him his pleasure. He wanted to see her cry and beg him to stop. Her dad stopped it in the end, saying he would break her and then where would he get his fun next time? Heh heh.

She is out of the gates and running down the pavement in the direction of the park. If she can get there she will be able to hide in the bushes. Unless they send dogs out for her. Or heat-seeking helicopters like she's seen on the TV.

If she jumps into the lake perhaps they will not be able to see her, or smell her. But then she might drown.

Footsteps behind her.

She manages to speed up a little. She has the body of an athlete, the PE teacher always says. He can't understand why she is so slow, why she tires so easily. He says she is unfit, eating the wrong things, staying up too late.

She risks a glance behind and her bladder loosens with a vicious burning sensation. It is the policewomen who are pursuing her — and they are fast. The older one, surprisingly, is in the lead, her cap wedged on her head, her arms and legs pumping.

The little girl whimpers and tries to increase her pace as she runs past the row of houses. But she is so tired, and her back hurts, and her bladder hurts, and there is no strength left in her legs. She is not strong, or clever, or brave. She is just tired. She just wants it all to stop.

There is a lorry coming. She feels its rumble through her feet before she hears its roar behind her, drowning out the sounds of her pursuit. Suddenly she knows what she must do.

The lorry is at her back, so huge and loud,

shaking the trees and rattling the windows of the houses.

It won't hurt at all; it will be so quick.

She stops and turns. The robot face of the cab stares at her impassively, the driver no more than a shadow behind glass that reflects the trees of the park. They would have found her in the park. They would find her anywhere. Find her and take her back to the men, and it would be so much worse than before, though she cannot imagine how.

She bends her knees, ready to leap.

The lorry is so close she can smell its dirty, gusty breath.

It is a metre away, half a metre; she launches herself upwards, into the air, closing her eyes, waiting for the huge, shocking impact.

Her feet leave the edge of the pavement. The lorry screams, the whole world shakes.

But just in time she is being yanked back. There is an impact, a painful one, as she lands on her back on the pavement, but the lorry thunders past and she screams in an agony of despair.

The grey-haired policewoman is kneeling beside her, holding her down as she tries to scramble up and throw herself into the path of the next vehicle, or the next, or the next.

Eventually she gives up, collapses on her torn back on the pavement and sobs.

Running footsteps skid to a halt and the other policewoman is beside her, but they make no attempt to drag her to her feet or handcuff her. Instead they lift her gently and, like doting parents with a new baby, gather her into their arms, not caring that bloody, acid wee is soaking through her threadbare school skirt.

They rock her like that, until she starts to calm down, and can make sense of the words they are repeating over and over, their voices merging together, as if they are singing a round.

'You're all right, now. You're all right. Nothing bad is going to happen to you again. Not ever. I promise. I promise.'

The little girl raises her head and looks from face to face. She is surprised to see that both women are crying.

Saturday 12 November

20

Mags

Spending the night with Daniel makes me feel so much better.

Though in fact, as the morning progresses and I'm still floating on a bubble, I wonder if it isn't the sex — which was fairly clumsy and unrewarding due to the fact that we were both drunk — and rather simply waking up with a nice warm human being beside me. I didn't realise how much being alone in this dark flat, in this shitty neighbourhood, in this freezing city, was getting to me. Plus there's no denying I felt safer with a big strong man in the house.

To make up for the last morning we spent together, where I basically told him to fuck off, I leave him sleeping and head out to get breakfast from the local bakery.

I won't go to the hospital today in case I bump into Jody. I still haven't had a chance to gather my thoughts about the stuff I saw on Abe's computer last night. It wasn't just photographs. I opened his history and logged onto a chat room he seemed to have spent a lot of time on. There was a thread from this Redhorse. Predictable enough stuff initially: *Can't stop thinking about the taste of your cock, I want to feel you inside me*, that sort of thing. Porn talk. But the later ones are more intimate. From Abe: *I've been thinking about you all day*. From Redhorse: *Not long before I see you*. Abe: *I'm counting the seconds. x*.

My first thought was: did Jody find out he was screwing around on her? He'd set the computer to remember all his passwords so it wouldn't have been hard for her. But that just brings me back to the idea that she pushed him, and I just don't buy it. Say she confronted him about it and there was a struggle, she isn't physically strong enough to overpower him. Plus, would you really confront a cheater on your way home from a night out? Wouldn't you do it in the privacy of your flat?

On my return from the bakery with my bag of croissants Jody is coming out of St Jerome's. When she sees me she stops dead on the path. I'm wearing Abe's parka — it's far warmer than the rain jacket — and maybe I look a bit intimidating. I pull the hood down and force myself to smile.

'Where are you off to?'

She blinks rapidly, then finds her voice. 'The hospital.'

'Oh.' My smile falters. Guiltily I screw the croissant bag up until it's as unobtrusive as possible. *While you're enjoying a nice slow morning screw she'll be conducting her bedside vigil for your brother*. I'm the bad guy yet again.

I could get rid of Daniel and go with her, but I'm not sure I can sit there beside her and pretend nothing's happened.

'Tell him I'll try and pop by later.'

She nods quickly, tucking her hair behind her

ear in that coy way she has. I don't think she even realises it looks flirtatious. The ring on her engagement finger has twisted around and now I can see the whole phrase that has been engraved into the stainless steel (as I now know it must be — not silver).

True love waits.

My eyes widen. Seriously, Abe? I know you didn't have much money but was this all you could manage for an engagement ring?

It's a purity ring, given to Abe and me by our parents when the first wisps of pubic hair appeared. I took mine to university with me, determined to be wearing it as I lost my virginity. Friends at King's came up with the nickname when they found it in my stuff. The name on my birth certificate is Mary Martha, but they thought that was inappropriate, so I became Mary Magdalene, the whore with the seven demons inside her.

But my friends didn't know their scriptures. The Magdalene may not have been the Virgin Mother, but she was by Jesus's side when he was crucified. Not a whore but a saint. Either way I played along and the name stuck. Magdalene. Mags.

At the end of the first term I laid my ring carefully down on the railway line and the fast train to London obliterated it. I can understand why Abe kept his — to laugh at, or hurl across the room when he thought of our parents — but why give a chastity ring to his lover? Ironically? Somehow Jody doesn't seem the type to get the joke.

As I watch her hurry away to be by his side, my wave of pity is accompanied by anger. How could he do this to her?

Piles of post sit on the table in the foyer. A pile for everyone who lives here. The bountiful friendship of the takeaway. Even the junkie in Flat Seven has a generous handful and she doesn't look as if she's eaten in years.

Abe's is held in an elastic band. A white corner is just visible between the menus of Bengal Kitchen and Pronto Pizza.

Sliding it out I open the blank envelope.

She was there

I pass my fingertips over the indentations from the biro, as if they will tell me something my eyes can't. I know for certain now that this note is meant for me.

And I'm pretty sure what they mean by *there*; they mean when he fell.

'*Who* was there?' I call up the stairwell, and the echoes of my voice seem to go on for long moments.

Then my door opens high above.

'Mags? You OK?'

'Fine. Be up in a minute.'

Tucking the paper into my pocket, I start climbing the stairs. On the second floor I pause, wondering whether to knock on the junkie's door. Clearly Abe was bisexual: could she have been one of his lovers too? She's probably capable of anything to get the money for her next hit, but I'm not sure Abe was that desperate, not with Redhorse on the scene. Unless she was before Redhorse, before Jody and now resentful of the love that she had with my brother? Resentful enough to try and set Jody up?

No. She said she was out that night and stuck with her story even when she thought there might be a reward for saying she'd seen something.

I pass the grumpy queen's door. Potentially another jealous lover?

I realise, suddenly, how desperate I am to try and prove that someone is jealous of Jody and is trying to drive a wedge between us, because the alternative — that this letter writer is telling the truth — is unthinkable.

Jody's lying to me.

She was there when Abe fell.

* * *

Daniel is standing by the window drinking coffee when I get back in. The flat always gets the morning sun and his blond hair is lit up with all the colours of the apostles' cloaks. He's wearing jeans and a T-shirt of Abe's — I said he could — which is a bit too tight because his muscles are just starting to turn to fat. He turns and smiles at me, and for a moment I feel a rush of emotion, like the release of some narcotic into my bloodstream. I bustle around the kitchen until it has passed, then lay out the croissants and the paper.

'Well, this is nice,' he says, sitting down at the

table. 'You'd make someone a lovely wife.'

'Piss off,' I say half-heartedly and pick up the money section, though my eyes skim across the page, unseeing.

I know he's got to go. He's promised to take his sons to the Warner Brothers Studio, but when he glances at his watch and sighs, I feel a sudden stab of desolation, and when I kiss him goodbye he says, 'Careful.'

'What?'

'Almost let some feelings show, there.'

'Yeah, the feeling of wanting you to piss off so I can get on with some work.'

He grins, his teeth Vegas-white.

I hear his footsteps all the way down the stairwell and then the creak of the door opening, and then silence. I'm alone again, and the oppressive atmosphere of the church returns, as if the lead roof is pushing down on me.

21

Jody

It was a beautiful evening because of the volcano. On the news they said it was something to do with the layers of ash in the atmosphere. As the afternoon wore on the sky became streaked with a million different shades of red. I tried to think of all the different names as I gazed out of the clear patch of window in my flat. Crimson and scarlet, vermillion, fuschia, cherry, burgundy, ruby, baby-pink, pillarbox, blood.

But it seemed as if this wonderful gift was all for me: the people scurrying down Gordon Terrace kept their heads bent, and the high road was as noisy as ever, with roaring bus engines, horns and sirens and the occasional shouting match. No one else had noticed what was spread out above them.

There was a tap at my door; it sounded hesitant, as if the person had come to ask a favour. I thought of my silent neighbours. Perhaps the woman next door wanted to borrow something. I decided not to answer. To become friends with her would mean having to have contact with her partner, and he frightened me. On the few occasions I'd actually seen him in the flesh he seemed to me, like most men who choose to look that way, more like an animal than a person.

The tap came again.

I sat still and silent by the window. Then I saw Mira walking across the waste ground in the direction of the high road.

If it wasn't her at my door, then who? What if it was her partner? He knew I lived alone. What if he —

'Jody?'

I sprang to my feet with a gasp and raced to the door.

You stood there, your body half turned away, as if you were about to leave, and my heart jumped into my throat — I came so close to missing you.

'Hi, sorry, I didn't hear the door,' I gabbled.

You turned back. You were all dressed up to go out — a big parka with a furry hood. My mind raced with the possible reasons for you coming round — could I take in a parcel for you, could I let you back in because you'd lost your key . . . ?

Then I noticed what you were carrying. A bottle of wine in one hand, and two plastic wine glasses in the other, a grey woollen blanket over your arm.

'I'm going to watch the sunset,' you said.

'I thought I was the only one who'd noticed it.'

Our eyes met. 'You're probably busy. And it's cold, but I was wondering if you wanted to co — '

'Yes!' I almost shouted, then was stricken with terror in case you were going with someone else and wanted me to lend you a corkscrew or something.

But you exhaled with relief. 'Great.'

I stepped out onto the landing and went to close the door.

'Erm, you might want to put something warmer on!'

That was just like you, Abe, to think about me and how I was feeling, before yourself. Your sister was right about one thing at least — you *can* love someone too much. You should have thought about yourself more, my darling. You should have told me how you were feeling. I could have helped you.

'Come in a minute. I won't be a sec.'

You came in behind me, shutting the door softly. I led you through to the living room and you went straight to the window. I watched you for a moment, silhouetted against the red sky. So beautiful.

Then I went to change. It felt strange as I hurried into my bedroom, knowing you were there, in my flat. A raw sort of feeling, as if I'd peeled off a plaster and the newly exposed flesh was throbbing in the air. Not painful, just sensitised.

I pulled off my jogging bottoms, my sweatshirt and T-shirt, sniffed my armpits, and then, on a whim, I changed my knickers, from the boy-shorts I've always worn, to a pink pair with lace at the front that one of the girls in the bedsit gave me. You can just glimpse my pubic hair through the lace and, looking at myself in the mirror, my heart started to pound. What was I doing?

I told myself to calm down, that I just wanted to feel confident and attractive with you.

I put on the rain dress, with a cable-knit Aran jumper over the top and a pair of pink cashmere socks — both Christmas presents from Helen. Then I pulled on my charity shop boots, brushed my hair, and at the last minute, put on some lipstick.

It would have to do, I told myself, as I slipped out of my bedroom to find you still by the window. You started when you saw me, then you smiled.

'My favourite dress.'

I looked away, the blush deepening.

'Well, come on then, rainwoman, this wine won't drink itself.'

When we got out of the flat I started going down the stairs because the best view would be around the back and we could sit on the wall that surrounds the car park.

'Where are you going?' You leaned over the banisters, grinning. The banister rail pressed into your hips and your T-shirt moved in the currents of air rising up through the stairwell.

I'd passed the door at the end of the landing every day since I moved in, but never thought to wonder where it led. Somehow you got it open and as soon as you pushed it wider and the cold air rushed in, I realised — it led to the roof.

Giving me the wine bottle to carry you took my other hand and began leading me up a flight of concrete steps. The door below us swung shut and for a moment we were in total darkness.

My grip on your hand tightened.

'It's all right,' you murmured. 'Don't be scared. I'm here.'

I squeezed your hand. 'I'm not.'

I remember thinking how soft your hands were. Not rough and coarse like a normal man's, but soft as mine.

It was so dark that I couldn't see where I was going and when you got to the door at the top I didn't stop in time and bumped into you. You held me to stop me from falling back and for a moment we were in each other's arms. The fur of your parka was against my cheek. It smelled warm and cosy, like fresh hay.

You held me just a second longer than was necessary and then you pushed open the door.

The lead was slippery with rotting leaves and the soles of my boots had no grip, so I didn't let go of your hand as you led me out into the middle of the roof.

A low crenellated wall surrounded the spire.

'We could sit here,' you said. 'The view's OK. Or,' you grinned and raised your eyebrows, 'we could go higher.'

I followed your gaze towards the little door set into the spire, and then up. Just before it narrowed to a point, there were two arched apertures, open to the elements.

'The view from there will be amazing, but it'll be pretty windy. Will you be OK?'

I held your gaze and said, 'As long as you keep me warm.'

It was shockingly brazen of me. You might even call it *provocative*, but I knew I could trust you, Abe, and I was right.

The little door opened onto a spiral staircase and a moment later we stepped out onto a stone floor covered in twigs and dry leaves, I guess from birds making nests higher up. Opposite the windows that looked towards the high road, there was another pair, looking west, into the sunset. For a moment I couldn't catch my breath.

'Careful,' you said, as I went over. 'We're pretty high now.'

The view was incredible, stretching right across London. I could see Hampstead Heath and the Emirates Stadium, as well as all the office blocks and cranes in the city.

You came to stand beside me, resting your arm on my shoulder. I could feel the stiffness in your muscles, as if you weren't sure you were doing the right thing, and I slid my arm around your waist to let you know that you were, that I was comfortable with you. More than comfortable — I was happy.

'I bet we've got the best view in the whole country right now,' you said softly.

I let my head fall on your shoulder. The fabric of the parka was cold and slippery under my cheek.

'Wait a minute.'

To my disappointment you took your arm away, but then I understood. You were taking your coat off and wrapping it around my shoulders. Now I was nestled into the sweet warmth of your body.

For several minutes we just watched the sky. The wind had blown up again and the bands of different colours were coiling and unravelling under high, gold-bellied clouds.

'What are you thinking about?' you said.

I looked into your eyes and my voice was barely audible when I spoke. 'You.'

I placed my hand on your chest. It was rising and falling so fast. You stayed perfectly still as I rose up onto my tiptoes and kissed you. Your lips were so soft. For a moment you didn't respond and I was terrified I had misunderstood your intentions, as I always do, but just as I was about to pull away you circled me with your other arm and kissed me back. Time seemed to stand still. Even the wind dropped. The sun must have come out from behind a cloud because my eyelids glowed red. I was overcome with such dizziness that if you hadn't been holding me up I think I would have tumbled out of the window.

After a few more minutes you pulled away and smiled at me, but I didn't want it to stop. And more than that, I wanted it to go further. I had waited so long.

I moved my hand across your chest and you caught your breath.

I started unbuttoning your shirt.

'Are you sure?'

I nodded and you kissed me again, and now I felt the tip of your tongue against my own, I responded, and you held me tighter, crushing me to your body. Something was rising in me that I had never felt before.

You lowered me gently down onto your coat and, with the dry leaves whispering around us, I lay back and closed my eyes.

Then you were inside me, moving in me. And then we were moving together. We were like one person, our breath, our heart-beats, the slow coming together and moving apart: all synchronised, as if our bodies knew one another already.

Something inside me opened then, like a flower blooming. For the first time I felt like a woman.

You whispered my name breathlessly and I murmured yours, moving my palms up and down your smooth back, feeling the energy that rippled through your body. Your lips were on my neck, on my shoulders, my breasts.

Heat flowed across my skin. All my nerve endings were alive, as if my whole body thrummed with electricity. In the core of my being I felt a bursting warmth, like a wellspring, bubble up and start to flow through my veins.

'Jody,' you said, your voice cracking. 'Jody.'

Afterwards, for those still seconds as you lay on top of me, your breath in my ear, your heart thudding against my ribs, I looked out at the last flare of the dying sun, and I knew that I had left it all behind, all the pain and fear and shame. Down there, where the rubbish whirled around the broken tarmac — that was my past. Up here, in the cold, clean eye of the wind, with the bloody sky stretched over me — this was my future. With you.

You pushed yourself onto your elbows and smiled at me. 'All right?'

I knew I wouldn't be able to tell you just how right I was, so I just smiled back at you, touched your face.

'I know this will sound strange because we hardly know each other,' you said, your hand cupping my cheek. 'But I think I love you.'

I love you too, Abe.

I love you.

I love you.

I love you.

I love you.

I love you.

I love you.

22

Mags

I decide to print out the messages from Redhorse and show them to Derbyshire. If this were my case there are several lines I'd follow. One: Jody found out Abe was having an affair and pushed him. Two: Abe tried to end the affair and his lover pushed him. Three: An as yet unidentified third person pushed him for unknown reasons, but someone witnessed this.

If I can't get Derbyshire to take any of these scenarios seriously then I'll either have to call in some favours with private investigator firms in the US — see if they can't trace the IP address where the chat room messages came from — or give up and go home. To be honest, the latter option is by far the most attractive, but I feel like I owe it to Abe to at least try.

I save the messages onto a stick and head to the Internet café on the high street. There's a queue for the printer so I order a latte and sit at a table to wait. The girl behind the counter is Eastern European; her skin has a greenish waterlogged pallor that makes me think of a drowning victim.

Outside a wall of buses inches by, slower than walking pace.

Sipping the latte I flick through one of the coffee-ringed magazines. It's filled with gleeful descriptions of celebrity break-ups. *Friends say Kelly just couldn't take his hostility towards her BFFs. Friends say he didn't like her leaving her dirty underwear around the bedroom.*

I toss the mag down in a pool of coffee. A whole industry based on *schadenfreude*, making their inadequate readers feel smug about their drab little lives and relationships. Celebrities break up because their egos are solid enough not to put up with other people's bullshit. The rest of us don't have the balls, because we're too insecure to be alone. Maybe you have to be as wet as Jody to be really happy.

She'll be at Abe's bedside by now, gazing into his grey face, clutching his flaccid hand, her lips moving in silent prayer.

How could he do it to her?

Was it simple cowardice? Easier just to let her believe that they had the perfect romance than admit that he was not only unfaithful, but also bisexual? That wouldn't really fit into Jody's dreamscape. If she'd found out it would have been such a horrible shock. Enough to make her do something entirely irrational and out of character? Something violent? Jody? It seems so outlandish.

I can see her with a knife, wild-haired and crazy-eyed, slashing with anguished abandon at the murderer of her dream, but pushing someone over a banister? That feels like a man's doing.

I recounted to Daniel Jody's story of her and Abe's first kiss, hamming up the scene of Abe stripped to the waist, beating down the flames with his sodden shirt while Jody swoons by the sink.

He said, 'Then did he sweep her off on his Arab stallion to his yacht in the Caribbean?'

I laughed. 'What do you mean?'

'Well, it's a bit Mills and Boon, isn't it?'

At the time I'd been more interested in teasing him about his knowledge of pulp romance plots, but now . . . I don't know. Is it all a bit *too* perfect? Is Jody making things up to make her relationship look rosier? If so, is she hiding more about herself? Is there actually a violent psychopath hiding behind the mask of pathetic little victim? Am I that bad a judge of character?

Christ. How much longer do I have to wait in this dump?

One guy's hogging a computer while he gabbles on his mobile in some obscure guttural language, Latvian or Ukranian or something. I feel like going up to him and tearing the phone out of his hand. I close my eyes and breathe deeply, while manic Latvian laughter ricochets around the café. I'm in the throes of that post-sex low when all I want to do is ring Daniel and ask him back for more. That's the last thing I should do, of course. I already said he could call me, but I'll just send him straight to voicemail. I certainly don't want to be embroiled in a relationship with a man who has kids and an ex-wife.

Finally the Latvian logs out and I print out the messages and leave.

Returning to the flat is a battle against the relentless wind. It tugs at my clothes, roars in my ears and up my nose, as if it's trying to get inside me. By the time I get to Gordon Terrace I feel like I've run a marathon, but as I turn towards St Jerome's it starts buffeting my back, thrusting me forward, making me stumble over the ridges on the path. They look like tree roots pushing up through the tarmac, but there isn't a single tree as far as the eye can see.

A woman is standing by the main doors. As I approach the building I see she's carrying what looks like a large basket wrapped in paper with a repeating pattern of teddy bears.

'Sorry.' I move past her to open the door.

'Oh, excuse me,' she says. Her middle-class accent is incongruous in these surroundings. From behind she looked younger but her face suggests she must be in her mid to late sixties. 'Do you know Jody Currie, on the fourth floor? It's just I have a parcel for her but she doesn't seem to be in.'

'I can take it.'

Then I catch sight of the label: *To Jody, with all my love, Helen*.

So, this is the aunt who was supposed to be Jody's guardian but kicked her out to grow up in a care home: now trying to buy off her own conscience with what smells like a hamper full of bath bombs.

Without catching her eye I hold out my arms to take the package.

'Actually,' she says, 'perhaps I could pop up? Maybe she hasn't heard the buzzer.'

'I saw her go out,' I say coldly.

When she doesn't hand over the present I look

up. Her eyes are hazel and heavily ringed with brown. She's breathing heavily, the silver locket on her bony chest bouncing the light as it rises and falls.

'She's told you, hasn't she?' she says.

I make my expression blank. 'I'll make sure she gets the parcel.'

'Well, it isn't true. None of it. You should know you can't believe a word she says.'

She is lying.

I glance down at the label again. It's in slanting block capitals where the notes were in a more looping lower case. But even if she was disguising her handwriting, I can't imagine what motive she would have to leave me them.

'What do you mean?'

'All the fairy tales about her past, her family. My son. The rape.'

I stare at her. The rape?

'No smoke without fire, they say, don't they?' Her eyes are a mass of tiny thread veins. 'He could never get away from it. What she said he'd done, him and his friend. Their mates stuck by them, of course, but there are enough people who insist you always have to believe the victim. They never took into account her background. She was so damaged . . . '

Her junkie son raped Jody and now she's accusing Jody of lying about it. Disgusted, I push past her and open the door.

'Goodbye.'

'We loved her,' she says.

'Clearly not enough.'

Her face twists in anguish. 'But we had a

responsibility towards our own flesh and blood, didn't we? If it had been just us, then we would have managed, but when she started lying about our son . . . Her accusations destroyed him.'

'Sounds like he destroyed himself. It was nothing to do with her.'

Helen's hands are still clasped around the present, gradually turning blue in the cold. There's a sudden snatch of wind and the label twists and tears itself free to flap across the grass, but her eyes don't leave my face.

'What did she tell you?'

So, Jody never mentioned this rape. Why should she? But still I can't quite bring myself to close the door.

'You promised to look after her, to bring her up as your own, and then because your son became a drug addict you decided you couldn't cope and kicked her out. And these *accusations* . . . ' I emphasise the words with contempt. 'If it comes down to Jody's word versus the word of a drug addict, I know who I'd trust. You had a responsibility to look after that girl. However difficult it was, you were her aunt. Her parents entrusted her to you.'

I hold Helen's gaze. I may not have been the best of sisters to Abe but I'm here, aren't I? I'm doing my duty. This woman abandoned hers.

'But I'm not her aunt.' Helen's bloodshot eyes are wide and bewildered. 'I was her foster mother.'

I stare at her for a moment, then I force the door closed.

'Felix turned to drugs because of what she

did!' Her voice is muffled by the glass. 'She cried rape against our son and his friend and it destroyed his life!'

I push through the inner door, my chest tight.

'She lies about everything. Everything.' Her voice is drowned by the wind. 'She's dangerous!'

Inside the sanctuary of the stairwell I stop and lean against the wall, breathing heavily. My head is spinning.

Jody lied to me about Helen.

The light through the stained glass is subdued, a dull red wash darkening the concrete.

What else has she lied about?

I run back outside. Helen is hobbling along Gordon Terrace now, hunched against the wind, the basket wobbling in her hands.

Dodging the piles of dog shit and ridges of bursting tarmac, I set off after her. I need her to tell me exactly what happened with this alleged rape. Did Jody make it all up, or are Helen's maternal instincts blinding her to the unpalatable truth about her son, the rapist junkie?

I step out onto Gordon Terrace, my Converses slipping on a discarded crisp packet — or maybe they're Abe's Converses: I've started wearing his clothes without a second thought, as if they are my own.

Helen is almost at the high road but to get there she must run the gauntlet of a gang of youths in low-crotch jeans and pulled-up hoodies. They've separated into two lines on the pavement to let her through, but they are clearly saying something to her. Their laughter's like the yapping of dogs. One of them reaches out for the parcel and she yanks it away, breaking into an awkward trot, making them laugh all the harder.

I think about calling out to her, but that will attract their attention.

What's the point anyway? I slump against a lamp post. Who's to say Helen will tell me the truth any more than Jody? What was it Daniel said? There's no such thing as truth, only what you can make someone believe. Helen will want me to believe Jody is guilty so her son can be innocent.

She reaches the end of the terrace and turns onto the high road, the awkward parcel still clutched to her chest. The youths have noticed me, so I turn and head back to the church. But the idea of passing another evening in gloomy, wine-fogged solitude is pretty unappealing, so against my better judgment I take out my phone to call Daniel.

I sit, revolving slowly on the roundabout, as we talk for almost an hour. He had a good time with his boys. So good that when he describes how they cried as he dropped them off, his voice cracks.

'They love you, Dan,' I say softly, cupping my hand over the receiver to protect it from the wind. 'That's why they're upset. Isn't it worth it to know that?'

It's not like me to issue words of comfort, but I'm desperate to keep him on the line.

He sniffs. 'Sage words from the mistress of the heart.'

I guess he feels he has to lighten the tone or risk spooking me.

'How was it with . . . your wife?' The pause is

almost imperceptible.

'Donna? She wants me to move back and give it another go.'

To my utter surprise my heart lurches and it's hard to catch my breath.

'Mags?'

'Sorry, I can't hear you very well over the wind.'

'Go inside then, you nutcase.'

'I will.' I glance over at the church, becoming blacker and more forbidding by the moment. If I don't go now I may just chicken out and book into the hotel again. 'In a minute. What did you say to her?' I keep my voice light.

'What, you mean did I say, *Sorry, I'm in love with someone else?*' His tone is mocking.

'Hey, look, I should really go. Good luck with everything. I hope it works out for you.'

'Mags.'

I hang up, feeling inexplicably wounded.

The roundabout makes one more slow revolution and I find myself face to face with the group of youths.

'You got a joint, sexy?' says the tall one at the front.

'No.' I stand.

'Can I have a look?' He fingers the zip of my bag.

'Like hell you can.'

'No need for that, bitch,' says one of the others. I glance back at the church but the windows are all dark. On Gordon Terrace the few functional street lamps are flickering into life. The street is deserted.

'Gimme the bag,' says the tall one conversationally.

'No.' It's hanging across my body. He will have to physically assault me to get it off. Unless he cuts the strap.

He produces a knife. It has a shiny blue handle with a spider graphic. The blade has holes down the blunt edge, and is jagged from halfway down the sharp side. I find myself wondering about the holes, then I realise. They're like the ones in a cheese knife — to prevent them from sticking inside the thing you are trying to slice up.

The tall one's eyes are black and pitiless. 'Gimme the bag.'

'Or he'll cut your tits off,' another adds.

I have no choice. No one is coming to help me. I'm about to lift the strap over my head when, from behind me, comes the characteristic creak of the church door opening. The security light blares on.

The boys' attention is diverted. This is my chance. I glance back to see whether whoever has come out will retreat rapidly when they see what's going on, or whether they might hold the door open for me to flee into the block.

To my surprise it's the old lady from Flat One. In the harsh light her hair is the colour of Fanta, moulded around her head like a crash helmet. Even from here I can see the splodges of rouge on her cheeks and the ragged slash of shakily applied lipstick. She's holding up a tablet in a fluffy pink case, the camera trained at our little group.

Over the wind her voice is quavering but

strong. 'Can you see, Martin? Are you recording? There are five of them.' She describes them in turn. 'Yes, I've already called the police. They're only in the high road.'

'She FaceTimin' us, isn't it,' one of the boys says.

'Should teach that old bag a lesson,' another says. He takes a few steps in the direction of the church. The old woman doesn't move. The pink fluff ripples in the wind, but her hair remains utterly still.

A police siren wails in the distance.

'Feds comin', man.'

The tall one flicks his chin contemptuously in the direction of the church. 'Bitch lives *here*. She ain't got no shit worth havin,' then he turns and starts walking back across the grass to the low fence that stops the Staffies getting in.

I have the satisfaction of seeing one of them trip over the fence. As he stumbles his hoodie rides up, revealing an expanse of white underpants and the crack between his pale buttocks.

I resist a jeering laugh — that might be pushing my luck — and set off quickly in the opposite direction.

The old woman holds the door open for me and we hurry into the foyer, making sure it's shut firmly before passing through to the stairwell, where we pause, panting.

She extends a hand. 'Lula Lyons. Pleased to meet you.'

'Mags Mackenzie.' Her hand is paper-dry but the grip is firm. 'I think my brother was your carer.'

She nods, her milky eyes bright. 'Do you know, the first time I saw you coming up that path, bumping your case over the bodies, your hair all tied back, I thought you were him. Gave me quite a turn.' She shakes her head. 'He was such a pretty boy.'

'Thanks,' I say, nodding at the tablet, hanging slackly from her grasp, 'for what you did. And please thank the person you were talking to as well, for recording them.'

'Martin?' She gives a wheezy laugh. 'Martin Scorsese was my cat. Died last year. And this bloody machine,' she waggles it contemptuously, 'hasn't worked for weeks. Your brother always sorted out my technology. Taught me how to FaceTime my friend in Catford. Not that we really want to see one another's faces these days. Come in and I'll give you something to settle your nerves.'

She moves haltingly, as if in pain, but her clothes are that of a woman sixty years her junior: a gold lamé top with sequined sleeves, a fitted black skirt with a thigh slit, royal blue fishnet tights, and a pair of crocodile skin, kitten-heeled ankle boots. I smile as I follow her in, then make my face serious again as she turns and asks me to sit down.

Her flat must have the same dimensions as Abe's, but you'd never know. The place is decked out like a Persian bazaar. Silk throws billow from the ceiling, studded here and there with silver lanterns, and what I took to be net curtains at the window are actually pieces of antique lace. The floor is layered with what look like flying carpets at rest. The sofa's a huge mahogany thing, piled with cushions and a slightly chilling rag doll with green paste jewels for eyes. I sit down in the opposite corner to the doll as Lula hobbles to the kitchen behind me. Again, it's similar in layout to Abe's, but instead of cabinets, there are rows of open shelves heaving with bric-a-brac: old tins and bottles, jars of multicoloured pulses, copper pans, stacks of old pudding basins, flowery jugs of utensils.

'Whisky? Or brandy? I think I've got some schnapps here somewhere.'

'Christ, yes, please. Whisky.'

On a mother-of-pearl inlaid table beside me is a lamp, draped with a fringed shawl. It casts an ethereal glow over the photograph next to it — a large version of the one in Abe's file. It looks like a film studio shot from the forties or fifties. The eyes of the woman in the picture are emerald green, like the doll's.

'Are you an actress?' I say.

'Was,' she snorts. 'Last job I did was a corpse in *Casualty*. But now I'm too old even for that!' She gives a wheezy laugh.

With her gnarled fingers she fills two greasy tumblers to the brim and brings them over, then sits down on the club chair opposite with a grunt of effort. Cataracts have turned her green eyes milkily opaque.

'What did you mean, when you said you saw me bumping my case over the bodies?'

She sips her whisky. I notice her lips pull up at each side, Joker style, perhaps from a primitive attempt at a facelift.

'This is a church,' she says. 'So, where's the graveyard?'

'I assume they moved it.'

'They moved the headstones, but left the bodies. Over the years, they've been gradually coming to the surface. Sometimes the dogs dig up a bone.'

I grimace, thinking of what I have been walking over every day and she laughs again. With the draperies muffling all sound, it seems eerily close to my ear.

'Isn't there anyone you want to call? It can really shake you up, that sort of thing.'

There is. I want to speak to Daniel so badly my chest aches, but that's precisely the reason I can't. Relying on someone else as an emotional crutch means I'll just end up like Jody.

'If you don't mind my just sitting here for a bit, I'll be fine.'

'Not at all, lovely. It's nice to have a visitor. Especially one who reminds me so much of my beautiful boy.' She sips her drink, leaving a scarlet semicircle on the glass.

'Did you see Abe often?'

She sighs. 'Every Monday evening he'd come down and have supper with me. Meatballs, or smoked haddock with mash, and liquor, half a bottle of whisky and sherry for afters. Lovely. I miss him.'

I smile at the thought of them getting pissed, then maybe dancing together to some music hall tune.

'He's not going to wake up, is he?'

My smile fades. I shake my head.

'I could tell when I saw him, that night. All that blood. I didn't want to look but you can't help yourself.'

'Did you see what happened?'

'I heard footsteps on the stairs and then that poor girl screaming. By the time I got outside she was kneeling by his side, all covered in blood like the one in that Stephen King film . . . *Carrie*.'

'I'm sorry,' I say. 'If there's anything I can do, then please — ' I bite my tongue immediately. I don't want to be spending my Monday nights down here eating smoked haddock and discussing Lula's past glories.

'He was *your* brother, darling,' she says gently. 'I'm sorry for you.'

'And Jody,' I say. 'She's the one really suffering. Abe and I barely knew one another.'

There's a beat of surprised silence.

'Jody?'

I can't read her expression. My heart pounds faster. Did Abe tell her something about their relationship? Some secret? But she called Jody *that poor girl*, so she can't think Jody had anything to do with his fall.

'He wasn't . . . violent towards her, was he?'

She raises her drawn-on eyebrows. 'Your brother didn't have a violent bone in his body. Though I wouldn't have blamed him.'

I put the tumbler down, half finished. 'Why do you say that?'

'Ah, she just wouldn't leave him alone.' She rolls her eyes and a flake of crusted mascara drifts down onto her blouse. 'Whenever he turned around, there she was, like a bad smell.'

Sounds like the old lady's jealous. Perhaps she was hoping those drunken Monday nights might turn into something else. A friendly fuck now and again, to remind her of the beauty she used to be.

She's watching me and there's something about that heavy-lidded gaze that makes goosebumps spring up on my arms. I'm not sure I like her, or the dismissive way she talks about Jody. I find myself bristling on Jody's behalf.

'She *was* his fiancée. It's not abnormal to want to spend time with the person you're about to marry.'

Lula laughs then, loud and ringing, like the crowing of a cockerel. 'Fiancée? Abe didn't have a fiancée! He didn't have a girlfriend at all, my love.'

The way she's staring at me, with an expression of pitiful disbelief, makes me feel like slapping her.

Abe didn't have a girlfriend, so what was Jody? A casual shag who got the wrong idea? No, no, it's more than that, it has to be. What about the ring? The photograph? He didn't tell Lula because he thought she'd be jealous.

'Don't tell me you didn't know?'

My heart is pounding now, with anger and with something else. I'm beginning to understand. Oh, God.

The sequins on her sleeves shimmer, making me dizzy. Her red lips open, a string of saliva stretching and breaking. 'Abe was queer as a nine-bob note!'

'No,' I say stupidly. 'No, he . . . she . . . ' But even as my brain struggles to process the information I know it's true.

I'm looking into his eyes on the parapet of Eilean Donan castle and I know he's different. The girls' cooing means nothing to him, but his hidden heart yearns to open up to someone the same way mine does. Was he trying to tell me then? I could have guessed, if I hadn't been so wrapped up in myself. I could have helped him, taken him with me. I could have stopped him having to pretend to be something he's not.

But he's in London now, with a million other men just like him. There's no reason for shame or fear. Our parents are five hundred miles away. What possible reason would he have to pretend?

'Then why . . . ' I begin, and my voice is thread-thin. 'Then why was he stringing Jody along? Why didn't he just tell her?'

'He wasn't stringing her along. She knew perfectly well, just pretended it wasn't happening. She's mad, of course. You're not going, are you? You haven't finished your drink.'

But I'm already up, hopping and stumbling over urns and Ali Baba baskets in my rush for the door.

I take the stairs two at a time, thumping the light switches as I go. Bursting out onto the fourth floor I hammer on Jody's door, hard enough for the splintered lock to crunch and give a little more.

'Jody! Jody! Open the damn door, now!'

23

Mira

I have got what I wanted. I have made the sister think it is Jody that killed Abe.

Bang, bang, bang, on the door.

She is shouting and swearing and threatening. I crouch in the hallway, praying for Jody not to open up. The sister is so angry she will hurt her, I am sure. And it will be my fault. I should have thought. I was trying to protect you, Loran, and now that poor mad girl will suffer for it.

There is a louder bang and the wall I am leaning on shudders. She has broken the door in.

I hear her footsteps pound up the hall and another bang as she kicks open the inner door.

I know I should keep away, look after the baby and you, but how could I live with myself if Jody was hurt because of what I did?

I open the door of the flat, just a crack, to listen.

There are no sounds of an argument. Perhaps Jody is out. Has the sister already gone away, or is she waiting inside the flat for her? Should I call the police and say there has been a break-in? Perhaps the sirens will frighten her off and I will have a chance to warn Jody.

But when I return from getting my phone I hear something that makes my heart squeeze up in fear.

Footsteps on the stairs.

I pray that I will hear them stop and one of the other flat doors open, but they continue on, up and up and up.

And then her face comes into view, behind the bars of the banisters.

She is pale and sad, like always. She has no idea what has happened. What is waiting for her.

She comes out on the landing. And still I am too much of a coward to open the door and stop her. I tell myself I am thinking of the baby, of my blood pressure, but it is just fear. Fear to admit what I have done. Fear of what will happen to you if I do.

Then there is no more time. As she passes in front of my door the breeze from her skirt wafts against my face and then she is gone, into the flat.

There was a time when the difference between right and wrong seemed so simple — before I met you. I was brave, then. I was brave because I was surrounded by people who loved me. Now they are very far away and all I have is you, and you do not love me.

I stand up.

I may have changed much from the girl I used to be. I may have become afraid, shameful, unlovable. But I will not let this happen. I will face whatever harm may come to me — and yes, the baby too — because I cannot live with the woman I have become. A woman who will allow others to suffer because of her own lies and cowardice.

I go out on the landing.

Now I hear voices

They are speaking too fast for me to understand. But I can clearly make out the harsh anger in the sister's voice and Jody's trembling, high-pitched responses, like the fluting of a tiny *kaval*. She is afraid. I have passed my fear onto her because I could not stand the burden of it. I must have the courage to take it back.

Jody's latch lies on the floor, the wood splintered where it was torn out. She must be strong, this woman, though she is as slight as her brother. Perhaps she would not be so easily hurt.

I creep over the threshold, cupping my belly as if a comforting hand will somehow make up for the pounding of my heart and the rushing of my blood. The baby will feel my fear and be afraid too. I am unworthy of him.

As I pass up the landing, the shouting goes on.

I reach the open door.

I have never seen inside this flat before. It is as shabby and poor as ours, but the walls are covered in pictures. They are, I think, supposed to be drawings of Abe. In the middle of the room is a table. A pair of scissors lie open.

The sister stands at the table, her face as white as *sultjash* as she snatches up scraps of paper and hurls them to the floor, shouting all the while. Jody is pressed against the wall, sobbing.

It does not seem that the sister plans to hurt her. I could go. They haven't seen me.

The sudden silence when the shouting ends is almost shocking. My ears ring with it and my pulse pounds in my head. The baby is so still inside me. Have I frightened him to death?

Abe's mild eyes look down at us from the pencil drawings.

She picks up the last piece of paper — or two pieces taped together. I cannot make out the image and I think she is going to toss it to the floor with the rest, but she doesn't. The next words she says are so cold and clear that I understand them perfectly.

'You were never his fiancée. You were his stalker.'

Now, with a curl of her lip, she brings both hands up to the top of the paper and starts to twist her thumbs. She is going to tear it in two.

Jody screams then. 'No, no, no!'

And before I can do anything, she lunges for the scissors.

She sits quietly, her homework on her lap, pencil poised, head bent as if she is reading it. But she isn't. She can't concentrate. Her brain is throwing off pulses of warmth that fizzle through her blood, making her fingertips and the ends of her toes tingle.

Her brother sits beside her. It's a double bed so there's plenty of room, but still they sit close — well, she sits close to him, close enough to feel his chest rise and fall at every sigh and every impatient jerk of his pencil, as if the movements were made by her own body.

The duvet is soft under her bare thighs. And so clean! They change the sheets every single week and pour a thick blue liquid into the tray of the machine that makes them smell like flowers. The carpets are a sort of fudgey beige and most of the furniture is white. Their father drinks lots of coffee from an expensive stainless steel machine, and so, from breakfast onwards, the house is filled with its rich musk. Their mother bakes. From scratch. Using free-range organic eggs with bright orange yolks. She has promised to teach her how to make a lemon drizzle cake. Which is easy, apparently.

Her brother huffs and drops his head back onto the headboard.

'Bollocking hell.'

She smiles. She would like to be able to help

him, but though they are the same age, he is working at a far higher level. She is dyslexic. Not thick and useless and a waste of space, but dyslexic. Her new parents suspected it and the teachers tested her. That is why she's so creative, they have told her. Dyslexic people are more imaginative than other people. They call it 'having an imagination', not lying, and she doesn't feel the need to lie any more. There is nothing to hide. Nothing to be ashamed of, they have told her. Nothing.

'Wanna play Grand Theft Auto?'

She screws up her nose. She promised their mother that she wouldn't let him distract her. She has been making good progress. She can read Harry Potter, now. She is on a special programme at school called Soundbites, and that has helped her with her spelling.

'Come on. I can't be arsed with this.'

He slides off the bed, scratching his lower back, making the T-shirt ride up. The skin on his back is completely smooth and blemish-free. His deodorant has a pleasant minty smell, and sometimes after a match he smells of sweat. But not the stale, fetid reek of poor hygiene: a fresh, young smell, that you could bottle and sell as a perfume. She can feel love creeping up on her, warming her up from deep inside all the way to her fingertips. But she's not afraid. This is not a love born of desperation, but one she can rely on forever. Her brother will always be there for her. They will grow old together. He will be a doting uncle to her children. She will have children. They've told her that she *still can, despite everything.*

He swigs from the litre bottle of Coke on the bedside table, the muscles in his throat rippling as the liquid passes through. When he has finished he hurls the empty bottle at the waste paper basket and misses, then wipes his arm across his mouth.

'Come on, dopey. What are you staring at?'

'What if Mum comes up?'

He snorts. 'Mother wouldn't dare come in here without asking. I might be wanking!'

He gives a barking laugh and she smiles.

They play the game for more than an hour and she is glad when he announces that he is hungry and tosses the controller onto the carpet. He stands and stretches. His stomach is muscley and a line of dark hair runs from his belly button into the low waistband of his sweatpants. Suddenly, for no reason at all, her throat tightens.

'I'll go and make you a sandwich,' she says quickly. 'What would you like?'

'Bacon. Loads of ketchup. Microwave the bread first so the butter melts. Oh, and a cup of tea, please, Sisterella.'

'Coming right up!' She smiles and tries to haul herself out of the beanbag, but it's too low and she stumbles back. He sniggers and grasps her arm, yanking her up so forcefully she stumbles into his chest. It is rock-hard from all the exercise he does. She can feel his blue eyes on her face but she doesn't meet his gaze.

'What do you fancy?' he says. His breath is sour from all the Coke.

'I'm not hungry.'

'You should eat more. Look at you. Flat as a pancake.'

'I will.'

'Good girl.' He smacks her bum as she goes out of the room.

24

Mags

I sit alone in a squalid little interview room. The table is pocked with cigarette burns, little black scoops I can just fit my fingertips into. I've been doing this for three quarters of an hour, waiting for PC Derbyshire to come and take my statement, all the time resisting calling Daniel, who is probably on *date night* with Donna.

Jody's attempt to murder me was laughable. The hand holding the scissors shook so much the light bounced off the blades like a disco ball. I leaped at her and snatched them from her hands and I might very well have used them on her if weren't for the sudden entrance of Mira. She thrust herself between the two of us, waving her hands and crying, '*Ndal! Ndal!*'

It was the sheer melodrama of the scene as much as anything else that chastened me, and I sat down at the table while she called the police and Jody wailed like a child and tried to gather up her papers.

There were several letters from Abe, asking her to stop, warning her that he'd speak to Peter Selby, letters she had cut words out of to make new ones that said what she wanted them to. *My Jody . . . I love you . . . I will be yours always . . .*

The walls of her flat were covered in awful pencil drawings of Abe. Abe smiling. Abe sleeping. Abe gazing into the distance, a Pierrot tear on his cheek. Laughable if they weren't so pathetic.

There were a few photographs too, that I assume she stole from his flat after the accident. One had been grainily enlarged on a copier and clumsily spliced together with a picture of her. Full-size it was obvious, but she had managed to shrink it down — perhaps at the Internet café — to create something vaguely convincing, if you didn't look too hard. I'd been convinced. It had been sitting there on the bedside table all that time and I hadn't given it a thought.

My cruel impulse to tear it to pieces in front of her eyes was what triggered her pitiful attack, but I'm glad it stopped me. This was evidence after all. Evidence that she was a psychopath who had stalked my brother and pushed him to his death when he rejected her advances.

How could I have been so stupid to believe all her stories? They might as well have been about fairies and unicorns.

The door finally opens and PC Derbyshire enters. 'Miss Mackenzie.' Her voice is clipped and professional. Hopefully she's feeling rather foolish for all that *not every tragedy is a crime* stuff.

'I've written out my statement,' I say, pushing the paper across to her.

It takes her a long time to read it. Finally she looks up at me. 'You seem convinced Miss Currie attempted to murder your brother.'

'Of course she did. And it won't be *attempted* murder when I have to switch his machine off.'

She shifts in her seat, making the swivel mechanism squeal, then places both hands palm down on the table. I wonder if this is in the police handbook under *placating gestures*.

'Miss Currie denies having anything to do with your brother's accident.'

'Oh, so it's an accident now? You do know she's a notorious liar? You told me she wasn't known to you in a criminal capacity. That's bullshit, though, isn't it? Eight years ago she was done for crying rape.'

'Charges were never brought, and I didn't feel it was relevant to this case.'

'Not relevant that she's a liar? You believed her story without making any attempt to investigate.'

Derbyshire inhales and exhales before she speaks again. 'The unfortunate fact is that there were no witnesses to your brother's fall, and the evidence is circumstantial at best. Aside from the caution over the rape claim, Miss Currie has never been in trouble before, she never threatened your brother with violence, and your brother never complained to us that he felt in danger. Plus I don't think she'd have the physical strength to overpower him.'

I fold my arms. 'Don't they call stalking 'murder in slow motion'?'

She blinks her porcine eyes, the lashes clogged with brown mascara. 'Your brother never reported a stalking incident to us. It seems to me that Miss Currie just had a very strong, unrequited attachment to him — and she does have a history of forming these powerful crushes.'

'What sort of history?'

'The boy she accused of rape — the foster brother — she had, by all accounts, developed a bit of a thing for him too.'

'*A bit of a thing?*' I stare at her. 'She accused him of rape when he rejected her. And now my brother rejects her and, lo and behold, he ends up brain-dead. You can't see any sort of connection there?'

'Shall I help you out?'

Her square cheeks are reddening now as she tries to stop herself saying the things she really wants to. She's not silly. She knows I'm a lawyer, can probably guess that I'm recording all this with my phone.

'Crying rape and murder are very different crimes, with very different criminal pathologies.'

'That boy turned to drugs because of her. A good life ruined.' I remember how cold I was to his mother, Helen, and guilt only fuels my anger. 'She should have gone to prison.'

The policewoman sighs. 'Jody Currie has only ever been a danger to herself, Miss Mackenzie. There's not enough evidence for us to charge her with your brother's attempted murder.'

'I want to see your superior officer.'

She breathes deeply, gathering herself. 'Your brother was a strong, fit young man. He'd been a member of Stone's Boxing Club for five years.'

I open my mouth to protest that this is the first I've heard of it, but remember in time that everything Jody told me about my brother is a lie.

'We've spoken to the manager, who said that

though he was lean he was a very powerful fighter. Apparently he had been the victim of homophobic attacks as a teenager and had learned to defend himself.'

An angry flush rises to my cheek. 'You knew my brother was gay and you never thought to mention it to me? You just let me believe that crazy bitch all along.'

She looks at me steadily. 'People's private lives are complicated, Miss Mackenzie. It only becomes our business when a crime has been committed.'

'Like pushing someone over a stairwell?'

'As I was saying, it's very unlikely Miss Currie would have the physical strength to overpower your brother.'

'She could have caught him off-balance.'

'If there had been a fight or argument someone in the building would have heard something.'

I snort. 'Those crazies?'

'We've spoken to her social worker who agrees that she is not a violent person. In her statement Miss Currie says that she came out of her flat after hearing a strange noise — sound carries in that place, as I'm sure you've noticed. It was then that she saw your brother lying on the concrete floor at the bottom of the stairwell. She went down to see if she could help him, then called 999 on her mobile. I can play you the recording if you like. She's hysterical.'

'So she's a good actress. She had her mobile with her, so either she'd just been out and was coming in or she was just about to go out, right?'

'Some people keep their mobiles in their pockets all the time. Or she looked over the stairwell, saw what had happened, and ran back inside to get it before she went down to Abe.'

I stare at her. Useless bloody British police. I could pursue a civil case: the circumstantial evidence is pretty strong.

'You are at least going to prosecute her for wasting police time, right? All the lies about him being depressed?'

'We've cautioned her.'

'Another caution? After everything she's done?'

'I accept you must feel that she made a fool of you, but injuring someone's pride is not a crime.'

I hesitate before answering, to make sure my voice is steady. 'So, if you're so convinced this psycho had nothing to do with it, how *do* you think my brother fell?'

There is a long beat of silence.

The policewoman's lipstick is seeping through the wrinkles around her mouth. Why do they do it, these women in positions of power? Why do they cling to these outdated conventions of femininity? She looks like an aging air stewardess.

'I still think he might have been depressed,' she says, and her voice is human again. 'I believe there's a history of mental illness in your family.'

I stiffen.

'We spoke to his GP in Scotland. As you probably know, Abe was treated for depression when he was fifteen.' She's carefully avoiding my gaze. 'He was working extremely hard and his employers admitted he was under a lot of pressure. I think he jumped, Mary,' she says gently.

I breathe slowly, tempted to correct her — *It's Miss Mackenzie to you* . . . but what's the point?

She gets up from the table, taking my statement, with all its allegations about Jody, and tells me I can stay in the room as long as I need to.

When she's gone I stare at the pale blue wall, with its single claw mark gouged into the plaster.

It's not my fault. I was a child, the same as Abe. We both did what we had to to survive. I'm not surprised he was depressed. He medicated with Prozac while I did so with booze and casual sex. The former cry for help always inspires more sympathy than the latter, of course.

In the silence I hear muffled voices and doors banging, the occasional laugh. Is she here? Have they finished with her already? Sent her home to spin some new tale in which I am the wicked sister who poisoned Abe's mind against her, or killed him to prevent them being together?

I could just go home. Back to my old life. Back to Jackson for some no-strings distraction. Leave Abe's future in the hands of the doctors. Leave Daniel to Donna.

I close my eyes and push my fingertips back into the scorch marks. The room smells of stale smoke. They can't have changed the carpets since the ban came in.

My phone, on silent, buzzes angrily in my bag. I force myself not to look, but can't repress the childish hope that it might be Daniel.

Perhaps Derbyshire is right. Maybe Abe jumped. He was overworked and exhausted. Jody was on his case the whole time, so he must have dreaded coming home. Perhaps the relationship with Redhorse came to a bitter end and left him heartbroken.

And yet the picture Lula painted of my brother was not of a depressive. He might have been as a closeted gay teenager, but now? The novel on his bedside table was half finished; I came to the page he had turned down, just after the plot twist when everything was thrown up in the air. It sounds silly, but wouldn't you find out what happened before you jumped? Unless it was on impulse. On the way back from a lonely night drowning his sorrows? Except that he hadn't drunk anything. And it happened early evening, only an hour or so after he'd got back from work.

I think of the bruises on Jody's clavicles and Abe's torn shirt, both signs of a struggle.

The fact that Abe was not wearing his jacket with his wallet in the pocket, as if someone had knocked on the door of the flat and called him outside for a moment.

I think about the fact that she was prepared to destroy the life of the last man who rejected her.

She did it, I'm sure of it, but with so little evidence what can I do?

Getting up from the table my legs feel like lead. I pick up my bag and walk back through the corridors of the police station, then ask to be buzzed out into the main reception area.

The woman who visited Jody on her birthday

is there, speaking quietly to the duty officer. She breaks off when she sees me.

'Excuse me, Miss Mackenzie. My name's Tabitha Obodom. I'm Jody Currie's social worker. Could I — ?'

I push past her. I don't want to hear whatever sob story she's going to concoct to make me drop my accusations. I'm done with stories.

It's dark outside and raining just heavily enough to be unpleasant but too lightly to make the pavements glitter. I must confront Jody, but the last place in the world I want to be is back at St Jerome's. I just don't have the energy to sort out a hotel. I'll go back, have an early night, and then tomorrow decide whether to stay and see this thing through or else give up and go home. I'm not usually a quitter but tonight I feel beaten. In fact, when a passing lorry throws up a cascade of filthy puddle water all over my legs, I feel like crying.

Instead I take out my phone to call a cab.

The missed call was from Daniel. He didn't leave a message. I don't call him back.

25

Mira

The sister told the police that they must arrest Jody for murder. They didn't, though. They didn't say, *You have the right to remain silent*, as I have seen on British police programmes. They just said they would take her to the station for a little chat. They treated her gently because she was so very upset, and for that I was grateful.

I don't know what she will tell them at the station. Will she carry on protecting you?

You are still not home when I start preparing our dinner. The smell of the onions makes me feel sick. I know that a piece of bread and butter will take away the feeling but I do not have any. You said that women use pregnancy as an excuse to become fat pigs.

Is that why you went with another woman?

You think I didn't guess?

The *kondomat* in your gym bag.

The late nights spent in the pub with the guys from the building site?

The times you crept out in the middle of the night, like you did a few nights ago.

The sudden smiles as you thought about something — about someone — that wasn't me.

It's as if you wanted me to find out, but I stubbornly refused to, didn't I?

Because where will I be without you?

I will have to go home. To the shame of being unable to keep my husband. The dishonour surrounding me like a bad smell, revolting all other men who might once have wanted me. The sullied woman with her bastard child.

They will think it strange, the people who gave us the flat, that I can just trot back to Albania where we were persecuted for being Roma. We are not Roma. Why did you tell them we were? Because it would be easier for us to stay, you said. Why do you not want to go home? My parents do not understand. *I* do not understand. You have cut all ties with our mother country as if you believed the lie you told the English authorities, that we were in danger of our lives, that we could never go back.

Could you bear to leave the baby, Loran? He, at least, you seem to love.

It is a miracle I am pregnant at all. We make love perhaps once or twice in a month when you are drunk. You want a son. Then, you say, you would not be sad to come home every night to a miserable house. You would call him Pjeter, which means rock, because this is how a man must be. A man must be strong because women are weak. It is your burden to look after me.

The burden of my mother and sisters rested lightly on my father's shoulders. Papa often laughed. But all men are different, and all women are different.

The day I left her Mama told me to be a good wife for you. She was crying as she said it. You stood by the car with your face as hard and cold as the ice on the puddles in the ditches. Papa told her to hush because she was upsetting you. He was grateful that you had chosen me because you were from a better family than ours. Mother said it was you who should be grateful, that I was the most beautiful girl in the whole of Tirana and could have any man I chose.

I chose you. Papa would not have forced me into marriage if it wasn't my desire. I did it for love: but not of you, of them. I could not bear to see them disappointed. And I thought I could make you love me. It had been so easy for me before you. I had thought men were simple creatures, that all you had to do was smile and wear pretty clothes, to paint your lips and listen to their woes, to make them supper and open your legs. I thought love would follow.

I could tell you did not desire me, even as you asked for my hand. This had not happened to me before. It excited me, the thought of bringing you round. I imagined the moment you would break: as you made love to me, suddenly love would come upon you like a wave crashing over your head. But you fuck me like my *pidhi* is just a hole in the mattress.

What is *she* like? Is she tighter? Softer? Does she smell better?

Is it Jody? Or the drug-addicted woman downstairs? If so, I shudder to think of what you bring back to our bed.

I swallow the nausea and chop the onions, then I wash and dry up, then I fold the towel and tuck it into the handle of the cooker so that the back and front edges are lined up. You like the house to be tidy.

I don't know what else I can do.

If I were at home I would feed the pigs now, or collect the eggs, or cut off the runners from the pumpkin plants. Or I might treat my hair with olive oil and simply lie by the window, reading Kadare, while the oil sinks into my hair to make it glossy and thick for a man to run his fingers through.

I am glad to hide my hair under a scarf now. Seeing it would only make me sad.

There is nothing else to do but wait, so I watch television — very quietly so that I can hear your footsteps echoing in the stairwell. I watch the programme where English people shout at each other because they have behaved badly towards one another. Sometimes it makes me laugh. Always it lightens my heart because it shows me that all people have troubles, that mine are not so great.

26

Mags

Daniel leaves me a message, asking how my brother is. It's a stupid thing to say and probably only born out of guilt over his rekindled domestic bliss, so I don't bother replying.

From the police station I head straight to the hospital.

Jody isn't there. The nurses don't seem aware of what's happened. If I told them she'd be barred from visiting him, but this is the one place I can guarantee to see her, so I don't.

I ask for Dr Bonville and they say he is coming to see another patient so I can catch him then. I'm intending to tell him I want Abe's machines turned off. But as I sit down beside my brother's bed, I find I cannot concentrate on the novel. I read the same page six times without taking it in, then finally close it with a sigh.

Abe's face is almost as pale as the pillowcase. The bruises are greenish yellow now, as if he's been painted with glow-in-the-dark paint.

I cannot fawn over this near-cadaver, like Jody, but I could at least hold his hand.

The rush of blood in my ears drowns out the monitors as I reach forward and take his limp hand in my own.

'Hello, Abe. It's Mags.'

It's the first time I have touched him since

Eilean Donan castle.

His fingers are cool and dry. I was afraid they would be clammy with death. Carefully I slide my palm beneath his until we are palm to palm, then I clasp my fingers around his.

All sounds recede as I close my eyes and focus on the connection between us. A channel of electricity, or magnetism, or whatever it is that makes up the human soul.

Good God. I can't breathe, or swallow, or open my eyes. I'm as paralysed as my brother. My beautiful, kind, self-sacrificing brother, who ate fish suppers with a worn-out starlet, who loved a man with a red horse on his hip.

Fingers clasp mine. Just for a moment.

Then the fingers slacken and the hand becomes limp again.

I open my eyes. My brother lies motionless, his eyelashes still against his cheek. The respirator sucks and blows, the heart monitor bleeps. Somewhere far away, in the bowels of the hospital, an alarm sounds.

I don't lean forward and beg him to give me a sign that he can hear me. I'm not under any illusions. It is a reflex, that's all.

But I don't let his hand go.

Dr Bonville arrives on the ward but I'm already pulling on my coat. I've changed my mind. Was I actually going to let my brother die just as a way of getting back at Jody? What's wrong with me? Have I spent so long repressing all human emotion that I've become inhuman?

Bonville is busy with a patient at the end of

the room, but as I pass him he looks up and moves towards me.

'Miss Mackenzie. I was intending to phone you but we've been very busy. May I have a word?' He gestures to the door, for me to follow him.

'We can talk here,' I say. 'It's not as if Abe's going to hear.'

He looks torn for a moment, then seems to give up. When he speaks again his voice is very low. 'Your brother developed a chest infection and by the time it was spotted it had become pneumonia.' He adds hastily, 'This wasn't down to anyone's negligence. These things happen with our very sickest patients.'

'So you've put him on antibiotics?'

He hesitates. 'We have.'

'Fine. You don't need to ask me about that sort of thing, right?'

'We don't.' He glances up at the door, as if seeking a way out, and I think again how young he is.

He inhales. 'Pneumonia is very dangerous for people in Abe's condition. Especially if the infection spreads to the bloodstream and causes sepsis — blood poisoning. I'm afraid the tests for sepsis have come back positive.'

He waits for me to process this fact — that my brother might die without my having to make any decision at all.

'I wanted to let you know, even though I'm not required by law to do so, that we've added a DNR to his notes.'

'Do not resuscitate?'

He nods, watching me warily.

I'm not sure whether to be angry or relieved. A wave of heat passes over my skin, followed by a wave of ice.

'There are several reasons for this, not just that his quality of life would be — '

'You don't need to explain,' I say. 'I under — ' To my surprise, my voice cracks on the last word. However hard I try, I cannot force it out. And then I break down in tears, there, in the middle of the ICU, surrounded by the bleep and whir and gasp of machinery.

He lets me cry for a moment, then he puts his soft young hand on my shoulder. 'I really am so sorry.'

I know it's just what everyone says and means nothing coming from a man who must parrot it every day, but I reach up and clutch his hand as if I'm drowning.

'I can remove the DNR, if that's what you want.'

It takes me a minute to get myself together again, and when I'm finally able to speak my voice is soft and high as a child's. 'No. Leave it. Probably the best thing. And call me Mags.'

I raise my eyes to meet his gaze.

'I knew when I first saw him that he wouldn't last long,' he says. 'I've seen people of Abe's age with lesser head injuries and I've seen them get better enough to go home. But when I look at them slumped in their chairs, dribbling, I think, if it were me, I'd rather be dead.'

I can tell he thinks he's made a mistake to talk to me that way. He fingers his name badge, blinking his long eyelashes. I want to tell him that it's OK. That the fight is seeping out of me by the day, that I can feel Abe's ghost creeping into me, softening me. But I can't speak.

'He sounded like a good man. I wish I'd had the chance to know him. And I promise you, Mags, I won't let him suffer.'

He squeezes my shoulder then and I reach up to clutch his hand, eyes closed, trying to suck enough oxygen into my lungs that I won't cry again.

When I'm sure I'm in control I open my eyes and smile at him.

'Thank you.'

He dips his head, then turns and walks away. A moment later the door swings shut and I am alone among the silently dying.

He has given me a gift. I won't have to make the decision to end my brother's life. I thought I could do it, but I am beginning to understand that I cannot trust myself any more. The person I thought I was, the person I made myself into, was a lie. And the lie is cracking like an eggshell, gradually exposing something new and white and clean.

I take the bus back to St Jerome's, hoping to compose myself during the journey. I can't confront Jody like this.

A work chat will sort me out. It's one-ish back home so Jackson will be back from his Saturday morning run. I take out my phone and hit the contacts, but on my way to his name I overshoot and Daniel's rolls up. I let my thumb hover over the number. Then I swipe up to Jackson's. Back down again. Back up. I am about to make the call to the US when the phone shudders in my hand and starts to ring. I almost drop it with shock. A number I don't recognise.

I answer.

'Miss Mackenzie?' A woman's voice, vaguely familiar.

'Who's calling?'

'It's Tabitha Obodom, Jody's social worker. The police have told me that you believe Jody pushed your brother over the stairwell. They said you're considering bringing a civil case against her.'

'And you want to try and talk me out of it.'

'Yes. Yes, I do.'

I tell her I'm not interested in hearing any bleating about Jody's difficult childhood. I tell her that England is too damn full of bleeding-heart social workers making excuses for criminals. I'm about to say that maybe the US states with the death penalty have got it right, but stop myself just in time.

'Please,' she says. 'Just a few minutes of your time. Just hear me out.'

It's nearly nine o'clock on a Saturday and this woman is prepared to come to St Jerome's for what is bound to be a hostile confrontation and then travel home again in the dark and cold. I think of Abe. I think of his kindness.

I say yes.

27

Mira

I am watching at the window when you finally come home, Loran. I see you park the car but you don't get out straightaway. You just sit there in the darkness. I wonder if you are looking at your phone, if you are texting *her*. But if you were on your phone I would see the glow of the screen.

After a while you get out. I watch you walk around the side of the building and then, just as you are about to disappear, you look up. I shrink back from the window in case you accuse me of spying on you, but you are not looking at me. You are looking at the flat of the man who fell, where the sister now lives. It cannot be her that you like — this affair was going on before she came. Before Abe fell.

Poor Abe. He always seemed such a kind, gentle man. The first time he spoke to me it was to offer help.

I was taking out the rubbish. Three big black bin bags, mostly clothes I knew I wouldn't wear again after the birth: tight dresses and short skirts, clothes from my life before. They weren't heavy, just awkward to carry, and it was taking me a very long time to get them down the stairs. Then I heard a door open on our floor.

'Let me help you with that,' he said.

'It's OK,' I said, but he was already taking two of the bags, his fingers brushing my own. I followed him down the stairs, a few paces behind so as not to have to speak to him. The only sounds were the rustling of the plastic and our quiet footsteps on the stairs.

We went out into the sunshine. It is not often sunny in England and when it is it makes me feel strangely sad. It makes me think of long summer evenings at home, lying out in the back field with my sisters, always barefoot, our feet nut brown and dirty as children's.

The sun on my skin was like a lover's caress and I paused for a moment with my eyes closed, letting its warmth spread across my face.

When I opened my eyes he was waiting for me at the corner of the building. I went to join him and we walked together to the bin store around the side. It's a disgusting place; in the summer I could not open the windows because the whole flat would smell of rotten meat and fermenting vegetables. As we threw down the bags there was an explosion of flies that made me jump back. I tripped on the handle of a toy pushchair and almost fell, but he caught my arm just in time and pulled me upright.

He smiled at me as I adjusted my scarf.

'Why do you wear that?'

'For modesty,' I said.

He laughed. 'You can be too modest, you know. Let the world see how beautiful you are. It doesn't last long.'

English people think it is funny to insult you,

and I thought it was funny too. I said, 'Actually, I am seventy-two.'

At first he just stared at me — it took him a very long time to realise an Albanian Muslim woman was making a joke. Then he laughed so loudly it echoed around the building. I liked the way he laughed, with his whole body. He was so long and lean it was like a violin bow bending back.

My face warmed up, and my heart beat faster, and I realised I was feeling something I hadn't felt for three years, not since we left Albania. I was feeling attraction for a man. I didn't want him to leave.

'You are my neighbour,' I said.

'Only for the past two years,' he said, and his brown eyes sparkled. 'When's the baby due?'

I told him. 'I hope you have — ' I gestured, poking something into my ears.

'Not at all,' he said. 'This place could do with a bit of life. Perhaps your husband will let me babysit.'

'Perhaps,' I said.

For a moment he just looked at me, and my smile faltered a little, because I knew the look was not one of desire, but of pity.

Then he looked over his shoulder. I followed his gaze and my blood turned cold.

You were watching us, Loran. You were standing by the corner of the building, with your work bag slung over your shoulder, and your face was white. Then you stepped to the side and were hidden by the building.

'I have to go,' I said, and hurried away, but

when I got to the main entrance you had already gone.

I heard Abe's footsteps behind me, and I was so scared that he would try and talk to me again and you would see that I set off at a run across the waste ground, and the children in the playground laughed to see a Muslim woman running, with her black dress flapping.

You were nowhere to be seen, and when I got to Gordon Terrace and glanced back, I saw Abe going back into the building. I stood there for some minutes, feeling foolish, until the boys in hoods came walking down from the high road and then I hurried back to the building.

You were very late back from the pub that night, and I lay awake, terrified that this time you would actually strike me for behaving like a whore with another man.

But you did not.

You came into the bedroom and undressed and showered, then you made love to me, silently, in the darkness. I was glad because I had made you jealous.

* * *

Your key in the lock makes me jump. I did not hear your footsteps on the stairs.

I wipe my face with a tea towel and run my fingers through my hair. It is short now, because you said it is more becoming for a married woman. I slip on my shoes too, because you say it is slovenly to go about barefoot, and I am smiling when you come into the room.

You do not know that I know.

'*Si je?*' I say.

'Mira.' You smile back at me, but the smile only reaches your lips. Your eyes are so sad and you drop your work bag onto the floor as if it was heavy as a dead body.

'How was your day?' I say.

You shrug. 'We are delivering cement.'

'Oh.' I smile, blinking, wondering how it can be so difficult to talk to a fellow human being. It is as though we are from different species. But the way you look back at me, I think perhaps you feel the same. You take out your phone.

'Look at these. They are cement silos. They store the powdered cement. We have to climb all the way up there.'

You point to a tiny ladder, as narrow as a zipper, running all the way up an enormous cylinder.

'You must be careful,' I say. 'If you fell you would break your neck.'

'Better that than falling into the silo. Then we would drown in cement powder and no one would ever know.'

'Don't say that. It's horrible.'

You are quiet then, frowning, lost in thought. Have I upset you again?

'I have made us *gjelle*,' I say brightly.

After a moment you raise your head. 'I can smell it. It smells good.'

'The baby makes it smell bad!' I say, and try to laugh.

Then, as if he is listening, he starts to kick.

'He is kicking, look.'

I know you would not want to see my bare belly so I press my dress against the bump and a tiny lump appears. The heel of the baby's foot.

I fear you might be disgusted, but now your smile goes to your eyes. You come over and cup your hand around it, and then it pushes out even further.

'He knows his father's touch,' I say.

For some reason tears start to my eyes, and then I see they are in your eyes also.

You stand with your hand on the baby's foot and we both know that you are only touching me to get to him.

'Does he hurt you?' you say.

'A little. Sometimes.' Your smile fades and I am sorry.

That night you do not go to the pub, but go to bed straight after dinner and cry yourself to sleep. When I thought you were happy with another woman it was bad, but this is worse. Loran, what are we to do?

28

Mags

Tabitha is short and overweight, but close up she has the face of a supermodel, with smooth, glowing skin and tranquil black eyes.

We sit down at the table, streaked blue by the street light through the stained glass. I've made her tea and myself a strong coffee. Even though the heating is up to maximum I feel cold and stiff, like an old woman, and I hunch over the steaming mug, trying to inhale some of the heat.

She sips her tea and then puts the mug down deliberately. 'Have you spoken to her yet?'

'I've tried, but she's not answering her door or phone.' I don't add that I'd been fully intending to break in if she hadn't had the locks changed. That must have been what she was busy doing when I was at the hospital.

'She'll be scared.'

'With good reason.'

'She didn't push Abe, if that's what you think,' she says. 'She was obsessed with your brother, but it was an infantilised thing. A pre-teen sort of crush. Had he shown the remotest interest in her sexually, she'd have been terrified. That's why she fixated on him. I think she knew underneath that he was gay and that she was perfectly safe from any adult involvement.'

'What makes you so sure she didn't do it?' I

say. 'Seems to me British social workers have a habit of giving their clients far too much benefit of the doubt.'

She looks down at her mug, then back up at me.

'Your brother suffered her attentions with very good grace. She was happy with the status quo and his gentle discouragement did her no harm. There was no reason for her to challenge things. Even if she wanted to hurt him — and these letters contain no suggestion of that — she's far too physically and mentally fragile. She could never have managed it.'

I experience a strong sense of déjà vu. 'She could have caught him off guard. There's evidence of a struggle. And who knows what she's capable of when she lies all the time.'

Tabitha sighs unhappily. 'I accept Jody is strongly self-delusional. She constructs these fantasies because the truth of her past is so unbearable. Her father was never in the forces, if that's what she told you. He didn't die in a plane crash and her mother didn't kill herself.'

I hold her gaze. Nothing she can say will surprise me.

'She died of a drugs overdose a year or two after Jody was taken into care.'

So what? Plenty of care home kids go on to lead fully productive lives and don't become pathological liars/stalkers, so if she's trying to make me pity Jody she's going to be disappointed.

'It wasn't because of the drugs,' the woman goes on. 'Why she was taken into care.'

I sigh. 'Go on then.'

'Her father was part of a paedophile ring. Jody and her stepbrother were traded back and forth between men on a farm in Surrey.'

Sickened, I look away. Then I look back and my lip curls. 'Sure this isn't one of her tall tales?'

'Five of them are serving life sentences in Woodhill Prison. Jody has severe internal injuries consistent with sustained abuse as a very young child.'

In the silence that follows I can hear the swings creaking in the playground below.

Abe and I used to play in the local playground next to our house. We had to be back in time for tea or my mother would shout across the little park to us: 'Daddy's waiting to say grace!'

It was humiliating when that happened and the kids mercilessly ribbed us for it, so most days we made sure we were back in good time. But one day when Abe called me, I ignored him and carried on playing in the sandpit.

He climbed in and yanked at my sleeve. 'C'mon Mary.'

'Git tae fuck!' I shouted and pushed him so hard he fell back and banged his head on the wooden wall.

The playground fell silent.

Then all around me I heard the little thunks of spades and buckets being dropped into the sand. Abe was crying but no one went to comfort him. I knew why. I could feel the presence behind me, huge and dreadful as a monster from a nightmare.

'MARY MACKENZIE.'

My father's voice was like a sonic boom echoing across the mountains behind the estate. Slowly the other children got up.

'I don't think she heard Mrs Mack calling,' one of them said shrilly. I never did find out who had tried to defend me, but Christ, they were brave.

He ignored them. His shadow turned the yellow sand grey.

'Up.'

I carried on playing, grimly pouring sand from a plastic watering can to make a little hill that never got any higher.

'Up.' His voice was getting quieter and quieter.

The last dregs of sand poured out. The watering can was empty. I stared at it until the bright colours smeared together. It was so quiet you could hear the hairdryers growling in the salon down the road.

'Git tae fuck,' I said quietly.

He yanked me up so forcefully my humerus fractured with an audible crack. I screamed, and carried on screaming as he dragged me by my broken arm across the cold concrete, through our side gate and up the steps to our kitchen. Only when I started vomiting did they call the ambulance.

The doctor at Inverness Hospital had used my father's roofing firm and wasn't inclined to question his account of how my arm 'twisted awkwardly when I pulled her up'. It didn't sound so bad, and it wasn't a lie. The other children told their parents what I'd said and they all thought I deserved it. At church the following

Sunday he received understanding pats on the back while I, the cross he had to bear, sat in the back pew with my arm in plaster, fiercely ignoring the shaken heads and pursed lips. For a time after that I thought about cutting his throat when he slept, but later I discovered boys and decided there were plenty of better ways to get my revenge.

The bastard ruined my childhood, but compared to Jody's he was Ned Flanders.

'I understand you're angry, Mags.' I bristle at her use of my first name. 'You feel humiliated and betrayed because you believed her. Don't be. She's very convincing. Of course she is. Even *she* believes everything she says, on some level at least.'

'So, what does she say when you pull her up on it?'

'We don't engage with her on those subjects. We let her tell herself the stories she needs to to keep herself strong.'

'You let her delude herself?'

'It's harmless. The whole thing about reliving your past to come to terms with it has actually started to be discredited. Everyone deals with trauma in a different way. Some people in Jody's position experience a complete fracturing of their personalities as they try to block out what happened to them. They become drug- or drink-dependent. Jody was a self-harmer for a while. But now she manages to keep her head above water with anti-anxiety medicine and a few harmless delusions — like the delusion that your brother was in love with her.'

'But they're not harmless, are they?' I get up and walk to the window. 'What about when she cried rape against those boys — her foster brother and his friend? That could have seriously affected their futures, and probably did do some real damage. Enough people still think there's no smoke without fire.'

When she says nothing I turn around. Tabitha is looking at me.

'Don't tell me you actually believe her.'

'Yes, I do.'

I laugh. 'Oh, get real. She's a fantasist! I feel sorry for her, I really do, but she's clearly mentally ill.'

Tabitha's expression hardens. 'Abuse often leads to mental illness, but does that mean we should never believe a victim? Of course not — although abusers have used it as a defence for years. You must have read reports of the more high-profile cases in the press, bringing up the troubled lives and suicide attempts of abuse victims. It's done deliberately to trigger doubt in our minds: they're unstable, their story's a fabrication. Poor politician, or TV presenter, or whatever, their lives turned upside down by these twisted liars.'

She pauses to take a breath. 'That's the line those boys' lawyer pulled. It's called discrediting the witness and it should have been made illegal years ago.'

'It's not discrediting if it's true.'

'She was covered in bruises. Her vagina was torn.'

'Self-inflicted.'

Tabitha shrugs. 'That's what the judge said. Anyway.'

She gets up and the dark velvet skirt falls from her wide hips like a theatre curtain. I never thought a fat woman could be so beautiful. She is wearing a wedding band and a diamond solitaire engagement ring. I bet her children adore her.

'You know what I've often thought,' she says softly. Her eyes are nearly black, with just the merest prick of light at their centre. 'If I were a rapist I'd choose someone just like Jody. A self-harmer. A fantasist. Someone no judge in his right mind would believe.' She picks up her tapestry bag and slings it over her shoulder. 'Easy meat.'

After she's gone, I open my first bottle of wine.

Sunday 13 November

29

Mags

I wake at midday with a pounding headache and a churning stomach that only eases after I've made myself sick in the toilet, the sounds of which presumably echo through my neighbours' pipework.

Every half hour throughout the day I knock on Jody's door, but either she's too scared to answer or she's managed to slip out without my noticing. Even though I know it's pointless, that she will just spin me some new line, I have to speak to her. I know she knows more about what happened to Abe than she's letting on, and I can't believe the police are letting her get away with flat-out lying to them.

By three I'm going stir crazy, so I decide to head to the Food and Wine to stock up on booze for later.

Letting myself out of the main door onto the deserted waste ground I'm suddenly convinced that Tabitha has already spirited her out of my clutches and I go around the side of the building and look up at her window. It reflects the darkening sky. I wouldn't even know if she was up there looking down at me.

I realise I'm not alone. The junkie from Flat Seven is standing behind the bins, smoking. She's wearing a short black lace dress and patent leather high heels. Heavy make-up masks the worst ravages of her face and I suspect whoever she's waiting for won't care that the wasted arms protruding from the lace sleeves are track-marked.

Turning to leave I tread on something slippery and, fervently hoping it's not a used condom, glance down. The flowers are incongruous among the fast-food wrappers and nappies. Some are still crisp and pink, the others, now brown and dying, must once have been — I catch my breath — white.

They're the flowers from the tumbler on Abe's bedside table.

The man who broke in must have thrown them down here on his way out of the building. His reasons for doing this are as mysterious as his reasons for taking them in the first place.

I go back the way I have come, then head across the waste ground towards Gordon Terrace.

Passing the playground I experience that familiar, unnerving sensation of being watched and turn, expecting to see the cat, or the junkie gazing balefully from the corner. But the grass is deserted and Lula's curtains are still.

I catch movement from the corner of my eye, but am only in time to catch a flicker of shadow disappearing down the other side of the building. Just someone on the way to the car park, perhaps, but in that case why didn't I hear them come out of the building?

Could it be Jody waiting for me to leave so she can creep back home?

I call her name, and my voice sounds lonely and small in the silence.

A minute passes.

Should I go after her? Assuming it is her. Assuming it's anyone.

No. I turn back and set off quickly for Gordon Terrace, where I'm relieved to see a mother with a double pushchair trundling towards the high road. I fall into step behind her.

Coming back from the Food and Wine half an hour later with two clanking blue carrier bags, I see a man waiting outside the main door. On the concrete beside him sits a large Amazon box.

As I come next to him I see the address label.

M. Ahmeti, Flat 11, St Jerome's Church

It must be something for the baby. A flatpacked cot or a baby bath, perhaps. She can't lug it all the way up the stairs on her own. And I can take the opportunity to thank her for preventing me murdering Jody.

'I'll take that up for her,' I say to the delivery man, averting my eyes from the piercings through his cheeks and eyebrow.

'It's OK,' he says. 'She's on her way.'

'In that case I'll wait and help her up with it.'

Opening the front door I see Mira's shadow through the frosted glass, coming down the last flight of stairs. After a glance at the piles of post — no telltale white corners protrude from the pizza menus — I hold the door open for her.

She doesn't catch my eye as she comes out.

'I'll help you,' I say loudly and clearly. 'With the parcel. Up the stairs.'

'No, no,' she mumbles. 'Is OK. I manage.'

She moves past me, wafting the baby-scent of talcum powder, and takes the electronic pen the delivery guy is holding out for her. I didn't plan to — though I don't know why I hadn't thought of it — but something makes me glance over her shoulder as she signs the screen.

The man slides the parcel through the doors and then goes. Mira bends to lift it and I take the other end.

We climb wordlessly. Past the first floor of the cripple and the bitter queen, past the junkie's flat and the fat man's, up to our floor. The alcoholic, the fantasist, the poison-pen writer.

She murmurs words of thanks and insists that she will be fine, but I just smile as she fumbles with her key, and say nothing. Perhaps she knows what's coming.

Once she has the door open she kicks the box through, slides through herself, and tries to close it again.

But I'm too quick for her.

The door twangs against my foot and I force it open, driving her back. We stand in the darkness of the hallway, both breathing heavily.

She says nothing. She knows why I'm here.

'You saw something, didn't you? The night my brother died. That's why you wrote the notes.'

I wait for her to deny it, but she doesn't. There is a rustle as she leans against the wall and takes a shuddering breath.

'I am a bad person,' she whispers. 'I think Loran is having affair with her.'

I sigh, disappointed. I thought she was going to tell me something real. Thought perhaps she might have seen what happened after all. Though Jody has managed to pull the wool over my eyes quite spectacularly, of one thing I am certain: she only had eyes for my brother. Is it possible that Mira, living right next door, could not have known that?

'I don't think so. Jody was in love with my brother.'

'I know, I know. I just stupid, imagining things.'

'Like what?'

'What?'

'What were you imagining you saw between Loran and Jody?'

And then, without warning, the door behind us opens and a man walks in. For a moment he is just a shape in the darkness, but at once the whole atmosphere in the room changes.

As the light flicks on Mira starts, her shadow jumping on the wall.

A big, Eastern European man, dressed in heavy work clothes, his black boots crusted with cement, a rucksack slung over his shoulder. He's in his early thirties, at a guess, with a broad, flat face and fair hair cropped so short he is almost bald.

How can he have come up without us hearing? Unless he was trying to be quiet, to catch us doing something wrong.

I glance at Mira. All colour has drained from her face.

But when I turn back to the man, I see that he too is as white as a ghost and he's staring at me with an intensity that makes my heart pound, his eyes passing down my body, then up to the parka's fur trim around my neck. Is he going to hit me?

He's blocking my exit and there's nowhere to go but back into the flat. I finger my phone in my pocket.

'What do you want?' His voice is low. The thick accent, so familiar these days, is suddenly threatening.

How much has he heard?

'I was just helping your wife with this heavy box.' My voice is steady and I force myself to meet his cold gaze. Surprisingly, he doesn't seem able to hold mine, and his grey eyes flick away. 'Can't be long now until the baby comes. You must be very excited.'

He glances at his wife and she stares back at him with wide dark eyes. She's afraid of him. Does the bastard beat her up?

'Thank you, thank you, it was very kind of you,' she gabbles, trying to herd me to the door but he doesn't move to let me past. The hands hanging by his side are large and rough.

I straighten my back to let him know that he isn't scaring me and the parka rustles softly.

Then something in him gives. His shoulders sink, his head drops, and we both press ourselves into the wall as, without another word, he stalks up the hall, wafting the smell of sweat and dust.

Kicking open the inner door, he crosses the room, dumps his bag on the sofa and kicks off his boots, sending chunks of cement skittering across the floor. Then, as if I'm not there, he starts undressing, dropping his bomber jacket where he stands, then pulling off his T-shirt before disappearing from sight. A moment later I hear the characteristic drone of the shower pump kicking in.

'Thank you,' Mira murmurs. 'He would be very angry if he knew what I had told you.'

'That's OK,' I manage, but I'm barely listening. I don't even glance at Jody's door as I make my way back across the landing and into the flat, where I close the door and lean against the wood, breathing heavily in the darkness.

I don't know what it means yet, but I know what I saw.

As Loran undressed I caught a glimpse of a tattoo just peeping from the waistband of his tracksuit bottoms. A tattoo of rearing hoof and the feathery tips of a mane, inked in red.

Loran is Redhorse.

The stereo is on too loud for her to hear what they're saying, so she just gazes out of the window at the terraced houses flashing past. Felix's friend is driving too fast, occasionally slugging from a can of lager. So is Felix, and he already seems drunk, though the other one doesn't seem to be affected by it. Occasionally he glances at her in the mirror and waggles his eyebrows. She thinks he's trying to be funny so she smiles, but when his eyes go back to the road she shuffles along the back seat, out of his line of vision. The car's in a disgusting state: fast-food cartons litter the footwells, the upholstery is stained and clotted with mud, and CDs and magazines are scattered on the seat and the parcel shelf. The front covers of the magazines either feature bare-chested muscular men or almost-naked young women. The lower half of one of these front pages has torn away to reveal an article entitled: 'Potting the Brown — honest, love, me knob slipped!' The picture is of a woman's buttocks in a G-string. It makes her feel sick.

This boy is Felix's oldest friend. They'd been at nursery school together and only separated when Felix got into the grammar school. The other boy's parents had sent him to a private boys' school specialising in sport, but the two of them play rugby every Sunday morning for the local club. He is much bigger than Felix, as if a cursor has been put in the corner of a normal nineteen-year-old and then dragged out a bit.

Today is Monday and Felix's parents won't be back from their long weekend in Whitstable until the following morning, so both boys have bunked off school for a bit of fun. They persuaded her to join them and for a while she was flattered, but now she just wishes she was at school. Back at the house they made no attempt to include her in the conversation; in fact, they positively excluded her, whispering and giggling in corners like seven-year-olds.

She gets out her phone to check the time.

'No phones!' the big one barks and thrusts his arm between the seats, beckoning with his fingers for her to hand it over.

She does so automatically and regrets it immediately. She has always been too biddable. To her horror he throws it straight into Felix's lap.

'Any dirty selfies?'

'Felix, please.'

She watches helplessly as he scrolls through the shots of the rooftops she took from her bedroom, the dead Red Admiral butterfly, and the unsuccessful attempts to capture the full moon. She squeezes her eyes shut and waits, but when it comes, the explosive jeer makes her start from her seat and the inertia-reel snaps tight across her chest.

'What are you, a fucking stalker or something?' Felix shouts over the music.

'She's probably got a pair of your skiddy pants

under her pillow!' the other one bellows.

She makes a dive for the phone but Felix whisks it out of her grasp. On screen is the close-up of his face that she took when he fell asleep on the sofa.

'Ahh, bless,' the big one coos. 'Ook at iddy biddy Fewix!'

'Piss off!'

Her breath is speeding up. In a minute she will cry, which will either antagonise or encourage them.

But then a worse thing happens. Felix has carried on scrolling through the pictures and now he comes to one she has forgotten she even took.

'Shit,' says Felix, holding the phone out for his friend, who snatches it and stares at it even though he should be concentrating on the road.

'Gross,' he says, tossing the phone back again. 'You got some disease or something?'

She had been trying to get a picture of the scars on her buttocks and thighs, to see if they would be visible if she wore boy-leg bikini bottoms instead of the Bermuda shorts she habitually wears on their family trips to the pool on Saturday mornings. They were. They fanned out from the inadequate strip of fabric, still livid purple, despite the doctors' promises that they would fade to white.

'Self-harmer,' Felix says, dismissively.

She is about to correct him — he knows at least some of the things that have happened to her — but changes her mind. Self-harming is something this other boy can get his head around. The other stuff isn't.

'Wow, you really are fucked-up, aren't you?' he says, craning his bull-neck to try and catch her eye in the mirror.

'Drop me off here,' she says suddenly.

She waits, with her fingers poised on the door handle, for the car to slow, but it does not.

'Let me out,' she says, her voice rising. 'Please!'

'Ah, come on,' Felix says softly. 'Let's just forget it, mate. Let her out.'

'Keep your hair on,' the friend says, but his tone is gentler. 'I'm only messing with you. Look, we're there now.'

And now it's too late — they're driving through the gates of the rugby club.

30

Mags

He was screwing my brother while his wife was pregnant.

And now she's trying to protect him.

At least, I assume that's what the notes were for, to deflect attention away from him and onto Jody. And there's only one reason I can think of that he needs to be protected.

If he pushed Abe.

Mira knew he was having an affair but she thought it was with Jody. Maybe Abe threatened Loran that he would tell her the truth. A heterosexual affair she might be able forgive, but a gay one? Especially as she's a devout Muslim. She would leave him. He would lose his child.

So he pushed Abe over the stairwell to keep him quiet, and Mira saw him do it.

She said he was at the boxing club the night Abe fell and I rack my brains to think of the name Derbyshire told me.

Stone's.

I look it up on my phone and find an address. North from here, in the no man's land between Crouch Hill and Hornsey, places I never knew existed before. I check my phone app and find a bus that runs from the high road.

Stop. Think.

Do I really want to get into this?

Jody's one thing — a mentally unstable, rather pathetic young woman, physically weak and easily intimidated. But Loran is something else. If he did push Abe then he's capable of anything, and clearly his wife is scared of him. He could simply kill me and then head back to Albania. Am I prepared to risk that just to find out the truth?

But have I ever risked anything for Abe? Isn't it about time I did right by him and ensure that whoever hurt him is caught and punished?

I scroll through my phone contacts for Derbyshire's direct line, but then I hesitate. With only a few porn photos and some sexy texts to go on, who's to say she'll do anything?

I need more.

I'll just have to be careful.

Checking the address one more time I let myself quietly out of the flat and head downstairs.

There's no sign of the gang and for once Gordon Terrace is busy with people returning from work. Under the dull orange street lights the faces look sallow and ill. No one gives me a second glance. A woman in the baggy, chequered trousers of a chef lets herself out of a front door and I hear a snatch of the boisterous family life going on within. She closes the door and walks down the path, looking tired.

I follow her to the high road.

The bus stop is crowded and I ease myself into a gap beside a pushchair with a listless toddler staring at a tablet.

It's getting colder. Under the canvas of the

Converses my feet are numb. I wiggle my toes and stamp my feet, smiling at the toddler who glances up at me with blank eyes. The eyes of her mother, who is arguing with someone on the phone, telling them that *it's not fucking acceptable*.

The rush hour traffic is heavy. Nose to impatient tail. Checking the bus app to see if it's close I see a text has come through from Daniel.

I've told Donna I can't try again because I've met someone I care about. I know it's tough with your brother but if we take it slow . . . ?

I'm seriously not in the mood.

I told you I didn't want a relationship.

The bus arrives and I get on. Instantly too hot in the parka, but feeling somehow protected by its bulk, I find a seat at the back. Heat is pumping from vents by my calves and the shudders of the engine pass straight through my spine, but at least no one can sit behind me.

I realise that I am scared. It's an unfamiliar feeling. An unpleasant one.

The buildings thin out as we turn off the main road and we speed up, past rundown housing estates and warehouses with all their windows broken. The few cars parked by the kerb are scratched and dented. Some have crude signs offering them for sale at a paltry few hundred pounds. The pavements start emptying out, leaving only the drunks and the elderly and a few hurrying schoolboys. The street lights cast their faces in a gritty orange glow.

As we draw nearer to the gym the bus and a car behind us are the only vehicles on the road. If I thought the area around St Jerome's was bleak, this place is infinitely worse.

Perhaps all this fuss I am making is for nothing and Derbyshire was right all along. Perhaps Abe did kill himself. If I lived here, I would.

I hear the rumble of trains before I see the bridge.

Stone's Boxing Club is set into the arches beneath. Surrounded by a concrete forecourt, the door is a slab of metal, its windows protected by metal shutters. The effect is almost comically macho. I presume Mira has never been here, because surely even *she* couldn't miss the fact that this is a gay gym. To my relief, just above the main door is a CCTV camera.

I get out and the bus roars away. The burgundy hatchback that was following us turns into a side street and I am left completely alone under a street lamp.

I take out my phone to photograph the place and see another text from Daniel.

Message received. Over and out.

I stare at it for a moment, wondering whether to reply. It's so quiet that the sudden thunder of a train passing overhead makes me start. The silence resumes, but for some reason the hairs on my back are prickling. I turn again, but the street is deserted in both directions. Then I notice a man smoking outside a pub a little way up the road. A squat, drab building with an ugly case of concrete fatigue, its incongruously pretty name is the Blue Mermaid. As I watch he tosses the butt into the road and goes back inside.

The sooner I'm out of here the better.

I take the photos and pocket the phone, then, pulling the coat tighter around me, stride up to the metal door and hammer on it with my fist.

An ugly teenager opens it. His vest and boxing shorts reveal a physique far too bulky for his years. Steroids probably. Perhaps he's hoping to distract attention from his underbite and acne, but the effect is just orcish.

He looks me up and down with an expression of distaste.

'I want to speak to the manager.'

'Stanley!' he shouts, then waddles away, his thighs so big his legs don't scissor properly. As he opens a door at the end of the corridor there's a sudden cacophony of animal noises — grunts and squeals and roars, added to the slaps and thuds of impact.

Did my brother come here? I wouldn't have thought it was his scene, but then a beautiful black man emerges from the same door and pads down the corridor to the water fountain. I avert my eyes from his buttocks, clad in shorts so tight they look painted on. *Fair enough, Abe*.

A wiry man who must be in his seventies at least emerges from a door to my right. His tracksuit is halfway between street style and PE teacher. He's even got a whistle.

'Can I help you?'

'I need to talk to you about my brother, Abe Mackenzie.'

We enter a little office that looks out over the rings. Beyond the glass men are sparring, pummelling punch bags, running at each other with giant plastic pillows, and dancing around like ballerinas, all the while looking intensely, aggressively serious. I stifle a laugh.

'Sit down, please.'

I lower myself into a rickety wooden chair with a cracked red plastic cushion, oozing foam. He sits behind the metal desk, his back to the glass.

'How can I help you?' His voice is expressionless. I'm already the enemy and I don't think feminine charm is going to cut it here.

'I need to see your CCTV footage from the night of my brother's accident.'

He looks at me steadily. 'Why?'

'I want to know if Loran Ahmeti was here at the time, as he claimed to be. I have reason to believe that he might have been involved in the accident.'

'You speak like the police — only you're not.'

'No.'

'So why should I hand over the footage?'

'Firstly, because I'm asking you nicely, and unless you're trying to protect him for a reason, I don't see why you'd be reluctant. Secondly, I'm a lawyer, and if I decide to bring a private prosecution against Mr Ahmeti you will be called as a witness. If you can't then produce the CCTV footage of the night in question, the judge will want to know why you've deleted it. I imagine they will want to look more carefully at your business.'

We stare at one another. It was just a punt but I'm pretty sure this place is not completely above board. There's the illegal steroids, for starters.

'What if I say it's not working?'

'Then I'd be inclined to call the police right now.'

He sighs and pushes his chair back, gazing across at the men slogging it out in the ring.

'Loran and Abe were close,' he says. 'I don't see why he would hurt Abe.'

'Lovers have rows. Don't tell me he's not capable of it.'

'Controlled aggression,' he turns back to me, 'is not the same as violence.'

'Do you have the footage or not?'

He hesitates, then gets up.

I follow him down the main corridor and he unlocks a door that leads into a storeroom. A black-and-white screen displays the front entrance of the gym, so still it might as well be a photograph.

The scene vanishes as he flicks out the disc from the machine on a shelf beneath.

'You're lucky.' He holds it out to me. 'There's fifty-four days on there, and it only goes up to sixty before we overwrite.'

'You're going to let me take it away?'

He holds my gaze. I guess his faded eyes must once have been a quite startling blue.

'Abe was one of ours. I don't believe anyone here would ever have harmed him, but if they did, I want them caught. Whoever they are. Tell me what you find.'

'I will. Thank you.' I slip the disc into my pocket and reach for his hand. His grip is crushingly firm.

It's only as I step back onto the cracked concrete and the metal door clangs shut behind me that I realise I should have stayed inside and waited until the bus was near. Pressing my back against one of the metal grilles I check the app. Twelve minutes. *Shit*.

There are no new messages from Daniel. That's that, then. It's what I wanted, I suppose. His fault if he's screwed up his chance with Donna.

Nervously I glance across at the Blue Mermaid. The strains of 'Babooshka' drift across the pavement. No one is outside, but now I notice that although the place is a complete dive, someone cares about it enough to decorate it with hanging baskets. The pink, yellow and white flowers draw my eye: they're the first flowers I've seen outside the hospital.

There's something familiar about the white ones, the single stems branching out into a knot of blossom.

No one's outside so I cross the road and approach the pub. The street lamp shining on the window means that while I can't see in, those inside will be able to see this lone woman approach. I must be quick.

It only takes a moment for my suspicions to be confirmed.

The white flowers are the same as the ones in

the jar on my bedside table.

Growing beside them are some yellow pansies and the same kind of cerise blossom I found around by the bins.

Is this the pub Mira mentioned? The one Loran went to after the gym?

Could it have been him who had my brother's keys?

Him who came in that first night and took away the dying white flowers?

Was he also planning to leave some fresh pink ones? The ones that ended up by the bins so that Mira wouldn't find them?

Were they a token of love? Or guilt? Or just flowers for the dead?

There's a burst of laughter from inside. If this is Loran's pub he could arrive at any time. I turn and start walking quickly, back the way I came.

But coming level with the gym I see there isn't a bus stop on this side of the road. The closest one is behind me, just past the Mermaid, but now three men are standing outside, lighting up. I will walk further up.

As I cross the side road I notice the burgundy hatchback that was following the bus, parked just down from the corner. The driver is still sitting in the driving seat.

It's Loran Ahmeti.

Our eyes meet.

I'm halfway across the road now, so I keep walking — perhaps he hasn't recognised me in my hood — but once I've passed out of sight around the corner I quicken my pace.

The hood against my ears muffles the sound,

so I pull it down. I'll be easier to spot but at least I'll hear any footsteps following me.

But why should he be following me? How can he know what I have discovered? He might just have been heading for the gym and stayed in the car to make a phone call, or wrap his hands, or whatever boxers do.

Unless someone at the gym has warned him I was there asking questions.

The sound of a car engine starts up.

I quicken my pace. Glancing behind I see the burgundy nose of the hatchback nudging out of the turning. From the angle I know that it's not going in the direction of the gym.

He's coming for me.

But I'm in luck. A van is coming down the main road and he must wait for it to pass before he can pull out.

I run.

Up ahead the road divides into two, but they're both dead straight. Whichever one I take he will see me. The bus stop is a few hundred metres down the left fork, but there's no one waiting and the houses that surround it are in darkness. Then I notice a little way down the right-hand fork there is a block of shadow. It must be the entrance to an alleyway. A place to hide.

There's no time to think of another plan.

With the brief shield of the passing van I sprint across the road and dive down the alley.

Broken bottles crunch under my feet and, as my eyes get used to the gloom, I make out high breeze-block walls that have been liberally graffittied. There's a faint glow coming from the other end of the tunnel. It must be a short cut linking the left and right forks of the main road. That's my escape route if he comes this way.

But I can't hear the engine any more.

I ease along the wall and peer out. There's no sign of the car. Maybe he wasn't coming for me at all. Maybe he just went home.

Deciding to wait a bit longer before emerging I retreat into the safety of the darkness, my ears pricked for any sound. A couple of cars go by, none, I think, the hatchback, and I back up further to escape the sweep of their headlights.

I have no idea where I am, or what time it is. I just know I don't want to be here when the pubs chuck out. I'll check when the bus is due and then dive out at the very last minute. I take out my phone and tap it into life, casting this small section of the alley in a cold pool of light.

Loran Ahmeti is standing a few feet away from me.

I try to run but he grabs the hood of the parka and yanks me back, throwing me against the wall. His grip on my shoulders is iron, thumbs driving into my clavicles. I scream, but the sound is swallowed by the high walls. No one is coming to help me.

31

Mira

You are back.

I hear the car engine and look out of the window. As you head towards the building you glance up again at Abe's window.

Is it the sister that you like? She is a fine, handsome woman. Handsome in a European way, like a man almost. She does not wear feminine clothes, or sparkly make-up, or curl her hair into the full waves of the women in magazines, but perhaps this is what you like. Perhaps England has spoiled you for farm girls like me.

When I hear the door go I check my appearance in the black mirror of the oven and am smiling when you walk in.

You do not look at me.

You go straight over to the sofa and open one of your fitness magazines. The knuckles on your right hand are bleeding and there is a cut on your forehead, just below your hairline. I thought you always wore gloves to box. I wonder whether to ask about it, or just to bring you a bowl of warm water and cotton wool. But your jaw is set and your brow is low, so I leave you alone and start slicing tomatoes for supper.

The flat is so silent that I hear when your breathing catches. I wait for you to cough. I will bring you a drink. But you don't cough. Your breathing shudders, and then you are sobbing.

The pages of the magazine flutter in your shaking hands, making the glossy brown flesh smear.

Drying my hands I go over to you and kneel down. It is difficult now that the bump is so big.

I take your injured hand and am glad when you squeeze it back. Your grip is so tight it hurts, and you look at me with red, hollow eyes. How long have you been crying?

I wonder if she has ended things with you. Or perhaps you ended it. For the baby. For us. I know what it feels like to lose someone you love, and though I should feel jealous, I just feel pity for you.

'It's all right,' I murmur in our language. 'When the baby comes it will be all right. I promise. We will love him. That is all the love we will need.'

You grip my hand so tightly the bones crunch together and the eyes you turn on mine are beseeching as a child's.

'I'm sorry,' you say. 'Mira, I'm so sorry.'

When you say my name I start to cry.

32

Mags

If I'd worn my stilettos I might have been in with a chance, but the Converses are too soft to hurt him as I kick out wildly.

The phone light went out when I dropped it and the darkness is filled with my snarling cries and his grunts as he tries to restrain me. I give up on the kicking and start trying to knee him in the groin, my brain spinning through all the ways he might kill me. A slash of broken bottle across my throat, those big hands strangling the life out of me. Or simply kicked and beaten and left to bleed in the darkness.

Then, a miracle. My knee makes contact and he grunts and loosens his grip.

I twist free and bolt for the light of the road, slipping on the remains of old takeaways, the blood rushing so loudly in my ears I can't even hear if he's coming after me.

The alley elongates impossibly, the road becoming more distant the faster I run. My thighs burn, my lungs ache, my veins are electric wires of adrenaline.

I've almost made it. I can see the bus stop.

Someone there! A stocky skinhead in a bomber jacket. If I scream loud enough he will surely hear me. I open my mouth.

A hand slaps over it and I am wrenched back into the darkness.

As I slam into his body I can feel the slabs of muscle moving against my back. He must be twice my weight, strong enough to lift me off the ground with one arm, until I'm thrashing through air, trying to bite the fingers clamped around my mouth.

In desperation I jerk back my head. There's an explosion of pain in my sinuses and a sickening crack. And then I am falling, free. Landing heavily I roll onto my back. Oblivious to the glass and food slime I kick out at him, aiming for any target that might come within range as I scramble crablike towards the road.

But this time he doesn't try to stop me. He stands back, holding his palms up. In the light from the street lamps I can make out his face more clearly. His lips are moving; he's saying something I can't make out over my screams.

Eventually my voice grows hoarse and still he has made no attempt to silence me or murder me. My adrenaline is subsiding, taking with it my last ounce of strength. Dragging myself to the wall I lean there, panting. For a brief moment of silence we just stare at each other.

His face is white. A black line of blood runs from his hairline to the bridge of his nose, but he makes no move to wipe it away. His big, pale hands hang by his side.

Then he speaks.

It takes me a beat to make out the heavily accented words.

'Will he be OK?'

Slowly, my eyes never leaving his face, I get to my feet, clinging to the wall for support.

'Abe. Will he live? Please tell me.'

I manage to get my breath under enough control to be able to speak. 'Didn't you see what happened?'

He tips his head back and closes his eyes. A single star is visible in the strip of night sky above us.

'I wait for him here, at Stone's. He never comes. I get back to the church and all I see is blood. Only . . . blood.'

His voice breaks.

I wait. After a few moments he lowers his head and crosses his arms over his chest, as if he's cold. 'Will he be OK?'

'No, Loran. He's going to die.'

He stares at me, his face a wax mask.

'Hey! What the fuck's going on?' The skinhead is standing at the entrance to the alley.

'It's all right,' I manage. 'I'm fine.'

I want to talk to Loran, to find out more about his relationship with my brother, but before I can stop him he has spun away from me and stumbles into the darkness.

The skinhead runs over and helps me up. He is about to race after Loran, but I manage to stop him, gabbling that I fell, that there was a misunderstanding, that it's complicated. He doesn't believe me, but clearly decides it's safer not to get involved in a domestic and contents himself with walking me to the bus stop.

When the bus finally arrives I barely have the strength to raise my foot to the step. My shoulders hurt where Loran gripped me and I slump into the first seat I come to, letting my head loll against the shuddering window and wondering if I'm going to be sick.

I thought he was going to kill me.

I thought he had tried to kill Abe.

It turns out he just loved him.

33

Mira

You lie with your back to me and I am curled behind you, holding you. You have stopped crying and we lie on the sofa in peaceful silence.

I think about touching you. Perhaps making love will ease your pain. But your body is not my possession: I must wait until you want it.

But our moment of intimacy is passing. I know you sense it too, because your body is gradually stiffening and your breathing becomes lighter.

'You must be hungry,' I murmur. 'I will finish dinner.'

You sit up to allow me to pass, but I can tell you are still heartsick. I sit beside you and take your hand. You look up at me and I know you want to speak. You are ready to confide in me. I am glad. This terrible thing has brought us so much closer.

I squeeze your rough hand and whisper to you in our language, 'You are a good man and I know you are suffering for what you did. But it's OK. I am glad. It shows you love me.'

A shadow passes across your face. 'What?'

'I know you pushed him.'

'Who?'

'Our neighbour. You pushed him down the stairwell.'

You stare at me a moment, then snatch your

hand back and shrink away from me. 'What?'

'You pushed him because you thought he wanted me.'

You shake your head, your eyes wide with shock.

'I saw you. It's all right. I understand. It was the only way you knew how to express your love.'

You jump to your feet. 'What are you saying?'

'Don't worry. I will never tell. It will be our secret. It will bring us closer to — '

He lunges at me, grasping me by the shoulders and shaking me. 'Shut up!'

'It's all right. It's all right.'

'Can't you understand? I could *never* have hurt him!'

He is shaking me so hard my head waggles. 'Stop. You will hurt the baby!'

'Fuck the baby!' he screams in English, loud enough for all the flats to hear. 'The baby is a lie. It was not made with love! We are nothing to one another, Mira, don't you see! I married you because it would make it easier to get to the west, where people like me can live without fear. The baby will be nothing too. Hollow. An empty shell.'

'Don't say that!'

'I cannot do this any more. I cannot do it to you, or to me.'

'Stop. Where are you going?'

'Goodbye. I am sorry.'

'No! No! Come back!'

He is making for the door but I fly after him, screaming. 'You cannot leave me! I will not let you!'

You have taken my looks and my spirit and now you will abandon me. My child will be a bastard.

I fling myself onto your back, my arms around your throat. If you want to leave you will have to kill me first.

34

Mags

After I've watched the footage I save it onto the iCloud and eject the disc.

Then I just stare at the black screen. I can't make sense of anything any more.

Abe was happy; he didn't want to kill himself.

Loran loved him, he couldn't have hurt him.

And yet Jody is lying.

And Mira is lying.

Why?

Why?

Once more I stalk out of the flat and cross the landing to Jody's.

'Answer the door!' I hammer it with my fist. 'You need to tell me what happened, Jody. Because I know you know a whole lot more about this than you're letting on.'

The flat is silent. The insolent spyhole stares me out. I kick the door, making a sound like a gunshot, and I think I hear a whimper on the other side.

'I know you're there,' I hiss into the gap between the door and the frame. 'You really don't want to go up against me in court, Jody. You don't stand a fucking chance.'

After a few more growled threats I head back to the flat and pour myself a drink, staring moodily down at the playground as I knock it back with grim determination, swiftly followed by two more.

I'm dozing on the sofa when the buzzer sounds. It's Jody's social worker, Tabitha. I ignore it, but she won't go away and eventually I buzz her in. Maybe she's trying to get through to Jody too. Maybe with her social worker for backup Jody will open up and speak to me.

For a woman of her size she's very fit because within a minute there's a sharp rap on my door.

'Can I come in?' she says curtly when I open it. Her lips are tight and her black eyes flash.

I move to let her pass and she marches up the hall, swinging her bag, and then turns on me.

'You need to stop harassing my client.'

'What?' I splutter.

'Jody. You need to leave her alone. You're making her ill.'

'You are kidding me, right? The only person doing any harassing around here was *her*. I just want her to tell the truth about how Abe fell. To go to the police.'

I walk past her to the kitchen and grab a beer from the fridge, smacking it open on the worktop, taking a chunk of MDF with me.

'The truth?' Tabitha laughs grimly. 'When no one believes you then it stops being truth and becomes slander. Last time, when she was raped — '

'She wasn't raped!' I cry, slamming the bottle down so that the froth surges up and over the counter. 'That was made up, the same as everything else!'

'When she was *raped*,' Tabitha goes on, 'they

threatened her with jail. And with all the things that would happen to her there. This is a girl who has been abused as far back as she can remember. What would you do under those circumstances, Miss Mackenzie? Would you agree that you had lied, or just been confused, and go back to trying to live a quiet life where no one bothers you? Or would you put yourself through the horrors of a trial and all that that would rake up? Would you face those boys, with their upstanding families and admiring teachers, across the courtroom and hear yourself branded a liar and a fantasist and worse? The truth costs, Miss Mackenzie. And that cost is too much for people like Jody.'

Her chest heaves.

I am about to say that I don't give a shit what happened before: this is about *my brother*, but she has turned away and is riffling through her bag. Hopefully she's getting her phone to call a cab. There's no chance of her helping me winkle Jody out of her hiding place now.

She straightens up and slams a blue document file on the table.

'What's that?'

'Proof. People like you need that sort of thing, right? Have a look, and then please tell me how you can live with yourself, bullying that poor girl.'

'She's a woman,' I say. 'Not a *girl*. She's responsible for her actions.'

But Tabitha isn't listening. Swinging her bag over her shoulder she turns and stalks out of the flat, and a moment later I see her squat black shape hurrying across the waste ground.

I stare at the file, swigging my beer, then I flip it open.

⋆ ⋆ ⋆

Ten minutes later I'm vomiting in the toilet. The mixture of beer and wine, alongside the medical reports and photographs contained in Tabitha's file, were too much for my now daily hangover.

Eventually, when the last of the bitter yellow bile is flushed away, I straighten up and wash my face, then I go back to the living room and tuck the papers back into the file.

Jody's haunted, seven-year-old eyes gaze at me from the last page I tuck back inside.

It was harrowing and pitiful reading, but it's not the proof Tabitha claimed it to be. I can believe Jody experienced all that, and I'm sorry for it. But that doesn't mean she's not capable of crying rape. In fact, I'd say the opposite is true. That kind of trauma could fracture a personality completely. She might have believed she was raped, like she believed Abe loved her. I'd been coming round to agree with Derbyshire that Jody wasn't physically capable of pushing Abe, but now I'm not so sure. Can psychosis give you unnatural strength?

Flashes of the demon child from *The Exorcist* pass through my mind and another wave of nausea washes over me.

All I want is another drink but instead I put two slices of freezer bread in the toaster and sit down at the table to plough through them, dry. I haven't bothered to put the light on and I'm glad I didn't when I see that the gang's back in the playground, their shadows darkening the bench, the roundabout. One is on a swing, the tip of his cigarette drawing a red line through the darkness. I suppose it would only have been four or five years ago since he was asking his mum for a push. What happened to him? To all of them?

I think of another darkened window. Above the front door of our semi in Scotland. I'm standing in the hall. My brother is close beside me. My mother's head is silhouetted against the window. I'm looking up at her from a long way down, so I must be very young. I'm wearing my favourite orange coat with the blue stitching and am too hot. I wish my mother would let us go outside but she won't. We are waiting. Waiting for my father who is sitting in front of the TV in the living room. Cold white TV ghosts loom and shrink across the threshold.

It wasn't dark when we put our coats on to go for a walk in the park, but me and my brother had been rolling a penny along the floorboards and had taken too long getting our shoes on, so my father decreed that if he had to wait for us, then we would wait for him.

I don't own a watch or know how to read a clock but it feels as if we have been waiting hours. My brother is crying: silently, because otherwise my father will punish him, but I can hear his thick, wet breathing.

I'm not crying. My five-year-old mind seethes with loathing.

When the theme tune of his programme

comes on and he's finally ready for what must now be a short stroll around the block, I refuse to move, and am carried upstairs, thrashing and scratching, and hurled onto my bed hard enough for the centre slats to splinter. My brother is taken out by my father to collect a fish-and-chip supper for the three of them. I'm given nothing to eat that night, and in an act of defiance I refuse to eat for the whole of the next day. By evening my mother is in tears but my father asserts that I will eat when I'm hungry, and of course he is right. As I tuck gratefully into my mother's macaroni cheese I despise him more than ever.

This was some years before the Great Conversion, while my father was just your average domestic tyrant, rather than one with God's stamp of approval. The Conversion (or *breakdown* as it was referred to in a doctor's letter I steamed open) happened halfway up a mountain on a volunteer's training exercise for the mountain rescue.

He came down from that exercise convinced that Jesus had spoken to him from the sky.

It's not fashionable to be a Christian fundamentalist any more. There's something a bit twee about arguing over the consistency, fleshly or otherwise, of the communion wafer when compared with the beheadings and immolations indulged in by other brands of religious lunacy. It's almost comforting. But growing up tiptoeing around the hair trigger of my father's rage was exhausting and terrifying.

He had always been a bully and now he had

God to back him up. Our home was run like a prison camp. If we showed any form of dissent we were locked in our rooms and starved until we begged forgiveness for dishonouring the Lord. Sometimes I think my father got mixed up who was God and who was the self-employed roofer, but, deluded as he was, he was clever enough to understand the concept of divide and conquer when it came to his children. When one of us was in disgrace the other was treated like a prince or princess, so we learned to view one another as the enemy.

Looking back, I was far worse than Abe. In fact, I was a monster. Like my father.

There's *my* proof, Tabitha. There's my excuse.

A scream shatters the silence.

I jump up from the table and run to the door, bursting out onto the landing.

'Jody!'

But the screaming isn't coming from Jody's flat. Mira's door is open I run inside. The door at the end of the corridor is open and I see on the floor, lit by that single harsh light bulb, a wide smear of blood, as if a body has been dragged across the room.

Has Loran lost his mind and killed his wife?

A woman sobs.

Mira.

Was it her all along? Were the notes to deflect attention from her own guilt rather than Jody's? Did she kill Abe because he was having an affair with her husband? And has she just killed her husband?

The sob becomes a moan of pain.

My thoughts fracture. I am losing their thread. Has Loran attacked *her* and then fled?

I burst through the inner door. Mira is bending over the back of the sofa. For the first time I am seeing her in normal clothes, without the abaya: a pair of cheap supermarket jeans and a flowery shirt I have seen for sale in the market in the high road. At first I think the jeans are black but then I notice that the cuffs are pale blue.

They're not black — they're drenched in blood.

She looks up at me, her face the colour of marble.

'The baby,' she says. 'Help me.'

The clubhouse smells of stale beer and the floor is sticky with spillages. It takes a moment for her eyes to adjust to the gloom because the two boys didn't want to open the curtains and risk being seen. They had left the fire exit open the previous day for the express purpose of coming here to get lashed for free and nobody had noticed.

Felix's friend is already behind the bar.

'What's your poison, Jody?'

'Coke, please.'

He gives a bark of laughter. 'Vodka and Coke it is.'

The boys play pool and drink steadily, pints of lager with Jack Daniels chasers. Every time the big one knocks one of these back he gives a violent shudder accompanied by a loud grunt. He's like an animal, she thinks, and as the afternoon wears away he starts to smell.

Felix does too. Sweat beads his forehead and upper lip despite the fact that the clubhouse is getting cooler as the sun goes down.

The next round of drinks takes two of them to prepare, and for a moment they stand with their backs to her, whispering.

Felix brings hers over, another vodka and Coke, and something makes her glance into the glass. There's just the slightly flat brown liquid and a slice of lemon from a glass jar behind the bar. She does what she did to the previous four drinks and pours sips of it into the pot of the ailing yucca plant when their attention has turned back to the pool table.

'Let's have a look at those scars, then,' the friend says when they've finished their game.

She stares at him. 'I . . . '

'Come on, I'll show you mine.'

His hands go to his belt and before she can say anything he has dropped his trousers. He's wearing tight white underpants that cling to the outline of his large penis, flopped over to one side. He hesitates a moment, his eyes on her face, then the corner of his mouth twists into a smile.

'Nah, not there, darling,' he says. 'My knee. Tore my cruciate ligament.'

There's a long scar running down from his lower left thigh, across the kneecap and down to his shin.

'Oh,' she says vaguely. 'That looks painful.'

'Felix has got one too.'

'I know.'

'Ooooooooooooh!' he jeers, falsetto, and Felix pretends to smile.

As the next game progresses she catches them glancing up at her frequently, then the friend says, 'How are you feeling?' His voice is thick with drink.

'Fine, thanks.'

'Fuck's sake,' he slurs. 'You some kind of bionic girl or what?'

She doesn't know what he's talking about.

Felix puts the cue down and straightens up.

'It's getting late. We should go back.'

The big one rounds on him. 'Don't bail on me now, you pussy.'

'I'm ready to go too, Felix,' she says.

'I'm ready to go too, Fewix,' the other one mimics.

She goes to the door and stands there.

The big one bangs the cue down on the table, making both her and Felix start. Then he grins. 'You love him, don't you, eh?'

She hesitates. 'He's my brother.'

'It's more than that, though, isn't it?'

She stares at him. She doesn't understand what he wants her to say. What can she say that will make him let her go?

'We know what happened to you, and that's really shit. Seriously.'

She blinks at Felix. He has told?

'But it's not like that normally. You should try it again. You'll like it. And I bet you've learnt stuff, haven't you? I bet guys would pay you for the shit you know now.'

He's coming towards her.

'Felix.'

Felix moves closer to her.

'Go on, mate. Show her. Show her how nice it can be. For both people. For all of us.'

Felix's Adam's apple bobs, then he turns to face her. His skin is waxy. 'It's all right,' he says. 'You love me, don't you?'

She blinks and nods. She does love him. And he loves her. He would never hurt her.

'Go on, mate,' the big one murmurs. 'My cock hurts.'

Felix pulls off his T-shirt and stands bare-chested in front of her, like one of the men from the magazines. Then he raises her hand and lays it against his chest. His skin is clammy with sweat and she can feel his heart throbbing beneath her palm.

'See, Felix here has never done it before. Not properly. He needs a girl with a bit of experience to show him how.'

Felix's eyes are closed. She wills him to open them and look at her. They can both leave. Go home. Eat bacon sandwiches in front of the TV.

'Jesus Christ, mate, just get out of the way!' Felix stumbles aside and it's the other one looming over her, his alcohol breath hot on her face.

She doesn't say anything when he squeezes her breast, sniggering like there's something comical about it, rubber fruit from a comedy sketch. She doesn't ask him to stop. She knows there isn't any point. He won't stop, whatever she says. She has seen that look in men's eyes before. The cold, glazed stare of a shark. He is drunk and aroused and Felix has told him everything about her. He knows a hundred men have fucked her. He probably thinks: what's one more?

'Come on, Jode,' says Felix. 'Don't cry.'

'Shut up and have another drink, you queer. It might give you some balls.'

Felix watches from the pool table, slugging from a bottle of Jack Daniels, as his friend pushes his slimy tongue into her mouth, right to the back, as if it wants to slither down her throat.

Automatically she makes her throat go flaccid. She learned a long time ago how to deal with the gag reflex.

Her mouth is stretched as wide as it can go without the corners of her mouth cracking. Then she feels a familiar burn in her nipple as he grips it between finger and thumb and twists.

'Enough,' says Felix. Tossing the bottle onto the table he staggers over, pushing his friend out of the way.

Her heart lifts. They will go home now. Felix will never see this monster again.

But he doesn't take her hand. There's a hard glitter to his eye, and as he leans into her she can feel the lump in his trousers.

'Way to go, Felix!' the other one crows.

She loves him, so she kisses him back, even as tears trickle down her cheeks. The sweetness of the Coke has turned bitter on his furred tongue. She wraps her arms around him, her palms on his warm back, pressing him into her as if to protect him from what is going to happen.

Then his body moves away from hers and she thinks he is going to stop. She will run out of the clubhouse, then, and along the residential streets until she finds a bus stop.

But he's only giving himself room to allow his fingers to slide inside the cups of her bra. They're cold and wet from holding glasses with ice.

She tries to push him away then, but he holds her tight, his fingers digging into her clavicles. There is a grim look on his face now. His friend watches hungrily.

She closes her eyes. The hands squirm inside her bra, and now she feels others at her back, slipping the hooks of her bra in one deft move.

The bra loosens and these hands, bigger and rougher than Felix's, move to the button of her jeans. One boy is behind her and one is in front.

Fingers worm into her knickers then push their way inside her.

'Shit, man, she's loose as an old granny!'

Felix stops then. The hands cupping her breasts go still, his tongue freezes in her mouth.

Then the other one is pushing him aside. 'My turn!'

Felix staggers away, dazed, his mouth glistening with spit. He stares, stupefied, as his friend propels her forward until she comes up against the pool table. The impact makes her fold at the waist, and with the hand at her back, she is forced face-down onto the table.

Her jeans are yanked down.

'No!' she shouts, but it becomes a grunt as air is forced from her lungs when he slams inside her.

He is big. It hurts. That dull ache in the cervix, the tearing caused by the friction of dry skin against dry skin. He should have used lubrication. If she had known this was going to happen she would have got some liquid soap from the toilet.

The rough baize scrapes up and down her cheek.

Let it be over, just let it be over.

But he is drunk. It will take ages.

She raises her eyes up above the ledge of the

table, to the soft gold light filtering through the thin curtains. Through a gap she sees an expanse of grass and she thinks of horses running, the wind in their manes, their tails flipping.

Then, to her surprise, he grunts and withdraws. The hand is lifted from her back and she tries to straighten but it comes down again with a slap. And then he is back inside her again, but now he's only semi hard. This is even worse. He'll never come like this. He will blame her.

It flops out. He has lost his erection.

'Out of the way, you gay. Let the real man finish off.'

Felix?

She stares, green suffusing her vision.

It was Felix?

The thrust is so hard her hips slam into the wooden table edge and she cries out with pain.

'That's it, bitch,' the friend pants, through gritted teeth. 'That's what you want, isn't it?'

The big hands slide underneath her chest and start twisting her nipples. Is he one of those who gets off on inflicting pain? Will he twist until he draws blood? He's growling like a dog.

Hurry up and come.

Hurry up and come.

Hurry up and come.

Someone is being sick. The warm splatter hits her foot. She manages to raise her head enough to turn it, her chin scraping the baize. Felix squats beside the pool table, his face ghost-grey, his eyes hollow as they stare back at her, unseeing.

A thrust so hard she screams: from the pain in

her hips, and a deeper pain inside. The sensation of something breaking open.

They said she might be able to have children.

Felix is sick again.

And then the heavy body slumps over her, squeezing the air from her lungs. She can't breathe. She will suffocate. She struggles and he moves away, his still-hard cock twanging out of her with a wet sucking sound.

As soon as she is able she straightens up and stumbles around the table, making it a barrier between them as she pulls her jeans up, refastens her bra and yanks her T-shirt down. She's panting like a dog. She must control her breathing or she will start hyperventilating. She needs to stay in control. She needs to get out of here in case they decide to do it again, or something worse.

Felix's friend is at the bar drinking his JD and Coke.

Felix is still crouched on the floor, like a trapped animal. His eyes are wide with pure, cold terror.

She runs to the fire door and they don't try to stop her.

Monday 14 November

35

Mags

By some miracle they save the baby.

A little girl. With no name because Mira was so sure she would be a boy.

Mira has lost several pints of blood, and for a while it looked liked she might not pull through, but she has. I sit quietly by the bed as she slumbers in the peaceful depths of the anaesthetic. Somewhere in the hospital my brother slumbers too. I will go and see him when I have the strength to get up. At the moment all I can do is drink my warm sweet tea and stare at the light from the traffic outside strobing across the bed sheet.

At first I thought it was Loran — that he had kicked her or pushed her across the room, thought her insistence that he had done nothing was just to protect him. Again. But the nurses said that the bleeding was caused by something called *placental abruption*, and was due to Mira's high blood pressure.

The door opens quietly and a nurse comes in carrying the baby swaddled in a white waffle blanket.

'Would you like to hold her?' the nurse whispers. 'I'm sure she'd like some human contact until Mummy's feeling better.'

And then, without warning, this tiny scrap of

flesh and bone is placed into my arms.

She is as light as a paper kite.

'Are you sure?' I stammer. 'What if . . . ?'

'What if you break her?' The nurse chuckles. 'You won't. She's a fighter, this one. You can lay her in the cot afterwards.' She gestures to the Perspex box by Mira's bed.

'What if she starts to cry?'

'She won't be hungry yet awhile. You're safe for a bit.'

And then she's gone, and it feels to me as if there's only me and this tiny girl in the whole world.

I pull back the blanket from the downy cheek. Her eyes are open a sliver so I hold my hand up to shield them from the light above the bed. They open a little wider, then a little wider, until I'm looking into a pair of huge eyes, as dark and glistening as pools of tar.

'Well, hello,' I murmur. 'I guess this is a first for both of us.'

She starts to wriggle and whicker like a pony. Afraid she's about to cry, I loosen the swaddling to give her a little more freedom of movement. A tiny arm shoots out, pink and skinny, the fingers splayed. I raise my forefinger and touch it to the little palm and the fingers close around it.

The grasp is so tight. *Never let me go.*

When she falls asleep in my arms it feels — and though I want to, I cannot find a better phrase than one of my father's — like a blessing.

I raise her up until she's resting on my chest so that I can hear the high breaths. Damp strands of black hair are plastered to her forehead, as if she has exerted herself forcing her way out into the world.

When I feel myself slipping into drowsiness, I lay her in the cot, tucking the waffle blanket underneath her. We all sleep.

The entrance of the nurse wakes me.

Mira is awake now too. She lies there staring at the ceiling, as the nurse quietly checks her blood pressure.

'Can you manage a little breakfast?' the nurse says to her.

There are slivers of morning light through the blinds. I glance at my watch: 10.15.

'A cup of tea, please,' she says meekly.

The nurse turns to me. 'And one for you?'

I thank her. My mouth is furry from yesterday's drinking.

The baby starts to squawk. An impressively assertive noise for such a tiny creature, and the nurse lifts her from the cot and places her in Mira's arms. As she looks down at her daughter, tears roll down Mira's cheeks.

The nurse tugs back her gown and pushes the baby to her breast. The little hand appears, batting at the air, and then it settles, comfortable against her neck, and the breaths become muffled. I feel a pang of envy.

The nurse beams. 'There. You're a natural.'

'Did he come?' Mira asks me when the nurse has left the room.

I shake my head and wait for her to crumple, but she doesn't. 'He wanted a boy. He will not like her.'

'More fool him. She's perfect. What will you do?'

'I will take her back to Tirana. We will be feminists together and all the men will fear us.'

They must hear my laugh halfway down the corridor. Mira laughs too, and with her ruffled crop, her flushed cheeks and bright eyes, I no longer see the faded, downtrodden, oppressed Muslim wife, but a strong and beautiful young woman.

Then her face becomes serious.

'I lied to you.'

'I know.'

'Then you must know why. It was weakness and cowardice. I hope you can forgive it.'

I think of Daniel. 'We're all cowards sometimes.'

'Let me tell you the truth. And this time I will swear it on Flori's life.' She looks down at the baby. 'That is my mother's name.' Then she looks up at me. 'Before the scream, before that terrible sound of your brother striking the concrete, before Jody crying, I heard voices.'

I breathe slowly, in and out, trying to calm myself.

'One of them is Jody. She sounds afraid. The other is a man's voice. Not Abe's. I cannot hear the words because of your brother's music. There are bumps and rustling sounds and I begin to think that Jody is struggling with someone. I go into the hall then, and pick up the baseball bat Loran keeps by the door in case of trouble.'

I try and picture her, eight months pregnant,

armed and ready to fight off Jody's attacker. She is certainly brave.

'There is a big bang and Jody screams your brother's name, and then he comes out. The music is louder now and I cannot hear what they are saying. There is a little light from Abe's flat, so I look out of the spyhole.'

As she looks down at Flori's head I hold my breath. This is it. This is the moment I will find out the truth. The rumbling traffic outside makes the windowpane rattle, like the chatter of teeth.

'What did you see, Mira?'

She looks up at me. 'I see Loran.'

I blink at her, trying to make sense of her words. 'This was at the time of the accident? Eight in the evening?'

She nods miserably. 'I watch them struggling, just black shapes in the dark. I see your brother bent backwards over the banisters — I can tell him by his build — and I hear Loran grunt as he pushes him, and then there is only one of them. There is a sound. A thud. Jody screams.'

She squeezes her eyes shut. 'Loran pushed Abe. And then he ran away.'

She turns her head away from me. The baby stops nursing and moves its head away from her breast, gazing up at her mother with those impenetrable dark eyes.

I take my phone out of my bag. 'He can't have done. Look.'

Logging in to my iCloud account I bring up the fuzzy black-and-white video of the railway arches, and fast-forward. Men go in and out of the metal door at high speed. The scene darkens as the clock in the corner of the screen ticks by, and then brightens again as the street lights come on. When a bald man cycles up and dismounts I stop fast-forwarding. He is folding his bike up when the door opens and Loran steps out onto the concrete. The two men pause to talk.

'Look at the clock.'

I close up on it. Mira reads out the date and time and then she frowns up at me. 'They must have changed it.'

'The guy gave it to me as soon as I asked for it. He had no time to doctor it.'

Back in full-screen mode the man on the bike walks into the gym and Loran moves away, out of picture, in the direction of the pub. I fast-forward to eight thirty, nine, ten o'clock when Loran finally emerges from the pub. Then I stop the video.

She raises a hand to her face and the tube from the catheter coming out of it starts to tremble.

'Did you see his face, Mira? Could it have been someone else?'

'He had his back to me but I knew him by his build.'

'Lots of men are built like that. Why assume it was Loran?'

'Because otherwise Jody would have had to let him into the building, and why would she if it was a stranger? And I did not hear the buzzer, which is very loud.'

I don't know the answer to this. 'What was he wearing?' I hazard.

Her eyes drift away from mine. 'It is true I do not recognise the writing on the back of his sweatshirt. It is *something something RFC*. I thought he had borrowed it.'

We sit in silence but for the rattling of the window and small rustles from the baby. I'm not sure how to put what I'm about to say.

'Mira, what motive did you think Loran had to hurt Abe?'

She shakes her head. 'I don't know. I wonder if he is jealous because your brother liked me.'

Even as she speaks I know she doesn't believe it. That she doesn't expect *me* to believe it. I don't think that what I am about to tell her will come as a shock.

'Mira. Loran and my brother were in love. Loran's gay.'

She stares at me.

'You understand? Gay. Homosexual. He loves other men.'

The swaddling blanket rustles as Mira's chest rises and falls. The baby watches her face.

Then she nods.

She looks down at the baby and then up at the ceiling, and then starts to cry, quietly first, then building and finally breaking into a sob.

I reach forward and take her fragile hand. 'I'm sorry.'

But when she looks up at me she's smiling through her tears. 'No, no, no. I'm happy. There is a *reason* why he cannot love me.'

'You need to tell the police what you saw.'

She nods, wiping her eyes. 'And Jody. You must get her to speak to them. She must know this man. This man who killed Abe.'

My lip twists automatically. 'Who's going to pay any attention to what Jody says?'

But before she can answer me the nurse comes back in. I turn around to smile at her. Only she and I know that, aside from the medical staff, I was the first person in the world to hold Flori. I am amazed to find it means something to me.

But she doesn't return my smile. 'Miss Mackenzie?'

'Yes.'

'They need you down at ICU. You need to get there quickly if you want to — '

But I'm already out of the door.

* * *

Jody is hunched, foetal position, on the chair beside the bed. Wrenching sobs shake her whole body, as if an invisible giant is punching her again and again.

I feel no anger towards her. I feel nothing. It is as if I am watching action unfold on a screen: action which I have walked in on halfway through, before I've had the chance to care about the characters.

The nurses move around the bed. Above the rustling of the clothes and sheets I can hear Abe's breathing. They have taken the ventilator away. It sounds as if he is choking.

'Shouldn't you be doing something?'

'It's too late for that, I'm afraid.' Dr Bonville is standing next to me by the bed. 'Abe is dying. Sit down, Mags. It can take a while for a person to pass.'

'So, do something! Is it because I agreed to the DNR? I take it back. I want you to save him!'

'His system is shutting down. There's nothing we can do — and it was nothing to do with the DNR. Don't blame yourself.'

I sit and the foam seat cushion gives a heavy sigh.

'His system was irreversibly compromised by the accident. He was never one of those patients that would linger for years. Better this way, I think. Don't you?'

The nurses move away and I see that all the monitors have been switched off. One by one they slip away through the blue curtains, but Bonville stays.

Minutes pass. Abe's rattling breaths become more and more spaced out and Jody is now crying quietly.

And then his breathing stops.

My eyes are fixed on his face, watching for the moment of death, to see if something tangible will leave his body. I realise then, in the cocoon of those blue curtains, that however far and fast I ran from our father's creed, I never quite left it behind.

I am watching for Abe's soul.

The sudden choked gurgle makes me cry out and Dr Bonville lays a hand on my shoulder. 'Not yet.'

I don't know how long it takes, but it's exhausting listening to those last agonised breaths. The light strengthens and shadows pass across the bed. The morning rush-hour traffic begins. Engines are revved bad-temperedly, horns are sounded.

My mind drifts back to Eilean Donan. I wonder if we were both thinking the same thing as we stood on that parapet. *Is it worth struggling on?*

It was, Abe. For you at least. You, of the two of us, made something worthwhile of your life. You were loved, and you gave love. Whatever I used to think, I am sure now that this is all that matters.

I realise that several minutes have passed since Abe last breathed. I glance up at Dr Bonville, who has stood sentinel behind me all this time.

He steps forward and takes Abe's wrist. A minute passes, then he raises his head. 'He's gone.'

⋆ ⋆ ⋆

When I was nine I borrowed *Peter Pan* from the school library and would read it, hidden under the covers, listening for my father's footsteps on the stairs. I remember so much of that forbidden book, with all its blasphemous magic.

As I gaze at the body on the bed I think of Peter, flown off into the night, leaving just his shadow, tethered with lines and tubes. I stand up and begin to pull them out, one by one — pushing back the bandages, peeling off the tape — and Bonville does not try to stop me.

Finally I can see my brother clearly. He looks like a boy asleep.

I taught you to fight and to fly, Peter says to Wendy. *What more could there be?*

And yet without me you learned so much more, Abe. You learned to care for people. That, I have never learned. And now you are not here to teach me.

I don't want to see death take hold of him. I don't want to see his lips go slack or his skin turn grey, the eyelids peel back to reveal eyes as dull as pond water. And yet I cannot tear my gaze from my brother's face.

A flash of memory . . . Abe asleep on the sofa when he should be reading his Bible. Me leaning over him, a delicious sense of anticipation blooming in my chest as I realise I have something I can tell on him for. I will be rewarded. Daddy will be happy with me. So happy he will let me beat Abe myself. I have come to enjoy my brother's tears and pleas for me to stop because they mean that my star is in the ascendant. My fingers itch to feel the slippery length of leather, the chill of the buckle that will leave such precise half-moon bruises.

I lean over and kiss Abe's lips. They are warm and soft, but no breath tickles my cheek.

I've lived in America too long to place any store by *I love you*s, but I wish I'd had time to say that I'm sorry. For all that I did to him. For leaving him alone there. To tell him that it wasn't fear of our father that made me leave, but fear of myself, of what I had become.

I think of the last, meagre words we shared. Words on flimsy Christmas cards, hastily inscribed, destined to arrive late and unlooked for.

From Abe. From Mags.

But perhaps, after all, they said all we needed to.

I know. I understand. I forgive.

Ah . . . I have to go.

When I've moved away Jody falls on his body and howls. I watch her for a moment, transfixed. This liar. This fantasist. God, how she loved him.

⋆ ⋆ ⋆

I travel back to St Jerome's in a daze, and as I let myself into the flat I can barely remember how I got there.

Somewhere in the depths of my subconscious I must have always believed we would be reconciled one day. Now I am gripped with a wild panic. He is gone and I must imprint all that remains of him onto my mind before the darkness takes him away from me forever.

I pull open drawers, looking for photographs, mementoes, anything that will let me glimpse the real him, even just for a moment.

In the bedroom I become Jody, riffling through his wardrobe, trying to catch a fleeting scent of him.

I upend the box in search of letters or email print-outs, something that will let me hear his voice again in my mind.

A Christmas card falls out.

On the front is a picture of the Eiffel Tower wearing a Father Christmas hat. It takes me a moment to realise that it is not the real thing, only the mini one from the Las Vegas Strip, and another to realise it was I who sent the card.

I sit down on the bed and open it.

Seasons Greetings from Sin City!

I can barely remember scrawling my name at the bottom, but it is clear that it was done with little care, knocked out from duty, a year late because his had only reached my desk that February.

From Mags.

I close it and run my fingers across the embossed image. Then suddenly I remember Abe's card. A snow-covered castle on an iced-over lake, a trail of ducks padding across the ice towards the distant horizon.

You and I stood on that parapet, wondering whether to drown ourselves. Whether life would ever be bearable.

But we didn't need to go under. We could walk across the water to freedom. I suppose you never meant it as a metaphor, just a reminder of that time, that single time, when we were truly brother and sister. And contained within that, the hope that we could be again.

The cards we sent were more than the flimsy paper they were printed on and the trite sentiment within.

They were a covenant. A promise we made to one another to forget the past, to do right by one another in future.

I know then what I have to do. I start packing my case.

The lady's expression is so hostile the girl wishes she didn't have to sit next to her. Her eyes flash with the light from the fluorescent strips above the table, as if there are torches shining out from behind the black irises.

'Let's just get this over with, shall we?' Her voice is dangerous.

The policeman stacks his papers on the table, as if he's not really interested. He's wearing a short-sleeved shirt and his arms are thick and shapeless as sausages. Gingery hairs sprout from the freckles. Ignoring the lady he looks up at the girl. His orangey brown eyes match his hair.

'Before we begin, I need to confirm with you that you understand the significance of an official caution and you have given informed consent to receive it.'

'The Goddards gave consent on her behalf and they're hardly disinterested parties. I should have been called way before this.'

The policeman turns on her. 'Firstly, they were her legal guardians up to today.'

Were? The girl stares at him.

'And secondly . . . ' Just for a moment he is the hissing, spitting bully who terrified her into saying she was lying — into thinking it too. 'You're lucky we decided to offer it at all. She could have gone to prison, you do know that?'

'Oh shut up, Kellan. You know as well as I do

that a jury would never have convicted her. The case would have been thrown out and you'd have been wiping egg off your face until next year. I've named you personally in my complaint to the IPCC.'

He smiles drily. 'And your complaint will be fully investigated. Now, if we could get back to the job in hand?' His amber eyes click back to her. 'Do you understand the proceedings up to this point?'

She's supposed to agree with this so she nods.

'Good. I'm cautioning you for wasting police time and making false allegations. These are very serious crimes. You do understand that?'

She nods.

'This caution will appear on any CRB certificate applied for, for a period of two years, and will then remain on your criminal record and may be used as evidence in a court of law should you commit another offence.'

'Another offence?' the lady explodes. The girl wishes she would just be quiet. She just wants it over and done with so that she can go home.

'If you continue to be disruptive I'll ask you to leave.'

'No you won't,' the lady replies. 'I'm her legal chaperone.' But she settles nevertheless, literally shaking herself down like a fat pigeon after a rainstorm.

The girl glances behind her at the door. Why didn't Mum come in with her?

'A condition of the caution is that you will issue an apology to the boys involved.'

'Oh fuck off.'

'Watch your language, Mrs Obodom.'

'She's not apologising.'

The policeman turns to the girl, smiling. 'You ever been to a prison, young lady?'

She shakes her head.

'Some of the women there, well . . . ' He looks her up and down and shakes his head.

'You aren't seriously doing this? You do know her history?'

His smile slips. 'I've read the file.'

'So don't threaten her, or it won't be just the IPCC I'll be going to. It'll be the press. They won't have forgotten her.'

The policeman pushes his tongue into his bottom lip. Finally he says, 'Because of what you've been through in the past we'll waive the apology, but if anything at all happens like this in the future — '

'Come on, sweetheart.' The lady gets up. 'We're done.'

Mum and Dad are waiting for her outside. Mum is crying. Dad's face is as grey as his suit. He steps forward as she emerges.

'Mr Goddard.' The lady shakes his hand. 'Mrs Goddard.' But her mum won't take the proffered hand. She presses her tissue to her eyes as if she can't bear to look at it.

'There's her stuff,' Dad says.

The girl looks in the direction he's pointing. She recognises the red suitcase they took to Majorca in the Easter holiday.

'If there's anything we've left out we'll send it on. Email us, please. No phone calls. I won't have my wife being upset.'

'You're doing a good job there, then.'

Multicoloured ribbons flutter from the handle of the case. She and Mum tied them on to make it easier to spot on the baggage conveyor belt.

'Let's not draw this out any longer than necessary. You got the papers I sent last week?'

He's asking the lady, but she's just staring at the girl.

'You did tell her, didn't you?'

Last week was when Felix was staying with his Auntie Carol and she was in the house alone with her parents. Up in her room mostly, listening to the creaks and ticks of the silent house, trying to breathe air that seemed to have less oxygen, as if it had all been used up by the shouting and crying.

'I'm . . . sorry . . . ' It seems like it's difficult for him to say the word, like his tongue has suddenly stiffened. ' . . . it had to end like this. Goodbye, Mrs Obodom, and we really do wish her the very best of luck in the future.'

She wonders who he's talking about, because he isn't looking at her.

He turns then and clasps his wife's arm, but she shakes him off.

'Let me talk to her, David, please.'

He makes an angry noise as she moves past him. Her face looks so weird, all red and swollen, and her eyes are crusty and half closed. She takes the girl's hands and the trembling passes into her own arm.

'We had to choose.'

'Mrs Goddard, please be mindful of — '

'We had to choose and he was our son.'

'I'm your daughter,' she says, her voice rusty from disuse. 'Aren't I? That's what you said.'

'Helen, let's — '

'GET OFF ME, DAVID!'

It's the first time she's ever heard Mum raise her voice. The foyer of the police station falls silent. Even the drunk on the bench stops humming to himself.

Her dad's face goes tight and pinched up. 'Tell her whatever you need to. I'll be waiting in the car.' And without a backwards glance, he stalks out through the police station door.

The lady waits until Mum has finished talking, and then she takes the girl's hand and helps her over to the bench. The girl watches her mum's blurry figure, haloed by sunlight, as she opens the door and follows her husband out into the warm afternoon. A moment later the car engine starts. The girl remembers the way the car always smelled of boiled sweets and Felix's trainers.

The lady hands her a tissue but her hands are so numb she can barely hold it.

Thursday 17 — Saturday 19 November

36

Jody

It's three days since you passed on. I can't bring myself to call it anything else. I have to believe that you are somewhere. That one day, perhaps, I might reach you.

Every minute takes forever to crawl by. I watch the shadows in the flat creep across the floor and then up the walls where they spread out like ink in water. I sit in the darkness. I'm still sitting there at dawn.

I haven't washed or brushed my hair or cleaned my teeth since I got back from the hospital. I suppose I must smell.

It's so quiet.

Nobody hammers on the door any more. Nobody orders me to open up and explain myself. Even the baby has stopped crying. They've gone, she and her mother, back to Albania. The husband never came home.

Your sister pushed a note under my door.

I'm returning to the US. Have the decency to go to the police and tell them what you know, for Abe's sake. Mags Mackenzie

I watched her through the spyhole as she let herself out of your flat for the last time, in her masculine suit and tight ponytail. Her high heels clip-clopped down the stairs and then the front door banged closed and a moment later there was the sound of a car engine. Her taxi? Or the man I had seen her with that night?

Now I'm alone up here at the top of the church. Tucked away in the space near the roof. If St Jerome's was still a working church the congregation would be sitting in their neat rows far, far below, heedless of my existence. Just the way I like it. You were the only one who ever made me want to be noticed. Now I will fade into the background again. A grey girl in a grey dress in a grey city, living a grey life until it's my turn to pass. Will I see you then?

Yes.

You loved me.

You gave your life for me, Abe, and if that's not love I don't know what is.

I didn't need the pills when I had my love for you to keep me grounded. Now I do. I took the ones I had left but they ran out and the next morning I felt like going up onto the roof and throwing myself off. I could see myself lying spreadeagled on the tarmac of the car park, utterly still and peaceful. The image brought such a sense of release. But what if it went wrong? I'd end up like you did, or worse. And with no one there to hold my hand.

Oh God, I can't keep thinking like this.

In desperation I call Tabby and she organises for a prescription to be left for me at the chemist on the high road.

It's so cold now. The sky a dense white with no sign of the sun. We must be nearly in December. Christmas is coming. The mere thought of that is enough to make me want to step out in front of the bus passing the corner of Gordon Terrace. Helen will send a card, maybe even a pair of slippers or a set of bath oils that I will give to the charity shop. Tabby will buy me chocolates and her daughter will bake me another cake that will only make me feel more alone.

I cross the road and go into the chemist.

The pharmacist starts to smile when I go up to the counter, but then she looks me up and down and her face falls. She asks for my name then hurries into the partitioned section where they dish out the pills.

I wait by the window, looking out over the high road.

Cosmo is busy with lunchtime office workers. I close my eyes and picture the two of us there, sitting at our favourite table at the back, where we can hold hands and kiss without anyone noticing. I picture us chinking glasses as we talk about the wedding. You want a big one — you are proud of me, you want to show me off — I want to keep it small. Just one or two special people, because really I can only think of inviting Tabby and her daughter and maybe the lady from the charity shop.

I am smiling when I open my eyes.

The smile freezes.

You are there.

You are inside Cosmo.

Standing in the shadows at the back, the customers and staff milling around you heedlessly. If I didn't know you by your build, by your clothes and hair, I would know from the intensity of your gaze. I'm pinned to the spot, unable to breathe.

'Miss Currie?' says the pharmacist. 'Are you OK?' I tear my eyes away from you. She's holding out a paper bag. Hurrying to the counter I snatch it from her and go out.

The road is as busy as usual and I'm stuck on the pavement, dancing from foot to foot as I wait for a break in the traffic.

Eventually I dive out between a bus and a minicab, ignoring the angry beeping.

The window of Cosmo is a flat reflection of the street scene in front of it. To see inside I have go right up to the glass, ignoring the strange looks of the customers on the other side as I peer in.

You've gone.

If I'd just mistaken you for someone else and that person had left the restaurant I would have seen them walking up the high road, but no one has left, I'm sure of it. Ours is the only empty table in the whole place.

I know what Tabby would say — that I imagined seeing you because you're on my mind so much.

Perhaps she's right.

I try to put it out of my head as I turn for home.

* * *

I wake up in the middle of the night.

This is wrong. The pills are supposed to help with insomnia. But I realise that something specific has woken me when I hear it again. A woman's cry, like a wail of grief.

It's coming from close by. One of the other flats on this floor, surely.

I sit up in bed, staring into the darkness.

Is it Mira? Has she come back from Albania? Has her husband found her?

But it's not a cry of distress. It's a woman singing the blues. Now I can hear the words.

I get up and go into the living room. Street light spills through the window to form orange puddles on my bare feet. I breathe shallowly, listening to the words I know so well.

It's one of your favourites. You would play it late at night, when you came back drunk. I'd watch you through the spyhole weaving unsteadily to your door, blundering your key against the lock. And then you'd go inside and the music would come on, lullabying me to sleep when the pills couldn't.

And now it's playing again. How?

I creep down the hall and, as quietly as possible, turn the latch and step out onto the landing. The woman's voice streams out of your flat, echoing down the stairwell.

I grip the banister rail and look over the edge. Unless another listener is standing down there in the darkness, I'm alone. Can no one else hear this? Is it all in my mind?

The lino is cold beneath my feet. My breathing is quick and shallow with fear, and something else. Excitement? Then it catches in my throat.

I can smell your aftershave.

The black spyhole on your door glitters as I pad across the landing.

I press my ear to the wood and listen.

But the music is too loud to hear if there is any movement inside.

Then it hits me. All the flats come with two sets of keys, but I only found one of yours: the one I gave to your sister, that she must have sent back to the housing association.

Who has the other?

Are they in there now? Playing your music? Wearing your aftershave?

'Hello?' I say clearly and loudly.

The music shuts off and the darkness throbs with the sudden silence.

For a moment I'm paralysed with fear. Was it all a trick? Is some stranger going to wrench open the door and drag me inside?

Is it *him?*

Somehow I get my feet moving. I run back to the flat, slamming the door behind me and curling up into a ball on the floor. I stay there until the sun begins to come up, but your flat remains utterly silent. Finally I creep back to bed.

* * *

Nothing's working any more.

My vision is blurry.

I keep tripping over and dropping things. I scalded myself with a cup of tea and a large blister, tight and red and fragile, has come up on my right thigh, making me too scared to wear jeans in case it pops.

My stomach churns and I have to dash to the

toilet three or four times a day.

I go into a room and forget why I went there.

My hair is falling out.

I went to the charity shop to buy some books to try and distract myself, but when the woman behind the counter tried to talk to me I felt the first stirrings of a panic attack and had to run out.

I know that some of these are side effects of the pills, but I'm scared that something else is happening to my brain.

I'm seeing things.

Ghosts.

Your ghost.

I try to tell myself that it's all in my mind, keep my head down, stay indoors, don't talk to anyone, don't look out of any windows. But sometimes I have to leave the flat for groceries, and it's then that you come.

Checking the instruction leaflet in the pills I see that in the side effects section under *Rare: fewer than one person per ten thousand* it says *hallucinations*.

But the thing is, unless I'm imagining this part as well, other people can see you too. I watch them adjust their course to let you by. I see them standing on a full bus when you are seated in front of them. I see them serving you in cafés — but when I go in you're gone.

You're always moving away from me. On a bus or walking just too fast and too far for me to catch you up.

Tabby phones. She says the pharmacist has been in touch with my doctor, saying they're worried about my state of mind. Am I all right?

'I'm fine.'

A pause.

'You must miss him a great deal.'

I cannot stop my breathing from thickening.

'Just remember, that though they're gone from our sight, the dead are always with us, Jody,' she says gently. 'In our hearts and our memories.'

How can that ever be enough?

'I think you should start going back to the group, and stick with the pills — they'll take a couple of weeks to work after a break. I think that will help any negative thinking or . . . delusions.'

I want to tell her that you aren't a delusion. If it was all in my head would people be moving aside to let you pass them? My whole life I've been told that what I'm feeling or thinking isn't real, that I can't trust myself. Tabby was the only one who ever believed me, and now she's doubting me too.

I get her off the phone, saying I've got an upset stomach. It's not a lie. Though I don't want to have to see the woman who's been reporting me to my social worker, I need to go back to the chemist for something to settle it. I walk quickly, head bent so low all I can see are people's feet, and even then my heart lurches every time a pair of Converses or brogues steps into my line of vision.

The pharmacist doesn't try to talk to me this time, or slip any more leaflets into the bag when she hands me the medicine. I'm trying so hard not to look in Cosmo that I open the door without looking and almost walk straight into a young woman. At her gasp I turn and apologise, and it's then that I see you. Standing on the corner of Gordon Terrace. From this far away it's hard to see properly but I think you're looking back at me. Waiting.

I've spent such a long time trying to convince myself that these visions are wrong, something to be ashamed of, something to fear, that for a moment I don't move.

Then a wave of love and happiness so powerful washes over me that I think I might collapse. A woman looks you up and down as she passes and then I know for sure. I'm not imagining you, Abe. You're here. You've come back to me to show me that, whatever anyone else says, our love was real.

I run.

But by the time I've crossed the road you're nowhere to be seen. I stand on the corner, panting, waiting for the dizziness to pass, my eyes pricking with tears.

As I trudge back to St Jerome's the youths peel out from the shadows, but do you know? I'm not scared at all any more. The worst thing has happened to me — I've lost you — and now I don't care about anything else.

They ask me the time but I keep walking. One of them steps out in front of me and says that his friend asked me a question so I should have enough respect to reply.

I stare at him, dull-eyed. *Do what you want.*

He looks me up and down and wrinkles his

nose in disgust. Then he lunges for my bag. There's nothing in it of value: the few pounds change from my shopping, my keys, a couple of tampons, but I cling to it like it's bursting with fifties. I squeeze my eyes shut. I know why you've come back, Abe — to tell me that you're waiting for me. Well, I'm ready.

'Give me the bag.'

I hold on tight.

'Let. Go. Of. The. Bag. Bitch.'

I hear the rasp of a knife being taken from a pocket, and the flash of light, reddened by my eyelids.

It's time. I'm ready.

'Hey!'

I open my eyes. A middle-aged man with a Staffie stands at the corner of the high road, his legs spread as wide as his dog's. He might have been muscly once but it has all turned to fat. Under the football shirt his stomach is broad and squarish, like the shell of a tortoise.

The youths smirk as they look him up and down, but when he sets off at a fast stride towards us, the dog loping along by his side, they disperse, catcalling and gesticulating as they go. The man stops at the edge of the grass, panting, perhaps with the adrenaline rush of a narrow escape, or maybe just because he's fat.

I set off across the waste ground to the church.

'Don't fucking thank me, then!' he calls after me.

I turn, suddenly angry. *Thank you for what? I was ready!* But he's already stomping back to the high road.

I watch from the window of the flat as the light drains away and the remaining street lights on Gordon Terrace buzz on. I've microwaved a plastic tub of lasagne, but find I have no appetite. Instead of the usual aroma of cooked meat and cheese, the food smells sour, like bad breath. I guess that's the pills again.

José the building manager arrives, lugging a new mattress, and disappears through the front door, so I guess someone new is going to move into the empty flat on the third floor.

A lonely night stretches ahead of me. At least I should sleep properly. I've forced myself to stay awake for the past three nights, in case you play your music again, but all has been quiet.

After throwing away the lasagne and rinsing the fork under the tap, I get into bed and put the radio on. My bedroom window glows pink. The man on the radio says there's been a sandstorm in the Sahara and for the next few days the sunsets will be beautiful.

I turn my face to the wall and try to sleep.

* * *

On Saturday morning I go out looking for you. I try Cosmo and the baker's and even the Food and Wine where you would sometimes get a newspaper, but it's as if I was imagining you all along.

I wander up and down the high road until my hands are numb with cold and I can't feel my feet.

The pharmacist comes out and asks if I'm all

right so I have to stay on the other side of the road after that.

The sun starts going down and I get scared then because it's Saturday. Match day. Buses are backing up along the high road, their windows reflecting the sky. I hurry back to Gordon Terrace and across the waste ground, pausing for a moment at the children's playground to gaze at the church spire silhouetted against the sky.

The radio was right. It's going to be the most amazing sunset. The sky is streaked with a million different shades of red. The clouds curl like petals. The wind has dropped. From up there you could look out on the whole city, all pink and glowing like it's fresh out of a hot bath.

I realise that this is the night I described to Mags. The night you and I spent together on the roof. And then, for a moment, I think I see a figure standing in one of the windows of the spire.

It resolves itself into a block of shadow and I sigh and make my way to the door.

There's no post for me, as usual, and I'm relieved, having half expected a letter with an American postmark, threatening to sue me. Passing through the inner door all the rich colours of the evening are dulled. The sunset is behind the building so the stained glass is flat and grey.

I stand for a moment, gazing at the tranquil lake of concrete that shows no sign of what it did to you. I can remember the chill hardness of it under my knees as I crouched beside you, whispering that everything would be all right as your blood seeped into my jeans.

You were looking at something far away that I couldn't see, but then you must have felt my presence because your eyes moved, locking onto mine. We held each other's gaze for a moment — a minute? An hour? — and then your eyelids fluttered closed and I never saw them open again.

He must have come down behind me, have passed by while I was kneeling beside you, but I never heard his heavy footsteps, or the crash of the door closing. He might have been just a bad dream.

Turning on the light I start climbing the stairs. On the first floor I pass the grumpy man's flat, and Brenda's husband's, and the man who plays the silent organ. When I get to the second floor I walk quickly. Here lives the junkie, who terrifies me because of what I might have been, and the man who is eating himself to death because his mother died. I have just stepped out onto the third floor when the light clicks off and I'm left in darkness.

But not total darkness.

A red light spills down the stairs from above, as if from an emergency generator.

I listen. I can hear a whispering moan. Like the sound you made as you lay dying. But this time it's just the wind. I feel it on my face, lifting my hair, pushing my skirt between my thighs.

I grip the banister and continue climbing.

When I step out onto the fourth floor — our floor — my heart catches.

The door to the roof is open.

Blood-red sunset spills across the landing. I stand at the edge of the puddle of light, afraid of what will happen if I step inside. Am I finally going mad?

The air smells of your aftershave.

Movement by the door catches my eye. A strand of wool is caught on the latch and blows in the wind.

I have to force myself to cross the landing and unhitch the strand. I run it through my fingers. Cashmere. No one else in this place wears cashmere. Only you. This thread is from the diamond pattern on your cardigan with the big collar.

There's a footprint in the dust at the bottom of the staircase. The Xs and diamonds of a Converse sole.

The dull grey concrete steps turn shimmering gold as they rise up before me, like a stairway to heaven.

They were *wrong*. They were all wrong. You *did* love me, and now you've come for me. You're waiting for me up there in the sunset.

I put my foot on the first step.

37

Mags

I admit I enjoyed it.

From my table at the back of Cosmo I could watch the street. So many times she missed me as she hurried past with her head down, but I only needed one moment. It was busy that day. I'd just sat down at the one remaining table and was glancing over the menu I knew quite well by then when I saw her. A drab figure merging with the dullness of the street, I only noticed her as she moved across the brash window of the bookie's on the other side of the road.

When she went into the pharmacy I stood up, willing her to see me. And then she did. I held her gaze, wondering how good my disguise was, whether this close she would see that my hair was straighter, my shoulders narrower. But I guessed from her expression that she was taken in and I experienced a shiver of delight as the colour drained from her face.

When she turned to the woman behind the counter I retreated to the toilets at the back of the restaurant to wait until she'd gone. I knew she wouldn't have the courage to come in.

I'd told Peter Selby that I had to stay on for a week or so to arrange the cremation. But when I explained that being around Abe's stuff was proving too painful, the sentimental old queen readily agreed to my moving into one of the unoccupied flats.

José met me at the Moon and Sixpence at the far end of the high road to give me the key and we got semi drunk. The new flat smelled of stale alcohol and urine. The alcoholic's mattress had been destroyed so he went for a new one straightaway, hoping, I suspect, he would have the chance to christen it. I pretended to be busy when he came back, tapping away on my computer, ostentatiously oblivious to his shirt-stripping and sweat-from-brow-wiping as he manoeuvred the mattress into position.

He was halfway out of the door when I called him back. He returned, grinning.

'Can I borrow the roof key?'

His face fell. 'Ah, is not safe up there. What you want it for?'

'There's been a huge sand storm in the Sahara and apparently it's going to be a beautiful sunset for the next few days. I thought there'd be a good view from up there.'

'Sounds fun,' he said, leaning on the door frame. His aftershave was so strong it gave me the same head rush as alcohol. 'Like some company?'

I smiled. 'If I do, I'll give you a call.'

He slid his hand into the pocket of his low jeans and drew out a key ring.

'Don't lose it or I will spank you.'

'Sod off, José,' I said good-naturedly, and he swaggered across the foyer and out into the dusk.

After that I just had to lay low and wait for the right moment.

This morning I had my first attack of conscience as I watched Jody cross and recross the end of Gordon Terrace, searching for her phantom lover, and decided that, for better or worse, it would have to be tonight.

As the afternoon wore away I dressed up in Abe's trousers, shirt and cardigan, gelling my hair into his careful waves in front of the mildewed mirror in the bathroom.

And now I stand here, by a windowsill strewn with dead flies, swigging from a quarter bottle of whisky to calm my nerves while I wait for Jody to come back.

I had wondered, as the barber on the high road cut my new style, whether Daniel would still fancy me with short hair, but it's an academic point, since I won't be seeing him again. He and Donna probably have a cosy night planned, with a Netflix boxset, a nice bottle of chardonnay and a takeaway curry, the kids slumbering peacefully upstairs, happy in the knowledge that Mummy and Daddy are back together.

I throw the remains of the bottle down my throat and grimace. Is my contempt just sour grapes? Christ, who knows? I *thought* I was jealous of Abe and Jody.

The glass stops halfway to the sill.

She's coming.

Her steps drag down Gordon Terrace, her shadow yawning behind her. I would have spotted her before but her colouring merges with the grey pavement.

She reaches the end of the street and steps

onto the grass. At any moment she could look up and see me. A part of me wants her to, wants her to guess what's going on so I don't have to go through with any of it. But she doesn't.

Grabbing Abe's aftershave and the key from the stained Formica dining table, I let myself out of the flat. I take the stairs two at a time and have made it to the third floor when I hear the creak of the main door opening. Swearing under my breath — I should have moved earlier — I drop to my haunches and crawl up the final flight of steps. But my luck's in — for some reason she's lingering downstairs.

I crawl past Abe's door, then Mira's and Jody's, to the other end of the landing. A sliver of grimy wind creeps under the threshold where a semi circle of dust and grit has formed. A stroke of luck.

Awkwardly lifting my knee while trying to keep my head down, I press the sole of Abe's shoe into the dust, then I slide the key into the Yale lock and turn it with the utmost care. There is the tiniest scrape of metal and then the door unfastens, swinging out towards me. I catch it, and let it out slowly, wincing as it creaks a little. I can hear her coming up the stairs. She must be on the first floor.

Careful not to disturb the footprint, I climb into the stairwell, spritz the aftershave a couple of times, then run lightly up the cement steps to the door at the top.

Opening it I am assaulted by the wind. Fortunately José has thought to leave a chunk of breeze block up here and once I've secured it I straighten up to look for a place to conceal myself.

For a split second I forget why I came and simply stare.

The sky is on fire.

Ribbons of gold and scarlet light stream west to east, studded here and there with fireballs of slowly revolving cloud. The buildings are charred black stumps, with an occasional window dazzlingly aflame.

My hands are red, and so are Abe's Converses, as if I have been paddling in blood.

Behind me faltering footsteps scrape the gritty surface of the cement steps.

She has fallen for it.

Where can I hide?

The church spire lances up into the roiling sky. There are louvred windows on both sides, one set looking out over the high road, the other looking back over endless council estates. A small door in the wall must lead to a staircase. Is this the place from her imagination? Where she and my brother had their first earth-shattering night of passion?

It's pitiful, laughable. But I don't feel like laughing any more. I dive across the lead roof and conceal myself behind the spire.

Dead leaves crackle under my feet as I peer around the edge of the wall.

I squint, unable to distinguish her shape from the pink shadows on the wall of the stairwell, until she steps out onto the roof.

Close up she is so frail I fear that the wind will buffet her straight over the edge. Beneath the scarlet wash of the sunset, her face is drained, her eye sockets dark-ringed.

I assume she is blinking because the sun is in her eyes but then the tears spill out, the sun catching them and turning them to livid scratches down her cheeks. Like some kind of martyred saint.

No. I mustn't allow myself to think she is the victim. I must go through with this. Or I will never know.

She takes a step forward, then another, catching her toe on the edge of the leadwork and stumbling, then righting herself.

'Abe?' Her voice trembles.

I step back as her eyes scan the rooftop.

'Abe, I'm here.'

She walks hesitantly towards the door in the spire; presumably she intends to climb up to the windows. I hear a rattling. The door is locked. Poor Jody. No ghostly lover waiting to enclose her in his cold embrace after all.

I ease myself around the spire to approach her from behind.

Her body is angled away from me so it takes a moment for her to register my presence.

Jerking around she cries out, her hands flying to her face, and I realise, with incredulity, that even now she thinks I'm Abe. It must be the tears clouding her vision — or the pills clouding her mind.

Either way she stands rigid with shock, which gives me enough time to get between her and the steps that lead back down to the fourth floor.

'Hello, Jody.'

The hands fall from her face. Her eyebrows contract, tilting upwards as her face crumples with disappointment. No, more than disappointment. Anguish. She gives a moaning exhalation, as if she has been kicked in the stomach, and actually bends a little at the waist.

'You've been avoiding me. What was I supposed to do?'

I sound like a Bond villain. If I were watching this tableau on a film, I'd be willing her to snatch up the brick door prop and hurl it into my face before making her escape. I set my jaw. It has to be done.

'I need to know what happened — and you're going to tell me.'

Her eyes are dull. Her face has slackened, as if a little bit of her soul has drained away. Even the wariness seems blunted because when I step towards her she makes no attempt to evade me.

'You're not getting off this roof until you do.'

Her eyes slide away from mine. It's the only signal that she is alive or conscious.

I am ready to hurt her. It is the natural progression of my behaviour over the past week — watching her, following her, deliberately setting out to unnerve her. I've been stalking her. Murder in slow motion.

'You're going to tell me what happened that night. Who pushed my brother over that stairwell?'

She is silent.

Striding up to her I slap her hard enough to make my palm sting.

Her head stays where the blow put it, angled

away from me, her wide eyes staring down at the lead.

'Say something, Jody, or I'm warning you . . .'

But I'm losing hope already.

She is a broken doll, propped awkwardly on spindle-legs, unstable, liable to collapse like pick-up sticks. I grasp her bony shoulders, my thumbs digging into her clavicles, and give them a sharp shake as if I'm trying to dislodge something that's sticking. Her head waggles stupidly, the hair falling in front of her face, hiding her eyes from me. I yank it back, hoping to see fear, but they are blank and glassy.

'You owe me this, Jody. You owe it to Abe.' I force this attempt at emotional blackmail out through gritted teeth. In a minute I'll stop trying to keep a lid on my fury. It's getting dark. Our figures will no longer be silhouetted against a treacherously bright sky. I can do what I like to her and no one will ever know.

No one would care.

No one would believe her.

'Answer me!'

I count down the seconds: three, two, one.

Time's up.

And now a part of me doesn't even want to hear her story, it just wants to hurt her, to punish her for everything that has happened. For Abe, for our shitty childhoods, for the wreck of my emotions, for her own weakness, which has allowed people to hurt and abuse her for her whole pathetic life. She is a rag doll upon whom others can take out their misery and pain. And now it's my turn.

I drag her to the edge of the roof, where the lead falls away sharply, then kick her legs out from under her. She lands on her back with a grunt and before she can roll to safety I drop down to straddle her, my hands around her throat, forcing her head back over the edge. Her hair streams out in the wind.

'This is what it was like for Abe!' I snarl. 'Hanging over that banister, wondering if the person that held him would let him go. Was it you, Jody?'

Thrusting my hips I nudge her forward, and now her shoulders are off the edge. I only manage to stop the inexorable slide of the rest of her by grasping the head of a nearby gargoyle. Its pointed tongue stretches down between its legs and I want to snatch my hand back, but if I do she will fall. Even holding onto it I feel unstable. Perhaps we will both go over.

Somewhere a police siren wails. The vertical wind pummels my chin, all scent of aftershave long gone, replaced by the gritty, musty perfume of the city.

The siren recedes and an eerie quiet descends.

She isn't screaming or crying or begging me not to hurt her. She doesn't make a sound. Her eyes simply gaze skywards, like a long-suffering saint seeking deliverance from heaven. Her chest brushes my thighs as it gently rises and falls. She is totally calm.

I ask myself again: who is she protecting?

Whoever it is it seems she'd rather die than betray them.

How far will I have to go to make her tell me?

A gull swoops down in front of us, red-backed. We must be above the bins. Jody's view, in all these years of living here, has been of people's rubbish being fought over by gulls and rats and foxes, and the occasional whore turning tricks. The gull swoops up again, a chicken bone between its beak. Eating it's own kind in its greed or desperation.

And then it hits me.

I have been so stupid.

What was it Daniel said? There's no such thing as truth. Only the story we choose to tell, to others, and ourselves. Jody is sticking to her story to the bitter end. Sticking to that fantasy in which her life is bearable, in which she is loved and needed, even though it might kill her. Because what on earth is the truth worth to Jody? Nothing. Nothing but pain and despair. But in her parallel universe she is about to follow her darling into eternity.

Who is Jody protecting? Herself, of course.

Look at my knuckles, white with the strain of gripping her. I thought I'd left that girl behind, the one who took pleasure in inflicting pain, but she's been here all along, hiding under my skin, waiting for the next victim. The next piece of easy meat.

I loosen my grip. Suddenly released, she starts to slip and I have to grab her by the arm and yank her back. For a few agonising seconds I wonder if we will both fall, but then she seems to wake up from whatever stupor she has been in, and with my help, hauls herself back onto the safety of the leadwork.

We crawl away from the precipice on hands and knees, and don't stop until we have reached the spire. Suddenly exhausted, I lean against the cold stone and let my head drop to my knees.

I have become a coward and a bully again, like my father. Was the truth worth that?

I hear a rustle and raise my head.

Jody has crawled over and slumps against the wall beside me.

The wind rises, catching the gelled wave of Abe's hair and tugging at it, as if to tear it away from my skull. The last sliver of sunset bleeds across the horizon.

'I'm sorry,' I say. 'It doesn't matter any more. He's dead. I don't care who — '

'Wait.' Jody's voice is strong and steady.

I blink my eyes clear.

'Listen.'

It's getting dark. Normally she's home before dusk. She doesn't like walking down Gordon Terrace after six, when all the residents are home from their cleaning or catering jobs and are safely tucked away in their concrete boxes, curtains closed like a charm against the packs of feral youth prowling outside.

It was the psych assessment. They were running late. Saturdays are our busiest time, *the receptionist said accusingly.* You should have booked for another day.

Sorry, *she said. She didn't even know she could; she just came when she was told to. Her appointment was for four but they didn't call her in until half past five and now it's almost seven.*

The bus sits in traffic for so long that when a more ballsy passenger punches the emergency door-opening button she slips out after him and crosses the high road. It's about a mile to Gordon Terrace, but there's no point hurrying — it's already dark. When she gets to the corner she will just have to pick her moment — a late commuter returning, a car pulling up — and sprint for St Jerome's.

Lights glare from the fast-food outlets. Someone is having an argument in the kebab shop. The manager of the Greek bakery is pulling down the shutters. They are covered with graffitied names: Toxo, Barb, Stika. Like alien planets instead of human beings. The man in the Food and Wine is shouting down the phone in a foreign language.

The traffic is solid all down the other side of the road and as she walks past a stationary bus she senses a face turn in her direction. Ducking her head she quickens her steps and is passing Cosmo restaurant when a voice calls out behind her.

'Hey.'

She turns around.

Her legs become matchsticks and she almost falls to her knees on the pavement.

'Hey,' he says again, holding up his hands, palm first. 'Hey, don't look so freaked out. I just wanted to say hello.'

If she could move she would run now, as he comes towards her. She would run in front of the traffic and be hit by the now moving bus rather than have him come close to her.

But her legs won't work and now he is so close she can smell the beer on his breath and that oh-so-familiar deodorant, with the reek of stale sweat beneath.

'Hey, Jody.' His voice is soft. 'How are you doing?'

He looks the same, only bigger, and with less hair. His eyelids are heavy. He is drunk. 'Cat got your tongue?'

'Sorry,' she says. 'I'm fine, thank you. How are you?'

'Yeah yeah, not bad. Just had a fixture against Hackney. Nailed them. On my way home to get ready for the club later — couldn't believe it when I saw you. What you up to?'

'I'm on my way home.'

His head rocks backwards and forward. He doesn't know what to say to her. If she stays quiet he will get bored and go.

'You hear about Felix?'

She presses her lip together and shakes her head.

'Mainlining heroin now, apparently. Completely fucked.'

The high-pitched sound in the back of her throat is lost in the traffic. Her beautiful Felix. Still beautiful in her mind, whatever he did to her.

'Lost half his teeth.'

'Stop,' she says. 'Please.'

'Oh yeah, I forgot you had a thing for him. Don't reckon he'd twist your lemon these days, sweetheart. He stinks.'

He's looking at her, waiting for a response. She tenses up, trying to think of something that won't agitate him.

'I'm sorry to hear that.'

And then suddenly all his affability is gone. 'You should be, though, right? I mean, all that shit with the police. That's what really sent him over the edge.'

It's like being punched in the stomach. He's saying that it was her fault, what happened to Felix. She can't catch her breath as his cold eyes drill into her.

And then he smiles.

'Hey, listen, no hard feelings, though, OK? I mean, all that shit you said could've really fucked my prospects, but it's water under the bridge now, right? I've got a nice accounting job. Good money. I'm not bitter. In fact . . . ' He grins. 'To prove it, why don't I walk you home? Make sure you get back safe.'

'Thanks, but it's not far.' Her smile is skull-like.

'Nah, it doesn't bother me.' His huge hand closes around her upper arm. 'Lead on!'

He walks very fast and sometimes she stumbles. Now she remembers how his normal breathing sounded like panting. Like a dog. Her arm is in the grip of its jaws.

They reach the corner of Gordon Terrace. The boys are there, sitting on one of the garden walls, smoking.

Hearing footsteps their heads turn as one.

They know her, know that her bag is unlikely to contain anything but a few coins and a second-hand paperback, but surely this middle-class white boy, with his bulging kit bag and expensive-looking watch, is more promising. If they accost him he'll have to let go of her to deal with them.

But his steps do not falter as they come level with the group.

'Evening, fellas,' he says and one of them actually grunts a response.

A moment later they have passed by. She turns her head and the youths gaze back at her with flat, dead eyes. Where is the shark's bite when you need it?

Up ahead, St Jerome's is a black spike against the dark sky. Perhaps he will simply leave her at the door. He's a grown man now, not a reckless teenager. Back then it was Tabby who insisted it was rape, but the judge said it was no more than raging hormones, that she had willingly taken part, until the sober light of day had brought with it a sense of shame at her own promiscuity. That she had made the accusation to assuage her own guilt. Did she give them some sign that she wanted them to do what they did to her? Over the years she has decided that she must have done, that if she had been clearer they would have stopped. It wasn't rape, just a failure of communication. Her fault.

They reach the door.

'Thank you,' she says.

'My pleasure.'

He makes no move to go. Hopelessly she slips her key from her bag and pushes it into the latch. The foyer door opens.

'Thanks,' she says again. 'Bye.'

She walks in. He follows. The door shunts closed.

They stand in the gloom of the foyer.

'I'll be all right from here.'

'A gentleman always sees a lady to her door.'

He holds open the inner door and she steps through. A sliver of light spills from Mrs Lyons' flat, illuminating a semi circle of concrete. Should she scream for Mrs Lyons to help her? To call the police?

But what would they say at being called out because someone had the temerity to try and see her safely home? She was warned before about wasting police time.

'What floor you on?' His voice is loud and intrusive. No one speaks loudly in St Jerome's. From upstairs she can hear the lilting murmur of Abe's music.

'The fourth.'

His heavy steps echo through the stairwell as he follows her up the stairs.

'This must keep you fit, eh?' he says. 'No wonder you're so skinny. I always liked that about you. If you just had some tits . . .'

'Thank you,' she murmurs.

'Wanna know the other thing I've always liked about you?'

They're on the third floor now.

She gives a wan smile. 'What?'

He gestures for her to go on and she starts the final ascent to the fourth.

'That you're so completely full of shit.'

She hesitates. Has she misheard him over the music?

'Seriously, you're fuckin famous for it. Who were your parents again? Not a pair of kiddy fiddlers who pimped you out to dirty old farmers? Course not.' He laughs.

She stares at him.

'Remind me how they died again? Wasn't your dad castrated by his cellmate? Oh no, my mistake. He was shot down over Iraq, wasn't he? A war hero. You must be so proud.'

Her chest cavity fills with ice.

'And your mum found God, didn't she? Said the devil had made her let men stick farming implements up her six-year-old daughter? Hell, Jody. What a life, eh?'

Her trembling hand makes the carrier bag rustle.

'Now.' He steps up onto the fourth floor. 'You owe me, for what you did to me back then. What you did to poor old Felix.'

'Sorry. I'm sorry. I'm — '

'And this is payback time. A good match always makes me horny, and what's one more cock for a slag like you? Now be a good girl and don't make a fuss, because you know what happens if you try to stop me, right? I fuck you anyway and then it's my word against yours. And what do you think your word's worth, Jody?' His bottom lip pokes out and he shrugs. 'You tell me, honey. I might just sue you this time, for defamation. I could have done before, but I let you off cos I like you.'

She backs towards her door, fumbling for her keys.

'Did you get that stitch Felix was on about? Hope so, cos seriously, it was like driving a minibus through the Grand Canyon.'

His laugh is a rifle shot ricocheting around the stairwell. Her fingers close around the keys. She won't have time to open the door, dash through and close it again, but if he attacks her out here, surely someone will come out.

On impulse she hurls the keys over the banister.

He grins. 'Nice try.' Then he lunges at her, shoving her up against the door so hard the wood splinters. His thumbs gouge her shoulders.

She should have screamed before, when she first saw him. She should have screamed and run and not stopped until she reached the sanctuary of the church. This is her chance. Her last chance to save herself. To be saved.

'ABE!'

He punches her. Her lip splits and warm blood flows into her mouth.

'WHAT DID I SAY?'

The lock is loose now. One more blow and they will be through, into the flat, and he will be able to do whatever he wants to her. He yanks her body into his then throws it back to ram the door. Her head rebounds off the wood but though the lock rattles it still holds.

'Hey!'

Two heads turn in the direction of Flat Ten.

Abe is silhouetted against warm light. The music is louder and the lemon scent of washing-up liquid drifts across the landing.

'Piss off, mate,' her attacker sneers.

'The hell are you doing?'

'None of your business. Piss off back inside.'

'Jody?' Abe takes a single step out onto the landing. 'You all right?'

'Seriously. Get lost.'

She watches him, holding her breath. If she has truly been imagining his love for her all this time, Abe will do as he's told and go back inside.

Another step. 'Jody? Answer me. Are you all right?'

Their eyes lock. She is dumb with fear, but she doesn't need to speak. They have such a powerful connection he can read the truth in her eyes.

His brown eyes harden. 'Let go of her. Now.'

Miraculously the other man does so. Then, in one fluid, muscular movement, he crosses the landing and throws a punch that sends Abe crashing back against the door frame.

For a moment he sways, unsteadily, but though he is slight, she knows that Abe goes to the gym under the arches every day. As the other man draws back his fist, Abe bends at the waist and powers forward, butting her attacker in the abdomen, driving him backwards until, with a hollow ring of metal, the bigger man's meaty back comes up against the banister rail.

He has to grip the rail with both hands to stop himself tipping backwards and is helpless to protect himself as Abe draws back an elbow and punches him once, twice. The bigger man's nose explodes with blood and he gives a gargle of surprise, then brings his hands to his face, swearing.

Abe turns to Jody. His face is flushed. The mop of fringe falls damply across his forehead. She can hear the whisper of his shirt against his skin as his chest rises and falls. 'You OK?' He reaches out for her with those long, elegant fingers.

She is so filled with emotion she cannot speak. He loves her. He loves her.

She reaches for him. Their fingers are almost touching.

Then the monster raises its head. Over Abe's shoulder she sees black eyes glaring from a blood-streaked face.

'No!'

Abe turns too late. The creature clamps its

thick arm around his neck and drags him to the banister. There is a sickening crunch as his spine makes contact with the metal handrail.

It all happens so quickly.

Abe's feet scuffle against the lino, and then the scuffling stops and he is kicking through air.

'No!'

He bends like a high jumper.

He is balanced on the small of his back, a human seesaw. Then the seesaw tips.

Her feet carry her to the banister and the rail crushes the air from her lungs as she strains forward, reaching for his flailing arm. She manages to grasp the fabric of his shirt sleeve, but the stitches give and it slips from her fingers.

For a split second he is frozen in time, arms outstretched like wings, an angel flying out of the darkness. Then he is gone.

38

Mags

We sit side by side against the wall of the tower. Through Abe's shirt I can feel the rough stone against my back. It is as cold as the lead beneath me, as cold as her hand resting on mine.

She has stopped speaking.

Blown by the wind her hair is a silver curtain across her face. I push it behind her ears so that I can look into her eyes. They are watery grey, red-rimmed with the loss of my brother and perhaps the loss of everything she has ever dared to value.

'Abe didn't love me in the way I wanted him to,' she says softly. 'But if he didn't care about me at least a bit, why would he have given his life to save mine?' Her eyes search my face. 'It's true. Please believe — '

I smile at her. 'I believe you.'

Then I tell her a story of my own.

My father had found my stash of the pill that I'd persuaded the doctor to give me without their consent. When I got home from school he dragged me to the bathroom and held me, fully clothed and bellowing, under the hot shower, as punishment. I was fighting him so much that he'd actually had to get in the bath with me and was suffering under the scalding flow as much as I was.

I suppose it must have been shortly after Eilean Donan. Something had changed in mine and Abe's relationship. If not actual affection, then something like a mutual respect had grown up between us. We were partners in misery after all.

Without warning, my silent, self-contained brother burst into the bathroom and started trying to pull our father off me. As he pulled and I pushed, the old bastard slipped, cracking his head on the tiles. It wasn't much of an injury, but Abe knew what he would get for it.

The beating he received for protecting me was the impetus for my leaving home. In case he tried to stand up for me again. Because as he stood there, back straight, stony-faced, while my father clambered out of the bath with blood dribbling down his scalp, I knew that he would. He may have been skinny and young and scared, but he was brave.

I don't have much faith in Jody, but I have faith in Abe. I think he would have come out of his flat to help her when she called his name. He would have taken on a bully who was bigger than him, stronger, crueller. For love. Not the love Jody's talking about, but because he cared about people. Funny. She believed in guardian angels, and in the end one came to her aid.

When I've finished speaking she is smiling. But then the smile falters and she starts to cry. 'I'm sorry. He did this for me. He saved me, but I was too scared to do something for him. I should have told the police. I just knew they'd never believe me.'

The wind buffets her narrow frame, snatching at her hair and the grey dress, with its cheap plastic beads. The things that I sneered at her for when we met — the frailty, that yearning to please, to be loved — now twist my heart.

I'm not a good person; I know this. I'm impatient, selfish, contemptuous of people more vulnerable than me. I can be cruel. But how could anyone get pleasure from hurting someone like Jody? Like a child torturing a kitten.

What kind of a man would do that? Someone who must dominate the weak to make him feel less inadequate? A calculating psychopath? Or just a mundane bully satisfying his most basic urges?

He thought it would be easy. Easy meat. He knew he would get away with it, just like he did before, because who would believe someone like Jody? They didn't the last time he raped her; they considered themselves lenient in letting her off with a caution, free to return to the wreckage of her life. But if she dared to cry rape a second time, to throw their clemency back in their faces, she would have to be taught a lesson. No wonder she was afraid.

I take her by the shoulders and look into her eyes.

'They might not believe you, Jody. But they'll believe me.'

Then I smile.

He won't know what hit him.

New Year's Eve

39

Mags

Jody and I have Christmas together. I intend for us to spend it quietly, in memory of my brother, but then I think screw it, and book us into Claridges for Christmas dinner and a room for the night. It costs more than my flights, but we agree that Abe would love the idea of us being here. He would appreciate the spa-brand toiletries, the Egyptian-cotton bedding and deep-pile dressing gowns Jody swans around in. I tell her that they, and the contents of the mini bar, come with the room and she should help herself.

Over breakfast on Boxing Day I explain my plans for New Year's Eve. She blanches, but I reassure her that everything will be all right and, trusting soul that she is, she believes me. Later she returns to St Jerome's in a cab with both our cases, while I head off to the Boxing Day rugby fixture.

⋆ ⋆ ⋆

My father loved rugby, my brother too. Just being at the ground takes me back to my childhood. A bit of sentimental nostalgia to ease my homesickness, which is always worse around Christmastime, especially since I've just lost my brother. I couldn't resist popping in on my way past, during one of the longs walks that I've taken to making, to clear my head and process the last month's sad events.

This is what I tell the old duffers propping up the bar at half-time. Clive and his merry band compete to impress me, with their stats knowledge and witty banter. They buy me drinks and I'm careful to explain that I'm off alcohol since my brother's accident *(we think drinking may have exacerbated his depression)*. Not that it stops me enjoying myself, I insist. New Year's Eve is my favourite night of the year, although this year it will be a lonely affair . . .

Right on cue they bring up the party I've known about ever since I started researching the team.

Would I like to come, as their special guest, in memory of my brother?

'I wouldn't know anyone.'

'You'd know us!' They link arms, all rivalry for my attention forgotten.

'That's very kind of you. I might just take you up on it.'

My day's task successfully fulfilled, I can concentrate on the rest of the match. I spot him at once, from Jody's description and the name printed on the back of his shirt, and am relieved to discover how easy it is to dislike him: the way he stalks the pitch as if he owns it, the casual violence against players that oppose him, the unsportsmanlike crowing over points scored, and petulant protests when they don't go his way.

He comes within touching distance of me as he clumps past on his way to the dressing room, his flesh red with the blood pumping around his muscles, nostrils flared, smelling of sweat and victory. So big, so powerful. He must think he's invincible.

I smile and wave at the old duffers making their way back to the bar, and then I head for the bus stop.

* * *

He's clearly not expecting the call. I sense wariness in the careful neutrality of his tone but he's still in London and agrees to meet me in a coffee bar in the city, near to the bank where he works. They're in the middle of a corruption scandal, he says — that's why he hasn't yet returned to the US — so he won't be able to stay long. A get-out clause if he needs one.

I arrive early and scroll through all the possible ways I can frame the request I plan to put to him, all my possible bargaining tools. Free legal advice with his divorce, sex, money. I know gangsters; I can feel that greasy aura that tells you a person has a price, and usually how much it is. Daniel never had it.

He arrives before I've come up with anything.

He looks better than last time. His skin is smooth and his hair has grown and started to curl. He wears a grey suit with an open shirt and a pink tie flops like a dog's tongue from his pocket.

Clearly married life is doing him good. I resist

the urge to say so as I rise from the low-slung armchair. It will sound like sour grapes.

He hugs me, without a word, and for a moment I let myself sink into him. His neck smells of soap. Lucky Donna.

Then I take a deep breath and pull away. He's not mine to enjoy.

'Another coffee?' he says.

'Latte, please. No sugar.'

He brings the drinks on a tray with two pieces of cake.

My lip twitches. 'No sugar, but I want a super-healthy chocolate brownie?'

'You could do with putting on a few pounds. You don't look great.'

'Gee, thanks. You do.'

Shit, *shit*. We both studiously look in opposite directions.

'I meant,' *oh, here I go*, 'you look well. You must be happy. I'm glad it's all working out. With Donna.'

He cuts his brownie in half and clears up the crumbs with a licked fingertip.

'So, what can I do for you? After some investment advice?'

I look away. I deserve the insult.

'Sorry,' he says quietly.

'Don't be. I asked for it.'

His face is as expressionless as it has been since he arrived, then, heart-breakingly, he smiles. 'You did.'

I sigh. Too late now.

'I wanted to ask you something. To *do* something for me. A small thing.'

He watches me steadily, sipping his coffee. He knows it isn't a small thing.

I feel myself redden. It's an odd sensation. I haven't blushed since I was a child. The truth is, I have no idea how to approach this. I could pretend it was all true. *Truth is what you can make people believe*. I could make him believe the lie.

Or I could tell him the real truth.

I close my eyes and step out into the abyss. 'Someone might ask you . . . about the nights we spent together.'

His mug pauses halfway to his lips.

I take a deep breath. 'There's a court case. I've made a rape claim against someone and they will want to imply that I enjoy casual sex.'

He pales. 'Bastard.'

I hold up my hand. 'It's not as simple as that, Daniel.'

He leans towards me, head bent, waiting for me to continue.

I sip my coffee and put it down on the table. It's too milky, half cold. My hand is trembling and the liquid sloshes over the untouched brownie. Sweat is breaking out around my hairline.

'I want you to say we didn't sleep together.'

He frowns. 'But I asked the concierge for condoms.'

'Say we changed our mind, decided to wait. You're the only potential weakness in the case.'

He swallows. 'What did he do to you? The rapist.'

Around us the noise of the café becomes

muffled, as if we are enclosed in a bubble, a quivering wall of tension surrounding us.

'He killed my brother.'

* * *

I tell him everything. When I've finished there is a horribly long silence during which I want to stab myself with the brownie knife. What a stupid thing to do. I came here to patch up a potential weakness in my case, and now I've blown a hole in it so big it might not even get to court. Daniel might march straight to the police and tell them everything. They'll assume I've been taken in by Jody's lies and may be lenient with me, given my recent bereavement, or they may just charge me with wasting police time. Otherwise it's perverting the course of justice. If Jody's called in, it might very well be prison for her this time — for both of us.

I blunder to my feet, knocking over the cardboard cup, spilling beige milk all over the table.

'I shouldn't have come. Please, please forget everything I told you. Please don't go to the police.'

'Mags — '

'He's a bad person, Dan. Don't feel the need to protect him. He *did* rape her, the first time. Her social worker is a hundred per cent sure. She had injuries and — '

'Mags! Calm down.'

My heart is pounding and I can barely catch my breath. I sink back into the chair. He reaches across the table, his sleeve dragging through the puddle of coffee.

I grip his hand like a terrified child, but after a moment he lets go, slipping his fingers from mine and getting up.

'Let me think, OK?'

And then he's gone. A dark shape in the huge mirror on the wall opposite me, as he passes the window and is lost in the crowd.

What have I done?

March

40

Mags

The Crown Prosecution Service team's junior lawyer, Rauf Chaudhry, is a young Pakistani who took his degree at some crappy college in the Midlands, but I suspect he's going to be seriously good one day. He's clearly rabidly ambitious and when he finds out I'm a lawyer myself, pumps me for information about the US legal system.

His official role is to keep me informed about how the trial's progressing, but I tell him that after I've given my witness statement I intend to sit in on proceedings. He does his best to persuade me to present my evidence via video-link, as is the right of all rape victims, but I tell him that I want to face my attacker. I add, with a meaningful smile, that I also want to make sure the CPS doesn't balls up the case.

I wear a navy suit to court, no make-up, tie back my hair, put on my reading glasses. My rapist will struggle even to recognise me. No one in the court will remember me after the trial ends.

I'm the first witness and the court falls silent as the heels of my low pumps tap up to the stand. I can feel eyes upon me, trying to penetrate my drab disguise, to see the trembling victim — or the drunken slut beneath.

The jury's an even split, male and female, many of them under forty, which is good. Old prejudices about women who ask for it — with their clothes or their alcohol consumption — die hard.

That's why I stuck to Coke on the night of the rape. *He said he'd tell you I was drunk*, I told the police, *that you wouldn't believe a word I said*. The blood and urine test proved I was stone-cold sober, giving credence to my assertion that I'd just gone outside for a wee because the toilets were blocked. You're not allowed CCTV in a toilet so there was no evidence of me stuffing the bowl with hand towels until the water rose to the rim. It wouldn't even occur to them to suspect.

I face the court, unsmiling, and when the judge tells me to begin I give my version of the events of New Year's Eve with quiet determination, studiously avoiding the accused's eye — no tears. I'll let the photographs and the DNA do the work for me.

His girlfriend sits, stony-faced, listening intently. Sophie, I think. A standard-issue blonde with orange make-up and heavy eyeliner. After a few minutes she gets up and walks out of the court without a backwards glance. It wasn't hard to convince her that her boyfriend was a rapist. This makes me feel better.

I talk for half an hour or so and by the end the jury are sipping their water and fanning their faces.

It's all gone very well, I think, and it's hard to stop myself smiling as the judge calls for a break in proceedings before my cross-examination by the defence barrister.

Rauf is waiting for me outside and escorts me to a witness room with a coffee machine and a sagging sofa pocked with cigarette burns.

'I think that went pretty well, don't you?' I say, flopping onto the sofa and kicking off the frumpy shoes. But my smugness falters when he turns back from the coffee machine and hands me my drink.

'What?'

He shrugs. 'It sounded like you were reading a script. The jury didn't warm to you.'

'Frankly, it doesn't matter whether they warmed to me or not. It's an open-and-shut case. Wait till they see the photos.'

He gives me a penetrating glance then, which makes me wonder if he's guessed I'm hiding something. I can't tell him the truth, or he'd have to drop the case, but I want him to understand.

'He's a violent rapist, Rauf. He deserves to go down. And he will.'

Rauf runs toffee-coloured fingers through his hair and leans against the door. 'You've been living in the US too long. The white British male is an endangered species. Look at him, standing there in his too-tight suit, big hands hanging by his side because he doesn't know where else to put them. It's pathetic. The older women will want to mother him, the white men will see themselves. It only needs one or two prepared to give him the benefit of the doubt — that he's too thick or too unreconstructed to know when a woman means no.'

Before I can say anything the tannoy system announces that the court is about to go back into session. Rauf opens the door for me and as I move past him he murmurs, 'I want us to win this case as much as you do so just try and be a bit more . . . victim-y, OK?'

I smile indulgently, but as I follow him out into the corridor I'm feeling rattled. And now I must face cross-examination. I know how brutal that can be. I've done it myself.

This time I study the jury more carefully. A young overweight woman squeezed into a wrap-dress casts fluttering glances at the defence bench as the accused whispers with his brief. Is she one of those insecure, desperate types who writes to serial killers in prison, and occasionally marries them? Perhaps she will resent me for being slim and successful. Beside her is an elderly Asian man. He's beardless and in Western dress, but may still frown upon women who go out to parties without a chaperone.

Then there's the token middle-aged white guy: bald, over-weight, tattoos peeping from his shirt cuffs, he looks like an older, poorer version of the man in the dock. Shit. Maybe Rauf was right. I need to get them onside. Show some vulnerability.

Except that's the one thing I'm no good at. I can shut down but I can't open up.

I should take some tips from Jody, who trembled so much when I said goodbye to her this morning that she resembled an out-of-focus photograph. I told her she didn't have to attend if she didn't want to, but as I take the stand again I spot her, shrinking behind a pillar on the back bench. She keeps her head down and worries at the buttons of her cardigan as the judge speaks. By contrast Mira, beside her, is straight-backed, her eyes alive with intelligence as she follows the proceedings as best she can. She came back from Albania ostensibly to give Jody and me some moral support. But I think it's more than that. I think she knows what's going on, though neither of us has told her. I think she's here for Loran. To bear witness as the murderer of the man he loved is brought to justice.

Or not.

Not if I can't play my part.

Come on, Mags, I tell myself as the defence barrister gets up. Make it up. Put on a show. Pretend you're trying to convince your dad how sorry you are for smoking at the bus stop so that he'll stop cutting up your *little tart's* clothes.

She's tall and rangy, with thin dark hair that looks painted to her head and a hooked nose like a bird of prey. As our eyes meet I think I see a flash of recognition, one lawyer to another, both of us determined to play the trial like a game of chess. I look away quickly, dipping my head, the picture of demure trepidation.

But she's not stupid.

'It takes guts to face your supposed rapist in court, Miss Mackenzie, especially since you could have given your evidence behind a screen.'

I straighten my back and swallow hard. Here we go.

'And only a very self-assured young woman would walk alone into a party full of strangers. Just as it takes a certain devil-may-care confidence to urinate outside when other women are waiting patiently for the toilet to be unblocked. And yet,' she glances at the jury, 'we are being asked to believe that you are a shrinking violet, unable to defend yourself against my client, too frozen with terror to be able to utter an audible peep to alert the partygoers to your plight.'

'The countdown to midnight was happening,' I say. 'It was too noisy.'

She ignores this. 'You are a lawyer, Miss Mackenzie. Last year you defended a man charged with evading forty million dollars' worth of US tax, a man who, in his twenties, was convicted of the murder of three business rivals.'

I scowl at the CPS barrister. *Come on: object*.

'I put it to you, Miss Mackenzie,' she says, 'that you are a strong, clever, calculating young woman who is, for reasons best known to yourself, trying to manipulate the court.'

'What possible reason would I have to put myself through this?' I say, but I know at once I have made a mistake. She has caught the flash of anger in my voice, even if the jury hasn't. She knows this line is worth pursuing.

'You left home at sixteen, am I right?'

'Yes.'

'Why was this?'

'My relationship with my parents broke down.'

'Your parents? Or just your father?'

'My parents.'

'Is it the case that you reported your father to the police for false imprisonment?'

I hesitate. 'Yes.'

'And that you claimed he had beaten you, but no charges were pursued because he said you had attacked him and fallen when he pushed you away.'

I nod. She has spoken to my teachers.

'So, there has been a history of making false claims in your past?'

'They weren't false claims.'

'Your father was retired from his role in the Mountain Rescue for health reasons. What were these?'

'He had a nervous breakdown.'

'And your brother recently committed suicide.'

'Yes.'

'Do you believe mental health problems run in your family?'

My lip curls. 'No.'

She gives a small smile, flagging up to the jury my reluctance to co-operate. The glances they shoot in my direction are closed and hostile. Rauf was right. They don't like me.

'I put it to you, Miss Mackenzie, that as a young girl with a bullying father, you developed a hatred for men that has continued to this day and has rendered you unable or unwilling to form meaningful relationships. This fact, and your strongly religious upbringing, has resulted in a very conflicted sexual identity. After sexual relations you are so disgusted with yourself you have to lay the blame on someone else — in this case, my client. I suggest that the mental health problems that dogged your father and brother also affect you; that they drove you to make false accusations in the past, and that the cry of rape against my client is just another of these.'

She gives a sad smile and I want to claw out her eyes with my fingernails.

'You are a traumatised young woman who deserves our sympathy, but whatever your problems, you must not be allowed to bring down an innocent man.'

The jury's faces turn to the accused, slumped in his chair, the patch of fluff between the two glossy stretches of his receding hairline stirring in the breeze from the air con. He looks up at them with wide, bewildered eyes. *What did I do?*

'No further questions.'

* * *

Rauf is honourable enough not to say *I told you so*.

'Surely,' I rant, 'the jury isn't gullible enough to believe such convoluted nonsense?' But it's a rhetorical question. Of course they are. Juries will believe any old shit — look at O.J. — and I've made my living out of this fact.

I should have swallowed my pride and squeezed out some tears. I'll make sure I put on a better performance in the public gallery, in case the jury looks over at me to gauge my reaction to any other evidence or witnesses, but Rauf doesn't need to tell me I've missed my chance.

And there's worse. Before we say goodbye I ask him why the defence barrister didn't grill me about my sex life. Rauf says they've made it much tougher to go down that route. I should have known this. I should have found it out before I went shooting my mouth off to Daniel. They wouldn't have been able to call him as a witness to my loose ways so there was no need to tell him anything.

Let alone everything.

Back at Abe's flat that evening I experience my first unpleasant stirrings of doubt. I sluice them away with a bottle of wine.

⋆ ⋆ ⋆

The next morning I tell Jody and Mira I'm having breakfast with Rauf so as not to have to travel to the court with them, trying to keep the mask of unshakable confidence in place. I've lost track of the days, but a free paper I pick up on the bus tells me it's a Thursday. Always my favourite day in Vegas. The weekend just starting to get going and the air throbbing with anticipation. Jackson has given me a six-month sabbatical. Others have taken over my cases and are, according to him, doing *awesomely*. That'll teach me to think so much of myself. A little humility might have saved me yesterday's car crash.

Over a lonely coffee in a greasy spoon around the back of the courthouse I resolve to try and let go a little, let the CPS do its job. I've done all I can, for better or worse: it's up to them now to make the case. When I spot Rauf as we file back into court I don't even ask for the sheet of the day's proceedings.

There are the sounds of people settling, the clunks of knee against pew back, coughs and rustles, the slosh of the jury's water bottles. Mira and Jody are already here, two pews behind me. I give them a quick smile, then sit down with my back to them.

The next prosecution witness is called, the policewoman who found me sobbing in the rugby club toilets, being comforted by a couple of the older wives.

After her it's Elaine, one of the wives, who makes perfectly plain her distaste for Rob. Then the medical examiner and his sheaf of photographs.

I hope Jody doesn't catch a glimpse of them as they are passed around the jury.

There are more photographs of the scene of the crime, my clothing and Rob's, complete with ragged tears, lost buttons and stains; a few more witnesses, and then the prosecution case is closed.

It's a strong one, and as we break for lunch, my confidence has bounced back enough to face Jody and Mira. We eat sandwiches in the little rose garden by the court and I pretend to fall asleep in the sun so as not to have to answer any of Mira's questions.

When we get back it's the accused's turn to give evidence. As he stumps up to the witness box, the material of his suit pulling into wrinkles across his slab of a back, I remember the tackiness of his skin, the taste of salt when I bit him.

He gives his own account of the events of New Year's Eve. It's all pretty unbelievable stuff. The most unbelievable thing about it, of course, is that it's the truth.

His barrister cross-examines him, and then it's the CPS's turn.

Is the court supposed to believe that I approached him out of the blue and asked him to have sex with me? And that I insisted we have intercourse outside even though the temperature was minus two degrees?

Is it correct that his girlfriend was present at the party?

How would he explain the fact that I was covered in mud and patches of my hair had been pulled out?

His answer, that I fell over crossing the pitch, and that my hair must have got pulled out because I'd asked him to be rough with me, draws gasps of disgust from the public gallery.

'What, *this* rough?' says the barrister, holding up the photo that made most of the jurors wince.

During the cross-examination the jury is increasingly restive, and by the end some of the women are glaring him with barely disguised loathing.

But all is not lost for him.

Next they bring in character witnesses who, to a man, insist upon what a great bloke he is. Salt of the earth. Heart of gold. Wouldn't hurt a fly. Etc. We hear about his trek across the Andes for Sports Relief, his commitment to the rugby team, how every Sunday he visits his granny in her nursing home.

And then we're done. It's all over apart from the summing-up by each side. The prosecution evidence was strong, but there's no denying we are left with the impression of a confused drunken fool who *made a terrible mistake*.

As we file out of the court I put on a brave face for Jody and Mira but all my doubts have crept back in.

Rauf is waiting for me outside and I tell Mira and Jody I will see them back at home. His expression makes my heart sink into my shoes. Silently he leads me back to the witness room and closes the door behind him. I collapse onto the sofa, in dire need of whatever caffeine is left in the rat's piss the machine provides.

And then his face splits into a grin.

I stare at him. 'What? How was that not bad for us?'

He laughs then, and I want to hit him. 'Seriously, Rauf. What the hell is so funny?'

'Ah, come on, Miss Mackenzie, you're the laywer!'

I fold my arms and glare at him.

'They've introduced bad character reference for you, trying to make out that you like rough sex. And good character reference for him: all that shit about Granny Elsa. So . . . ?'

'So, *what?*' I still want to slap him.

He grins. 'So we get to introduce *rebuttal* evidence.'

My heart lifts, just for a moment, until I remember that I don't change my gran's nappy every weekend and my friends are not about to hop on a plane to tell the court what a truly saintly corporate lawyer I am.

'Do we have any?'

His infuriating smile becomes sly. 'Leave it with me.

* * *

So I do. I leave him to it and go off to enjoy my weekend, which is all the time the seriously pissed-off judge has given us to patch holes in our case. We should have given him warning in the prosecution case statement that we intended to do this, and he's within his rights to refuse to allow it. But Rauf and the CPS barrister somehow work their charms on him, to the disgust of the defence barrister, who's spitting blood as we file out of court.

Jody, Mira and I head straight for the station and by five o'clock we're lying by the pool in a spa hotel in the middle of Kent. Mira, who must have sensed what the trial is taking out of us, has banned all conversation to do with the case, and we simply read, eat good food, drink decent wine, and watch videos of a laughing Flori, smeared in fruit puree.

Arriving back at St Jerome's on Sunday night, my mood has lightened enough to gift me an unbroken night's sleep, but when my alarm goes off on Monday morning the dread descends once more. Aside from whatever rebuttal evidence Rauf has managed to dredge up, we are left with the summing up and my victim impact statement. I must somehow manage to make myself cry. I have to. Or he walks.

On the way to the station I buy a pack of tissues — I can press them to my face to hide any lack of tears.

The courtroom is quieter this morning. The rubberneckers have heard all the titillating stuff and moved on to the child-murderer in the court next door.

The first few minutes are taken up by the CPS barrister's creeping apology to the judge, who laps it up like a fat, cantankerous cat. All the while the defence barrister twitches in irritation, tapping her pen on the desk, and Rob sighs and shifts in his chair.

Finally the CPS barrister is ready to proceed.

He clears his throat, waiting for his audience to give him their full attention, and I remember the thrill of power this part always gave me. I can't wait to get back to work. I hope Jackson has a decent case lined up for m —

'I'd like to call Daniel Stillmans.'

There's a moment's silence, then white noise fills my head, so loud that I don't hear the clip of his footsteps as he walks past me to take the stand. The chair creaks as he sits down, his blond hair catching a bar of light slanting in from the high windows.

No, oh no. He's going to tell them everything I told him in the café. He's been biding his time, waiting for the perfect moment to blow the case apart.

'Would you please tell the jury how you met Miss Mackenzie.'

Wait. Calm down. The *prosecution* has called him, not the defence. And there's no way they could have known about him unless Daniel got in touch with them. Which means he's on my side. Isn't he?

'On the plane from McCarran.' His voice is clipped and professional.

The court is completely silent. All my attention is fixed on his face, willing him to turn and see the desperation in my eyes. Just get up and walk away. Please don't do this to me.

'We hit it off straightaway and over the ensuing weeks became close.'

I blink. He's making it sound like we had a relationship. His handsome face is drawn. The fat girl in the wraparound dress can't take her eyes off him.

'And what happened after the night of the New Year party?'

This is it. I wince. *Please*, I think. *Please don't tell them* . . .

He sighs and runs a hand through his hair. 'I'd planned to go with her, but Mags insisted I spend the night with my children.'

I stare at him.

'They love New Year's Eve and, well, it's been a long time, what with my divorce.'

Rauf glances at me and I can almost feel him thrumming with excitement. *Leave it with me*.

One of the jurors shifts in his seat. Perhaps I'm not such a hard bitch after all.

'If I'd gone it would never have happened. She would never have been . . . raped.'

For the first time he looks across at me. His

eyes are shining. Christ, he's good. Though I know it's all an act, my heart balloons in my chest.

His head drops in an Oscar-worthy demonstration of shame and when he speaks again his voice is soft. 'After that, things were different. She didn't want me anywhere near her. What happened to her made her afraid to be close to someone again. She doesn't trust anyone any more. She's built a wall, I think, to protect herself.'

I feel the eyes of the court fixed on me, and for the first time since I set foot in the musty-smelling building, the emotion that blurs my vision is genuine. I let my eyes well up and over, before dabbing the tears away with my sleeve.

The defence brief sighs at this cheesy attempt at emotional manipulation and, in one fell swoop, alienates the whole courtroom.

'So you noticed a definite change in Mary's character,' the barrister says, 'before the attack, and after?'

I nod and the transcriber's fingers fly.

'And prior to this, you never saw any evidence of mental health problems or this 'hatred of men' my opposing counsel implied?'

'No. Of course not. That's nonsense.'

'The defendant has tried to convince us that Miss Mackenzie liked *rough sex*.' He speaks the words with distaste. 'I'm sorry to have to ask such an intimate question, but did Miss Mackenzie ever ask you to hurt her in any way during your lovemaking? To beat, or scratch, or bite her? Anything that might have caused the injuries you see here.'

He walks across the room and hands Daniel a sheaf of photographs. For a moment I'm glad Rauf didn't tell me, because I'd never have allowed Daniel to see me that way.

He gives a sharp intake of breath.

Now that he knows how far I have gone in this deception, the depths I have stooped to, will he be disgusted? Will he feel he has to speak out?

He isn't looking back at me.

Not the merest rustle of paperwork or murmur of breath disturbs the silence. It's so quiet we all hear the slap of the photographs as he tosses them onto the floor. They fan out, face down with shame.

'That,' he says softly, 'isn't making love. It's torture.'

The barrister walks over, picks them up and tucks them back into the file. Then he turns back to Daniel. 'So that we can all cast these unpleasant aspersions aside, I must ask you to confirm whether Miss Mackenzie ever asked you to . . . hurt her in the ways depicted in the photographs.'

Daniel's lip curls. 'No. She did not.'

'Thank you, and my apologies, but it is, alas, all too common in these cases, even in the twenty-first century, to see the victim branded as a liar, or *mad*, or an indulger in 'rape fantasies'.' His air speech marks perfectly communicate the contemptibility of this idea. 'My commiserations that what happened that night destroyed this fledgling relationship, a normal, healthy relationship based on affection and respect, not sadistic torture fantasies.'

'Thank you. I still hope that . . . ' Daniel swallows. 'That one day, after all this is over, maybe we can start again.'

I can't look away, even as I hear the fat girl sniffle. The defence brief is muttering to her client. Jody's knee glances my own. The sun through the thin window is a blade of light bisecting the room between us.

And then tears are spilling down my cheeks. I fumble in my bag for the tissues, but it falls to the floor with a clunk. I am stripped bare in front of all these people. Eventually Jody hands me a handkerchief and I press it to my face, letting my hair fall like a curtain.

'Thank you, Mr Stillmans,' the barrister says. 'Your witness.'

'No questions, Your Honour.'

* * *

At the end of the day there's a delay as both counsels speak to the judge about something. I have arranged to meet Rauf in the coffee bar around the corner and, to his credit, my thunderous expression when he walks in does not give him pause. He stands by the counter, taking his own sweet time stirring his latte, letting me stew.

'Why the hell,' I hiss as he slides into the booth, 'didn't you tell me?'

He shrugs. 'Worked, though, right? You looked

one hundred per cent human, for once.'

'Quite a risk, don't you think? He might have said . . . anything. And how could you have predicted my response?'

He sips his coffee, then licks the froth from his shapely top lip. 'Mr Stillmans and I had a long conversation, from which I gleaned that your feelings for him might be worth exploiting.'

'You're a shit, Chaudhry. Don't ever do that again.'

He inclines his head, smirking.

'I'm glad you find my discomfort amusing.'

Now he grins openly. 'I can make it up to you.'

I place my own cup down on the sticky table and fold my arms. 'Please try.' As he speaks in his quiet, silky voice I realise I was wrong about him. He isn't *going* to be good. He *is* good.

But before I can think of a way to respond in a way that doesn't exacerbate his unbearable smugness, my phone rings. Its position, face up on the table, means that both of us can read the name on the display.

Rauf waggles his ridiculously bushy eyebrows. 'My pleasure,' he says, and gets up and walks out of the café.

I hesitate for the briefest moment, then draw my finger across the name to take Daniel's call.

41

Rob

Kathy says it'll be time for the summing up soon and then the jury will retire to try to reach a verdict. She says our chances are fifty-fifty. The bitch doesn't seem to give a shit either way, mind, and she's started taking calls about other cases, as if I don't matter a toss.

I've been making eyes at the fat girl and it was going quite well until that cheesy bastard with the perfect teeth opened his mouth. Kathy said that was a bad moment, that the lying slag's tears looked genuine enough to *convince* the jury. I tell her that maybe she decided to make it up because otherwise he'd have thought she was screwing around on him. Not that I should have to come up with this stuff. That's her job. My parents are paying her enough.

Without looking at me she gives a noncommittal 'hmm' and I know she's not convinced by my version of what happened at New Year. That's probably why she's not giving it 100 per cent. After all this is over I'm going to complain about her to the barrister's association or whatever. Maybe I'll sue her.

Their bloke walks in, bald head shining under the lights, heels clicking against the wooden floor. Smart shoes. Expensive. Another couple of years at the firm and I'd have been able to afford shoes like that. Course that's never going to happen if we lose. I glance at Kathy, but she's looking down into her lap. The bitch had better not be texting someone.

Before we came in this morning she told me to prepare myself for a *worst-case-scenario custodial sentence*. Which means I might go to prison. Apparently they've got a new witness. Kathy tried to stop it but because we'd already called witnesses to my good character, they're allowed.

Across the aisle, in the public gallery, the lying slag sits down. As she tucks her skirt under her arse she catches my eye, just for a moment, and gives me a flicker of a smile. I look at the jury to see if they've clocked it, but they're looking at their papers.

It was this smile, when someone handed her a glass of water on the first day, that made me finally recognise her as the girl from the bleachers. It was a horrible moment and I'm not ashamed to admit that my balls shrank right up into my pelvis. I didn't understand it then and I don't understand it now, but I'm not going to feel sorry for myself. She's a fucking nutcase and Kathy's bit about it being inherited from her nutcase family should work if the jury's got a single brain cell between them.

Baldy stands up. 'I'd like to call my next witness.'

The door opens behind me and footsteps shuffle up the aisle. I'm expecting a geriatric. Some old cow whose car I scratched or whose letterbox I shat through, but it's a skinny guy of about fifty or sixty who takes ages to clamber up to the stand. When he turns around people grimace. He looks like he's got terminal cancer.

I have no idea who he is, but a smell that fills the court makes me gag. I swivel in my seat and give Kathy a look. *Seriously*? But her eyes stay fixed on the cancer guy.

'Felix Goddard, you were a childhood friend of the accused.'

There's a high-pitched gasp behind me, as if someone recognises the name, but for a moment it doesn't register with me. Then it's like being hit by a train.

Felix?

Felix?

A *prosecution* witness?

'Yes.' His voice rasps, like it hurts to talk. I guess his vocal chords have been shredded by crack.

'We were mates since, like, four or five, right up to . . . I don't think I'm allowed to say, am I?'

'Correct. Please stick to answering the questions I ask you.'

Kathy shifts in her seat. Her jaw's tight. She looks like she's about to jump up and shout 'objection!', but she stays put.

'How old were you when the friendship ended?'

'Seventeen.'

'That's very specific. Clearly you remember the incidents surrounding the break-up very well.'

'Objection! Counsel is encouraging the jury to make negative inferences towards my client.'

'Sustained. Change your line of questioning.'

'Don't worry,' Kathy breathes as she sits

down. 'They're not allowed to bring up the other rape trial because you were acquitted.'

'Tell me about the nature of your friendship up to that point, please, Mr Goddard.'

I lean back in my seat, staring at him as he lists all the shit we got up to as kids. It sounds bad when you put it the way he's putting it. He's making out that guy's heart attack was solely caused by us playing our music on his front wall and dropping rubbish onto his lawn. He mentions the caution we got for squeezing the au pair's tits at the bus stop and I whisper to Kathy to see if he's allowed to bring it up. She nods tightly. The fat girl on the jury has stopped looking over at me, and the stocky tattooed bloke's just staring at the ceiling, as if there's something nasty playing on the telly and he doesn't want to watch.

'From what you describe am I to understand that the pair of you had little respect for women, seeing them only as objects for sex?'

Felix nods.

'Is that a yes?'

'Yes.'

'Did you ever feel remorse for these acts, Mr Goddard?'

Felix looks down. 'Yeah. Later on. After we did some . . . worse stuff.'

Kathy huffs and taps her pen on the table.

'I felt really bad. I wanted to blot it out and drink seemed to help that, and then drugs did too, and now,' his voice cracks, 'look at me.' He holds his arms out like a broken Jesus on the cross.

'Your witness.'

Kathy stands up. 'Perhaps we could spare the self-pity, Mr Goddard, and stick to the facts.'

Over the next half an hour or so she tries to make out that it was all Felix's fault, leading me astray, using the fact that he became a junkie while I straightened out. But even I can see it's not enough.

It's coming up to lunchtime when Kathy wraps up.

'Thank you, Mr Goddard, no more questions.'

But he doesn't go anywhere. My eyes burn into him, willing him to feel the hatred I'm firing in his direction, the fucking Judas. His hollow eyes are scanning the court. Then his body gives a jolt.

'Jody,' he says, his voice cracking.

I turn around, following his gaze. And whatever bullet just passed through Felix now passes through me, making my heart judder to a halt.

Jody Currie is sitting in the back row of the court.

'I'm so sorry, Jody. So very, very sorry.'

'Strike that from the record,' the judge snaps. 'Leave the stand now, Mr Goddard.'

He walks past me but my vision has fuzzed over like someone's just tackled me too hard.

Finally I get it.

The lying slag with the dead brother. The brother who didn't kill himself, like the papers all said. Who lived next door to Jody. She and Jody, somehow . . . To get back at me . . .

I need to tell Kathy.

But how can I? *Actually, I'm not a rapist, only a murderer*.

I'm trapped.

42

Mags

Jody, Mira and I waited for the verdict in the rose garden. The plants were still just dry-looking stalks, but at the end of each twig was a tiny, tight bud, as hard as wood but already snaked with the tiny fault lines from which the blooms would detonate. By June the place would be glorious.

Mira called home while Jody and I shared a packet of crisps, too nervous for anything more substantial, though I kept insisting it would all be fine.

The finger-whistle made us all jump.

Rauf stood on the steps of the court, his white grin glittering in the sunshine. He gave me a thumbs up. The jury was back already, which could only mean one thing.

As we approached the gate my eye was caught by a brass plaque on the wall. The garden was a memorial for twenty-nine people killed by a direct hit from a German V-2 bomb in 1944. Beside the list of names an angel hung his head in sorrow. I thought of another angel, drifting down through jewelled light, never landing.

* * *

Nine years.

It was never going to seem enough. Not for

murder, or for rape. But it will well and truly screw his life chances when he gets out. The three of us stood up as he was taken down, for the three lives he had torn apart: Jody's, Abe's and Loran's.

* * *

We cremated Abe on a spring morning before Mira went back home.

It'll be time for me to go back soon, too. Daniel's there already. Not exactly waiting for me — just waiting-and-seeing.

But before I go there's one more thing I need to do.

Something I should have done years ago, when I'd finally found the life I wanted to lead. I should have shuffled off my bitterness about my past back then, not let it become the baggage I always criticised other women for displaying with such martyred zeal.

Because however misguided they were, our parents thought they were doing what was best for us. And though I may dismiss it as a childish nursery tale, they considered their faith as a truth to live by.

So, this last loose end, I must tie up.

I must go home to tell my parents that their son is dead.

Father Archibald is long gone, but the church secretary promises me that Father Chinelo will call me back. When he does he tells me, in a booming Nigerian voice, that both my parents are alive and well and still worshipping at the same church.

'I didn't know they had a daughter,' he says, but I just thank him and say goodbye. I wonder how my father feels about an African priest leading him in worship.

On the long train journey from King's Cross I have plenty of time to think. I believe that after the initial shock and grief they will be satisfied with the manner of Abe's death. He died a hero. Protecting the weak.

I will tell them that the man who killed him is in prison and will be there for many years.

I will tell them Abe was loved, and if they ask the name of his partner, I will give them Jody's name.

I will not tell them he was homosexual. They're too old to overcome their prejudices now. Let them imagine the grandchildren they might have had (and perhaps will have one day, after all).

I will tell them that he kept the ring, and let them believe, in the end, that Abe found his way to Jesus, and that given a little more time, he would have found his way home. As I have done.

In the end they will consider it a good way to go.

At Edinburgh I change onto the local train and chunter through endless purple valleys threaded with silver streams and waterfalls, and the odd loch that crisply reflects the landscape around it — a looking glass that Abe has stepped through.

The names of the stations are so familiar: Crianlarich, Tydrum, Loch Awe, Bridge of Orchy.

To get to Eilean Donan I will have to travel further north. I wonder how many will have gone before me when I stand on the parapet and scatter Abe's ashes into Loch Alsh. I wonder how long it will take them to get to the sea.

The train slows, passing my old school, and the bridge that Maisie Ross jumped off for a laugh, into the burn that swept her away, never to be seen again. We pass the road that leads up to the old people's home where my nana died, and the pub where the wake was held, and where my father wouldn't let us join in the ceilidh dancing.

I start to feel sick as we pull into the station and my arms are so weak a man has to help me get my case down. Clutching it before me like a shield, I step off the train into the station I left fourteen years ago, vowing never to return.

The station concourse is bitterly cold, and every time the entrance doors slide open, there's a blast of icy wind. But it brings no litter, just a few fallen leaves.

I walk out of the station and emerge onto a roundabout. Even here, on the busy main road that runs up to Inverness, the air is different. Clean and mineral-tasting, like fresh water. To my right the loch sparkles. Whatever the weather, the surface of the water is always black. It has kept its secrets for ten thousand years.

This is my father's country. My mother was a second-generation Irish immigrant, but this land moulded him. And me, perhaps. We were two stones clashing together. No wonder there were sparks.

Do they regret the way they treated us, or are they still deluding themselves that they did the right thing and it was we who were in the wrong? The lie would have been easier to bear, but if nothing else, my father was a brave man. He brought seven half-dead climbers off that mountain, in weather that would have given the hardiest Sherpa pause. Of anyone, he might have had the balls to face up to his mistakes.

Ach! — I punch the button for the pedestrian crossing — what does it matter any more. Their child is dead. That's punishment enough.

I cross the road and enter my hometown.

The shops have changed: local independents have been replaced by the franchises you see in every other British town. The greasy spoon where I gossiped with the Proddy girls when my father thought I was at netball has become a Starbucks.

I walk past the war memorial, the rumble of my case wincingly loud in the silence. Away from the main road the street is empty of traffic, and the few pedestrians hurry on their way, their heads bent against the wind.

It was always so windy at St Jerome's. Was it Abe trying to tell me something? *Go home, Mags*.

I'm glad to be wearing his parka. The sweet smell of him still lingers in the fur. It's a smell I remember from long evenings of Bible reading on the sofa at home. I would slip a paperback into mine, but Abe never did. I used to sneer at him for his apparent devotion, but his eyes, though they gazed dutifully at the text, were always distant, as if fixed on some other reality. I wonder what you were thinking about then, Abe. Or who. Was it Dougie Kennedy, the cheesy football champ who most of the girls fancied? Or did you have my taste in boys — Pete Goldring, for instance, the dark, clever one who was never afraid to pass sardonic comment on the behaviour of the class morons, and got his face bashed in a few times for it?

How strange that I can picture them all so clearly.

The once grand Royal Highland has become a budget-chain hotel. It was the best I could find, and at least there's a bar. The place has that thick-carpeted hush of all provincial hotels, and the air smells of over-stewed vegetables. KFC for dinner, then.

A boy I went to school with is on reception. He doesn't recognise me and I give a different surname so as not to have to make conversation. Currie. It's the first name that pops into my head.

He gives me my room fob and I go up. The room is huge and bare, with a white-sheeted bed and a cheap-looking armchair. For company I turn on the TV as I run a bath.

Afterwards I feel different. I have been baptised in the waters of home, and it has washed something away. My confidence? My self-esteem? My sense of security?

No, I'm just afraid.

To steel my nerves I order up a gin and tonic and sip it by the window. If I'd raised my head as a child I would have seen that this place is breathtakingly beautiful. No wonder they get religion up here and never lose it. The place looks like it has been moulded by the hand of God.

I'd like a second drink, but I don't want to arrive at dusk so, slipping Jody's silver charm into my pocket, I head out.

The wind has died down and the loch is as still as glass as I walk down the steps.

I could make the short trip up the hill with my eyes closed. The road snakes up past the funeral director, the hairdresser, and the house with all the china dolls on the windowsill — all unchanged.

I take a deep breath and turn the corner.

There is the playground where he broke my arm. The sandpit replaced by a wooden castle surrounded by a blue rubber moat.

And there is my house. White walls, green door, slate roof, roses snaking up next-door's garage wall.

There is my bedroom, the rainbow sticker still in the window, its colours faded, almost transparent.

There is my father, digging in the rose bed. My fingers close around the guardian angel in my pocket.

The soles of my trainers are silent on the pavement as I walk the last few yards to the garden gate.

He was a big man, the muscles gone to seed when he stopped the mountain rescue, but always there; now he's leaner, and the coarse grey hair has thinned and become wispy. I never saw my father in jeans before. Jeans, a plain sweatshirt, and grey slip-ons that are caked in mud. His moustache is gone, revealing full lips like his son's. The ice-blue eyes are framed in square bifocals.

Oh, Daddy.

When he sees me he straightens up and frowns for a second, as if trying to remember. Then the trowel goes limp in his hand and the soil tumbles like confetti onto the multi-coloured petals below.

The high road is busy.

It's a mild evening, the Indian summer seems to be going on forever, and people on their way home from work pause to browse the vegetables outside the Lebanese supermarket, or take a tiny paper cup from an aproned young man standing in the doorway of the new coffee bar. His silver tray reflects a pink and yellow sky.

The Cosmo waiters are laying out the tables for the evening rush. It got into some London restaurant guide and now the locals can't get a table for love nor money, or so she has heard.

She waves at the woman in the pharmacy who gestures a cup tipping at her mouth: coffee? She holds her hand to her ear, finger and thumb extended: I'll call you. After shaky beginnings the two women have become friends. The pharmacist's mother has dementia and sometimes she needs a shoulder to cry on.

She crosses at the lights and turns into Gordon Terrace.

An explosion of colour halfway down marks out the house of the new family from Syria. They held a street party to celebrate their arrival and have since been filling their little front garden with flowers that would never grow under the harsh Arabian sun. There are roses and hydrangeas, a Californian lilac, hanging baskets of fuchsias, window boxes of lavender, and some pointed red and yellow blossoms that look like flames.

For a moment he flashes into her mind. The man she thought she loved. The man who saved her life. In all the lies and confusion she thought she had lost any notion of what was real and what wasn't from that crazy time, but the memory has come back so clear and so strong that she knows it must really have happened. Her grill pan caught fire and he rushed over and put it out with a wet tea towel. That was all.

She understands now why it affected her so powerfully. It made her feel cared for. And it felt good. She wanted more of it, but she was knocking at the wrong door. It took her too long to realise. She is sorry now, but her friend Mags says not to have any regrets, because he saved her life. And that was an act of love even if love wasn't in his heart.

Mags insists on telling her she is loved. Once, when she'd been drinking in a bar with Daniel, she told her, 'I love you.' The thought makes Jody smile.

Then Daniel took the phone and said he loved her too and when was she coming to visit them in Vegas. She promised she would, in the summer, if she had enough money.

They said forget the money, they would pay, but it's important to her to pay her own way. It helps with her self-esteem. Marian says that by next year she will be ready for a management role and there's a charity shop two miles away with a vacancy coming up because the manager's retiring. She's not afraid of the journey. The *youths who used to hang around Gordon Terrace have moved on because the police patrol it now, and the face she feared to see on every bus that passed her will be very changed by the time the prison sentence is over.*

At the end of Gordon Terrace she steps onto the path that runs up to St Jerome's. The new girl from Flat Three and her toddler are working on the community vegetable garden with Dale and Sara. Tessy the terrier skitters about, snapping at the white butterflies that have been disturbed by the presence of the gardeners. It's time to harvest the beans and she has promised to help, so she tells them she will just go and change out of her work clothes.

As she passes the ground-floor window she murmurs hello to Mrs Lyons. She's in a home now and doesn't recognise Jody, but sometimes she visits and sits on the sofa watching the Carry On films that still make the old lady laugh uproariously.

Dale's wheelchair has left muddy tracks through the foyer. José will be livid.

On the table is a postcard from Mira who is visiting relatives in Budapest with Flori. Jody smiles wistfully. She had hoped to see Flori grow, but of course it was right for Mira to return home to her family. If Jody had family she would have done the same. She will have to save up for those flights too, as the invitation to Albania is an open one.

She passes through the door into the stairwell, aglow with the colours of the stained-glass window. There's a young man sitting on the stairs, drawing with pastels. He is so thin she knows it must be the recovering anorexic who has moved into Flat Ten. He's absorbed in what he's doing and only looks up when her shadow falls across the page. He starts and drops his crayon.

'Sorry,' she says, picking it up.

'Hello,' he says. 'I'm Benno.'

They shake hands. His fingertips leave coloured spots of chalk on her skin.

'I'm Jody. Flat Twelve. We're almost next-door neighbours.'

They talk about the logistics of the flats: the unreliable availability of hot water, the dodgy tumble dryer in the basement that shrinks socks, the ambulance that arrived in the middle of the night to take the woman in Flat Seven to hospital. Benno says he was lucky to get a flat in such a beautiful place and did she know the window was by Thomas Willement? For a moment they gaze at Jesus. His mild brown eyes gaze back at them. Willement was good, Jody thinks; she cannot look away.

'Well, nice to meet you,' she says, finally. 'If you get bored, we'll be outside picking runner beans for the next three hours!'

Benno laughs. 'With these artist's fingers? I'll think about it.'

She goes upstairs and lets herself into the flat, dumping her bag and kicking off her shoes. She's got time for a cup of tea.

She listens for the sounds of the church: the distant gurgle of plumbing, the heartbeat of the organist's foot, the sighing of the wind around the spire, then she puts on the radio.

The kettle boils and she takes her tea to the table by the window. She used to think this view was terrible: looking down over the bins. But you don't have to look down. You can look up.

Above the shabby flats opposite the gulls soar through the sudden afternoon sunlight, their backs ablaze with gold and red. She watches them a moment, wheeling through the blue, never to land, then she goes to get changed.

Abe

It's cold and the wind's whipping my jacket around like mad but I don't go inside. I like the feel of the salt spray on my face and the boom when the ferry bucks through a wave. I'm standing at the front, right up by the chain, as far as they let you go, and I can't help the feeling that if I look back I'll see my da striding over the water to fetch me back.

Mam found me packing. I thought she'd try to stop me, or go and fetch my da from the prayer meeting, but she didn't. She just stood in the doorway watching me stuff a few pairs of pants and socks and some toiletries into my case. She must have seen the mobile Pete gave me when he got his iPhone, sitting on top of the pile, but she didn't say nothing. I didn't take many clothes. When I can afford it I'm going to buy new ones. Tight ones that cling to my body: *like a little tart*. It's not just girls that can be tarts, Daddy.

I met a man online who lives in Dublin. He's older than me and I don't fancy him much but he says he'll put me up, help me find a job, get me on my feet. Like a real dad should. I'll do whatever I need to to pay for it. I think I know what that'll be, because one of the other boys showed me a video online.

'Goodbye, Mam,' I said.

'Goodbye, Abraham.' When she said it her lips hardly moved.

I'd left myself only a couple of minutes to spare before the bus left, but they were the longest moments of my life as I waited in the wind for my da to come striding down the slope.

Even as I got on the bus I didn't believe it. Even as it pulled away and went rocketing down the motorway. Even as it clunked onto the ferry and the metal doors went down and the engines roared.

Still I was looking for him, not quite believing I'd made it.

I've not much to thank you for, Mary. And you've not much to thank me for — we were real bastards to each other, weren't we? But this one thing, I'll be grateful to you for my whole life.

You showed me it was possible to leave.

You laid a trail of white pebbles for me, and here I am following them. Da always said I was weak. Well, you've taught me to be brave, to fight for my dreams.

I know yours will come true. You were always clever and strong; you didn't take any shit from Da even when you were little. I always admired you. Even when you were grassing me up, or whipping my arse with that belt. I hated you, but I admired you. I reckon you'll be something really special. And I reckon that when you don't have to fight to stay alive any more you'll be a decent person.

I'd like to meet you then.

Seagulls have followed us all the way from Liverpool. Sometimes they're high up in the sky like white confetti, other times they fly really low to the water, and you can see the rippled reflections of their bodies. If I came back as an animal I'd like to be a bird. But all that reincarnation guff is heresy, Da, isn't it? When I die I'll be gathered into the bosom of the Lord Jesus, eh? Hope he's as hot as he looks in *Stories for Young Believers*. Big strong arms from carrying that bloody great cross, a decent tan, black wavy hair like Pete Goldring.

The ferry terminal comes in sight and I turn back to see if Da's made it across the Irish sea yet.

A man on the other side of the ferry is watching me. He's probably ten years older than me, short and stocky with a broad nose and a wide mouth. When I catch his eye I smile.

And then I laugh.

I am free.

I am free.

As the ferry starts to slow down a gull's white wing skims the surface of the water beside us, throwing up an arc of spray that catches the sunlight, and for a moment I can see a rainbow.

Acknowledgements

As ever, outpourings of gratitude to my agent Eve White, for her constant support, advice and championing. Without her TATTLETALE would never have been written. Also thanks to her trusty sidekick, Kitty Walker, who answers my pedantic questions with promptness and grace.

The team at Trapeze are a joy to work with. Along with her searing narrative insight, my editor, Sam Eades, seems to possess the energy of a five-year-old mainlining E-numbers. With epic publicist Ben Willis beside her, the world will know my name.

Thanks also to those working so hard on TATTLETALE'S behalf behind the scenes at Trapeze, including Laura Swainbank in marketing, Susan Howe and the rights team, Rachael Hum (who when we last met was sucking her thumb but now is, apparently, rather good at export sales), Ruth Sharvell in production, Loulou Clark in design, Sara Griffin and Katy Nicholl.

On a personal note, thanks to hotshot lawyer Jane MacDougall, who helped me tread the line between accuracy and drama in the trial scene.

Huge gratitude to all the early readers, the bloggers and authors who liked the book enough to get a buzz going on social media. And finally, of course, thanks to my stalwart first reader and biggest fan, my ma, Jill Smith.

We do hope that you have enjoyed reading this large print book.

Did you know that all of our titles are available for purchase?

We publish a wide range of high quality large print books including:

Romances, Mysteries, Classics
General Fiction
Non Fiction and Westerns

Special interest titles available in large print are:

The Little Oxford Dictionary
Music Book
Song Book
Hymn Book
Service Book

Also available from us courtesy of Oxford University Press:

Young Readers' Dictionary
(large print edition)
Young Readers' Thesaurus
(large print edition)

For further information or a free brochure, please contact us at:

Ulverscroft Large Print Books Ltd.,
The Green, Bradgate Road, Anstey,
Leicester, LE7 7FU, England.
Tel: (00 44) **0116 236 4325**
Fax: (00 44) **0116 234 0205**

Other titles published by Ulverscroft:

KILLING KATE

Alex Lake

Kate returns from a post-break-up holiday with her girlfriends to news of a serial killer in her home town — and his victims all look like her. It could, of course, be a simple coincidence. Or maybe not. She becomes convinced she is being watched; followed, even. Is she next? And could her mild-mannered ex-boyfriend really be a deranged murderer? Or is the truth something far more sinister?